ELECTRONIC DEVICES
Circuits and Applications

WILLIAM D. STANLEY

Eminent Professor, Old Dominion University

ELECTRONIC DEVICES
Circuits and Applications

PRENTICE HALL, Englewood Cliffs, NJ 07632

6/96

#18814665

Library of Congress Cataloging-in-Publication Data

STANLEY, WILLIAM D.
    Electronic devices.

    Includes index.
        1. Electronic circuits.   2. Electronic apparatus
and appliances.   I. Title.
TK7867.S77 1989   621.381   88-32190
ISBN   0-13-248949-X

Editorial/production supervision and
    interior design: Tom Aloisi
Cover design: Wanda Lubelska
Manufacturing buyer: Bob Anderson

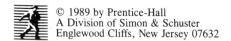 © 1989 by Prentice-Hall
A Division of Simon & Schuster
Englewood Cliffs, New Jersey 07632

Printed in the United States of America

10   9   8   7   6   5   4   3   2   1

ISBN 0-13-248949-X

Prentice-Hall International (UK) Limited, *London*
Prentice-Hall of Australia Pty. Limited, *Sydney*
Prentice-Hall Canada Inc., *Toronto*
Prentice-Hall Hispanoamericana, S.A., *Mexico*
Prentice-Hall of India Private Limited, *New Delhi*
Prentice-Hall of Japan, Inc., *Tokyo*
Simon & Schuster Asia Pte. Ltd., *Singapore*
Editora Prentice-Hall do Brasil, Ltda., *Rio de Janeiro*

Contents

2 Semiconductor Diodes 67

3 Applications of Diode-Resistor Circuits 94

4 Bipolar Junction Transistors 136

Contents

10   BJT Small-Signal Amplifier Circuits   403

11   FET Small-Signal Amplifier Circuits   451

Contents

20 Power Supplies 718

Appendix A: Resistor Values 751

Appendix B: Capacitor Values 753

Appendix C: Hybrid Parameters 754

Appendix D: Data Sheets 759

Index 801

Preface

The primary objective of this book is to provide a comprehensive treatment of electronic devices and circuits at a level suitable for first- and second-year students in electronic and electrical technology curricula. The large number of example problems and exercises should also make it very useful for self-study, review, and in support of more advanced courses. The approach in this book differs somewhat from that of other books and is believed to offer some significant advantages.

Most basic electronic texts offer several chapters on semiconductor physics. The treatment is quite superficial and is mostly incomprehensible to students at the first- and second-year levels. It has traditionally been included because authors probably felt that the coverage would be considered otherwise incomplete.

Since the vast majority of electronic technicians and technologists are *users* of electronic devices, and only a few are actually involved in device fabrication, the treatment of semiconductor physics here is limited to a few short sections in which some essential terminology is introduced. Rather, the major focus throughout the book is on device *modeling*, circuit *operation* and *analysis*, and applied *design*.

Basic electronic devices are introduced in the first third or so of the book, and this includes diodes, bipolar junction transistors (*BJT*s), and field-effect transistors (*FET*s). In each case, a chapter dealing with the particular device category and its models is first presented, and then a chapter dealing with simplified applications follows. The applications at this point include *only resistive components*, and circuits requiring reactive components are delayed until the reader has developed more sophistication. Included in the applications, however, are such important circuits as electronic switches, the emitter follower, the source follower, voltage-controlled resistors, and stable biasing circuits. Note that field-effect transistors are introduced in parallel with bipolar junction transistors, an approach that differs from most books at this level.

Prior to developing transistor amplifier circuits, a general development of amplifiers from a block diagram point of view is given. Concepts such as gain, input resistance, output resistance, stage loading, and interaction are developed with simpler models before the student becomes involved with specific circuit details.

Small-signal amplifiers, including all basic *BJT* and *FET* types, are developed in detail. The emphasis in circuit modeling is that of the simplified low-frequency *hybrid-pi model*. This represents a significant departure from most books at this level. However, it is believed by this author to be the best and most timely approach. It has several distinct advantages: (1) the forms of the *BJT* and *FET* models are essentially the same; (2) the basic model can readily be extended to cover high-frequency circuits; and (3) it can readily be adapted to computer modeling. With regard to advantage (1), the various equations developed first for *BJT* amplifiers are quickly adapted to the corresponding FET amplifiers, and this results in a significant correlation between the two types of amplifier circuits.

A few additional comments about small-signal models of transistors are in order. Most books at this level employ either the standard hybrid parameter model or the r'_e model. It is this author's strong opinion that early introduction of hybrid parameters acts more to confuse beginning students than to serve any real educational purpose. Their use is meaningful only after considerable intuition for the practical operation of electronic amplifiers is established. For instructors wishing to cover hybrid parameters, Appendix C is devoted to their coverage.

The r'_e model is quite simple and easy to apply, but it has the disadvantage that the approach is more formula oriented than circuit-model oriented. In fact, the value of r'_e is the reciprocal of g_m as used in the hybrid-pi model. Consequently, from one point of view, the approach of this text is nearly equivalent to that of the r'_e model if interpreted properly. However, the hybrid-pi model can be readily adapted to computer modeling and to high-frequency effects.

A complete chapter is devoted to the differential amplifier and its integrated circuit evolution to the operational amplifier. This sets the stage for the next several chapters, which are predominantly oriented toward integrated-circuit applications.

Negative feedback is developed in detail through the application of op-amp circuits. Various op-amp circuit applications (both linear and nonlinear) are considered. Included are numerous amplifier circuits, active filters, holding circuits, clamping circuits, and comparators.

Remaining topics near the end of the book include positive feedback and oscillators, timers, high-frequency amplifiers, power amplifiers, and power supplies. Placing the chapter on power supplies near the end of the book represents a significant departure from most other texts. However, this author believes it to be appropriate. The usual treatment of power supplies in a very early chapter causes difficulty with many students because of the variety of waveform concepts, definitions, and nonlinear functions encountered.

The book is suitable for a two-semester sequence in electronic devices and circuits. Such a sequence could start after the first course in circuits or basic electricity, or it could be taught in parallel with minimal modification in teaching style. The first chapter of the book provides a "minicourse" in basic dc circuit analysis, with specific emphasis on techniques useful in electronic circuit analysis. This chapter could be covered very quickly for students having at least one term in dc circuits. An important feature is that *ac circuit analysis is not required until about the middle of the book*.

DISCUSSION OF END-OF-CHAPTER PROBLEMS

At the end of most (but not all) chapters, there are five sets of problems. These problems are designated as Drill Problems, Derivation Problems, Design Problems, Troubleshooting Problems, and Computer Problems. The nature of these categories is discussed here.

Drill Problems

This set usually has more problems than the other categories and will represent the major focus for much of the material. Drill problems have been designed to solidify and reinforce the major principles presented in the text. Drill problems range from very simple types to more challenging problems representing situations somewhat different from example problems in the text.

Derivation Problems

The derivation problems are those in which the reader is asked to prove or develop some result in general terms. In some cases, these are simply verifications of rules or theorems given in the text, while in other cases, they are extensions or generalizations of the text results. The outcomes of most derivation problems are formulas or equations rather than numerical values. A few of the derivation problems require calculus, but such problems are identified as such. Such exercises represent the only points in the book where calculus is required.

Design Problems

Design problems are selected exercises in which the reader is expected to determine a specific circuit to meet a prescribed objective. In most cases, the actual circuit configuration will be one of those studied in the chapter, but the design aspect of the problem will involve the determination of element values for the various components. In some cases, it will be expected that component values can be rounded to standard element values as tabulated in Appendixes A and B. Design problems do not necessarily have a single correct solution.

Troubleshooting Problems

Troubleshooting problems are exercises in which the reader is expected to display judgment in selecting the probable cause of a malfunction or in predicting the expected response of the instrumentation to typical faults. Because a variety of faults can be responsible for an observed malfunction, most of the troubleshooting exercises have multiple-choice answers based on the *most likely* problem.

Computer Problems

For this purpose, the term *computer* will be interpreted to mean either a mainframe computer, a minicomputer, a "personal" microcomputer, or a sophisticated programmable calculator. Problems have been shown to illustrate how programs can be created to accomplish specific objectives in the area of circuit analysis or design. Many of the problems in this category specify *input data* and *output data* as part of the problem statements. The input data requirements are incorporated in the program so that the user may assign element values during a given execution. Without this feature, the program would be useful only for a specific set of values. The output data represent the desired program output variables. Because of the variation of languages and programming structures, these problems have intentionally been left loosely specified in many cases.

SEMICONDUCTOR DATA SHEETS

A concept heavily promoted throughout this book is the use of manufacturers' data or specifications sheets for obtaining information about semiconductor devices. To assist in that endeavor, samples of typical data sheets are provided in Appendix D. These sheets will be referred to throughout the book as the subject matter is developed and as individual specifications become relevant.

Device specifications are necessary in performing any type of design or special system implementation where components must be selected for a particular application. Without such specifications, one is "working in the dark," and what may appear to be a satisfactory design could be totally unworkable for the given set of conditions. Worst yet, if the maximum ratings of components are exceeded, the component may be destroyed.

The preceding discussion no doubt seems logical to the perceptive reader, but a few further comments are in order. First, there are literally thousands of different types of semiconductor devices available from a large number of different manufacturers. Some common devices are made by a number of different companies, and the identifying number may be the same or it may be different. The situation is often confusing even to experienced engineers and technicians, so it can appear to be overwhelming to the uninitiated.

The best advice to give to the beginner in this regard is not to worry about the fact that there are thousands of such devices in the field. In your entire career, you will probably have a direct encounter with only a small percentage of the available semiconductor devices, and you will tend to become quite familiar with these. When new devices are encountered, the first thing to do is to obtain data and/or applications sheets for these devices, and you can study these for further information. This author is personally acquainted with many outstanding engineers and technicians, and they all feel overwhelmed by the same situation. The fact is that we really do not need so many available types of devices, but the semiconductor industry has proliferated, no doubt by the profit motive, to produce the existing situation.

The next comment is in regard to the availability of data and applications sheets on individual components. Many semiconductor manufacturers publish one or more data manuals providing compilations of data sheets on devices. Such data manuals are constantly being revised with new devices added and older devices discontinued. As time passes and integrated circuits have become the major focus of most design, more diodes and transistors are being discontinued, and it may be difficult to find specifications on certain components. Some services provide coverage "across the board" and include data from a number of manufacturers. Of course, you need not be too concerned about this now, while you are learning the basics of electronics. Once you are on the job and have the need to know, help should be available in breaking through the information barrier.

The final comment is in regard to the specifications themselves. It is an interesting irony that a simple-looking three-terminal transistor may have two or more pages of specifications with dozens of graphs and several tables of data. This is enough to scare the impressionable person right out of the field! Fear not, however, for all is not yet lost. Such detailed specifications tend to provide about everything one would ever want to know about the device (and perhaps more!). Most general applications require the detailed consideration of only a fraction of the available information. The additional information is there for special applications and situations in which critical specifications are being "stretched to the limit."

In virtually any application, some of the most important data are the *maximum ratings*. Depending on the device, the maximum permitted values of such basic quantities as voltage, current, and power must be carefully monitored. Additional specifications relate to the types of applications to be performed. Throughout the book, we frequently extract pertinent data and apply it as the need arises.

I would like to express my sincere gratitude to Mrs. Estelle B. Walker for patiently typing this voluminous manuscript. I truly appreciate her fine work more than words can express.

William D. Stanley
Norfolk, Virginia

CHAPTER 1

Basic Circuit Analysis*

OBJECTIVES

After completing this chapter, the reader should be able to:

1. Define the basic circuit quantities and state the symbols and units used to represent them.
2. Define the basic active and passive circuit models and show their schematic forms.
3. State and apply Ohm's law.
4. State and apply Kirchhoff's voltage law.
5. State and apply Kirchhoff's current law.
6. Determine the equivalent resistance of a passive circuit containing only resistors.
7. Determine the current, voltages, and power in a single-loop circuit.
8. Determine the voltage, currents, and power in a single-node-pair circuit.
9. State and apply the voltage-divider rule.
10. State and apply the current-divider rule.
11. Analyze certain complete resistive circuits by applying voltage- and current-divider rules along with the equivalent resistance concept to achieve a step-by-step solution.
12. Define the notation used for voltages between two points, for voltages with respect to ground, and for current.
13. Apply the concept of the "voltmeter loop" to determine the voltage between two points.
14. Apply source transformations to change the forms of source models.
15. Approximate realistic sources as ideal source models, and identify conditions in which such approximations may be used.
16. Define the four types of controlled (or dependent) sources.
17. Determine the Thévenin and/or Norton equivalent circuits for a given circuit.
18. Apply the principle of superposition in the analysis of linear circuits.

*Portions of Chapter 1 have been adapted from the author's text *Network Analysis with Applications* (Englewood Cliffs, N.J.: Reston, 1985).

To acquire proficiency in the analysis and design of electronic circuits, it is essential that the reader have an understanding of basic circuit laws and theorems as well as proficiency in basic circuit analysis techniques. Many readers will already have established such a background from separate courses and/or texts. However, it is the experience of this author that most readers need some help in translating the various circuit theorems from their "purest" forms as given in circuit analysis texts to the forms encountered in practical electronic circuits.

This chapter has the primary objective of presenting the basic circuit laws and analysis methods that will be essential in studying electronic circuits throughout the remainder of the book. For readers who have a strong background in circuit theory, much of this chapter should serve primarily for review and strengthening. However, it is possible that new topics may be encountered, particularly those relating to dependent sources and the application of certain circuit theorems to circuits with dependent sources. Thus it is recommended that even readers with extensive circuit theory take some time to cover this chapter before proceeding to the remainder of the book. However, readers in this category will be able to cover portions of the chapter rather quickly.

For readers who are simultaneously studying basic circuit analysis from a different text, or for readers with little or no background in circuit theory, this chapter will be more challenging. Although this single chapter obviously cannot replace a complete text in circuit analysis, virtually all the basic circuit laws and analysis methods that will be required for much of the early study of electronics are provided here. Readers in this category should strive to master these basic concepts at this time, which should facilitate the study of electronics later in the book. As additional circuit laws are studied from a circuit analysis text, a more complete picture of the situation will emerge.

1-2 BASIC ELECTRICAL QUANTITIES

An electronic circuit is an interconnection of various components arranged to perform a specific function, such as switching, amplification, control, or signal processing. Components may be classified as either active or passive, according to their energy status. In the operation of electronic circuits, an *active* device is one that either supplies energy to the circuit (e.g., a battery or power supply) or converts energy from one form to another (e.g., a transistor). A *passive* device is one that either only absorbs energy (e.g., a *resistor*) or absorbs and later returns the energy to the circuit (e.g., a *capacitor* or *inductor*).

A complete electrical circuit consists of a combination of branches and nodes. A *branch* consists of one particular component with two terminals, but the definition is sometimes extended to include more than one component when the combination is to be viewed as a single component. A *node* is a junction point between two or more branches.

The concepts of branches and nodes are illustrated in Figure 1-1. This circuit

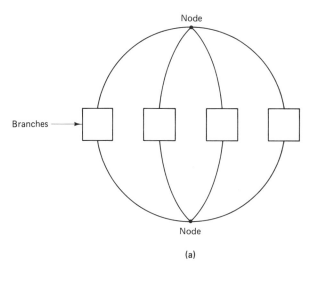

Node

Branches →

Node

(a)

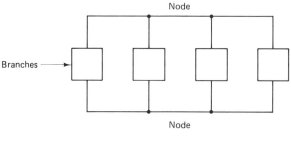

Node

Branches →

Node

(b)

Figure 1-1 Branches and nodes drawn two ways.

consists of four branches, each of which contains one component, and two nodes. In (a) the branch terminals are brought together at both the top and bottom to illustrate the two nodes. For complex circuits, it may be awkward to show the schematic diagram with all connections to a given node at a single point. An alternative arrangement is shown in (b), and this form is used most often. However, the circuit still has only four branches and two nodes because the additional connections along the top and bottom do not contain any components.

The various quantities used in most electronic circuit analysis will now be discussed. Each quantity is represented by one or more symbols, a unit of measurement, and an abbreviation for the unit. The symbols are used in various mathematical equations to refer to the quantity, and the units (along with abbreviations) are used to specify the value of the quantity.

The most common quantities used in electronic circuit analysis, together with their symbols, units, and abbreviations for units, are summarized in Table 1-1. The most common prefixes used in circuit analysis are summarized in Table 1-2. The units and their abbreviations are used in accordance with the International System of Units (abbreviated as SI). Not all quantities given in these tables will be discussed or

TABLE 1-1 MOST COMMON QUANTITIES USED IN ELECTRONIC CIRCUIT ANALYSIS

Quantity	Symbols	Unit	Abbreviation of unit
Time	t, T	second	s
Voltage	v, V, e, E	volt	V
Current	i, I	ampere	A
Power	p, P	watt	W
Resistance	R	ohm	Ω
Conductance	G	siemens	S
Capacitance	C	farad	F
Inductance	L	henry	H
Impedance	$\bar{Z}$	ohm	Ω
Reactance	X	ohm	Ω
Admittance	$\bar{Y}$	siemens	S
Susceptance	B	siemens	S
Frequency (cyclic)	f	hertz	Hz
Frequency (radian)	ω	radian/second	rad/s

TABLE 1-2 MOST COMMON PREFIXES USED IN ELECTRONIC CIRCUIT ANALYSIS

Value	Prefix	Abbreviation
10^{-15}	femto	f
10^{-12}	pico	p
10^{-9}	nano	n
10^{-6}	micro	μ
10^{-3}	milli	m
10^{3}	kilo	k
10^{6}	mega	M
10^{9}	giga	G

used at this point in the book. The reader may refer back to these tables later as the need arises.

In the second column of Table 1-1, note that both lowercase and uppercase symbols are listed for many of the quantities. As a general rule, *lowercase* symbols are used wherever a quantity *is* or *could be varying with time*; that is, it is considered as a *time-varying* or *instantaneous* function. *Uppercase* symbols are generally used for either a *direct-current* (abbreviated as dc) quantity or a *constant value* associated with a time-varying quantity (e.g., the peak value of a time-varying quantity). However, there are exceptions to this rule, so the reader should not interpret these guidelines as rigid. Subscripts may be added to these symbols as required when a number of quantities of the same type appear in a problem.

When it is desirable or necessary to emphasize that a quantity is time varying, functional notation of the form $v(t)$ or $i(t)$ can be used. The quantity t in parentheses represents time, and the notation indicates that the quantity is a function of time. To simplify expressions, the functional notation will be omitted in most equations, but if the quantity has a lowercase symbol, it should be interpreted as *possibly* being a time-varying quantity. For example, the voltage v and the current i are variables that could be varying with time.

Some of the quantities that will be encountered early in the book will now be briefly discussed. The reader should refer directly to Table 1-1 to support the discussion that follows.

Starting at the top of Table 1-1, the first quantity listed is *time*. Many electronic circuit variables vary with time in their instantaneous form. When time is considered as an arbitrary independent variable, it is denoted by lowercase t, as already noted.

Voltage is the electrical pressure between two points in a circuit. The unit of voltage is the *volt* (abbreviated as V). Voltage is always measured between two points, and the term *potential difference* is often used as an alternative term for voltage. By tradition, the symbols v, V, e, and E have all been employed to represent voltage. The major symbols used in this book will be v and V.

Current is a measure of the rate of flow of electrical charge through a circuit. The symbols for current are i and I, and the unit is the *ampere* (abbreviated as A). This book will assume *conventional current* flow within a circuit. Conventional current flow is the most widely used convention, and it is opposite to electron flow.

Current and voltage are the two most common variables that are measured and calculated in electronic circuits. Note that *voltage* is always measured *across* a portion of a circuit, while *current* is measured *through* a portion of a circuit. A *voltmeter* is an instrument used for measuring *voltage*, and an *ammeter* is an instrument used for measuring *current*.

Power represents the rate of performing work or the rate of change of energy in the circuit. The symbols for power are p and P, and the unit is the *watt* (abbreviated as W).

The instantaneous power p associated with a component having an instantaneous voltage v across it and an instantaneous current i through it is given by

$$p = vi \qquad \text{(1-1)}$$

Depending on the component and the remainder of the circuit, this power could either be *absorbed* or *delivered*. The two possible situations are shown in Figure 1-2.

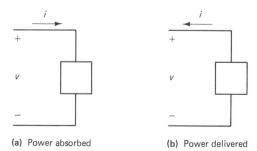

(a) Power absorbed   (b) Power delivered

Figure 1-2 Directions of voltage and current for power absorbed and power delivered.

If the current i is flowing into the positive terminal of the device as shown in Figure 1-2(a), the device is *absorbing* power from the remainder of the circuit not shown. If the current i is flowing out of the positive terminal as shown in Figure 1-2(b), the device is *delivering* power to the remainder of the circuit.

If either voltage or current (but not both) has a negative value, the assumption of absorbed or delivered is reversed by negation. For example, an absorbed power of -5 W implies that $+5$ W of power is actually delivered.

Resistance is the opposition to current flow present in all conducting materials. A lumped package of resistance is called a *resistor*. The symbol for resistance is R, and the unit is the *ohm* (abbreviated as Ω).

An alternative way to characterize resistance is through the concept of *conductance*. The symbol for conductance is G, and the unit is the *siemens* (abbreviated as S). The relationship between conductance and resistance is

$$G = \frac{1}{R} \qquad (1\text{-}2)$$

Thus the greater the resistance, the lower the conductance, and vice versa.

Discussion of the remaining quantities in Table 1-1 is deferred until they are encountered later in the text.

1-3 CIRCUIT MODELS

The functional operation of most complex electronic circuits can be investigated by the use of various equivalent circuits. Such equivalent circuits are used to display various voltage and current properties of the complete electronic circuit, and they are very important in both analysis and design. These equivalent circuits employ various *circuit models*.

A wide variety of useful electronic circuits may be represented as a combination of two active models and three passive models. The two active models are (1) the *ideal voltage source* and (2) the *ideal current source*. The three passive models are (1) *resistance*, (2) *capacitance*, and (3) *inductance*.

The models for the two ideal active devices are illustrated in Figure 1-3. The ideal voltage source model is shown in (a), and it is recognized by the presence of the $(+)$ and $(-)$ signs. The $(+)$ terminal is the assumed positive reference terminal, and the $(-)$ terminal is the assumed negative reference terminal. However, this does not necessarily mean that the $(+)$ terminal always has a higher potential than the $(-)$ terminal since this will depend on the functional form of v_s. For example, suppose

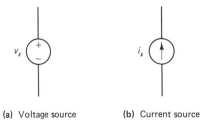

(a) Voltage source   (b) Current source

Figure 1-3 Active-circuit parameter models.

that at some value of time, $v_s = 5$ V. In this case, the actual positive terminal is the assumed positive reference. However, suppose that at some other value of time, $v_s = -5$ V. In this case, the assumed positive reference terminal is lower in potential than the assumed negative reference terminal.

The ideal current source model is shown in Figure 1-3(b), and it is recognized by the presence of the arrow. The arrow denotes the assured direction of positive conventional current flow. As in the discussion in the preceding paragraph, the current at some point in time could have a negative value, which would imply current flowing in the direction opposite to the given reference direction.

Either an ideal voltage source or an ideal current source may be further classified as an *independent* source or a *dependent* source. An independent ideal source is one whose specified value (voltage or current) is independent of any other variables in the circuit. A dependent source is one whose specified value is a function of some other variable in the circuit. As we will see later, the models of transistors and other active devices require dependent sources. Dependent sources will be introduced in Section 1-14.

The three passive circuit parameters are *resistance*, *capacitance*, and *inductance*. The schematic symbols for the three parameters are shown in Figure 1-4. Two possible symbols for capacitance are shown, but the one on the right using parallel lines is employed in this book. The actual devices exhibiting these three properties are denoted as *resistors*, *capacitors*, and *inductors*, respectively.

Of the three passive circuit parameters, only resistance appears in the basic discussion of electronic devices in the first seven chapters. Capacitance and inductance first appear in complete electronic circuits starting in Chapter 8. Thus at this point in the text, all discussions of circuit laws focus on resistive circuits.

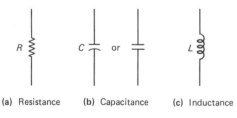

(a) Resistance   (b) Capacitance   (c) Inductance

Figure 1-4 Passive-circuit parameter models.

1-4 RESISTANCE AND OHM'S LAW

Consider the resistance R shown in Figure 1-5 as having an instantaneous voltage v across it and a current i flowing through it. Ohm's law states that

$$v = Ri \qquad (1\text{-}3)$$

or alternatively,

$$i = \frac{v}{R} = Gv \qquad (1\text{-}4)$$

where G is the conductance defined by (1-2).

Figure 1-5 Resistance with voltage and current conventions indicated.

From Figure 1-5, observe that *the most positive terminal of the voltage across a resistance is the terminal at which the current is assumed to enter*. This assumption is compatible with the fact that power is always absorbed by a passive resistance. (A few specialized active devices exhibit a phenomenon referred to as *negative resistance*, in which the devices deliver power, but that concept is not considered in this book.)

In many circuits, the actual direction of the current or the polarity of the voltage will not be known, and an arbitrary assumption must be made. However, when either the current direction or the voltage polarity is assumed, the other quantity must be chosen to be compatible with it according to the preceding convention. If the unknown quantity turns out to be negative, the literal use of Ohm's law for the other variable will also result in a negative value for that variable.

In defining Ohm's law, no mention was made of whether v or i was a source. It does not matter whether a voltage source is causing a current to flow or whether a current source is causing a voltage to exist. For most cases, neither v nor i for a given resistance is a source; they are simply particular variables representing a small part of a much larger circuit.

The instantaneous power absorbed by a resistance is given by

$$p = vi \tag{1-5}$$

If $v = Ri$ is substituted in (1-5), the power expressed in terms of the current and resistance is

$$p = i^2 R \tag{1-6}$$

Finally, if $i = v/R$ is substituted in (1-5), the power expressed in terms of the voltage and resistance is

$$p = \frac{v^2}{R} \tag{1-7}$$

Current Sampling Resistor

A useful and practical property of Ohm's law for instantaneous voltage and current will be discussed. From (1-3), the voltage is always a constant times the current, so the waveforms of voltage and current for a resistor have the same forms. Frequently, it is desired to monitor the waveform of a certain time-varying current on an oscilloscope. However, an oscilloscope is a voltage-measuring device. If the circuit contains a resistance (normally with one terminal grounded) through which the current of interest flows, the voltage across it will have the same waveform as the current through it, and the oscilloscope can be made to respond to current indirectly. In some cases, a small sampling resistor may be inserted in the circuit for this purpose. Obviously, the resistance must be sufficiently small that it does not disturb the circuit. The oscilloscope voltage scale can be calibrated so that the current level can be determined by a simple application of Ohm's law.

Example 1-1

It is desired to monitor the waveform of current in a certain electronic circuit by means of a standard oscilloscope. It is determined that the insertion of a 10-Ω precision sampling resistor presents negligible disturbance to the circuit. The resistor is inserted at a point where one side is connected to the common ground, so that the grounded lead of the oscilloscope may be maintained at the circuit common ground. A satisfactory display is obtained when the oscilloscope vertical calibration factor is set at 20 mV/cm. Determine the equivalent calibration factor for the measured current.

Solution

By Ohm's law with $R = 10\ \Omega$ and $v = 20\ \text{mV}$, the current is $i = 0.02\ \text{V}/10\ \Omega = 2 \times 10^{-3}\ \text{A} = 2\ \text{mA}$. The calibration factor on this scale for current is thus 2 mA/cm.

1-5 KIRCHHOFF'S LAWS

Two fundamental circuit laws discussed in this section are Kirchhoff's voltage law and Kirchhoff's current law. For brevity, these laws are often denoted as KVL and KCL, respectively.

Kirchhoff's Voltage Law

Kirchhoff's voltage law states that the algebraic sum of the voltages around a closed loop is zero. This can be stated mathematically as

$$\sum_n v_n = 0 \qquad (1\text{-}8)$$

where $\sum_n$ represents a summation of all voltages v_n around the loop. To apply KVL, an algebraic convention must be applied to all voltages. Voltage drops can be assigned as positive, and voltage rises can be assigned as negative, but the opposite convention will work just as well provided that it is consistent. It is also possible to restate KVL as follows: The sum of all voltage drops is equal to the sum of all voltage rises, with all terms being positive, but with rises and drops on opposite sides of the equation. Algebraically, all the forms are equivalent.

To maintain a consistent approach in this book, the first form is employed for all cases in which a formal loop equation is required. Thus, in formulating KVL equations, drops are assigned positive values, and rises are assigned negative values.

At this point in the text, clockwise loops are used for consistency. Later, when we study electronic circuits, we will see that counterclockwise loops are sometimes more convenient. As we progress through the book and acquire more experience in this matter, certain simple loops will be analyzed in a more intuitive, shortcut manner. However, when in doubt, readers are strongly urged to return to the fundamental KVL process for analyzing a loop.

To illustrate KVL, consider the loop shown in Figure 1-6. The blocks could represent any circuit components, and lowercase notation for the voltages implies instantaneous quantities. To apply KVL, we pick a point and "walk around the loop" until we return to the initial point. In accordance with the convention we have adopted, if we enter a positive terminal and leave a negative terminal, there is a drop in potential, so that the term is assigned a positive value. Conversely, if we enter a minus terminal and leave a positive terminal, there is a rise in potential, so that the term is assigned a negative value. This convention has the easily remembered property that the sign of the term is the same as the polarity of the terminal at which we enter. For the circuit of Figure 1-6, we start at point "×" and form a clockwise loop. The KVL equation is

$$-v_1 + v_2 + v_3 - v_4 + v_5 = 0 \qquad (1\text{-}9)$$

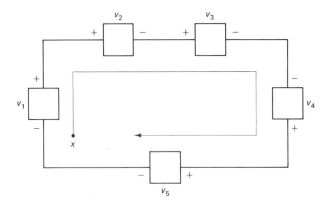

Figure 1-6 Loop used to illustrate Kirchhoff's voltage law (KVL).

Kirchhoff's Current Law

Kirchhoff's current law states that the algebraic sum of the currents at a node is zero. This can be stated mathematically as

$$\sum_n i_n = 0 \qquad (1\text{-}10)$$

where $\sum_n$ represents a summation of all currrents i_n at a node. To apply KCL, an algebraic convention must be applied to all currents. Currents leaving can be assigned as positive, and currents entering can be assigned as negative, but the opposite con-

10   Basic Circuit Analysis   Chap. 1

vention will work just as well provided that it is consistent. It is also possible to restate KCL as follows: The sum of all currents leaving a node is equal to the sum of all currents entering the node, with all terms being positive, but with currents leaving and entering on opposite sides of the equation. Algebraically, all the forms are equivalent.

To maintain a consistent approach in this book, the first form is employed for all cases in which a formal equation at a node is required. Thus, in formulating KCL equations, currents leaving are assigned positive values, and currents entering are assigned negative values.

To illustrate KCL, consider the node shown in Figure 1-7. The lowercase notation for the currents implies instantaneous quantities. In accordance with our convention, the KCL equation is

$$-i_1 + i_2 - i_3 - i_4 + i_5 = 0 \qquad (1\text{-}11)$$

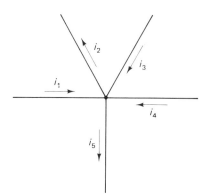

Figure 1-7 Node used to illustrate Kirchhoff's current law (KCL).

Example 1-2

A portion of some dc circuit of interest is shown in Figure 1-8. Determine the voltage V and the current I. (Uppercase notation is used since the circuit is a dc circuit.)

Figure 1-8 Circuit of Example 1-2.

Solution

We do not see the complete circuit, nor do we know the values of all compo-nents shown. However, after some initial inspection, we do recognize that there is a closed loop shown for which three of the four voltages in the loop are known. Thus, by KVL, we should be able to determine the remaining voltage V.

Starting at the negative terminal of the 36-V source, a clockwise KVL loop reads

$$-36 + V + 12 - 6 = 0 \qquad \text{(1-12)}$$

Solving for V, we obtain

$$V = 30 \text{ V} \qquad \text{(1-13)}$$

Since I is the current through the 10-kΩ resistor, for which the voltage is now known, it may be determined by Ohm's law. Thus

$$I = \frac{30 \text{ V}}{10 \times 10^3 \ \Omega} = 3 \times 10^{-3} \text{ A} = 3 \text{ mA} \qquad \text{(1-14)}$$

Example 1-3

A portion of some dc circuit of interest is shown in Figure 1-9. Determine the current I and the resistance R.

Figure 1-9 Circuit of Example 1-3.

Solution

As in Example 1-2, not all the circuit is evident. However, in this case, there is a node shown for which three of the four currents are known. Thus, by KCL, we should be able to determine the remaining current.

Assuming positive currents leaving the node, the KCL equation is

$$3 \times 10^{-3} - 15 \times 10^{-3} + 4 \times 10^{-3} + I = 0 \qquad \textbf{(1-15)}$$

Solving for I, we obtain

$$I = 8 \times 10^{-3} \text{ A} = 8 \text{ mA} \qquad \textbf{(1-16)}$$

Since I is the current through R, and the voltage across R is known, the value of R can be determined by Ohm's law as

$$R = \frac{16 \text{ V}}{8 \times 10^{-3} \text{ A}} = 2 \times 10^3 \ \Omega = 2 \text{ k}\Omega \qquad \textbf{(1-17)}$$

1-6 EQUIVALENT RESISTANCE

The equivalent resistance of some basic combinations of resistors is considered in this section. The formulas to be stated are derived in virtually all basic circuit texts and are stated here without proof.

The equivalent resistance of a large number of practical resistive circuits can be determined by the process of *successive series and parallel reduction*. Two rules are required to use this procedure, and they will be discussed next.

Series Resistance

Consider the n resistors connected in the fashion shown in Figure 1-10. These resistors are said to be connected in *series*. The equivalent resistance R_{eq} as viewed from the external terminals is

$$\boxed{R_{eq} = R_1 + R_2 + \cdots + R_n} \qquad \textbf{(1-18)}$$

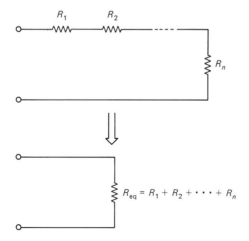

Figure 1-10 Equivalent resistance of n resistors connected in series.

Parallel Resistance

Consider the n resistors connected in the fashion shown in Figure 1-11. These resistors are said to be connected in *parallel*. The equivalent resistance R_{eq} is determined from the relationship

$$\boxed{\frac{1}{R_{eq}} = \frac{1}{R_1} + \frac{1}{R_2} + \cdots + \frac{1}{R_n}}$$ (1-19)

Alternatively, by working with conductances and letting $G_{eq} = 1/R_{eq}$, $G_1 = 1/R_1$, $G_2 = 1/R_2$, and so on, we have

$$G_{eq} = G_1 + G_2 \cdots + G_n$$ (1-20)

As a special case of *two* resistors in parallel, the following useful formula can be readily deduced:

$$R_{eq} = \frac{R_1 R_2}{R_1 + R_2}$$ (1-21)

A common notation that is often used in complex electronic formulas for two resistances in parallel is $R_1 \| R_2$. This notation should always be interpreted as the equivalent resistance of R_1 and R_2 in parallel. The notation can be extended to any number of resistors if desired. Thus the equivalent resistance of R_1, R_2, and R_3 in parallel can be expressed as $R_1 \| R_2 \| R_3$.

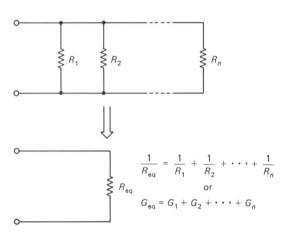

$$\frac{1}{R_{eq}} = \frac{1}{R_1} + \frac{1}{R_2} + \cdots + \frac{1}{R_n}$$

or

$$G_{eq} = G_1 + G_2 + \cdots + G_n$$

Figure 1-11 Equivalent resistance of n resistors connected in parallel.

Example 1-4

Using successive series and parallel reduction, determine the equivalent resistance at the input terminals for the circuit of Figure 1-12.

Figure 1-12 Circuit of Example 1-4.

Solution

The process consists of starting on the right-hand side of the circuit and working back toward the input, making appropriate series and parallel combinations at each step. The various steps are illustrated by the different parts of Figure 1-13. The calculations will not be given here since they consist of simple applications of the formulas of this section, but the reader may want to verify the results.

First, the series combination of the 9-kΩ and 6-kΩ resistances is determined as shown in (a) of Figure 1-13, and this result is 15 kΩ. Next, the parallel combination of the 15-kΩ equivalent resistance and the 7.5-kΩ resistor is determined, and the result is 5 kΩ, as shown in (b). Finally, the series combination of the 5-kΩ equivalent resistance and the 3-kΩ resistor is determined, as shown in (c). The result is the equivalent resistance, and thus $R_{eq} = 8$ kΩ.

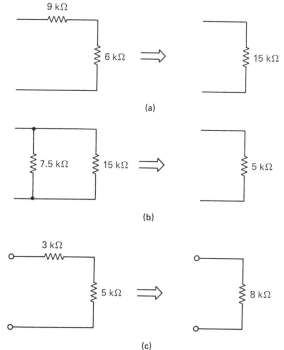

Figure 1-13 Successive steps in determining the equivalent resistance of the circuit of Example 1-4.

In this section the complete solution of two important practical configurations is discussed. The circuits are designated as (1) the single-loop circuit and (2) the single-node-pair circuit. Both circuit configurations arise frequently in the analysis of electronic circuits.

Single-Loop Circuit

The first special configuration is the single-loop circuit, whose basic form is illustrated in Figure 1-14. First, the situation with an arbitrary number of voltage sources, but no current sources, is considered. The presence of a current source will be considered later in this section. All elements in the single-loop circuit are said to be connected in *series*.

The procedure for analyzing the circuit under these conditions is based on the fundamental property that the *current through all elements in a single-loop circuit is the same*. The voltage drops across all resistors may be expressed in terms of the single current variable with the sign convention carefully noted; that is, the terminal of a resistor at which the current enters must be assumed to be the most positive. Kirchhoff's voltage law may then be applied to the loop, and the algebraic sum of all the voltages around the loop is equated to zero. As long as there is no more than one unknown quantity in the loop equation, it may be determined from the equation.

Formulation of a single-loop equation is illustrated by the resistive circuit shown in Figure 1-15. The circuit contains three voltage sources and four resistors. All elements in this illustration are labeled with symbols, so no particular quantities will be assumed to be known or unknown. Rather, the goal here is simply to express the KVL equation around the loop in terms of the labeled quantities.

For clarity, the current i is shown at a number of separate points around the circuit. Observe that the positive terminal of the voltage drop across each resistor is assumed at the point where the current enters. However, the signs of the source voltages are a function of the manner in which the sources are connected and have nothing to do with the assumed current direction.

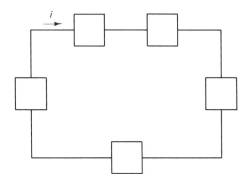

Figure 1-14   Basic form of the single-loop circuit.

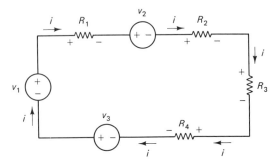

Figure 1-15   Representative single-loop circuit containing dc voltage sources and resistors.

With the circuit properly labeled, we pick a point in the loop at which to start and then we "walk all the way around" to complete a closed loop.

For the circuit of Figure 1-15, we start at a point just below the negative terminal of the source v_1 and take a clockwise path. Following the convention discussed previously, we obtain the following KVL equation for the loop:

$$-v_1 + R_1 i + v_2 + R_2 i + R_3 i + R_4 i - v_3 = 0 \qquad \textbf{(1-22)}$$

The reader should carefully study the sign pattern of this equation and relate it to the convention discussed.

An interesting pattern for the single-loop equation results when i is considered as the unknown; an expression for it can be determined in terms of the other parameters. The reader is invited to verify that the result is

$$i = \frac{v_1 - v_2 + v_3}{R_1 + R_2 + R_3 + R_4} \qquad \textbf{(1-23)}$$

This result may be generalized to the form

$$i = \frac{\text{algebraic sum of voltage source values}}{\text{net loop series resistance}} \qquad \textbf{(1-24)}$$

In the interpretation of (1-24), voltages of voltage sources are positive when they contribute to current flow in the assumed direction, and they are negative when they contribute to current flow in the opposite direction. For this form to be valid, *all the voltage sources must be independent sources.*

The reader may prefer a shortcut approach such as (1-24) for a problem such as this, and indeed, the author frequently uses this form. However, one has to exercise caution with this seemingly simpler approach because it is easier to make a sign error in the process. Further, this simpler approach is *not valid* when there is one or more *dependent voltage sources* in the loop. Therefore, it is recommended that the more fundamental approach of "walking around the loop" be retained as the basic method for analyzing a general single-loop circuit, and that any shortcut procedures be used cautiously and only after a clear understanding of the process is achieved.

Next, we consider the case where a single ideal current source appears in the loop. *The presence of an ideal current source in a single loop forces the current through all elements in the loop to be the same as that of the source.* Thus the current is immediately evident if the value of the current source is known. Any number of ideal voltage sources may be in the loop, but they do not affect the current. The various voltage drops across resistors may be determined directly from the known current and the resistances.

One quantity that requires some computation to determine is the voltage across the current source. An unknown voltage (say v_0) is assumed across the current source. By application of KVL to the loop, v_0 can be determined. The polarity may turn out to have the positive terminal at the point where current flows out of the current source, or vice versa, depending on the nature of any voltage sources in the loop.

Thus far, we have assumed only one ideal current source. The assumption of two or more ideal current sources with different currents in series results in a contradictory situation and is impossible. In a real-life situation where two or more current sources with different values are connected in series, the actual models of the current sources containing internal resistances would have to be employed to predict the results, and this would require a more complex model than the one under consideration.

Single-Node-Pair Circuit

The second special configuration is the single-node-pair circuit, whose basic form is illustrated in Figure 1-16. First, the situation with an arbitrary number of current sources, but no voltage sources, is considered. The presence of a voltage source is considered later in this section. All elements in the single-node-pair circuit are said to be connected in *parallel*.

The procedure for analyzing the circuit under these conditions is based on the fundamental property that *the voltage across all elements in a single node pair is the same*. The currents through all resistors may be expressed in terms of the single voltage variable, with the sign convention carefully noted; that is, the current flows into the terminal of a resistor at which the potential is assumed to be the more positive. Kirchhoff's current law may then be applied to one of the nodes (usually, the assumed positive node), and the algebraic sum of the currents leaving that node is equated to zero. As long as there is no more than one unknown quantity in the node-pair equation, it may be determined from the equation.

Formulation of a single-node-pair equation is illustrated by the circuit shown in Figure 1-17. The circuit contains three current sources and three resistors. All elements in this illustration are labeled with symbols, so no particular quantities will be

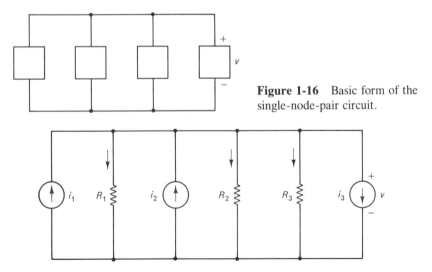

Figure 1-16   Basic form of the single-node-pair circuit.

Figure 1-17   Representative single-node-pair circuit containing current sources and resistances.

assumed to be known or unknown. The goal is to illustrate the formulation of a KCL equation at one of the nodes.

The voltage across the node pair is labeled on the right, but this voltage is the same across all the elements because they are all connected in parallel. Observe that the current through each resistor is assumed to be flowing into the positive terminal. However, the directions of the current sources are a function of the manner in which the sources are connected and have nothing to do with the assumed voltage polarity.

With the circuit properly labeled, we select one of the nodes at which to write a KCL equation. It is usually easier to work with the node at which the positive terminal of v is assumed, and thus the upper node in Figure 1-17 will be used. (Incidentally, either node produces the same result, so only one independent equation is possible for a node pair.)

The sign convention employed here and in most places in the book is as follows: Currents leaving a node are expressed with a positive sign, and currents entering are expressed with a negative sign. For the circuit of Figure 1-17, an expression of KCL for the upper node is as follows:

$$-i_1 + \frac{v}{R_1} - i_2 + \frac{v}{R_2} + \frac{v}{R_3} + i_3 = 0 \qquad \textbf{(1-25)}$$

The reader should carefully study the sign pattern of this equation and relate it to the convention discussed.

An interesting pattern for the single-node-pair equation results when v is considered as an unknown and an expression for it is determined in terms of the other parameters. The reader is invited to verify that the result is

$$v = \frac{i_1 + i_2 - i_3}{\dfrac{1}{R_1} + \dfrac{1}{R_2} + \dfrac{1}{R_3}} \qquad \textbf{(1-26)}$$

$$= R_{eq}(i_1 + i_2 - i_3)$$

where R_{eq} is defined as the equivalent parallel resistance of the three resistances as determined from the equation

$$\frac{1}{R_{eq}} = \frac{1}{R_1} + \frac{1}{R_2} + \frac{1}{R_3} \qquad \textbf{(1-27)}$$

This result can be generalized to the form

$$v = \begin{pmatrix} \text{net node-pair} \\ \text{parallel resistance} \end{pmatrix} \times \begin{pmatrix} \text{algebraic sum of} \\ \text{current source values} \end{pmatrix} \qquad \textbf{(1-28)}$$

In the interpretation of (1-28), current sources are positive when they contribute to the assumed positive terminal of v, and they are negative when they contribute to the opposite polarity of v. For this form to be valid, *all the current sources must be independent sources.*

The reader may prefer a shortcut approach such as (1-28) for a problem such as this. However, caution must be exercised with this approach because of possible sign errors. Further, this approach is *not valid* when there is one or more *dependent current sources* across the node pair. Therefore, it is recommended that the more fundamental approach of summing currents be retained as the basic method for analyzing a single-node-pair circuit and that any shortcut procedures be used cautiously for applicable circuits.

Next, we consider the case where a single ideal voltage source appears across the node pair. *The presence of an ideal voltage source across a pair of nodes forces the voltage across all elements to be the same as that of the source.* Thus the voltage is immediately evident if the value of the voltage source is known. Any number of ideal current sources may be connected across the node pair, but they do not affect the voltage. The various currents through resistors may be determined directly from the known voltage and the resistances.

One quantity that requires some computation to determine is the current through the voltage source. An unknown current (say i_0) is assumed through the voltage source. By application of KCL to a node, i_0 can be determined. The direction may turn out to be either out of or into the positive terminal of the voltage source, depending on the nature of any current sources across the pair of nodes.

Thus far we have assumed only one ideal voltage source. The assumption of two or more ideal voltage sources with different voltages in parallel results in a contradictory situation and is impossible. In a real-life situation where two or more voltage sources with different voltages are connected in parallel, the actual models of the voltage sources containing internal resistances would have to be employed to predict the results, and this would require a more complex model than the one under consideration.

Example 1-5

For the dc resistive circuit of Figure 1-18(a), determine (a) the loop current, (b) the voltage drops across all resistors, and (c) the power dissipated in the three resistors.

Figure 1-18 Circuit of Example 1-5.

Solution

As given initially, the circuit has no labels for the unknown current and voltages desired in order that such labels may be determined as part of the analysis. First, intuition tells us that the loop current will be clockwise since both the 24-V source and 18-V source would tend to cause current to flow clockwise, and their combined effect is more powerful than that of the 12-V source. Thus a clockwise current I is assumed. Once the current direction is assumed, the various voltage drops across resistors are chosen in the proper directions as indicated in Figure 1-18(b). For convenience, each voltage is also labeled with an appropriate defining subscript. (*Note:* Uppercase symbols are used for the current and voltages since this is a dc resistive circuit.)

(a) Starting at a point just below the 24-V source and moving clockwise, the KVL equation is

$$-24 + 200I - 18 + 500I + 12 + 800I = 0 \qquad \text{(1-29)}$$

This simplifies to

$$1500I = 30 \qquad \text{(1-30)}$$

or

$$I = 0.02 \text{ A} = 20 \text{ mA} \qquad \text{(1-31)}$$

(b) Once the current is known, the voltage drops across the resistors are determined as follows:

$$V_{R_1} = R_1 I = 200 \times 20 \times 10^{-3} = 4 \text{ V} \qquad \text{(1-32)}$$

$$V_{R_2} = R_2 I = 500 \times 20 \times 10^{-3} = 10 \text{ V} \qquad \text{(1-33)}$$

$$V_{R_3} = R_3 I = 800 \times 20 \times 10^{-3} = 16 \text{ V} \qquad \text{(1-34)}$$

(c) Let P_{R_1}, P_{R_2}, and P_{R_3} represent the three power levels dissipated in R_1, R_2, and R_3, respectively.

We have

$$P_{R_1} = R_1 I^2 = 200(20 \times 10^{-3})^2 = 0.08 \text{ W} = 80 \text{ mW} \qquad \text{(1-35)}$$

$$P_{R_2} = R_2 I^2 = 500(20 \times 10^{-3})^2 = 0.2 \text{ W} = 200 \text{ mW} \qquad \text{(1-36)}$$

$$P_{R_3} = R_3 I^2 = 800(20 \times 10^{-3}) = 0.32 \text{ W} = 320 \text{ mW} \qquad \text{(1-37)}$$

Incidentally, the 24-V and 18-V sources are delivering power, and the 12-V source is absorbing power. The net power delivered must equal the net power absorbed, and the reader is invited to verify this fact.

Example 1-6

For the dc resistive circuit of Figure 1-19(a), determine (a) the node-pair voltage and (b) the currents through all resistors.

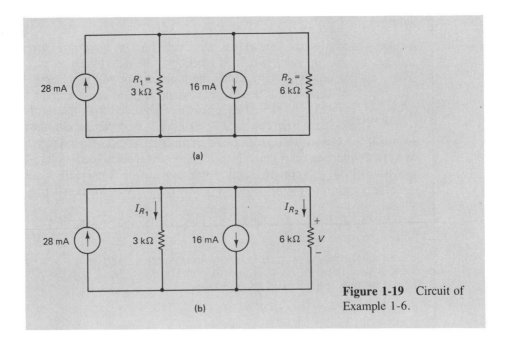

Figure 1-19 Circuit of Example 1-6.

(a)

(b)

Solution

As given initially, the circuit has no labels for the unknown voltage and currents desired in order that such labels may be determined as part of the analysis. First, intuition tells us that the more positive terminal of the voltage is at the top since the 28-mA source is more powerful than the 16-mA source. With this direction of V chosen, the directions of the two resistive branch currents are then determined in accordance with the standard sign convention as shown in Figure 1-19(b). For convenience, each current is labeled with appropriate defining subscripts.

(a) A KCL equation may be written at the upper node, and it is

$$-28 \times 10^{-3} + \frac{V}{3 \times 10^3} + 16 \times 10^{-3} + \frac{V}{6 \times 10^3} = 0 \qquad \textbf{(1-38)}$$

This simplifies to

$$5 \times 10^{-4}V = 12 \times 10^{-3} \qquad \textbf{(1-39)}$$

or

$$V = 24 \text{ V} \qquad \textbf{(1-40)}$$

(b) Once the voltage is known, the currents through the two resistors are determined as follows:

$$I_{R_1} = \frac{V}{R_1} = \frac{24}{3 \times 10^3} = 8 \times 10^{-3} \text{ A} = 8 \text{ mA} \qquad \textbf{(1-41)}$$

$$I_{R_2} = \frac{V}{R_2} = \frac{24}{6 \times 10^{-3}} = 4 \times 10^{-3} \text{ A} = 4 \text{ mA} \qquad \textbf{(1-42)}$$

Example 1-7

For the dc resistive circuit of Figure 1-20(a), determine **(a)** the node-pair voltage, **(b)** the currents through the resistors, and **(c)** the current through the voltage source.

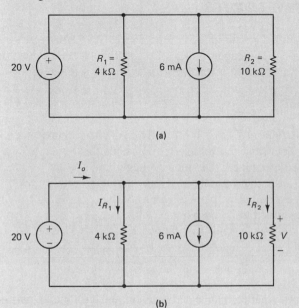

(a)

(b)

Figure 1-20   Circuit of Example 1-7.

Solution

The circuit with appropriate labels is shown in Figure 1-20(b). Note that a current I_o is shown through the voltage source.

(a) The voltage V is simply the value of the voltage source and is

$$V = 20 \text{ V} \qquad \textbf{(1-43)}$$

(b) The currents through the resistors are

$$I_{R_1} = \frac{V}{R_1} = \frac{20}{4 \times 10^3} = 5 \times 10^{-3} = 5 \text{ mA} \qquad \textbf{(1-44)}$$

$$I_{R_2} = \frac{V}{R_2} = \frac{20}{10 \times 10^3} = 2 \times 10^{-3} = 2 \text{ mA} \tag{1-45}$$

(c) To determine the current I_o, KCL is applied to the upper node. We have

$$-I_o + I_{R_1} + 6 \times 10^{-3} + I_{R_2} = 0 \tag{1-46}$$

This leads to

$$I_o = I_{R_1} + I_{R_2} + 6 \times 10^{-3} = 5 \times 10^{-3} + 2 \times 10^{-3} + 6 \times 10^{-3}$$
$$= 13 \times 10^{-3} \text{ A} = 13 \text{ mA} \tag{1-47}$$

1-8 VOLTAGE AND CURRENT DIVIDERS

Two very useful simplified circuit analysis rules for resistive circuits are discussed in this section. They are the voltage-divider rule and the current-divider rule. These rules are derived in basic circuit texts and are stated here without proof.

Voltage-Divider Rule

Consider the simple circuit of Figure 1-21(a). The voltage v_i represents a known *input* voltage to the divider, and the voltage v_o represents the *output* voltage of the divider, which is to be determined. The input voltage could be the voltage of an applied voltage source, or it could simply be a known voltage determined by the effect of one or more sources from other parts of the circuit not shown. The point is that v_i is known, and it is assumed to be the voltage that causes v_o.

The *voltage-divider rule* states that the voltage across R_o is

$$\boxed{v_o = \frac{R_o}{R_o + R_1} v_i} \tag{1-48}$$

Note that v_o is always smaller than v_i, a result of the fact that a portion of v_i is dropped across R_1, and the remainder appears across R_o.

The voltage-divider rule is readily extended to a series connection of several resistors, as shown in Figure 1-21(b). In effect, all resistances except R_o are treated as a single equivalent resistance, and the voltage-divider rule becomes

$$v_o = \frac{R_o}{R_o + R_1 + R_2 + \cdots + R_n} v_i \tag{1-49}$$

An important intuitive process for the voltage divider with two resistors may be deduced by dividing both sides of (1-48) by v_i and expressing the ratio v_o/v_i as

$$\frac{v_o}{v_i} = \frac{R_o}{R_o + R_1} = \frac{R_o/R_1}{R_o/R_1 + 1} \tag{1-50}$$

where the last form of (1-50) was obtained by dividing numerator and denominator of the preceding expression by R_1. This function has the form shown in Figure 1-22 with the ratio v_o/v_i as the dependent variable and R_o/R_1 as the independent variable.

(a)

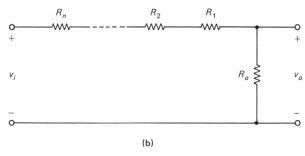

(b)

Figure 1-21 Circuits used to illustrate the voltage-divider rule.

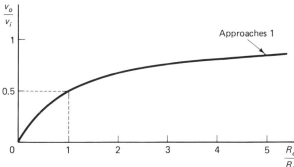

Figure 1-22 Voltage-divider ratio as a function of R_o/R_1 based on circuit of Figure 1-21(a).

From the plot of Figure 1-22, it is noted that for $R_o/R_1 \ll 1$ (i.e., $R_o \ll R_1$), the ratio v_o/v_i is very small. In this case, most of the input voltage is dropped across R_1 and very little appears across R_o. Conversely, if $R_o/R_1 \gg 1$ (i.e., $R_o \gg R_1$), v_o/v_i approaches the value 1. In this case, very little of the input voltage is dropped in R_1, and most appears across R_o. In some applications, a voltage divider is designed to reduce the level of a signal deliberately, in which case R_o and R_1 are selected to achieve the required reduction. However, in many applications, a voltage-divider process appears because of interactive and coupling effects, and it is desirable not to reduce the signal any more than necessary. In this case the ideal circuit design is to force R_o to be as large as possible relative to R_1 so that most of the available input voltage appears across the output.

As an easily remembered rule of thumb, if the output resistance R_o is at least 100 times as large as R_1 (i.e., $R_o \geq 100R_1$), the loss in input voltage across R_1 will be less than 1%. Verification of this rule is left as an exercise for the reader (Problem 1-53).

Sec. 1-8   Voltage and Current Dividers

25

One particular point on the curve of Figure 1-22 deserves attention. If $R_o = R_1$, it is readily shown from (1-50) that $v_o/v_i = 0.5$. Thus when the two resistances are equal, half of the available voltage appears across each resistor. This property is useful in performing certain laboratory tests.

Current-Divider Rule

Consider the simple circuit of Figure 1-23. The current i_i represents a known *input* current to the divider, and the current i_o represents the *output* current of the divider, which is to be determined. The input current could be the current of an applied current source, or it could simply be a known current determined by the effect of one or more sources from other parts of the circuit not shown. The point is that i_i is known, and it is assumed to be the current that causes i_o.

The *current-divider rule* states that the current through R_o is

$$i_o = \frac{R_1}{R_o + R_1} i_i \qquad \textbf{(1-51)}$$

Note that i_o is smaller than i_i, a result of the fact that a portion of i_i is shunted through R_1, and the remainder flows through R_o.

If the current-divider rule of (1-51) is compared with the voltage-divider rule of (1-48), it is "the other resistor" that appears in the numerator of the current-divider rule, while it is the resistor across which the output voltage is measured that appears in the numerator of the voltage-divider rule. Because of this difference in forms, some caution must be exercised in applying the current-divider rule to a divider containing more than two resistors in parallel. This can be achieved by first determining the equivalent parallel resistance of all resistances except the output resistance R_o. The parallel equivalent resistance of all the other resistances is treated as R_1 in (1-51), and the rule can then be applied as in the case of two resistances.

An intuitive process for the current divider may be deduced by dividing both sides of (1-51) by i_i and expressing the ratio i_o/i_i as

$$\frac{i_o}{i_i} = \frac{R_1}{R_1 + R_o} = \frac{1}{1 + R_o/R_1} \qquad \textbf{(1-52)}$$

where the last form of (1-52) was obtained by dividing the numerator and denominator of the preceding expression by R_1. This function is plotted in Figure 1-24 with the ratio i_o/i_i as the dependent variable and R_o/R_1 as the independent variable.

From the plot of Figure 1-24, it is noted that for $R_o/R_1 \ll 1$ (i.e., $R_o \ll R_1$), the ratio i_o/i_i approaches the level 1. In this case, most of the input current flows through R_o, and very little is diverted through R_1. Conversely, if $R_o/R_1 \gg 1$ (i.e., $R_o \gg R_1$), the ratio i_o/i_i is very small. In this case most of the input current is shunted through R_1, and very little flows in R_o. In many applications, a current-divider process appears because of interactive and coupling effects, and it is desirable not to reduce the signal any more than necessary. In this case, the ideal circuit design is to force R_o to be as small as possible relative to R_1 so that most of the available current reaches the output.

26   Basic Circuit Analysis   Chap. 1

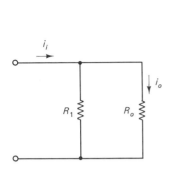

Figure 1-23 Circuit used to
illustrate current-divider rule.

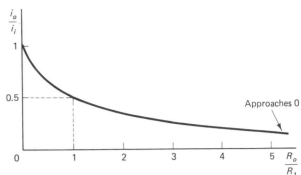

Figure 1-24 Current-divider ratio
as a function of R_o/R_1 based on
circuit of Figure 1-23.

As an easily remembered rule of thumb, if the resistance R_1 is at least 100 times as large as R_o (i.e., $R_1 \geq 100R_o$), the loss in input current through R_1 will be less than 1%. Verification of this rule is left as an exercise for the reader (Problem 1-54).

As in the voltage-divider rule, the case $R_o = R_1$ is worth noting. From (1-52) it is readily determined that when $R_o = R_1$, $i_o/i_i = 0.5$. Thus the current divides equally when the two resistances are equal.

Comparing the voltage- and current-divider rules again, note that the conditions for maximum output in R_o (voltage for voltage divider and current for current divider) are opposite. *To maximize voltage across R_o in a voltage-divider configuration, R_o should be large compared with R_1, whereas to maximize current through R_o in a current-divider configuration, R_o should be small compared with R_1.*

Example 1-8

Use the voltage-divider rule to determine the three voltages V_1, V_2, and V_3 in Figure 1-25.

Figure 1-25 Circuit of Example 1-8.

Solution

The applicable relationship is (1-49) since there are three resistors. However, the notation is different here, and three separate voltages are desired. This should illustrate that the process involved in a formula should be visualized rather than a rote memorization without understanding. The voltage across a given resistor is determined by first placing the resistance of the particular resistor in the numerator and dividing by the sum of the three resistances. This ratio is then multiplied by the source voltage of 16 V.

Applying this process to three resistors, we have

$$V_1 = \frac{1000}{1000 + 2000 + 5000} \times 16 = \frac{1000}{8000} \times 16 = 2 \text{ V} \qquad \text{(1-53)}$$

$$V_2 = \frac{2000}{8000} \times 16 = 4 \text{ V} \qquad \text{(1-54)}$$

$$V_3 = \frac{5000}{8000} \times 16 = 10 \text{ V} \qquad \text{(1-55)}$$

It can be readily verified that the sum of these three voltages is 16 V, the value of the source voltage.

Example 1-9

Use the current-divider rule to determine the current I_o in Figure 1-26(a).

(a)   (b)

Figure 1-26 Circuit of Example 1-9.

Solution

Since there are three resistors, to find the current through the 4-kΩ resistance, it is first necessary to determine the equivalent resistance of the 3-kΩ resistance and the 6-kΩ resistance in parallel. This equivalent resistance is readily determined to be 2 kΩ, and a modified equivalent circuit is shown in Figure 1-26(b). Application of the current-divider rule yields

$$I_o = \frac{2000}{2000 + 4000} \times 9 \text{ mA} = \frac{1}{3} \times 9 \text{ mA} = 3 \text{ mA} \qquad \text{(1-56)}$$

1-9 SIMPLIFIED CIRCUIT ANALYSIS

By combining the equivalent resistance simplifications with basic laws and the voltage- and current-divider rules, it is possible to analyze a number of complete circuits by a step-by-step approach. At each step of the process, the circuit is represented in both directions by a simplified form, allowing the use of one of the basic circuit laws or simple rules stated earlier. This method can be classified as an intuitive method as opposed to a more formal mathematical approach. Consequently, it is difficult to state any precise rules that always work. Although this method does not work for all circuits, it does work in a sufficient number of practical electronic circuits to justify coverage.

The circuits under consideration are generally those in which the circuit as viewed by a source can easily be represented by a single equivalent resistance. The current flowing from a voltage source or the voltage across a current source can then be determined. By moving away from the source and successively determining resistive equivalents at various points, use of Ohm's law forms or voltage- and current-divider rules may lead to a complete analysis. The method is best illustrated with a representative example.

Example 1-10

Consider the dc resistive circuit of Figure 1-27. The passive part of the circuit is the same as that of Example 1-4, but a 48-V dc source has been connected to the input. Using a step-by-step approach, determine all branch voltages and currents in the circuit.

Figure 1-27 Circuit of Example 1–10.

Solution

The equivalent resistances at various points in the circuit were calculated in Example 1-4, so the reader may wish to refer to that example (including Figure 1-13). In addition, some equivalent circuits with certain voltages and currents identified are shown in Figure 1-28 to aid in the steps that follow.

First, the equivalent resistance "seen" by the voltage source is 8 kΩ, so the current leaving the 48-V voltage source is readily determined by Ohm's law as $48 \text{ V}/8 \times 10^3 \ \Omega = 6 \times 10^{-3} \text{ A} = 6 \text{ mA}$. This is illustrated in Figure 1-28(a), with the resistance represented as the series combination of the 3-kΩ

Sec. 1-9   Simplified Circuit Analysis   **29**

Figure 1-28 Steps involved in analyzing circuit of Example 1-10.

resistor and the additional 5-kΩ equivalent resistance. The voltage across the 3-kΩ resistor is determined as $3 \times 10^3 \times 6 \times 10^{-3}$ A = 18 V, and the voltage across the circuit to the right of the 3-kΩ resistor is determined as 5×10^3 Ω $\times 6 \times 10^{-3}$ A = 30 V. Alternatively, the voltage-divider rule could have been used twice to determine the two voltages.

With the 30-V voltage across the 5-kΩ equivalent resistance known, this resistance is next represented as the parallel combination of the 7.5-kΩ resistance and the additional two resistors in series, as shown in Figure 1-28(b). The current through the 7.5-kΩ resistor is determined as 30 V/7.5 $\times 10^3$ Ω = 4×10^{-3} A = 4 mA. The current through the series combination of the two remaining resistors is then determined as 30 V/15 $\times 10^3$ Ω = 2×10^{-3} A = 2 mA. Alternatively, the current-divider rule could have been used twice to determine the division of the 6-mA current through the two paths.

With the 2-mA current to the right now known, the two remaining voltages may be calculated by Ohm's law as 9×10^3 Ω $\times 2 \times 10^{-3}$ A = 18 V and 6×10^3 Ω $\times 2 \times 10^{-3}$ A = 12 V.

All the currents and voltages calculated in the preceding steps are identified on the original circuit diagram in Figure 1-28(c). The reader may readily verify that KVL is satisfied around all loops and KCL is satisfied at all nodes.

The perceptive reader may have observed some possible variations in some of the analysis steps that could have been taken in "moving" through the circuit. The key with circuits of this type is to be able to simplify the circuit successively at different points so that the intuitive rules may be applied. To apply either of the divider rules, the loading effects of all resistors must be considered in the form of equivalent resistances at different points.

Basic Circuit Analysis   Chap. 1

1-10 VOLTAGE AND CURRENT VARIABLES AND NOTATION

Some basic conventions concerning voltage and current are established in this section. In a relatively simple circuit with only a few voltages and currents, no special notation may be necessary. However, as the circuit complexity increases, special notation and conventions are necessary so that one can keep track of the many variables within the circuit. The convention will be established using uppercase symbols, which generally imply dc quantities. However, the concept may be applied to time-varying quantities as well. In the discussion that follows, the term "point" in a circuit will usually denote the same meaning as the term "node."

Voltages Between Nodes

Consider two nodes designated as node A and node B, respectively. The two-point notation V_{AB} is defined as the voltage at node A with respect to node B. Thus node A is considered as the *positive reference* for the voltage. However, V_{AB} may be either a positive or a negative value. If V_{AB} is positive, the voltage at node A is indeed more positive than the voltage at node B. If, however, V_{AB} is negative, the voltage at node A is actually lower than the voltage at node B. In some cases it may be desirable to reverse the positive reference in the latter case. If we move the positive reference node to B, the voltage should be expressed as V_{BA}. The two voltages are related by

$$V_{BA} = -V_{AB} \tag{1-57}$$

This concept is illustrated by the examples of Figure 1-29. The blocks could represent any circuit components. In (a), the voltage at node A is 5 V higher than the voltage at node B. In this case, $V_{AB} = 5$ V. In (b), the voltage at node A is 6 V lower than the voltage at node B, so $V_{AB} = -6$ V. Alternatively, we can reverse the reference and say that $V_{BA} = 6$ V.

Voltages to Ground

Most electronic circuits contain a common ground from which most of the voltages are referred. Schematic diagrams often show many of the typical voltages with re-

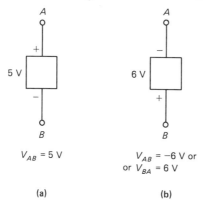

$V_{AB} = 5$ V

(a)

$V_{AB} = -6$ V or
or $V_{BA} = 6$ V

(b)

Figure 1-29 Examples of two-point voltage notation.

spect to ground that can be measured if the equipment is functioning properly. This allows one to determine probable faulty components when troubleshooting. The voltmeter common terminal is connected to the system common ground point, and the other lead is simply moved around to various nodes. In many cases, the common circuit ground is also connected to the chassis of the unit.

One could label the ground node as G and express the voltage at a point A with respect to ground as V_{AG}, if desired. However, when two-point quantities are being used in an analysis, we will usually simplify the notation by omitting the G subscript and having it understood. Thus a voltage V_A will be interpreted as the voltage at point A with respect to ground unless otherwise indicated.

This concept is illustrated by the circuit of Figure 1-30. This circuit is assumed to be part of a larger circuit, and the blocks could represent any components. Note that two ground symbols are shown, but they are at the same potential. To simplify schematic diagrams, ground symbols are typically shown at many points. In the actual wiring of the circuits, the grounds may be connected at different points, or in some critical cases, they may be brought to a single point to minimize difficulties with a phenomenon called *ground loops*.

From the diagram, point A is 8 V above ground, which can be expressed as $V_A = 8$ V. Point B is 4 V below ground, so it can be expressed as $V_B = -4$ V. If it were necessary to express the latter voltage as a positive value, the ground symbol G would have to be introduced, and we could write $V_{GB} = 4$ V.

Currents

Consider a single branch connected between nodes A and B, and assume that there are no separate paths for current between A and B. This restriction assures that there is only a single current to consider. The current I_{AB} is defined as the current leaving node A and entering B through the branch. The actual current may be flowing from A to B or from B to A. If I_{AB} is positive, the current is actually flowing from A to B. If, however, I_{AB} is negative, the current is actually flowing from B to A.

In some cases it may be desirable to reverse the assumed direction. If we reverse the assumed direction, the current should be expressed as I_{BA}. The two currents are related by

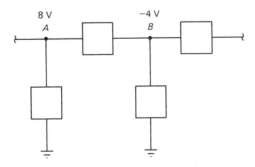

$V_A = 8$ V (with respect to ground)
$V_B = -4$ V (with respect to ground)

Figure 1-30 Examples of voltages with respect to ground.

Basic Circuit Analysis   Chap. 1

Figure 1-31 Examples of two-point current notation.

A ——3 A——→ B
$I_{AB} = 3$ A

(a)

A ——2 A←—— B
$I_{AB} = -2$ A or
or $I_{BA} = 2$ A

(b)

$$I_{BA} = -I_{AB}$$
(1-58)

This concept is illustrated by the examples of Figure 1-31. The blocks could represent any circuit components. In (a), the current is actually flowing from A to B, so $I_{AB} = 3$ A. In (b), the current is flowing from B to A, so $I_{AB} = -2$ A. Alternatively, we can say that $I_{BA} = 2$ A.

The general guidelines given here will be followed as necessary and appropriate throughout the book. The node symbols will often be those that represent certain terminals of a semiconductor device. For example, a bipolar junction transistor has emitter, base, and collector terminals, so the corresponding node symbols will be E, B, and C. With many transistor currents, however, it may be desirable to employ only one subscript.

1-11 USEFUL VOLTAGE RELATIONSHIPS

Two useful concepts for determining voltages between points in a circuit are discussed in this section. Both of these concepts are special applications of Kirchhoff's voltage laws. However, they are sufficiently useful that they deserve special consideration.

Voltmeter Loop

Frequently, it is necessary to determine the voltage between two points in a circuit separated by a number of branches, for which the voltage across each branch is known. As a simple example, consider the series connection of Figure 1-32(a), in which the voltage V_{AB} across the combination is desired. The two branch voltages are 10 V and 5 V, and the polarities are aiding. Most readers will probably deduce immediately by an intuitive process that the net voltage between the two points is 15 V and that point A is the most positive node.

If that example seems too easy, look at Figure 1-33, in which the voltage between points A and B is to be determined. Some readers might be able to deal with this circuit on an intuitive basis, but the chances of error in determining the magnitude and the sign of the net voltage are much higher in this case.

A technique which this author refers to as "forming a voltmeter loop" is the best way to deal with this situation. To apply this concept, there must be at least one path between the two points along which all branch voltages are known. We then assume that an ideal voltmeter is connected between the two points. The voltage measured by the voltmeter is the unknown voltage to be determined. The combination of the path mentioned previously and the voltmeter constitute a closed loop, and KVL may be applied.

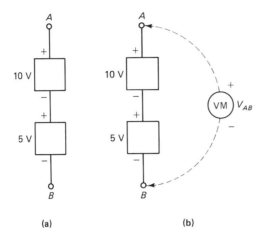

Figure 1-32 Technique for forming a voltmeter loop with a simple circuit.

(a)   (b)

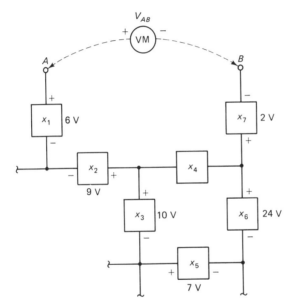

Figure 1-33 Voltmeter loop formed in a more complex circuit.

The concept is illustrated first by the simple circuit of Figure 1-32(a). An ideal voltmeter is assumed to be connected between points A and B as shown in Figure 1-32(b). The positive reference terminal of the voltmeter is connected to the assumed positive reference terminal of the voltage. Thus, to determine V_{AB}, the positive voltmeter terminal is connected to node A, and the negative voltmeter terminal is connected to node B. The voltage across the voltmeter is labeled as V_{AB}. An ideal voltmeter draws no current, so none of the branches are assumed to be disturbed by the voltmeter's presence.

The closed loop that results is now evident, and KVL may be applied. Starting at B and moving clockwise, we have

$$-5 - 10 + V_{AB} = 0 \tag{1-59}$$

or

$$V_{AB} = 15 \text{ V} \qquad \text{(1-60)}$$

which agrees with our earlier intuitive observation.

Consider next the more complex circuit of Figure 1-33. Observe that the path from A to B through x_1, x_2, x_3, x_5, x_6, and x_7 has all branch voltages known. Note that the voltage across branch x_4 is not known without a separate calculation, so that branch will be avoided in the path. (The voltage across x_4 could be calculated if desired. Can the reader supply this value?)

Consider that point A is the assumed positive reference. In this case we start at point A and move clockwise. Application of KVL yields

$$V_{AB} - 2 + 24 - 7 - 10 + 9 - 6 = 0 \qquad \text{(1-61)}$$

or

$$V_{AB} = -8 \text{ V} \qquad \text{(1-62)}$$

This result means that the voltage at point A is actually lower than the voltage at point B, so we write

$$V_{BA} = 8 \text{ V} \qquad \text{(1-63)}$$

A good practical measurement reminder will be inserted here. Assume that this was an actual circuit and that the voltmeter had been connected as shown. If the voltmeter had been a digital meter, the negative or minus indication would have been generated, and the meter could remain as originally connected to take the reading. However, an analog voltmeter would have experienced a negative deflection, and possible damage could result. When the polarity and approximate level of a voltage are unknown, start with the highest available meter scale. A quick initial check can be made to determine the polarity. If the deflection is negative, the meter terminals can be reversed so that a positive deflection results. The range switch can then be adjusted downward until an accurate measurement can be made.

Voltage Between Two Ground-Referred Points

Frequently, it is desired to determine the voltage between two points in a circuit, for which the voltage is known at each point with respect to ground. Consider the two nodes A and B in Figure 1-34(a), and assume that the respective voltages with respect to ground are V_A and V_B.

Figure 1-34 Process of determining the voltage between two ground-referred points using a voltmeter loop.

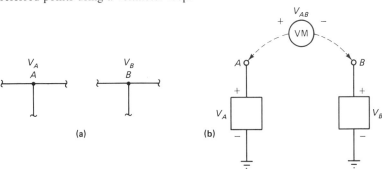

To determine the voltage between A and B, we can model the circuit as shown in Figure 1-34(b). Although there can be any number of branches between points A and B and ground, the voltages V_A and V_B can be visualized as shown. An ideal voltmeter is connected between A and B to form a loop according to the procedure discussed previously. Note that the positive references of V_A and V_B are at nodes A and B, respectively. However, either V_A, V_B, or both could be negative.

Starting at ground and moving in a clockwise direction through the left-hand part of the circuit, the KVL equation is

$$-V_A + V_{AB} + V_B = 0 \qquad \textbf{(1-64)}$$

or

$$\boxed{V_{AB} = V_A - V_B} \qquad \textbf{(1-65)}$$

Once understood, this relationship is very simple to apply. The voltage between any two points is the voltage of the assumed positive reference node with respect to ground minus the voltage of the assumed negative reference node with respect to ground.

As an example, refer back to Figure 1-30, in which $V_A = 8$ V and $V_B = -4$ V. The voltage V_{AB} between points A and B, with A the assumed positive reference, is

$$V_{AB} = V_A - V_B = 8 - (-4) = 12 \text{ V} \qquad \textbf{(1-66)}$$

1-12 SOURCE TRANSFORMATIONS

When the effects of inductance and capacitance are negligible, an actual electrical source of energy can be modeled by either an ideal voltage source in series with a resistance or an ideal current source in parallel with a resistance. As long as the source internal resistance is finite and nonzero, either form may be used to represent a real-life source. Some devices tend to be associated with one particular form. For example, it seems more natural to represent a battery by the voltage source model. As we will see later, the action of a transistor is more naturally represented by a current source model. However, theoretically, either model could be used for either case.

Some techniques of analysis are more easily performed when all sources are voltage sources, and other techniques are more easily performed when all sources are current sources. For that reason, the source transformations to be given here are useful in allowing conversion between the types of sources.

First, consider the voltage source having an instantaneous voltage in series with R_s, shown on the left in Figure 1-35. As far as the external terminals of the source are concerned, the circuit may be replaced by a current source of value v_s/R_s in parallel with a resistance R_s as shown on the right. Said differently, the current source model on the right will produce the same external effect as the voltage source model on the left.

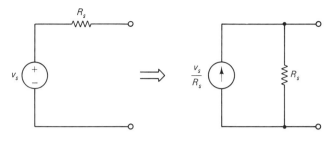

Figure 1-35 Transformation of voltage source to current source.

Figure 1-36 Transformation of current source to voltage source.

Next, consider the current source having an instantaneous current i_s in parallel with R_s, as shown on the left in Figure 1-36. As far as the external terminals of the source are concerned, the circuit may be replaced by a voltage source of value $R_s i_s$ in series with a resistance R_s as shown on the right. As in the previous case, the new circuit predicts the same external behavior as the circuit it replaces.

To convert a voltage source model to a current source model, a series resistance must be present with the voltage source model; and to convert a current source model to a voltage source model, a parallel resistance must be present with the current source model. This means that if a source is assumed to be ideal and is standing alone (i.e., not connected to a circuit), it is not possible to perform a conversion. However, by associating an assumed ideal source with external circuit components, it may still be possible to perform a source conversion if some care is used in interpreting the results.

Readers familiar with Thévenin's and Norton's theorems will recognize that the transformations given here may be considered as special cases of these theorems and may be interpreted as such. A general development of Thévenin's and Norton's theorems is given in Section 1-15.

Example 1-11

For the circuit of Figure 1-37(a), determine the voltage V_o two ways: **(a)** by analyzing the circuit directly as given and **(b)** by first performing a voltage-to-current source transformation to the left of terminals a–b and analyzing the modified circuit.

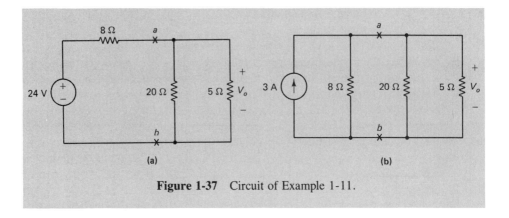

Figure 1-37 Circuit of Example 1-11.

Solution

(a) By treating the parallel combination of the 20-Ω resistor and the 5-Ω resistor as a single equivalent resistance, the voltage-divider rule may be applied directly to this circuit. We have

$$V_o = \frac{20\|5}{20\|5 + 8} \times 24 = \frac{4}{4 + 8} \times 24 = 8 \text{ V} \tag{1-67}$$

where it was determined that the parallel combination of 20 Ω and 5 Ω was 4 Ω.

(b) A source transformation applied to the left of terminals a–b yields a 3-A current source in parallel with 8-Ω as shown in Figure 1-37(b). This resulting circuit is of the single-node-pair form, for which a KCL equation at the upper node may be written as follows:

$$-3 + \frac{V_o}{8} + \frac{V_o}{20} + \frac{V_o}{5} = 0 \tag{1-68}$$

Simplification of this equation leads to

$$\frac{3}{8}V_o = 3 \tag{1-69}$$

$$V_o = 8 \text{ V} \tag{1-70}$$

as in part (a).

For this particular example, it took at least as much work to solve the circuit with the transformation as with the original circuit. However, the intent was to demonstrate the equivalence of the transformation process and to demonstrate an alternative visualization of the circuit layout. In many complex electronic circuits, significant simplification and visualization may be obtained through the source transformation process.

Both ideal and nonideal source models are considered in the text. In this section, the following three questions are investigated: (1) Under what conditions can a real source be represented by the ideal voltage source model? (2) Under what conditions can a real source be represented by the ideal current source model? (3) Under what conditions must the internal source resistance be retained in the model? These questions will be answered by considering two particular circuit conditions, denoted as the low internal resistance case and the high internal resistance case.

The *low internal resistance* case represents the situation in which the source resistance is small compared with the external load resistance connected to it. It is convenient in this case to use the voltage source model shown on the left in Figure 1-38(a). The resistance R_s represents the source resistance, and R_L represents the *net* load resistance. Since $R_s \ll R_L$, the voltage drop across R_s is very small compared with the drop across R_L. As a good approximation under many conditions, the circuit may be represented by the model shown on the right. The extent of the approximation depends on the degree of the inequality. As one particular criterion, if $R_s \le 0.01R_L$, the voltage or current computed in any external resistance will differ by no more than 1% from the true value. Under this condition the source is sometimes referred to as a *stiff* voltage source.

Some conclusions from this discussion are that (1) the voltage source model is more convenient when the internal source resistance is small compared with the load resistance; and (2) *if the internal source resistance is sufficiently small, it may be neglected altogether and the ideal voltage source model may be used.*

The *high internal resistance* case represents the situation in which the source

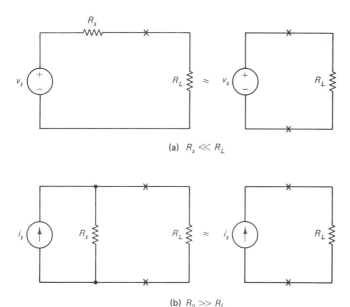

(a) $R_s \ll R_L$

(b) $R_s \gg R_L$

Figure 1-38 Realistic source approximations for very low and very high internal resistances.

resistance is large compared with the external load resistance connected to it. It is convenient in this case to use the current source model shown on the left in Figure 1-38(b). The resistance R_s represents the source resistance, and R_L represents the *net* load resistance. Since $R_s \gg R_L$, the current through R_s is very small compared with the current through R_L. As a good approximation under many conditions, the circuit may be represented by the model shown on the right. The extent of the approximation depends on the degree of inequality. As one particular criterion, if $R_s \geq 100R_L$, the voltage or current computed in any external resistance will differ by no more than 1% from the true value. Under this condition, the source is sometimes referred to as a *stiff* current source.

Some conclusions from this discussion are that (1) the current source model is more convenient when the internal source resistance is large compared with the load resistance; and (2) *if the internal source resistance is sufficiently large, it may be neglected altogether and the ideal current source model may be used.*

It should be noted that neglecting the internal resistance in the ideal current source model means replacing it with an *open circuit*. Thus *the internal resistance of an ideal current source is infinite*. This is in sharp contrast to an *ideal voltage source, in which the internal resistance is zero*.

When the internal source resistance and the load resistance do not satisfy inequalities of the types just discussed, the internal source resistance must be retained in the model and the choice of the voltage source in series with a resistance model versus the current source in parallel with a resistance model will depend on a number of factors, particularly the method of analysis employed.

1-14 DEPENDENT SOURCES

The voltage and current source models introduced in Section 1-3 represent the common independent source models. Independent sources are those whose values are independent of the levels of any voltages or currents in the circuit. The values of all voltage and current sources encountered thus far have either been independent constant values (corresponding to dc sources) or independent time-varying quantities.

Dependent (or *controlled*) source models are a special class in which the voltage or current is dependent on some other voltage or current within the circuit. Such sources arise from complex physical interactions in electronic devices. For example, the base current of a bipolar junction transistor causes a much larger collector current to flow, and this process is modeled by a dependent source, as we will see later.

A dependent source, like an independent source, may be either a voltage source or a current source. However, it may also be controlled by either a voltage or a current. This results in four different combinations for dependent source models, and these are illustrated in Figure 1-39. The symbol used for the dependent source models is in accordance with an increasing portion of the literature, although earlier texts employ the same symbol (circle) as for independent source models. Each of the four forms will now be discussed briefly.

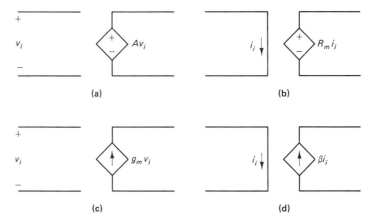

Figure 1-39 Four possible models of ideal dependent (or controlled) sources.

Voltage-Controlled Voltage Source

The model for the voltage-controlled voltage source (VCVS) is shown in Figure 1-39(a). An independent control voltage v_i is assumed to exist across certain control terminals. The voltage controls a dependent voltage whose value is Av_i. The quantity A is dimensionless and corresponds to a *voltage gain* when the model represents a voltage amplifier.

Current-Controlled Voltage Source

The model for the current-controlled voltage source (ICVS) is shown in Figure 1-39(b). An independent control current i_i is assumed to be flowing in certain control terminals. This current controls a dependent voltage whose value is $R_m i_i$. The quantity R_m, relating the dependent voltage to the controlling current, has the dimensions of ohms and is called the *transresistance*.

Voltage-Controlled Current Source

The model for the voltage-controlled current source (VCIS) is shown in Figure 1-39(c). An independent control voltage v_t is assumed to exist across certain control terminals. This voltage controls a dependent current whose value is $g_m v_i$. The quantity g_m, relating dependent current to the controlling voltage, has the dimensions of siemens and is called the *transconductance*.

Current-Controlled Current Source

The model for the current-controlled current source (ICIS) is shown in Figure 1-39(d). An independent control current i_i is assumed to be flowing in certain control terminals. This current controls a dependent current whose value is βi_i. The quantity β is dimensionless and corresponds to a *current gain* when the model represents a current amplifier.

1-15 THÉVENIN'S AND NORTON'S THEOREMS

Thévenin's and Norton's theorems are very important in analyzing electronic circuits. They permit the reduction of complex circuits to very simple forms whenever the primary focus of interest is on the external behavior of the complex circuits. First, a brief statement of each of the theorems as applied to resistive circuits will be made, and explorations of the procedures for determining the equivalent circuits will then be given.

Thévenin's theorem as applied to resistive circuits states that all effects of any linear circuit external to two reference terminals can be completely predicted from a model consisting of a single ideal voltage source in series with a single resistor. The concept is illustrated in Figure 1-40. The voltage v_t will be referred to as the Thévenin voltage, and for reasons that will shortly be seen to be logical, R_{eq} will be referred to simply as the equivalent resistance.

Norton's theorem as applied to resistive circuits states that all effects of any linear circuit external to two reference terminals can be completely predicted from a model consisting of a single current source in parallel with a single resistor. The concept is illustrated in Figure 1-41. The current i_n will be referred to as the Norton current, and R_{eq} is the equivalent resistance.

For a given circuit and a given set of terminals, R_{eq} is the same for either the Thévenin or the Norton model, and for that reason, the terms "Thévenin resistance" and "Norton resistance," which are used in some books, are not used here. Such terms could give the impression that the resistance is different for the two models, which is not the case.

The reader should recognize that if either the Thévenin or the Norton model is known, the other may readily be determined with one of the source transformations

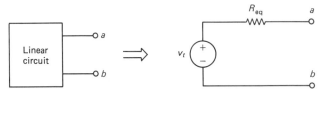

Figure 1-40 Thévenin model of an electrical circuit at a set of reference terminals.

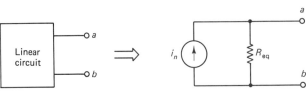

Figure 1-41 Norton model of an electrical circuit at a set of reference terminals.

of Section 1-12. For example, if the Thévenin model is known, the Norton current is readily determined as

$$i_n = \frac{v_t}{R_{\text{eq}}} \tag{1-71}$$

Further, if the Norton model is known, the Thévenin voltage is determined as

$$v_t = R_{\text{eq}} i_n \tag{1-72}$$

At the outset, a fundamental point concerning the external nature of those models must be stressed. When a complex circuit is replaced by a Thévenin or a Norton model, the model predicts correct results *external* to the reference terminals only. The internal action of the original circuit has been "lost," and any internal calculations in the Thévenin or Norton model are generally meaningless. For example, power delivered by the sources v_t or i_n generally is not equivalent to the power delivered by the source or sources in the original circuit. If any internal calculations are required, the original circuit must be reconstructed in the analysis process. This is a limitation frequently misunderstood by beginners in circuit analysis.

The Thévenin and Norton models are most useful when there is a portion of a circuit that remains fixed for which there is little interest in the internal behavior but in which the effect on an external circuit is to be studied under varying conditions. Rather than perform a complete circuit analysis for each load, the static part of the circuit is replaced by its Thévenin or Norton model, and the effect for each load condition is then more easily studied.

A few comments about the practical nature of Thévenin's and Norton's theorem are in order. Although laboratory experiments to verify these theorems are widely performed in electrical circuits courses, their major utility is in "paper" analysis. Most real-life practical circuits are wired a certain way to perform a job and, occasionally, the use of one of these models may suggest a better way to design the circuit. However, the major benefit from the theorems is in analyzing the circuits on paper and in simplifying the calculations. Thus, in using an expression such as "replacing a circuit with a Thévenin equivalent," we will be referring to the analysis of these circuits rather than to their construction. Indeed, some of the operations may be totally impractical or unrealistic when viewed from a practical point of view. For example, reference will be made to determining the "short-circuit current" in certain steps that follow, and this may be totally impractical (or even dangerous) in a realistic situation.

Having established the concepts of Thévenin's and Norton's theorems, we now turn to the necessary task of determining the values of the Thévenin voltage, the Norton current, and the equivalent resistance. The problem is complicated by the fact that the easier methods to apply do not work in the most general cases, and the more general methods tend to be clumsy when applied to the simpler circuits. No doubt these facts contribute to the difficulty that many students encounter when determining Thévenin's and Norton's equivalent circuits.

For the purpose of Thévenin and Norton simplification, it is convenient to divide circuits into two categories: (1) circuits containing only independent sources and (2) circuits containing one or more dependent sources.

Independent Sources

Refer to Figure 1-42 in the discussion that follows. When a circuit contains only independent sources, the following procedure may be used:

1. Determine the open-circuit voltage v_{oc} across the two reference terminals. The value of this open-circuit voltage is the Thévenin voltage, that is,

$$v_t = v_{oc}$$ (1-73)

2. Deenergize all internal sources, and determine the equivalent resistance as viewed from the two reference terminals. This resistance is the equivalent resistance R_{eq}.

To deenergize sources as indicated in step 2, ideal *voltage sources* are replaced by *short circuits* and ideal *current sources* are replaced by *open circuits*. Since non-ideal sources contain resistances, these resistances will remain in the circuit models when the ideal source portions of the actual models are deenergized by the preceding steps.

Although any Thévenin model could readily be replaced by a Norton model by use of a circuit transformation, a separate statement of a general approach for determining the Norton model is needed for completeness. Furthermore, it is easier in many cases to determine the Norton model directly from the given circuit.

A general approach for determining the Norton equivalent circuit for a resistive circuit *when the circuit contains only independent sources* is illustrated in Figure 1-43. The procedure consists of two steps:

1. Assume that a short circuit (on paper) is placed across the two reference terminals. Determine the current i_{sc} that would flow through the short. The value of this short-circuit current is the Norton current i_n, that is,

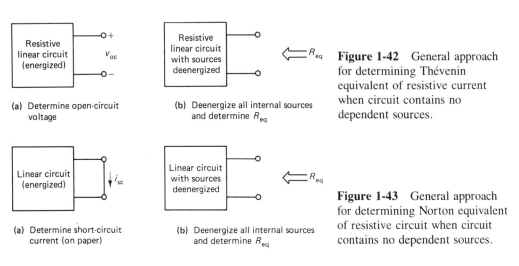

(a) Determine open-circuit voltage

(b) Deenergize all internal sources and determine R_{eq}

Figure 1-42 General approach for determining Thévenin equivalent of resistive current when circuit contains no dependent sources.

(a) Determine short-circuit current (on paper)

(b) Deenergize all internal sources and determine R_{eq}

Figure 1-43 General approach for determining Norton equivalent of resistive circuit when circuit contains no dependent sources.

$$\boxed{i_n = i_{sc}}$$ **(1-74)**

2. The second step is the same as for determining the Thévenin equivalent circuit; that is, deenergize all internal sources, and determine the equivalent resistance R_{eq} as viewed from the two reference terminals.

It can now be stated that the source transformations of Section 1-12 were nothing more than special cases of Thévenin and Norton equivalent circuits. Reviewing Figure 1-35, the value of the current source in the right-hand model represents the current that would flow through a short circuit connected across the terminals of the left-hand model. In Figure 1-36, the value of the voltage source in the right-hand model represents the open-circuit voltage that would be measured in the left-hand model. In all these cases, the equivalent resistance, $R_{eq} = R_s$, is the resistance that would be viewed from the output terminals when the internal sources are deenergized according to the procedures delineated earlier.

Although these procedures may, in theory, be applied to any circuit containing only independent sources, they are often clumsy and inefficient. Consider, for example, a circuit containing a number of meshes and nodes for which a Thévenin model is desired at a set of terminals on one side of the circuit. To determine the open-circuit voltage or the short-circuit current, the solution of several simultaneous equations may be required. Although this may be necessary if all else fails, a simpler step-by-step approach is often possible.

A large number of circuits that the "average" technical person will need to analyze in detail can be reduced to Thévenin or Norton form by successive application of the source transformations discussed in Section 1-12, coupled with combining resistances by series and parallel equivalents. Although this approach does not work in all cases, it works in enough important and useful cases to justify a major focus of our efforts in that direction. It is difficult to identify a simple "formula" for recognizing such a circuit, but with some practice, an intuitive process will quickly evolve. As a general guideline, if the circuit is one in which an equivalent resistance could be determined by successive series and parallel combinations if all sources were deenergized, the technique under consideration will usually work.

Dependent Sources

The method discussed here is quite general and may be employed when the circuit contains dependent sources. The procedure is discussed in the steps that follow and is referred to as the *external generator method*.

First, *either* the open-circuit voltage or the short-circuit current is calculated as was done for circuits having only independent sources. The variable that is calculated could depend on which is easier to determine or whether the Thévenin or Norton circuit is the form desired. Actually, it is usually not critical which is determined because eventually a source transformation can be used to change the form after the equivalent resistance is determined.

The procedure for determining the equivalent resistance is illustrated in Figure 1-44. All *independent sources* are first deenergized. This means that all ideal inde-

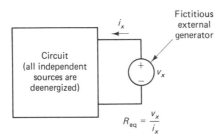

Fictitious external generator

Figure 1-44 Application of fictitious external generator to determine Thévenin and Norton equivalent resistance.

pendent voltage sources are replaced by short circuits, and all ideal independent current sources are replaced by open circuits. A fictitious external generator is then connected to the output terminals. Although the source may be either a voltage source or a current source, we will assume a voltage source v_x for this discussion. The current i_x flowing from this source is calculated. The equivalent resistance is then

$$R_{eq} = \frac{v_x}{i_x} \tag{1-75}$$

Alternatively, a current source could be chosen and the voltage could be calculated. The relationship of (1-75) is still the same.

It should be stressed that while all independent sources are deenergized, any dependent sources that are functions of circuit variables that would not be zero when the external generator is applied must be retained in the analysis.

Example 1-12

Determine the Thévenin equivalent circuit for the dc circuit of Figure 1-45(a) at the terminals x-y two ways: **(a)** by determining V_{oc} and R_{eq} directly and **(b)** by successive internal reduction.

Figure 1-45 Circuit of Example 1-12 and the development of the Thévenin equivalent circuit.

(a)

(b)

(c)

(d)

Solution

(a) This particular structure occurs very frequently in practical problems, which justifies the decision to use two different approaches for illustration. Reference should be made to Figure 1-45 in this part. First, the open-circuit voltage V_{oc} at terminals x–y must be determined, and this process is illustrated in Figure 1-45(b). An important point is that the open-circuit voltage across the output is the same as the voltage across the 6-Ω resistor. The reason is that no current flows through the 5-Ω resistor under open-circuit conditions, and there is no voltage drop across it. In determining the open-circuit voltage, imagine that one is using an ideal voltmeter, which would theoretically draw no current from the circuit to which it is connected. The type of situation illustrated here occurs often in performing open-circuit voltage measurements, so it is important to understand this concept fully.

Since the open-circuit voltage is the same as the voltage across the 6-Ω resistor, a simple application of the voltage-divider rule to the left-hand loop may be used to determine V_{oc}. We have

$$V_{oc} = \frac{6}{6+3} \times 12 = 8 \text{ V} \tag{1-76}$$

To determine the equivalent resistance, the 12-V independent voltage source is replaced by a short circuit as shown in Figure 1-45(c). The resistance "looking in" from terminals x–y is then determined. Note that the 3-Ω and 6-Ω resistors turn out to be in parallel as far as R_{eq} is concerned. The parallel equivalent resistance is 2 Ω, which is in series with 5 Ω. The net equivalent resistance is thus

$$R_{eq} = 5 + 2 = 7 \text{ } \Omega \tag{1-77}$$

The Thévenin equivalent circuit is shown in Figure 1-45(d).

(b) The various steps involved in a successive reduction process are delineated in Figure 1-46. First, the 12-V voltage source in series with a 3-Ω resistor is converted to 4 A in parallel with 3 Ω, as shown in (a). This circuit is then connected to the 6-Ω shunt resistor, and the parallel combination of the two resistors is determined as shown in (b). The resulting 4-A current source in parallel with 2 Ω is converted to 8 V in series with 2 Ω, as shown in (c). The additional 5 Ω is connected as shown in (d) and the net series resistance of 7 Ω is determined. The resulting circuit is the Thévenin equivalent circuit.

For this particular circuit form, the author personally prefers the first method because it is more direct. However, the next example will illustrate a circuit in which it would be very difficult to determine the open-circuit voltage directly, but for which successive internal reduction may readily be applied.

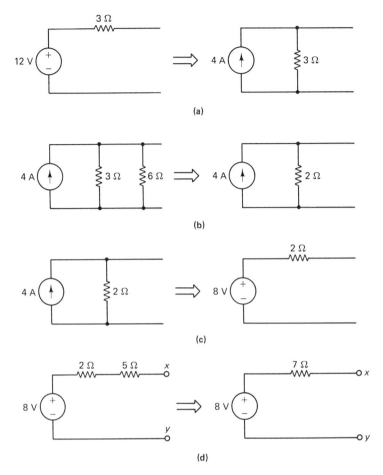

(a)

(b)

(c)

(d)

Figure 1-46 Alternative development of the Thévenin equivalent circuit in Example 1-12.

Example 1-13

Determine the Thévenin and Norton equivalent circuits at terminals x–y in Figure 1-47.

Figure 1-47 Circuit of Example 1-13.

Solution

The process of determining V_{oc} directly with this circuit would require a somewhat detailed evaluation, so a step-by-step approach using source transformations will be used. In the steps that follow, reference should be made to the
circuit diagrams of Figure 1-48.

First, the circuit to the left of the 11-Ω resistor in Figure 1-47 is reduced
to a 20-V source in series with 4 Ω, as shown in (a) of Figure 1-48. The details of this reduction are not shown, but it can be achieved by calculating the
open-circuit voltage across the 12-Ω resistor and then determining the parallel
combination of the two resistors after the source is deenergized. Next, the 11-
Ω resistor and 40-V source are combined with the results of the preceding
step, and the circuit is simplified as shown in (b).

To combine this circuit with the 7-A shunt current source in (d), the
voltage sources and series resistor are first converted to a current source in
parallel with a resistor as shown in (c). Then, when the shunt current source is
added, the two current sources may be combined as shown in (d).

The next step consists of converting back to a voltage source form, as
shown in (e). The additional 5-Ω resistance is included and combined with the
15-Ω resistance as shown in (f), which is the desired Thévenin form.

The final Thévenin model is shown again in Figure 1-49(a). A direct application of a source transformation leads to the Norton model shown in (b).

Figure 1-48   Successive reduction of the circuit of Example 1-13.

Figure 1-49 Thévenin and Norton models for Example 1-13.

Example 1-14

The circuit of Figure 1-50(a) represents a field-effect-transistor common-drain amplifier, whose details will be studied much later in the book. At this time, the so-called small-signal equivalent circuit will be given and is shown in Figure 1-50(b). The quantity v_i is the input signal voltage, and v_o is the output signal voltage. The parameter g_m is the transconductance of the transistor. Apply Thévenin's theorem to determine a new equivalent circuit in which the source is not a function of the output voltage v_o.

Figure 1-50 Common-drain field-effect transistor amplifier and equivalent signal model of Example 1-14.

Solution

The source in the model is a dependent source and is a function of both v_i and v_o. First, the open-circuit output voltage will be determined, and this quantity is denoted as v_{ooc}. The process for "measuring" this voltage is shown in Figure 1-51(a). Note that since the output voltage for this condition is $v_o = v_{ooc}$, the value of v_o in the expression for the dependent source is also replaced by v_{ooc}.

From the output portion of the equivalent circuit, we have

$$v_{ooc} = g_m(v_i - v_{ooc})R_s \qquad \text{(1-78)}$$

This expression may seem strange in that v_{ooc} is expressed in terms of itself! However, the situation is resolved by regrouping all terms containing v_{ooc} on the left side of the equation, which results in

$$v_{ooc}(1 + g_m R_s) = g_m R_s v_i \qquad \text{(1-79)}$$

Solving for v_{ooc}, we have

$$v_{ooc} = \frac{g_m R_s v_i}{1 + g_m R_s} = v_t \qquad \text{(1-80)}$$

Thus the Thévenin voltage v_t is now a function only of the independent input signal v_i; it is not a function of the output voltage v_o.

Next, the external generator method will be used to determine the equivalent output resistance. The steps involved are shown in Figure 1-51(b). The input signal v_i is deenergized (i.e., replaced by a short circuit). A fictitious generator of value v_x is connected across the output terminals as shown, and a current i_x is identified as flowing into the output. Note that since $v_i = 0$ and $v_o = v_x$, the value of the dependent source is $g_m(v_i - v_o) = g_m(0 - v_x) = -g_m v_x$. (The direction could be reversed if desired and the value would then be be $g_m v_x$.)

(a)

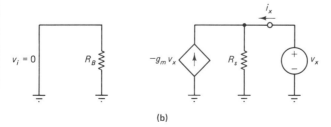

(b)

Figure 1-51 Steps involved in the determination of the Thévenin equivalent circuit of Example 1-14.

Application of KCL to the upper node of the circuit yields

$$-(-g_m v_x) + \frac{v_x}{R_s} - i_x = 0 \qquad \textbf{(1-81)}$$

Rearranging terms, we have

$$v_x \frac{1 + g_m R_s}{R_s} = i_x \qquad \textbf{(1-82)}$$

The equivalent resistance is

$$R_{eq} = \frac{v_x}{i_x} = \frac{R_s}{1 + g_m R_s} \qquad \textbf{(1-83)}$$

The resulting Thévenin equivalent circuit is shown in Figure 1-52.

Note that the effective output resistance is $R_s/(1 + g_m R_s)$, which is quite different from the value of R_s appearing initially in the circuit. This example illustrates how the presence of dependent sources can strongly affect the net circuit properties.

Figure 1-52 Final equivalent circuit of common-drain amplifier showing voltage gain and output resistance.

1-16 SUPERPOSITION

The *principle of superposition* states that any voltage or current response in a linear circuit resulting from several voltage and/or current sources may be determined by first considering the response produced by each source individually and then algebraically combining the individual responses. As the effect of each source is considered, all other sources are deenergized. To deenergize an ideal *voltage source*, it is replaced by a *short circuit*. To deenergize an ideal *current source*, it is replaced by an *open circuit*. Any internal resistances assumed in the source models are retained in the circuit when the ideal sources are deenergized. The use of superposition will be illustrated with an example.

Example 1-15

Consider the circuit of Example 1-13 (Figure 1-47), for which Thévenin and Norton equivalent circuits were established at terminals x–y. The circuit was simplified in Example 1-13 by using successive source transformations. Use superposition to compute V_{oc} from the original network structure.

Solution

For convenience, the circuit is repeated in Figure 1-53(a), and the desired variable V_{oc} is indicated. Since there are three sources in the circuit, there are three contributions to the voltage. Let

$$V_{oc} = V'_{oc} + V''_{oc} + V'''_{oc} \tag{1-84}$$

where V'_{oc} is the response due to the 30-V source, V''_{oc} is the response due to the 40-V source, and V'''_{oc} is the response due to the 7-A source.

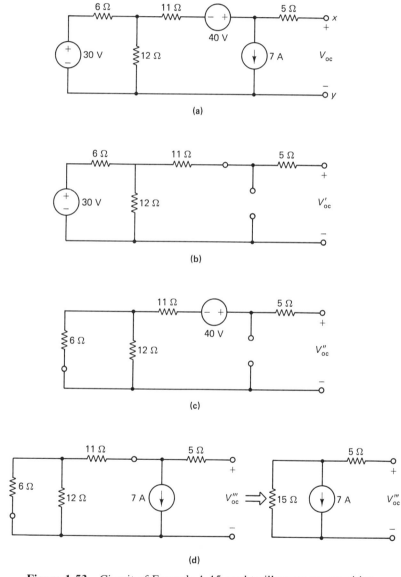

(a)

(b)

(c)

(d)

Figure 1-53 Circuit of Example 1-15 used to illustrate superposition.

To determine V'_{oc}, we deenergize the 40-V source and the 7-A source as shown in Figure 1-53(b). Note that the 40-V source is replaced by a short circuit and the 7-A source is replaced by an open circuit. Since no current flows through the 11-Ω and 5-Ω resistors, V'_{oc} is the voltage across the 12-Ω resistor, and this is determined by the voltage-divider rule as follows:

$$V'_{oc} = \frac{12}{12 + 6} \times 30 = 20 \text{ V} \tag{1-85}$$

To determine V''_{oc}, we deenergize the 30-V source and the 7-A source as shown in Figure 1-53(c). No current flows in this case, so the open-circuit voltage is the same as the source, that is,

$$V''_{oc} = 40 \text{ V} \tag{1-86}$$

To determine V'''_{oc}, we deenergize the 30-V source and the 40-V voltage as shown on the left in Figure 1-53(d). To simplify this circuit, the resistance in parallel with the 7-A source is determined as 15 Ω, as shown on the right. The open-circuit voltage will have its more positive terminal at the bottom, and V'''_{oc} is

$$V'''_{oc} = -7 \times 15 = -105 \text{ V} \tag{1-87}$$

The net voltage from (1-84) is

$$V_{oc} = 20 + 40 - 105 = -45 \text{ V} \tag{1-88}$$

This result is in agreement with the outcome of Example 1-13 as shown in Figures 1-48 and 1-49.

DRILL PROBLEMS

1-1. For each of the branches shown in Figure P1-1, determine the power and whether it is being delivered or absorbed. (The components are not identified.)

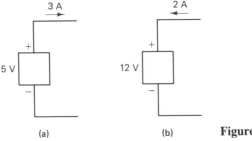

(a)   (b)   **Figure P1-1**

1-2. For each of the branches shown in Figure P1-2, determine the power and whether it is being delivered or absorbed. (The components are not identified.)

(a)   (b)   **Figure P1-2**

1-3. The dc voltage across a 2.7-kΩ resistor is 12.6 V. Determine the dc current and the power.

1-4. The dc voltage across a 68-kΩ resistor is 30 V. Determine the dc current and the power.

1-5. The dc current through a 3-kΩ resistor is 12 mA. Determine the dc voltage and the power.

1-6. The dc current through a 150-kΩ resistor is 8 μA. Determine the dc voltage and the power.

1-7. The dc voltage and current associated with a certain resistor are 30 V and 2 mA. Determine the resistance and the power.

1-8. The dc voltage and current associated with a certain resistance are 60 V and 3 μA. Determine the resistance and the power.

1-9. A certain resistor has a resistance of 2 kΩ. Determine the conductance.

1-10. A certain resistor has a conductance of 8 μS. Determine the resistance.

1-11. A certain resistor dissipates 0.4 W when the voltage is 20 V. Determine the resistance.

1-12. A certain resistor dissipates 0.2 W when the current is 50 mA. Determine the resistance.

1-13. In the circuit of Figure P1-13, the source voltage and the voltage at the right-hand node with respect to ground are known. Two resistance values are known. Determine V_1, I_1, I_2, and R.

Figure P1-13

1-14. In the circuit of Figure P1-14, the two source voltages and the voltage across one resistor are known. Determine I_1, I_2, and R.

Figure P1-14

1-15. For the circuit of Figure P1-15, determine V_o and R.

Figure P1-15

1-16. For the circuit of Figure P1-16, determine I_1, I_2, and V_o.

Figure P1-16

1-17. Using successive series and parallel reduction, determine the equivalent resistance at the input terminals for the circuit of Figure P1-17.

Figure P1-17

1-18. Using successive series and parallel reduction, determine the equivalent resistance at the input terminals for the circuit of Figure P1-18.

Figure P1-18

1-19. For the circuit of Figure P1-19, determine **(a)** the loop current I, **(b)** the voltage drops across all resistors, and **(c)** the power associated with each element and whether it is absorbed or delivered.

Figure P1-19

1-20. For the circuit of Figure P1-20, determine **(a)** the loop current I, **(b)** the voltage drops across all resistors, and **(c)** the power associated with each element and whether it is absorbed or delivered.

Figure P1-20

1-21. For the circuit of Figure P1-21, determine **(a)** the loop current I, **(b)** the voltage drops across all resistors, and **(c)** the voltage V_o across the current source.

Figure P1-21

1-22. For the circuit of Figure P1-22, determine **(a)** the loop current I, **(b)** the voltage drops across all resistors, and **(c)** the voltage V_o across the current source.

Figure P1-22

1-23. For the circuit of Figure P1-23, determine **(a)** the node-pair voltage V and **(b)** the currents through all resistors.

Figure P1-23

1-24. For the circuit of Figure P1-24, determine **(a)** the node-pair voltage V and **(b)** the currents through all resistors.

Figure P1-24

1-25. For the circuit of Figure P1-25, determine **(a)** the node-pair voltage V, **(b)** the currents through all resistors, and **(c)** the current I_o through the voltage source.

Figure P1-25

1-26. For the circuit of Figure P1-26, determine **(a)** the node-pair voltage V, **(b)** the currents through all resistors, and **(c)** the current I_o through the voltage source.

Figure P1-26

1-27. Use the voltage-divider rule to determine the voltages V_1 and V_2 in Figure P1-27.

Basic Circuit Analysis   Chap. 1

Figure P1-27

1-28. Use the voltage-divider rule to determine the voltages V_1 and V_2 in Figure P1-28.

Figure P1-28

1-29. Use the current-divider rule to determine the currents I_1 and I_2 in Figure P1-29.

Figure P1-29

1-30. Use the current-divider rule to determine the current I_1 in Figure P1-30.

Figure P1-30

1-31. Assume that a 100-V dc source is connected across the input terminals in the circuit of Figure P1-17. Determine all branch voltages and currents.

Drill Problems **59**

1-32. Assume that a 120-V dc source is connected across the input terminals in the circuit of Figure P1-18. Determine all branch voltages and currents.

1-33. Determine the voltage V_{AB} in Figure P1-33.

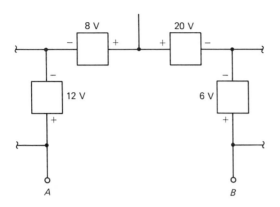

Figure P1-33

1-34. Determine the voltage V_{AB} in Figure P1-34.

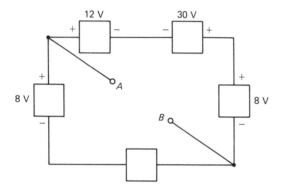

Figure P1-34

1-35. The voltages referred to ground at two points A and B are $V_A = -6$ V and $V_B = -18$ V. Determine V_{AB}.

1-36. The voltages referred to ground at two points A and B are $V_A = 6$ V and $V_B = -14$ V. Determine V_{AB}.

1-37. Perform a source transformation on the circuit of Figure P1-37 to obtain an equivalent current source.

Figure P1-37

1-38. Perform a source transformation on the circuit of Figure P1-38 to obtain an equivalent current source.

Figure P1-38

1-39. Perform a source transformation on the circuit of Figure P1-39 to obtain an equivalent voltage source.

Figure P1-39

1-40. Perform a source transformation on the circuit of Figure P1-40 to obtain an equivalent voltage source.

Figure P1-40

1-41. The load to be connected to the circuit of Figure P1-41 will always be greater than 5 kΩ. Based on a maximum 1% error, determine the form and value of an ideal source approximation.

Figure P1-41

1-42. The load to be connected to the circuit of Figure P1-42 will always be less than 2 kΩ. Based on a maximum 1% error, determine the form and value of an ideal source approximation.

Figure P1-42

1-43. The load to be connected to the circuit of Figure P1-43 will always be less than 5 kΩ. Based on a maximum 1% error, determine the form and value of an ideal source approximation.

Figure P1-43

1-44. The load to be connected to the circuit of Figure P1-44 will always be greater than 200 kΩ. Based on a maximum 1% error, determine the form and value of an ideal source approximation.

Figure P1-44

1-45. Determine the Thévenin equivalent circuit for the dc circuit of Figure P1-45 at the terminals x-y.

Figure P1-45

1-46. Determine the Thévenin equivalent circuit for the dc circuit of Figure P1-46 at the terminals x-y.

Figure P1-46

1-47. Determine the Thévenin equivalent circuit for the circuit of Figure P1-47 at the terminals x-y.

Figure P1-47

1-48. Determine the Thévenin equivalent circuit for the circuit of Figure P1-48 at the terminals x-y.

Basic Circuit Analysis   Chap. 1

Figure P1-48

1-49. The open-circuit output voltage of a certain electronic circuit is 8 V. When a resistance of 5 kΩ is connected across the output, the voltage drops to 4 V. Determine the Thévenin equivalent circuit. (*Hint*: Look ahead to Problem 1-55.)

1-50. The open-circuit output voltage of a certain electronic circuit is 6 V. When a 120-Ω load is connected across the output, the voltage drops to 5.5 V. Determine the Thévenin equivalent circuit. (*Hint*: Look ahead to Problem 1-56.)

1-51. Use the principle of superposition to determine the open-circuit voltage in the circuit of Figure P1-47 by considering the effects of the individual sources separately.

1-52. Use the principle of superposition to determine the open-circuit voltage in the circuit of Figure P1-48 by considering the effects of the individual sources separately.

DERIVATION PROBLEMS

1-53. Consider the voltage-divider circuit of Figure 1-21(a). Verify that if $R_o = 100R_1$, $v_o/v_i > 0.99$, which means that the loss in input voltage through the divider is less than 1% of the available input voltage.

1-54. Consider the current-divider circuit of Figure 1-23. Verify that if $R_1 = 100R_o$, $i_o/i_i > 0.99$, which means that the loss in input current through the divider is less than 1% of the available input current.

1-55. The determination of the Thévenin or Norton equivalent circuit of a signal source in the laboratory usually requires a procedure completely different from the "paper" procedures discussed in the chapter. A procedure that works well for small-signal sources with moderate to high values of internal resistance is illustrated in Figure P1-55. First, the open-circuit voltage v_{oc} is

(a) Measure v_{oc} (b) Measure v_L **Figure P1-55** measured, and we know that $v_t = v_{oc}$. Next a variable resistance R_L is connected across the output, and the resistance is adjusted until the load voltage v_L is

$$v_L = \frac{v_{oc}}{2}$$

At this point R_L is disconnected and measured. Show that the equivalent resistance R_{eq} is

$$R_{eq} = R_L$$

1-56. The procedure of Problem 1-55 for determining the equivalent resistance of a source is generally not suitable for low internal resistance sources due to the excessive loading when $R_L = R_{eq}$. Referring again to Figure P1-55, an alternative procedure is to adjust R_L for only a moderate decrease in the terminal voltage. For a terminal voltage v_L with load, show that the value of R_{eq} is

$$R_{eq} = \frac{(v_{oc} - v_L)R_L}{v_L}$$

1-57. A requirement that frequently arises in laboratory testing is to provide a simple voltage divider to attenuate (reduce) the level of a signal without changing the output resistance. Referring to Figure P1-57, assume that a given laboratory generator has an open-circuit voltage v_g and an output resistance R_g. The resistances R_1 and R_2 are added externally, and the following requirements are imposed:

$$\frac{v_o}{v_g} = \alpha \qquad (\alpha < 1)$$

R_{eq} = Thévenin equivalent resistance looking back from output = R_g.

Show that the resistances R_1 and R_2 are

$$R_1 = \frac{1 - \alpha}{\alpha}R_g$$

$$R_2 = \frac{R_g}{1 - \alpha}$$

Figure P1-57

1-58. An ideal voltmeter would have infinite resistance and would present no loading to a circuit. Practical voltmeters have a finite resistance, of course. In general, the resistance should, if possible, be very large compared with the Thévenin resistance seen at the point of measurement. With passive voltmeters not containing electronic isolation, this is sometimes difficult to achieve with high-resistance circuits.

The actual resistance of many passive voltmeters is a function of the scale range. Assume that two voltage measurements, based on two different ranges, are made. On a range in which the voltmeter resistance is R_1, the voltage is measured as V_1, and on a scale in which the voltmeter resistance is R_2, the voltage is measured as V_2. Show that the true open-circuit voltage V_{oc} and Thévenin equivalent resistance R_{eq} are given by

$$R_{eq} = \frac{R_1 R_2 (V_2 - V_1)}{V_1 R_2 - V_2 R_1}$$

$$V_{oc} = \frac{V_1 V_2 (R_2 - R_1)}{V_1 R_2 - V_2 R_1}$$

(*Hint*: A voltage-divider model may be constructed in which the measured voltage is the drop across the voltmeter resistance.)

DESIGN PROBLEMS

1-59. The range of an ammeter may be extended by placing a shunt resistance R_s in parallel with the ammeter. The resistance is selected to absorb the current in excess of full-scale meter current. Assume that a given 0–1 mA ammeter is to be extended to cover the range 0 to 10 mA. If the internal resistance is 50 Ω, determine the required value of the shunt resistance.

1-60. A classical approach to designing dc voltmeters is illustrated in Figure P1-60. A basic D'Arsonval galvanometer is used as the indicating device. The series resistances are chosen such that the full-scale current I_{fs} flows when the maximum voltage for a given range is applied. For the circuit illustrated, assume that the four full-scale desired voltages are 3 V, 10 V, 30 V, and 100 V. Assume that $I_{fs} = 100 \ \mu A$ and that the meter contains an internal series resistance $R_m = 1 \ k\Omega$. Determine the required design values of the four resistances R_1, R_2, R_3, and R_4.

Figure P1-60

1-61. Using the results of Problem 1-57, assume that a given laboratory generator has an output resistance of 50 Ω and that it is desired to reduce the output voltage by a 10 : 1 ratio (i.e., $\alpha = 0.1$). Design the attenuation circuit.

1-62. Repeat Problem 1-61 if $\alpha = 0.01$.

TROUBLESHOOTING PROBLEMS

1-63. Assume that the circuit of Figure P1-27 is not working as expected and that you measure the following voltages: $V_1 = 4.6$ V and $V_2 = 15.4$ V. Which one of the following is the likely trouble? **(a)** 6-kΩ resistor open; **(b)** 6-kΩ resistor shorted; **(c)** 5-kΩ resistor open; **(d)** 20-kΩ resistor open; **(e)** incorrect value of dc input voltage.

1-64. Assume that the circuit of Figure P1-28 is not working as expected and that you measure the following voltages: $V_1 = 51.4$ V and $V_2 = 8.6$ V. Which one of the following is the likely trouble? **(a)** 18-kΩ resistor open; **(b)** 36-kΩ resistor open; **(c)** 6-kΩ resistor open; **(d)** 6-kΩ resistor shorted; **(e)** 2-kΩ resistor open; **(f)** 2-kΩ resistor shorted.

COMPUTER PROBLEMS

1-65. Write a program to determine the current and voltages in a single-loop circuit with up to four resistors and up to four dc voltage sources.

Input Data: R_1, R_2, R_3, R_4, V_{s1}, V_{s2}, V_{s3}, V_{s4}
Output Data: I, V_{R_1}, V_{R_2}, V_{R_3}, V_{R_4}

1-66. Write a program to determine the voltage and currents in a single node-pair circuit with up to four resistors and up to four dc current sources.

Input Data: R_1, R_2, R_3, R_4, I_{s1}, I_{s2}, I_{s3}, I_{s4}
Output Data: V, I_{R_1}, I_{R_2}, I_{R_3}, I_{R_4}

1-67. Consider the ladder network shown in Figure P1-67, containing N resistors arranged in the systematic form shown. Write a computer program to determine the equivalent resistance R_{eq}.

Input Data: N, R_1, R_2, . . . , R_N
Output Data: R_{eq}

(*Note:* Depending on whether N is even or odd, the last resistor R_N may be either a series resistor, as shown, or a shunt resistor.)

Figure P1-67

CHAPTER 2

Semiconductor Diodes

OBJECTIVES

After completing this chapter, the reader should be able to:

1. State the three classifications of materials in terms of their relative conductivity.
2. State the two most common types of semiconductors.
3. State the two types of charge carriers in semiconductors.
4. Explain briefly the concepts of conduction band and valence band.
5. Classify the types of impurities and the types of charge carriers they produce.
6. Explain the difference between *p*-type material and *n*-type material.
7. Discuss the construction of a *pn* junction diode.
8. Discuss the depletion layer and its effect.
9. Draw the symbol of a diode, and identify the anode and cathode terminals.
10. Show the directions of "easy" and "hard" current flow.
11. Explain the significance of a terminal characteristic and indicate the independent and dependent variables.
12. Sketch the form of the terminal characteristic of a *pn* junction diode.
13. Discuss the properties of the diode terminal characteristic for the forward-biased region and for the reverse-biased region.
14. Construct the circuit form and the terminal characteristic of the ideal diode model.
15. Construct the circuit form and the terminal characteristic of the constant voltage model for the diode.
16. Construct the circuit form and the terminal characteristic of the constant voltage source plus resistance model of the diode.
17. Analyze simple dc circuits containing a diode using one of the models of objectives 14, 15, or 16.
18. Utilize specification sheets to determine diode parameters.
19. Define *peak reverse voltage* (or *peak inverse voltage*).
20. Explain the behavior of a zener diode and draw its schematic diagram.
21. Define for a zener diode the *knee current,* the *maximum reverse current,* and the *test current.*

22. Define the zener impedance, and apply it to predict voltage variations.
23. Define the temperature coefficient of the zener voltage, and apply it to predict voltage variation.
24. Discuss the general operation of optoelectronic diodes, including light-emitting diodes, photodiodes, and optocouplers.
25. Discuss briefly the general operation of other miscellaneous diodes, such as Schottky diodes, varactor diodes, constant current diodes, tunnel diodes, varistors, step-recovery diodes, and back diodes.
26. Explain the procedure for testing a diode with an ohmmeter.

2-1 APPROACH

An initial discussion of the approach for dealing with electronic devices in this chapter and in succeeding chapters is appropriate. For the purpose of this discussion, let us classify the behavior of electronic devices at two levels: (1) internal device physics, and (2) external terminal characteristics.

At the most fundamental level, the internal device physics are very important and must be understood by those who design and manufacture semiconductor devices and integrated circuits. However, this segment of the electronics industry is much smaller than those segments that utilize electronic devices in the numerous application areas. There are books, courses, and complete college curricula devoted to semiconducutor theory at a level appropriate for component and integrated-circuit design. Because the approach at that level is quite different from that of circuit design and applications, books attempting to cover all aspects of the subject equally well are quite rare.

The vast majority of engineers, technologists, technicians, and instrumentation personnel tend to be "component users"; that is, they purchase components or integrated circuits from a manufacturer and adapt them to fit their needs. In most cases, they are not particularly concerned with what is happening inside the component, but they want to know about the terminal characteristics and how it relates to their requirements.

What we are going to do in this book is concentrate on the terminal characteristics of devices and how they can be modeled for analysis and design. The internal mechanisms will be considered only qualitatively and to the extent that they aid in the visualization of circuit operation and in reading specifications. This is the approach taken by the majority of electronic component and integrated-circuit users in their day-to-day encounters with the complex world of electronics.

Although no substantive treatment of device physics is given, it will be necessary to define a few semiconductor terms as they occur. Such "buzzwords" appear in the labels of many semiconductor devices, and users must be able to recognize them in specifying various devices.

2-2 SEMICONDUCTOR CONCEPTS

A brief overview of some aspects of semiconductor theory as it relates to diodes will be given. As explained in the preceding section, the purpose is to provide enough insight to allow qualitative understanding and to develop appropriate terminology.

From the standpoint of electrical conductivity, materials may be classified as (1) conductors, (2) insulators (*or nonconductors*), and (3) semiconductors. *Conductors* are materials having very low resistance or high conductivity. Examples are copper and aluminum. *Insulators* are materials having very high resistance or very low conductivity. Examples are dry wood and plastic.

Semiconductors are materials that lie between conductors and insulators in their conductivity; that is, current can flow but not as easily as in conductors. Two examples of semiconductor materials are *silicon* and *germanium*. Silicon is the most widely used in modern electronic components and integrated circuits. Semiconductors are very useful in electronic components because their atomic structures may be modified to create various types of voltage and current control functions.

Conduction in semiconductors may occur as a result of the movement of either (1) electrons or (2) holes. *Electrons* are negatively charged particles and they move in a direction opposite to that of conventional current flow. Electron flow is the type of conduction occurring in ordinary conductors, and the flow occurs in the so-called *conduction band* of the atom.

Holes represent the absence of an electron in an atom, and they are positively charged units. Holes move in the direction of conventional current flow. The movement of holes is peculiar to semiconductors and this flow is said to occur in the *valence band* of the atom.

Semiconductor materials are processed for device construction by the process of *doping*, in which *impurities* are added. The impurities are materials classified as either donors or acceptors. *Donor* materials result in an excess of *negative* charge carriers (*electrons*), and *acceptors* result in an excess of *positive* charge carriers (*holes*).

A semiconductor material having an excess of positive charge carriers (holes) is called a *p-type* material, and a material having an excess of negative charge carriers (electrons) is called an *n-type* material.

2-3 JUNCTION DIODE

A basic semiconductor *pn junction diode* is formed from the combination of a section of *p*-type semiconductor material with a section of *n*-type semiconductor material, as illustrated in Figure 2-1(a). As a result of a certain interaction of the charge carriers, a neutrally charged layer devoid of carriers forms at the surface between the regions. This region, called a *depletion layer*, inhibits current flow. It is difficult for current to flow in the direction shown by the dashed line in Figure 2-1(a). However, a positive voltage applied to the *p*-terminal with respect to the *n*-terminal is capable

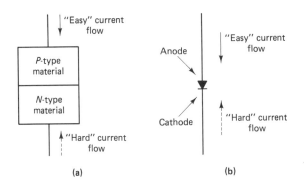

"Easy" current flow

P-type material

N-type material

"Hard" current flow

(a)

Anode

"Easy" current flow

Cathode

"Hard" current flow

(b)

Figure 2-1 Composition of *pn* junction diode and the circuit symbol.

of overcoming the effect of the depletion layer, and the result is possible current flow in the direction of the solid line in Figure 2-1(a). Thus the junction diode acts as a type of one-way valve for current flow.

The schematic symbol for the diode is shown in Figure 2-1(b). Note that the arrow on the symbol points in the direction of "easy" positive conventional current flow. The terminal on the *p*-side is called the *anode*, and the terminal on the *n*-side is called the *cathode*. These terms date back to the era of the vacuum-tube diode, but they are still widely used. In summary, current flows easily from anode to cathode but not from cathode to anode. Diodes come in a variety of sizes, with current levels ranging from a fraction of an ampere to thousands of amperes.

2-4 TERMINAL CHARACTERISTICS

Before developing circuit models for the diode, it is important to understand and be comfortable with certain basic concepts of linear and nonlinear relationships. Consider the block shown in Figure 2-2, having a set of terminals. This set of terminals is also called a *port*. An arbitrary voltage v is assumed across the terminals, and a current i is assumed flowing into the upper terminal (which means that i will be flowing out of the lower terminal, but it is usually not shown on the lower terminal). For the purpose at hand, we assume that the port has no reactive elements (i.e., no inductance or capacitance), and is passive (i.e., there are no energy sources inside).

Assume that the voltage v is varied through some range of values and that i is measured at each value of v specified. In a sense, we are considering v as the *independent variable* and i as the *dependent variable*. However, we could also vary i through a range of values and measure v at corresponding points. In this sense we would be considering i as the independent variable and v as the dependent variable. A curve showing i as a function of v (or v as a function of i) is called a *terminal characteristic*. In many cases, it is also possible to express the terminal characteristic as an equation relating i to v, or vice versa.

i

+

v

−

Passive circuit (no *L* or *C*)

Figure 2-2 Block used to discuss terminal characteristic.

Semiconductor Diodes   Chap. 2

Let us illustrate the concept by observing the terminal characteristic of a simple resistor. Consider the resistance of Figure 2-3(a) with voltage v and current i. If we consider v as the independent variable and i as the dependent variable, we have

$$i = \frac{v}{R} = \frac{1}{R}v \qquad (2\text{-}1)$$

The terminal characteristic is shown in Figure 2-3(b). Observe that the independent variable is shown as the horizontal axis (abscissa) and the dependent variable is shown as the vertical axis (ordinate). The slope of this curve is $1/R$.

If current is considered as the independent variable, we have

$$v = Ri \qquad (2\text{-}2)$$

This curve is shown in Figure 2-3(c), and its slope is R. In both curves we are assuming that the resistor is operated below its maximum power rating for the curve to be valid.

Either form of the terminal characteristic of an ideal resistor is a *linear characteristic*; that is, the dependent variable changes linearly with changes in the independent variable. Although this rather trivial example has not taught us anything new, it should help us to visualize the concept of terminal characteristics before proceeding to more complex situations. Further, by visualizing the characteristic for an ideal linear resistor, it can serve as a basis of comparison with other types of characteristics.

As an example of a nonlinear characteristic, consider the circuit of Figure 2-4(a), in which a nonlinear resistor is shown. Although there are a number of differ-

Figure 2-3 Ideal linear resistor and two forms of linear terminal characteristics.

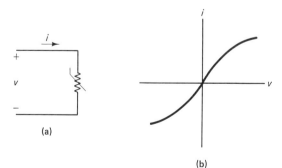

Figure 2-4 Nonlinear resistor and possible terminal characteristic.

ent types of nonlinear resistors, this particular one is assumed to have a resistance that increases with increasing voltage. The resulting terminal characteristic might be of the form shown in Figure 2-4(b). Many nonlinear characteristics are not easily described by equations, so the graphical form is often the only way to characterize the device.

Diode Terminal Characteristic

The terminal characteristic of the junction *pn* diode will now be investigated. Consider the diode shown in Figure 2-5(a) with voltage V_D and current I_D as labeled. The assumed voltage polarity and current direction is referred to throughout as the *positive reference direction* for the diode.

Assume that the diode is connected in the circuit of Figure 2-5(b). An experiment is to be performed for the purpose of obtaining a terminal characteristic. By varying the dc voltage source, the voltage V_D across the diode and the corresponding current I_D through the diode can be measured so that enough points can be determined to plot a curve. The voltmeter measures the voltage across the diode, and the ammeter measures the current through it. The only purpose of the resistor is to limit the maximum current through the diode. As we will see shortly, slight changes in the diode voltage in one direction can result in large changes in current.

Refer to Figure 2-5(c) for the discussion that follows. This curve represents a typical junction diode terminal characteristic, which is discussed in relationship to

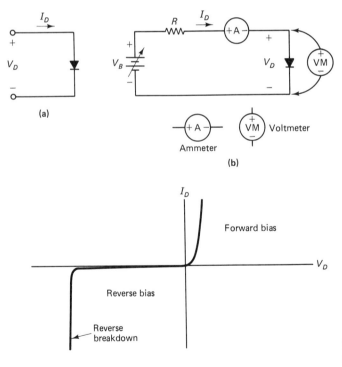

(a)

(b)

(c)

Figure 2-5 Semiconductor diode and the form of the terminal characteristic.

the experiment. First, the portion of the characteristic corresponding to positive voltage and positive current is considered. This condition is called *forward bias*. As the power supply voltage is varied in positive steps from zero, there is little or no measurable current flow for V_D less than a few tenths of a voltage. This is the range in which the depletion layer within the diode is blocking current flow. However, once V_D reaches a certain voltage level (to be quantified later), the current starts to flow and increases rapidly for small changes in voltage, as shown. Without the resistor it would be difficult to adjust the voltage without creating an excessive current.

The dc voltage is next reduced to zero and arrangements are made to obtain the portion of the characteristic corresponding to negative voltage and negative current. This can be achieved by reversing the terminals of the power supply and the polarities of the ammeter and voltmeter. (Alternatively, one can simply reverse the diode.) All subsequent measurements will then be based on negative voltage and current with respect to the assumed positive direction.

The condition of negative voltage and current with respect to the original direction is called *reverse bias*. The voltage and current in this condition are also called *reverse voltage* and *reverse current*.

As the reverse voltage is increased from zero in the reverse-biased region, the current would ideally be zero. In practice, a very small *reverse current* does flow, but it is normally much smaller than the forward current measured previously. In many applications, the reverse current will be ignored, but some discussion of this phenomenon is included in Section 2-6.

If the magnitude of reverse voltage were allowed to reach a certain level, a *reverse avalanche breakdown* would occur; then the current magnitude would experience a pronounced increase, as shown. *Zener* diodes are designed to utilize the nearly constant voltage properties of this region, and such diodes are considered in Section 2-7. However, ordinary diodes should *not* be operated in the reverse breakdown region since they may easily be damaged.

2-5 DIODE CIRCUIT MODELS

We consider next the process of modeling diodes for the purpose of analyzing their behavior in circuits. The diode is basically a nonlinear device, and an exact analysis is rather difficult. However, some very good approximations have been developed that permit simplified analysis to be performed.

Ideal Diode Model

The first and simplest approximation is the *ideal diode model*. Consider the diode shown in Figure 2-6(a) with the positive references for voltage and current indicated. The form of the ideal diode model is shown by the two circuits of Figure 2-6(b). This model assumes that the diode is either a perfect short circuit or a perfect open circuit, as shown. It is considered as a perfect short circuit if the direction of current flow is positive (in the direction of the arrow), and it is considered as a perfect open circuit if the voltage is negative (i.e., the cathode is more positive than the

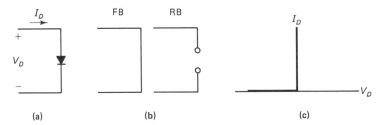

Figure 2-6 Ideal diode model and terminal characteristic.

anode). The perfect short circuit corresponds to ideal *forward bias* (indicated as FB) and the perfect open circuit corresponds to ideal *reverse bias* (indicated as RB).

The terminal characteristic of the ideal diode model is shown in Figure 2-6(c). In effect, the characteristic becomes the positive vertical axis and the negative horizontal axis. This means that the voltage cannot become positive and the current cannot become negative under these ideal conditions. This curve is a graphical representation of the following conditions: *The voltage is zero when the current is positive, and the current is zero when the voltage is negative.*

Constant Voltage Model

The second model to be considered will be referred to as the *constant voltage model.* It is based on the fact that when the diode is forward biased, very little current flows until the forward voltage reaches a few tenths of a volt. [Refer back to Figure 2-5(c), if necessary.] However, once current starts flowing, very little change in voltage occurs over the much larger range of operating current. As a very good approximation for many purposes, therefore, the forward-biased diode can be approximated as an ideal voltage source.

Refer to Figure 2-7, in which the diode references are shown again in (a). The constant voltage source model is shown in (b). The voltage V_o is denoted as the "turn-on" voltage, the "cut-in" voltage, or the "knee" voltage, depending on the reference.

The value of the knee voltage V_o depends on whether the diode is germanium or silicon. The value for germanium diodes typically varies from about 0.2 to 0.3 V, and the value for silicon diodes varies from about 0.6 to 0.7 V for small-signal

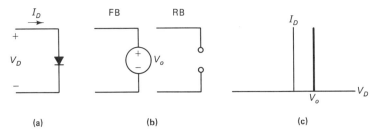

Figure 2-7 Constant voltage diode model and terminal characteristic.

diodes. For diodes designed to handle large current levels, such as in power supplies, the voltage may be of the order of 1 V or so.

In most circuit analysis problems throughout the text, the value $V_o = 0.7$ V will be assumed. This will represent a sort of "standard" valued based on silicon diodes. However, the reader should understand that it is not an exact value and will vary both with operating current and from one diode to another.

Understand that although the circuit model shows a voltage source, the diode is a passive device and cannot cause current to flow. The voltage source simply accounts for the forward-biased voltage drop, and this source does not "come alive" until the diode is forward biased. The terminal characteristic of the constant voltage model is shown in Figure 2-7 (c).

Constant Voltage Source Plus Resistance Model

The third model that we will consider is the *constant voltage source plus resistance model*. For the diode shown in Figure 2-8(a), the pertinent model is shown in (b). In this model, a resistance r_D is added to the constant voltage source model. This resistance allows the net voltage across the diode to increase as the current increases, which makes the result closer to a real situation. The corresponding terminal characteristic is shown in Figure 2-8(c).

The major difficulty with this model is in determining the value of the resistance to be used in the model. There is no simple value that will work over the entire range; that is, the resistance is heavily dependent on the operating range. For that reason, the constant voltage source plus resistance model has limited value in practical circuit analysis.

Shockley Model

The last method to be considered will be mentioned briefly, but is not considered within the scope of this book. This model utilizes the so-called Shockley equation, in which an exact mathematical voltage–current relationship utilizing an exponential function for the diode is provided. This method is capable of providing virtually exact results, but it cannot be adapted easily to a simple circuit model.

The reader is no doubt confused at this point and is asking: "Which model do I use?" The answer is that any of the models can be used provided that the answers are

Figure 2-8   Constant voltage source plus resistance diode model and terminal characteristic.

interpreted in terms of their real-life meaning. We must understand that with the exception of the Shockley equation, none provide *exact* answers. However, acceptable answers can be obtained from these models, and they can produce meaningful results if interpreted properly.

The ideal diode model is used when a very quick analysis of a circuit is desired and where high accuracy is not required. For example, when one is making a quick overview of a circuit and "ballpark" estimates are being used, ideal diode models are frequently assumed.

The constant voltage drop model is probably the most widely employed model, and it is capable of providing results sufficiently close for a variety of applications. Although one does not always know the exact value of voltage drop, an assumed value of 0.7 V for silicon or 0.3 V for germanium usually results in acceptable answers for most routine applications. This model is used extensively throughout the book.

As mentioned previously, the ideal voltage source plus resistance model is more difficult to use because of the possible wide variation in resistance, so we will not utilize it very often. We illustrate its application in Example 2-3.

Forward or Reverse Bias?

One basic question that must always be answered in analyzing a circuit with a diode is to determine whether the diode is forward biased (FB) or reverse biased (RB). When there is only one diode in a circuit, it is usually straightforward to predict the condition. Based on your intuition (which will develop with practice), assume momentarily one of the two possible conditions (FB or RB). If you assumed FB, replace the diode with the appropriate FB model (usually the constant voltage source), and calculate the current through the diode. If the current turns out to be in the direction of the diode arrow, your guess was correct, and the assumed model may be retained. If, however, the current turns out in the opposite direction (which the diode won't allow), your guess was wrong and you must replace the model with the RB model.

Suppose your intuition told you initially that the diode was probably reverse biased. In that case, replace the diode with the RB model (open circuit), and calculate the voltage across the diode. If the voltage at the cathode turns out to be more positive than the voltage at the anode, your guess was correct, and the assumed model may be retained. If, however, the voltage turns out in the opposite direction, your guess was wrong and you must replace the model with the FB model.

With a little practice, the reader should develop a quick intuitive approach for dealing with diode circuits having one diode. However, when the circuit contains more than one diode, the situation starts to get complicated, particularly when there are several diodes. The reason is that the assumption of one diode condition can affect the condition of other diodes. We defer this situation for awhile.

The examples that follow are designed to illustrate the various levels of modeling based on the preceding developments.

Example 2-1

Consider the circuit shown in Figure 2-9(a). Verify that the diode is forward biased. Calculate the current I and the two voltages V_D and V_R based on **(a)** the ideal diode model and **(b)** the constant voltage drop model. The diode is assumed to be a silicon type with $V_o = 0.7$ V.

(a)

(b)

(c)

Figure 2-9 Circuit of Example 2-1.

Solution

Forward bias is readily verified by observing that the direction of current flow would have to be clockwise in the loop, that is, out of the positive terminal of the source and back into its negative terminal. This current flow is in the direction of the diode arrow, so the diode must necessarily be forward biased. With a little practice, the direction of forward bias can be quickly determined in a simple situation such as this.

(a) The equivalent circuit with the ideal diode is shown in Figure 2-9(b). In this case, the diode is replaced by a short circuit, and we readily note that

$$V_D = 0 \tag{2-3}$$

Thus the entire 10-V source would appear across the resistance in this case and

$$V_R = 10 \text{ V} \tag{2-4}$$

The current is determined by Ohm's law as

$$I = \frac{10 \text{ V}}{1 \text{ k}\Omega} = 10 \times 10^{-3} \text{ A} = 10 \text{ mA} \qquad \textbf{(2-5)}$$

(b) The equivalent circuit with the constant voltage diode model is shown in Figure 2-9(c). We immediately note, of course, that

$$V_D = 0.7 \text{ V} \qquad \textbf{(2-6)}$$

By KVL we have

$$-10 + V_D + V_R = 0 \qquad \textbf{(2-7)}$$

or

$$V_R = 10 - V_D = 10 - 0.7 = 9.3 \text{ V} \qquad \textbf{(2-8)}$$

The current is

$$I = \frac{9.3 \text{ V}}{1 \text{ k}\Omega} = 9.3 \times 10^{-3} \text{ A} = 9.3 \text{ mA} \qquad \textbf{(2-9)}$$

As explained earlier, neither solution is necessarily correct, but the second solution is probably quite close to the actual result and *could* be the exact solution. (It *could be* because there is a point in which the voltage drop across the diode is *exactly* 0.7 V, but we do not usually know when this condition is met.)

Example 2-2

Consider the circuit shown in Figure 2-10(a). Verify that the diode is reverse biased, and determine the voltage across the diode.

(a) Note that V_D is negative with respect to the assumed direction

(b)

Figure 2-10 Circuit of Example 2-2.

Solution

Current flow around the circuit *would* be clockwise (if it could flow), and this would be opposite to the diode arrow, which is impossible. Thus the diode is reverse biased, and it is replaced by an open circuit as shown in Figure 2-

10(b). Observe that we have chosen the positive reference terminal of the voltage across the diode in accordance with the basic convention established earlier. Since no current flows in the circuit, there is no voltage drop across the resistor, and an application of KVL gives

$$-10 - V_D + 0 = 0 \qquad (2\text{-}10)$$

or

$$V_D = -10 \text{ V} \qquad (2\text{-}11)$$

Thus the voltage across the diode with respect to the reference direction is negative, meaning that the cathode (left-hand terminal) is actually more positive than the anode (right-hand terminal). This validates our assumption of reverse bias.

We could, of course, have assumed the diode voltage drop as plus to minus from left to right, and the answer would have been positive. Later, we'll take more liberties in choosing directions, but for the moment, we'll adhere to the basic reference directions established earlier. With that convention, a reverse-biased junction will always yield a negative voltage.

It should be emphasized that although there is no current flow, the reverse voltage clearly exists across the diode. In circuit design it is sometimes easy to overlook the fact that very large negative voltages may appear across reverse-biased diodes. From the actual diode characteristic of Figure 2-5, recall that junction breakdown may occur if this reverse voltage is too large. More will be said about this later.

Example 2-3

This example will illustrate the effect resulting from the constant voltage source plus resistance model. Consider the circuit of Figure 2-11(a), and as-

Figure 2-11   Circuit of Example 2-3 for different input conditions.

sume that the diode is represented by a constant voltage source of value 0.6 V in series with 50 Ω as shown in (b). Calculate the current I, the diode voltage drop V_D, and the resistive voltage drop V_R for each of the following values of V_i: (a) 10 V; (b) 20 V.

Solution

(a) The equivalent circuit with $V_i = 10$ V is shown in Figure 2-11(c). The KVL equation for the loop is

$$-10 + 50I + 0.6 + 10^4 I = 0 \qquad (2\text{-}12)$$

Solving for I, we obtain

$$I = \frac{9.4}{10,050} = 0.9353 \times 10^{-3}\text{A} = 0.9353 \text{ mA} \qquad (2\text{-}13)$$

The resistive voltage V_R is simply

$$V_R = 10^4 I = 10^4 \times 0.9353 \times 10^{-3} = 9.353 \text{ V} \qquad (2\text{-}14)$$

The diode voltage V_D consists of a contribution from the assumed internal resistance plus the constant value of 0.6 V. By forming a "voltmeter loop," we obtain

$$V_D = 0.6 + 50I = 0.6 + 50 \times 0.9353 \times 10^{-3} = 0.647 \text{ V} \quad (2\text{-}15)$$

Alternatively, the diode voltage is simply the difference between the source voltage and the resistive voltage.

(b) The equivalent circuit with $V_i = 20$ V is shown in Figure 2-11(d). The KVL equation is now

$$-20 + 50I + 0.6 + 10^4 I = 0 \qquad (2\text{-}16)$$

Solving for I, we obtain

$$I = \frac{19.4}{10,050} = 1.930 \text{ mA} \qquad (2\text{-}17)$$

The voltages V_R and V_D are now determined as

$$V_R = 10^4 \times 1.930 \times 10^{-3} = 19.30 \text{ V} \qquad (2\text{-}18)$$

and

$$V_D = 0.6 + 50 \times 1.930 \times 10^{-3} = 0.697 \text{ V} \qquad (2\text{-}19)$$

Let's see if this example can serve to illustrate a few points. The source voltage in (b) is twice the value in (a). In a perfectly passive linear resistive circuit, all voltages and currents would also double. However, the current actually increased by a factor $1.930/0.9353 = 2.064$. What is happening here is that the diode drop has less effect at the higher source voltage, so the current more than doubles. The resistive voltage V_R increases in the same proportion as the current since that voltage drop varies linearly with current. The diode voltage, however, only increases from 0.647 V to 0.697 V. Of course, if the simpler constant voltage source model had been used, the result would have shown no increase in diode voltage. A diode voltage actually does increase with increased current, so this model predicts the actual behavior more closely. Yet it is still approximate because the assumption of a constant voltage plus a fixed resistance is usually valid only over a very narrow range of operation.

2-6 READING DIODE DATA SHEETS

A typical set of specifications for some representative diodes is provided on the first page of Appendix D. Although we are not yet able to relate to all these specifications, some of the information may be understood at this point. This particular sheet provides specifications on the family consisting of seven diodes, denoted as 1N4001 through 1N4007. Note that they are called *rectifiers* on this sheet. As we will see in Chapter 3, a common application of a diode is as a rectifier, so the term is often used interchangeably with diode. Observe the band on one end of the diode, which represents the *cathode* side. This is a widely employed convention.

The various maximum ratings are provided in a table. The significance of some of these ratings may not be clear at this point, but a few will be meaningful. The first two maximum ratings relate to reverse bias and provide data on the *peak reverse voltage*, which is also referred to in some references as the *peak inverse voltage*. The first rating is based on a repetitive situation, which means that the diode would be reverse biased on a regularly occurring basis. Note that in all cases, this voltage rating is less than that of a nonrepetitive situation. Note also that the peak reverse ratings vary considerably for different diodes in the family. The maximum value of the average rectified forward current is given as 1 A. However, the diodes can withstand a surge of 30 A for 1 cycle.

A table providing certain electrical characteristics is also given. A typical value of forward voltage drop is given as 0.93 V at 1 A. Note that this is somewhat greater than our assumption of 0.7 V. However, the value of 0.7 V is more accurate at the much lower current levels typical of electronic signal levels. The value of 1 A used here is more typical of certain power supply levels.

A typical reverse current is indicated as 0.05 μA at a junction temperature (T_J) of 25°C. However, the typical value increases to 1 μA for a junction temperature of 100°C. The maximum values for the two cases are 10 μA and 50 μA, respectively.

2-7 ZENER DIODES

In the discussion of the semiconductor diode characteristic in Section 2-4, reference was made to the reverse avalanche breakdown process, which can occur in any diode if the reverse voltage is excessive. By increasing the doping level, it is possible to decrease the magnitude of the reverse voltage and to control the process in an orderly fashion.

In the higher range of voltages, the breakdown is due to the reverse avalanche effect, as already noted. In the lower range of voltages, a separate effect called *zener breakdown* occurs. However, all diodes in which specified reverse breakdown voltage levels are established are collectively referred to as *zener diodes*. Thus, whereas reverse breakdown is undesirable for an ordinary diode, it is the primary objective sought in a zener diode. The variety of available zener diodes is illustrated in Figure 2-12.

The schematic symbol of a zener diode is shown in Figure 2-13(a). The normal positive reference voltage and current for a diode are labeled in Figure 2-13(b). Although intended operation of the zener diode is in the reverse region, the normal voltage and current convention is compatible with standard semiconductor diodes.

The terminal characteristic of a typical zener diode is shown in Figure 2-14. Generally, it has the same form as an ordinary diode, as given in Figure 2-5. In the forward-bias region, a zener diode behaves like an ordinary diode, and the same circuit models used earlier may be applied to the zener diode in that region.

As the zener characteristics are studied in the remainder of this section, it may be helpful for the reader to refer to Appendix D. The second set of specifications provided there relate to several families of zener diodes. Included are the 1N746 through the 1N759, the 1N957A through the 1N986A, and the 1N4370 through the

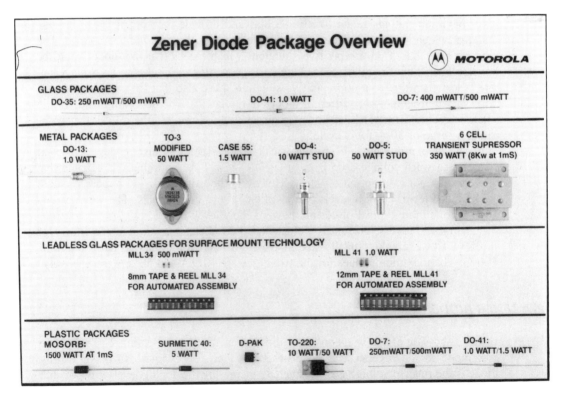

Figure 2-12 Typical packages for zener diodes. (Courtesy of Motorola, Inc.)

Semiconductor Diodes   Chap. 2

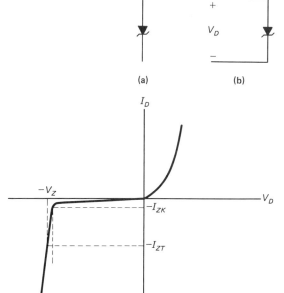

(a)   (b)

Figure 2-13 Schematic symbol and vaiables for zener diode.

Figure 2-14 Zener characteristic with parameters identified.

1N4372. These are all 500-mW diodes. However, these are only a small sample of the available zener types, and the available power levels vary widely.

The notation on Figure 2-14, along with other parameters to be discussed, has been formulated to be compatible with that of the zener diode specifications in Appendix D.

The distinguishing feature of the zener diode is that reverse breakdown occurs at a specified voltage $-V_Z$, and the value of V_Z is usually much smaller than the reverse breakdown voltage of ordinary diodes. Once breakdown is reached, the voltage varies very little over a wide range of current. The zener is thus capable of providing voltage regulation and in establishing voltage references.

Some important parameters of a zener diode are identified in Figure 2-14. Strictly speaking, some of these quantities have negative values when referred to the basic voltage and current convention of Figure 2-13(b). However, as is customary with data manuals and with references, the definitions below are given in terms of positive values.

The nominal *zener voltage* is designated as V_Z. Since V_Z does vary slightly, it is customary in data manuals to state the value of V_Z at some reference *test current*, which is denoted as I_{ZT}.

An important parameter for a zener diode is the maximum power rating. In general, the power dissipation P_D is given by

$$P_D = V_Z I_Z \qquad (2\text{-}20)$$

where I_Z is the zener current. Since the value of V_Z is known, the value of I_Z corresponding to the maximum power dissipation could be computed. However, since V_Z will vary somewhat from one unit to another and with different operating conditions, a separate maximum current rating is provided for a given zener type. The maximum current rating provides some leeway to ensure operation below the maximum power level and will be discussed shortly.

The current I_{ZK} is referred to as the *knee current*. It is the smallest value of current at which regulation of the voltage can be assumed to occur.

The current I_{ZM} is the *maximum* reverse current at which the diode can operate safely, and it provides sufficient leeway to ensure operation below the maximum power. Thus all circuit designs must be checked to ensure that operation is between I_{ZK} and I_{ZM} under all conditions.

Since there is some variation of voltage in the zener region, the *zener impedance* is specified. It is actually a resistance, but a common symbol for the quantity is Z_{ZT}. Let ΔI_Z represent a change in current in the zener region, and let ΔV_Z represent the corresponding change in voltage. Then Z_{ZT} is defined as

$$\boxed{Z_{ZT} = \frac{\Delta V_Z}{\Delta I_Z}} \tag{2-21}$$

Assume that the voltage V_Z and impedance Z_{ZT} at a given current I_Z are known. If the current then changes to $I_Z + \Delta I_Z$, the corresponding zener voltage changes to $V_Z + \Delta V_Z$, and

$$\Delta V_Z = Z_{ZT} \Delta I_Z \tag{2-22}$$

The new value of zener voltage V_Z' can then be estimated as

$$V_Z' = V_Z + \Delta V_Z = V_Z + Z_{ZT} \Delta I_Z \tag{2-23}$$

The value of Z_{ZT} varies with the operating point. From the zener data of Appendix D, note that the maximum value of Z_{ZT} is specified at the test current I_{ZT}. For some of the diodes, the maximum value of Z_{ZT} is also specified at the knee current I_{ZK}. The value of the impedance at the knee current is quite large, indicating that regulation is very poor at that level of current. The value of the impedance at the test current is more typical of the impedance in the desired operating range.

The variation of the zener voltage with temperature can be determined from the *temperature coefficient of the zener voltage*, which is denoted in Appendix D as ΘV_Z. This quantity is usually expressed in mV/°C and may be positive or negative, depending on the breakdown process. In general, at zener voltages below about 6 V, breakdown is due to the zener effect and the temperature coefficient is *negative*. However, at zener voltages above about 6 V, breakdown is caused by the avalanche effect, and the temperature coefficient is *positive*.

Assume that V_Z and ΘV_Z are known at a certain reference temperature. If the temperature then changes by ΔT, the change in voltage ΔV_Z is

$$\Delta V_Z = \Theta V_Z \Delta T \tag{2-24}$$

The new value of voltage V_Z' is then estimated as

$$V_Z' = V_Z + \Delta V_Z = V_Z + \Theta V_Z \Delta T \tag{2-25}$$

Note that ΘV_Z may be either positive or negative, so V_Z' may be larger or smaller than V_Z.

On the fourth page of the zener specifications in Appendix D, curves providing temperature coefficients are provided. Note that the two curves represent a *range* of values, so the actual coefficient for a given diode could lie anywhere within the range. Note also how the temperature coefficient changes from a negative value at low voltages to a positive increasing value as the voltage increases.

Example 2-4

A 1N753 zener diode has a specified value of $V_Z = 6.2$ V at $I_{ZT} = 20$ mA. The maximum zener impedance is specified as 7 Ω. Estimate the maximum value of zener voltage at $I_Z = 60$ mA.

Solution

The change in current between the reference point and the point of interest is

$$\Delta I_Z = 60 \text{ mA} - 20 \text{ mA} = 40 \text{ mA} = 0.04 \text{ A} \tag{2-26}$$

The change in voltage is

$$\Delta V_Z = Z_{ZT} \Delta I_Z = 7 \times 0.04 = 0.28 \text{ V} \tag{2-27}$$

The new voltage is

$$V_Z' = V_Z + \Delta V_Z = 6.2 + 0.28 = 6.48 \text{ V} \tag{2-28}$$

Note that the basic units of ohms and amperes were used in the multiplication of (2-27).

Example 2-5

A possible value of the temperature coefficient of a 1N753 zener diode is +3 mV/°C. If $V_Z = 6.2$ V at 25°C, estimate the zener voltage at 75°C.

Solution

The temperature difference is

$$\Delta T = 75 - 25 = 50°C \tag{2-29}$$

The change in voltage is

$$\Delta V_Z = 3 \times 10^{-3} \times 50 = 0.15 \text{ V} \tag{2-30}$$

The new voltage is

$$V_Z = 6.2 + 0.15 = 6.35 \text{ V} \qquad \textbf{(2-31)}$$

Note that the temperature coefficient was expressed in the basic units of volts /°C in (2-30).

2-8 OPTOELECTRONIC DIODES

The field of *optoelectronics* is a rapidly developing segment of the electronics industry. This term refers to those devices and applications that combine *optical* and *electronic* technology. Some of the most common optoelectronic devices combine the action of a *pn* junction diode with light-sensitive materials. In this section we consider three of the most common optoelectronic devices: light-emitting diodes, photodiodes, and optocouplers.

Light-Emitting Diodes

The *light-emitting diode* (widely referred to as an LED) is a special *pn* junction diode that emits light when forward biased. In an ordinary *pn* junction diode, charge carriers crossing the junction radiate energy in the form of heat. In the LED, special materials are used that convert a portion of this energy to light. Some of the elements used are gallium, arsenic, and phosphorus. A variety of colors are available in LEDs. In addition to visible radiation with common colors, some LEDs produce invisible infrared radiation. Invisible radiation could be useful in applications such as burglar alarm systems, for example.

The schematic symbol for an LED is shown in Figure 2-15. The arrows obviously indicate light *leaving the diode*.

The voltage drop for a forward-biased LED is less predictable than that of an ordinary diode. Typical values range from 1.5 to 2.5 V, and the current range is typically from 10 to 50 mA. A reasonable estimate for the voltage when the exact value is not known is 2 V.

A number of LEDs can be combined to form an electronic display. One of the most common electronic displays is the *seven-segment indicator*. The form of the seven-segment display is shown in Figure 2-16(a), and the schematic diagram is

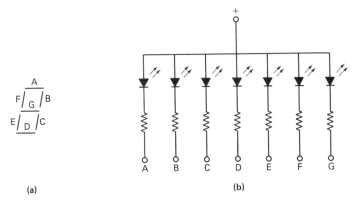

(a)

(b)

Figure 2-15 LED schematic symbol.

Figure 2-16 Seven-segment LED indicator and schematic.

shown in Figure 2-16(b). The seven segments of the display are denoted as A through F, respectively, and each represents a separate diode and associated dropping resistor. The form shown in Figure 2-16 is referred to as a *common-anode* form since all the anodes are connected to a common point. If a positive voltage with respect to ground is connected to the common-anode form, an individual segment is activated by grounding the appropriate branch. These indications are also available in *common-cathode* form.

By activating various combinations of the seven segments, all integer numbers can be formed. Certain letter combinations can be formed as well.

Photodiode

The *photodiode* is opposite to the LED in that it responds to impinging light. The schematic symbol for a photodiode is shown in Figure 2-17. In contrast to the symbol for the LED, the arrows point *toward* the diode in the photodiode symbol.

The photodiode employs the principle that the reverse current in a reverse-biased *pn* junction is a function of any light radiation received by the junction. The energy received from the light increases the reverse current. A photodiode is built to enhance this effect, and the light is channeled to the junction by a special window. Thus the normal mode of operation of a photodiode is with reverse bias.

Optocoupler

An *optocoupler* is a combination of an LED and an optically aligned photodiode in a single package. The schematic diagram is shown in Figure 2-18. The LED is on the left, and the photodiode is on the right.

Assume that a signal is connected on the left and that a load is connected on the right. Both portions of the circuit will require appropriate biasing circuitry, of course. The signal will cause variations in the light intensity produced by the LED, and this will result in corresponding variations in the current through the photodiode. The result is that the signal and source are completely isolated from each other electrically. The signals are, instead, said to be *optically coupled*.

This type of application is particularly useful where a small signal is to be transferred between two points without the presence of undesired electrical coupling such as high voltages or ground loops.

The term "optocoupler" includes similar coupling components employing transistors instead of diodes. Larger output levels are achieved with transistors as a result of the amplification process (to be discussed later).

Figure 2-17   Photodiode schematic symbol.

Figure 2-18   Optocoupler schematic symbol.

Example 2-6

An LED is to be connected through a resistor to a 15-V dc supply as shown in Figure 2-19(a). If the LED voltage is known to be 2 V and the desired current is 20 mA, determine the required resistance R.

Figure 2-19 Circuit of Example 2-6 and equivalent loop.

(a)   (b)

Solution

The circuit form shown in Figure 2-19(a) is the form that probably appears on a schematic diagram. The circuit is redrawn in an equivalent single-loop form in Figure 2-19(b), in which the 15-V supply is represented as a battery. An application of KVL for the loop reads

$$-15 + 20 \times 10^{-3}R + 2 = 0 \tag{2-32}$$

or

$$R = \frac{15 - 2}{20 \times 10^{-3}} = \frac{13}{20 \times 10^{-3}} = 650 \ \Omega \tag{2-33}$$

In the expression of (2-33), note that the value $15 - 2 = 13$ V represents the voltage absorbed by the resistance, and this value is divided by the known current in accordance with Ohm's law. With reasonable practice, a simple circuit such as this can be analyzed directly. However, one can always return to the more "formal" approach when necessary.

2-9 OTHER TYPES OF DIODES

In this section, several other types of diodes are discussed. These include Schottky diodes, varactor diodes, and others.

Schottky Diodes

When an ordinary pn junction diode is forward biased, holes are injected from the p-type material across the junction into the n-type material, and electrons from the n-type material are injected into the p-type material. These charges on both sides are momentarily stored before recombining with charges of the opposite side. To change the forward current or voltage, it is necessary that the stored charge change. This phenomenon is a capacitive effect and is referred to as *diffusion capacitance*.

Because of the stored charge, it takes some time to terminate current flow when the diode is changed from forward to reverse bias. This time is called the *reverse recovery time*. It is typically of the order of a few nanoseconds for a small-signal diode. Its effect is negligible at low frequencies, but it can become significant above 10 MHz or so.

The *Schottky diode* is of a special construction that reduces the forward diffusion capacitance to a negligible value, which in turn reduces the reverse recovery time considerably. This is achieved by utilizing a thin layer of aluminum on lightly doped *n*-type silicon, which forms a rectifying junction between the metal and the semiconductor.

The schematic symbol of a Schottky diode is shown in Figure 2-20. These diodes are usable well into the microwave frequency range. They are also used in high-speed digital switching circuits.

Figure 2-20 Schematic symbol for Schottky diode.

Varactor Diodes

All junction diodes exhibit a capacitance effect when the diode is reverse biased. This effect is referred to as *reverse depletion capacitance*. This capacitance varies with the magnitude of the reverse voltage across the junction.

Diodes manufactured specifically to employ the capacitive effect are referred to as *varactor diodes*. The term "varactor" refers to *voltage-variable capacitance*. Such diodes are also called varicaps. Among the applications of varactors are the tuning circuits of many high-frequency receivers. The schematic symbol is shown in Figure 2-21.

Figure 2-21 Schematic symbol for varactor diode.

Other Diodes

Other diodes include constant current diodes, tunnel diodes, varistors, step-recovery diodes, and back diodes.

The *constant current diode* is capable of providing a constant current to a variable load. It is thus opposite in its effect to a zener diode. Constant current diodes are related very closely to field-effect transistors (discussion is deferred to Chapter 7).

A *tunnel* or *Esaki diode* exhibits a *negative resistance* region in its terminal characteristic. This means that over a portion of the terminal characteristic, the current decreases as the voltage increases. Tunnel diodes are used in certain high-frequency and microwave circuits.

A *varistor* is similar to two zener diodes connected in series with opposite directions and with a high breakdown voltage in either direction. These devices are used as transient suppressors in power systems. Sudden line voltage changes, switching loads, lighting, and other effects can produce devastating overvoltages in some power systems. Varistors are used to suppress these effects.

A *step-recovery diode* has a doping profile that results in a continued current flow for a portion of a reverse-biased interval, followed by a sudden termination of the current. This effect is called *reverse snapoff*, and step recovery diodes are sometimes referred to as *snap diodes*. Step-recovery diodes are used in frequency-multiplier circuits to generate frequencies that are integer multiples of the input frequency.

A *back diode* is one in which the doping level is increased to the point where reverse breakdown occurs at approximately -0.1 V. By employing this diode in a reverse direction and for signal levels below 0.7 V, the output is reduced only by the 0.1-V drop.

2-10 TESTING A DIODE WITH AN OHMMETER

An *ohmmeter* may be used to provide a quick check on the probable condition of a diode. Current actually flows out of one of the terminals of an ohmmeter through the resistance being measured and back into the other terminal. Check the manual for the ohmmeter so that you will know the actual positive direction of current flow through an external load. *It is absolutely essential that all power sources be turned off for components being tested.* An ohmmeter can easily be damaged when connected in a "hot" circuit.

The diode being tested should not be shunted by any additional resistive load. Consequently, it may be necessary to disconnect at least one of the diode terminals from any circuit in which it is connected.

Connect the ohmmeter leads to the diode in the direction that will result in forward bias and note the approximate value of resistance. Next, reverse the ohmmeter leads so that reverse bias is established, and note the resistance. A good diode should exhibit a large reverse resistance compared to the forward resistance. A ratio of 1000:1 is typical. In fact, on some ohmmeter scales, the reverse resistance of a good diode appears to be an open circuit ($R = \infty$).

The actual value of forward resistance that would be read on a typical ohmmeter depends on the scale and can vary from a few hundred ohms to thousands of ohms. The resistance read on the ohmmeter will change as the scale is changed. Consequently, this process is more of a quick check than a quantitative measure.

Some ohmmeters can produce enough current on the very lowest ranges to destroy a small-signal diode. Consequently, when testing a very small diode, avoid using the lowest ranges.

A final point is that not all ohmmeter scales of many modern multimeters necessarily provide sufficient voltage to forward bias diodes. With such instruments, it is intended that only certain scales be used to check diodes. Sometimes these ohmmeter scales are identified on the instrument. In any event, the operating manual will be your best guide.

DRILL PROBLEMS

2-1. Consider the circuit shown in Figure P2-1. Verify that the diode is forward biased. Calculate the current I and the two voltages V_D and V_R based on **(a)** the ideal diode model and **(b)** the constant voltage drop model. The diode is assumed to be a silicon type with $V_o = 0.7$ V.

Figure P2-1

2-2. Consider the circuit shown in Figure P2-2. Verify that the diode is forward biased. Calculate the current I and the two voltages V_D and V_R based on **(a)** the ideal diode model and **(b)** the constant voltage drop model. The diode is assumed to be a silicon type with $V_o = 0.7$ V.

Figure P2-2

2-3. Consider the circuit shown in Figure P2-3. Verify that the diode is reverse biased, and determine the voltage across the diode.

Figure P2-3

2-4. Consider the circuit shown in Figure P2-4. Verify that the diode is reverse biased, and determine the voltage across the diode.

Figure P2-4

91

2-5. For the circuit of Problem 2-1 (Figure P2-1), assume that the diode is represented by a constant voltage source of 0.6 V in series with a resistance of 10 Ω. Determine I, V_D, and V_R.

2-6. For the circuit of Problem 2-2 (Figure P2-2), assume that the diode is represented by a constant voltage source of 0.56 V in series with a resistance of 20 Ω. Determine I, V_D, and V_R.

2-7. A 1N755 zener diode has a specified value of $V_Z = 7.5$ V at $I_{ZT} = 20$ mA. The maximum zener impedance is specified as 6 Ω. Estimate the value of zener voltage at $I_Z = 50$ mA.

2-8. A 1N759 zener diode has a specified value of $V_Z = 12$ V at $I_{ZT} = 20$ mA. The maximum zener impedance is specified as 30 Ω. Estimate the value of zener voltage at $I_Z = 30$ mA.

2-9. A possible value of the temperature coefficient of a 1N755 zener diode is +4 mV/°C. If $V_Z = 7.5$ V at 25°C, estimate the zener voltage at 75°C.

2-10. A possible value of the temperature coefficient of a 1N759 zener diode is +8 mV/°C. If $V_Z = 12$ V at 25°C, estimate the zener voltage at 75°C.

2-11. A possible value of the temperature coefficient of a 1N4372 zener diode is −2 mV/°C. For $V_Z = 3$ V at 25°C, estimate the zener voltage at 75°C.

2-12. A certain zener diode has a specified value of $V_Z = 10$ V at 25°C and $I_{ZT} = 20$ mA. The zener impedance is 20 Ω, and the temperature coefficient is +6 mV/°C. If the impedance and temperature effects are assumed to be additive, estimate the zener voltage for $I_Z = 30$ mA and at a temperature of 75°C.

2-13. In Example 2-6, the resistance was determined on the assumption that the diode drop was 2 V. If the diode drop actually turns out to be 1.5 V, determine the actual LED current that would result from the use of the 650-Ω resistance.

2-14. In Example 2-6, the resistance was determined on the assumption that the diode drop was 2 V. If the diode drop actually turns out to be 2.5 V, determine the actual LED current that would result from the use of the 650-Ω resistance.

2-15. (a) For the LED indicator of Figure 2-16, indicate which segments should be grounded to form the number "2." (b) If the current in each segment is 20 mA, what is the total current for this character?

2-16. (a) For the LED indicator of Figure 2-16, indicate which segments should be grounded to form the letter "A." (b) If the current in each segment is 20 mA, what is the total current for this character?

DESIGN PROBLEMS

2-17. An LED is to be connected through a resistor to a 5-V dc power supply as shown in Figure P2-17. (a) If the LED voltage is estimated as 2 V and the

+5 V

R

Figure P2-17

desired current is 20 mA, determine the required resistance R. **(b)** Based on the value of R determined in (a), estimate the possible range of current if the actual LED voltage can range from 1.5 to 2.5 V.

2-18. An LED is to be powered from a power supply whose voltage is -15 V with respect to ground. The LED voltage is estimated as 2 V, and the desired current is 15 mA. **(a)** Draw the schematic diagram and determine the value of a dropping resistance R. **(b)** Based on the value of R determined in (a), estimate the possible range of current if the actual LED voltage can range from 1.5 to 2.5 V.

TROUBLESHOOTING PROBLEMS

2-19. The voltages across the resistor and diode in Figure P2-1 are measured as $V_R = 0$ and $V_D = 16$ V. What is the likely problem?

2-20. The voltages across the resistor and diode in Figure P2-1 are measured as $V_R = 16$ V and $V_D = 0$. What is the likely problem?

COMPUTER PROBLEMS

2-21. Consider a series circuit consisting of a dc voltage source of value V_B, a diode, and a resistance R. The diode direction is forward bias. Assuming the constant voltage model, write a program to determine the current I.

> Input data: V_B, R, V_O
> Output data: I

2-22. Consider the same situation as in Problem 2-21 except that the constant voltage source plus resistance model will be employed. The assumed diode voltage and resistance will be denoted as V_O and r_D, respectively. Additional output data desired are the resistive voltage V_R and the net diode voltage V_D.

> Input data: V_B, R, V_O, r_D
> Output data: I, V_R, V_D

Computer Problems

CHAPTER 3

Applications of Diode-Resistor Circuits

OBJECTIVES

After completing this chapter, the reader should be able to:

1. Describe and illustrate the form of a transfer characteristic.
2. Show the difference between a linear and a nonlinear transfer characteristic.
3. Define the process of rectification.
4. Draw the circuit forms and transfer characteristics, explain the operation of, and perform the analysis of several half-wave rectifier circuits.
5. Explain the process of amplitude limiting.
6. Explain the function of a clipping circuit.
7. Explain the operation of an OR gate employing diodes and resistors.
8. Explain the operation of an AND gate employing diodes and resistors.
9. Draw the circuit form, explain the operation of, and perform the analysis of a zener regulator circuit.
10. Perform worst-case computations in a zener regulator circuit.
11. Design a zener regulator circuit to derive a fixed reference voltage from a larger available voltage.

3-1 OVERVIEW

The basic semiconductor diode and several special-purpose diodes were introduced in Chapter 2. Methods of analyzing circuits containing diodes were developed. However, up to this point, no specific applications of these devices have been considered.

In this chapter a number of basic applications of diodes in practical circuits are considered. The intent is to provide immediate application of the analysis skills of Chapter 2 to some useful circuits. The reader will also begin to start developing a "mental collection" of the types of circuits used in the field.

Of the three basic linear circuit parameters (R, L, and C), only resistance is considered in this chapter. The resulting circuits containing resistors and diodes will be classified as *diode-resistor* circuits. Thus circuits involving capacitance and/or inductance will be delayed to a point in the book where the reader's circuit analysis and mathematical skills are sufficiently developed. These diode-resistor circuits may contain fixed dc voltages as required, and their application with time-varying signal sources is also considered.

The specific types of applications considered include rectifiers, limiters, clippers, logic gates, and zener regulators. Some of the fundamental circuits in each category are considered.

Some basic advice to the reader is in order. This advice applies to this chapter and many others throughout the book. As you start looking at the various circuits, you will soon realize that there are many different types of functional operations that can be performed with electronic components, and there are often many ways to achieve the same operation. No one (including this author) can store the complete details of all these circuits in his or her memory.

Obviously, a reasonable amount of memory work is required when learning a new subject, and periodic review is required. However, it is more important to develop the necessary approach to analyze or develop the form of a given circuit than to rigidly memorize the circuit details without understanding. As you progress through various sections, hopefully you will see the similarity in analysis procedures as they are applied to individual circuits.

3-2 TRANSFER CHARACTERISTICS

Early in Chapter 2 the concepts of both linear and nonlinear *terminal characteristics* relating voltage and current at a set of terminals were introduced. Consider now a circuit having a set of *input terminals* (also called the *input port*) and a set of *output terminals* (also called the *output port*) as illustrated in block form in Figure 3-1. In many circuits it is possible to obtain an equation and/or a graph relating an output variable to an input variable. Such a relationship is called an *input/output characteristic* or *transfer characteristic*. Both terms are used in this book.

Depending on the circuit and the application, either the voltage or the current at the input port could be considered as the input variable, and either the voltage or the current at the output port could be considered as the output variable. In most situations the input variable is considered as the *independent variable* and is shown on

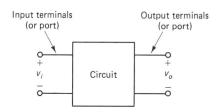

Figure 3-1 Circuit with input and output ports.

the *horizontal* axis, while the *output variable* is considered as the *dependent variable* and is shown on the *vertical* axis.

The most common type of application is where the input voltage is considered as the independent variable and the output voltage is assumed as the dependent variable. All applications in this chapter assume that situation. The most common symbol for the instantaneous input voltage is v_i, and the corresponding output voltage is denoted as v_o, as illustrated in Figure 3-1. In a few places, however, it will be necessary to modify that notation.

First, we consider a *linear transfer* characteristic. A simple example of a circuit that will produce a linear transfer characteristic is the *voltage-divider* or *attenuator* circuit of Figure 3-2, which can be used to *attenuate* or reduce the level of a signal. The output voltage v_o is readily expressed in terms of the input voltage as

$$v_o = \frac{R_2}{R_1 + R_2} v_i \tag{3-1}$$

This equation describes a linear equation with slope $R_2/(R_1 + R_2)$ and zero intercept. The term *zero intercept* means that the line passes through the origin (i.e., $v_o = 0$ when $v_i = 0$). The transfer characteristic is shown in Figure 3-2(b). It is assumed, of course, that operation is below the maximum power ratings of the resistors.

A basic property of a linear transfer characteristic with zero intercept is that the shape of any signal is preserved (i.e., no *distortion* occurs). For this attenuator circuit, the level would be reduced. For example, assume that $R_1 = R_2 = 10$ kΩ, which results in $R_2/(R_1 + R_2) = \frac{1}{2}$. For the waveform of Figure 3-3(a), the output would be of the form shown in (b), but the exact shape is preserved so that no distortion occurs.

An example of some arbitrary nonlinear transfer characteristic is shown in Figure 3-4. A signal passing through a circuit such as this would undergo a modification; that is, the output would be *distorted* with respect to the input. Distortion is undesirable in amplifiers, as we will see later, but in some circuits, such as will be considered in this chapter, possible alteration of the signal shape may be a design goal.

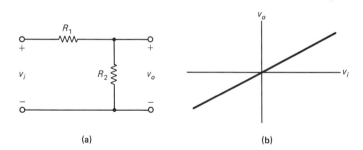

(a)   (b)

Figure 3-2 Simple attenuator circuit and transfer characteristic.

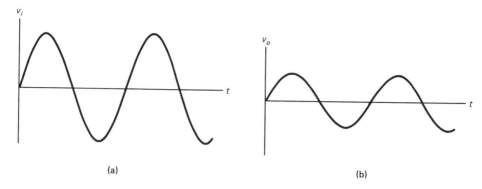

Figure 3-3 Input and output waveforms of attenuator circuit with $R_1 = R_2 = 10 \text{ k}\Omega$.

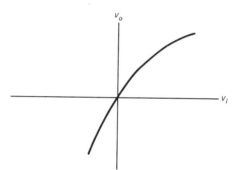

Figure 3-4 Example of a nonlinear transfer characteristic.

Piecewise Linear Characteristics

Many of the circuit applications involving diodes can be classified as having approximately *piecewise linear* characteristics. An example of a piecewise linear transfer characteristic is shown in Figure 3-5. Strictly speaking, piecewise linear characteristics are nonlinear. However, operation in certain regions of some piecewise linear

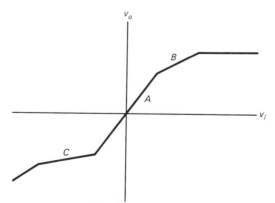

Figure 3-5 Example of a piecewise linear transfer characteristic.

characteristics may result in linear input–output conditions. With the characteristic of Figure 3-5, for example, a signal whose peak level is sufficiently low to keep operation along segment A would result in a linear input–output relationship, and no distortion would occur. However, if the peak signal moved to segments B and C, nonlinear operation and distortion would occur.

Linear Equations

Many transfer characteristics of electronic circuits may be closely approximated by linear or piecewise linear functions, as noted previously. To model or analyze such circuits, it is important to understand the mathematical and graphical forms of straight-line functions.

To relate this material to the reader's basic background, let us first review the form of a *straight-line* or *linear equation* as studied in most mathematics texts. Assuming x as the independent variable and y as the dependent variable, the most common form of a linear equation is

$$y = mx + b \qquad (3\text{-}2)$$

where m is in the *slope*.

A representative graph of a linear function is shown in Figure 3-6. When $x = 0$, $y = b$, and this value of y is called the *vertical intercept*. In this particular example, the vertical intercept is positive. When $y = 0$, $x = -b/m$, and this value of x is called the *horizontal intercept*. In this particular example, the horizontal intercept is negative. Note that in the common form of the straight-line equation as given by (3-2), the vertical intercept appears explicitly, but the horizontal intercept must be determined by setting $y = 0$ and solving for x.

The *slope m* provides a measure of how fast y changes as x changes. If m is very small, the linear function is nearly horizontal and displays very little change in y as x changes. Conversely, if m is very large, the linear function approaches a vertical characteristic, and y changes rapidly as x changes. If m is positive, y increases as x increases, while if m is negative, y decreases as x increases. In the example of Figure 3-6, the slope is positive.

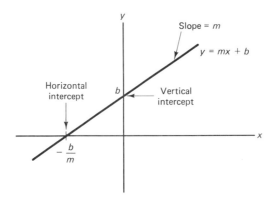

Figure 3-6 Basic linear (straight-line) function.

Applications of Diode-Resistor Circuits   Chap. 3

Common Linear Forms

We will now observe some of the possible forms that linear functions may assume as transfer characteristics for electronic circuits. The notation will change to correspond to typical electronic circuit terminology. Specifically, the input (independent) variable will be assumed to be an instantaneous voltage v_i, and the output (dependent) variable will be assumed to be an instantaneous voltage v_o.

A perfectly linear transfer characteristic with zero intercept is shown in Figure 3-7(a). This function can be expressed as

$$v_o = Av_i \qquad\qquad\qquad \textbf{(3-3)}$$

where A is the slope. Any purely resistive circuit with no sources could be described in this manner. For example, the voltage-divider circuit of Figure 3-2 has a characteristic of this form, and from (3-1), the slope is $A = R_2/(R_1 + R_2)$. In this case, $A < 1$. As we will see later, ideal amplifiers also have a linear characteristic with zero intercept. However, most amplifier characteristics have slopes whose magnitudes are greater than unity.

Some electronic circuits display an offset in the output voltage. An example is shown in Figure 3-7(b), and this characteristic is expressed as

$$v_o = Av_i + V_o \qquad\qquad\qquad \textbf{(3-4)}$$

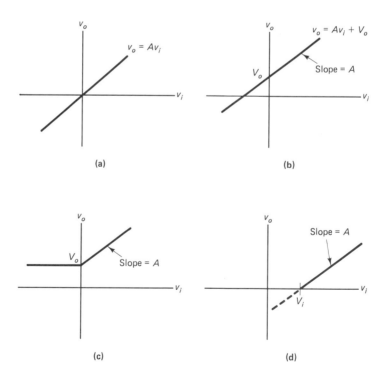

Figure 3-7 Some examples of linear and piecewise linear functions discussed in the text.

The output offset V_o is also the vertical intercept. Thus $v_o = V_o$ when $v_i = 0$.

Certain circuits display a constant value of output voltage when the input is in one range, and display a different behavior when the input is in a different range. A representative example is shown in Figure 3-7(c). This characteristic is a piecewise linear function. For this example, the output v_o can be expressed

$$v_o = \begin{cases} V_o & \text{for } v_i < 0 \\ V_o + Av_i & \text{for } v_i > 0 \end{cases} \tag{3-5}$$

Note that the slope is zero for $v_i < 0$, and the slope is A for $v_i > 0$.

In the preceding example, the vertical intercept for the right-hand segment was immediately evident, so the linear equation could be written by inspection. A different type of piecewise linear function in which the horizontal intercept is immediately evident is shown in Figure 3-7(d). In this case $v_o = 0$ for $v_i < V_i$ by inspection. To determine the straight-line equation applicable for $v_i > V_i$, we write

$$v_o = Av_i + b \tag{3-6}$$

where b is to be determined. We then substitute $v_o = 0$ and $v_i = V_i$ in (3-6) and solve for b. It is readily determined that $b = -AV_i$. Putting the preceding information together, we can write

$$v_o = \begin{cases} 0 & \text{for } v_i < V_i \\ Av_i - AV_i = A(v_i - V_i) & \text{for } v_i > V_i \end{cases} \tag{3-7}$$

Note in the second form of the second equation of (3-7) how the quantity $v_i - V_i$ appears as a multiplier for the slope. Some persons remember this form when the horizontal intercept is given.

The preceding linear and piecewise linear characteristics represent only a few of the possible types that will be encountered in the text. However, the basic straight-line equation form is always applicable to each segment of a piecewise linear function. Different equations are determined for the various segments, and the ranges of independent variables applicable to each are noted.

3-3 RECTIFIERS

The first application that will be considered for diode-resistor circuits is that of *rectification*. Rectification is the process of converting a voltage having both positive and negative polarities at different times to one having only one polarity. A circuit that performs rectification is called a *rectifier*. As noted in Chapter 2, diodes are sometimes called rectifiers.

Rectification is a fundamental process within a power supply, in which ac input is converted to dc output. However, a power supply requires a transformer, rectifiers, filter components, and fusing. A full treatment of power supplies is given in Chapter 20. The simple rectifier circuits to be considered here are those intended to function with electronic signals, and various voltage, current, and power levels are chosen on that basis.

Only simple half-wave rectifier circuits will be considered in this section. A *half-wave* rectifier circuit is one that passes the signal when it has one of the two possible polarities, but rejects the signal when it has the opposite polarity.

The first rectifier to be considered is a *series diode half-wave rectifier* circuit, of which one form is shown in Figure 3-8(a). When the input voltage v_i is less than about 0.7 V, the diode is reverse biased, and the equivalent circuit is as shown in Figure 3-8(b). There is an open circuit between the input and output under this condition, and the negative portion of the signal does not appear at the output. (It is assumed here that the peak value of the negative voltage is less than the reverse breakdown voltage for the particular diode used.)

When the input voltage exceeds about 0.7 V, the diode is forward biased, and the equivalent circuit is as shown in Figure 3-8(c). A loop equation reads

$$-v_i + 0.7 + v_o = 0 \tag{3-8}$$

or

$$v_o = v_i - 0.7 \tag{3-9}$$

The complete operation of the circuit can now be expressed in idealized form as

$$v_o = \begin{cases} 0 & \text{for } v_i < 0.7 \text{ V} \tag{3-10} \\ v_i - 0.7 & \text{for } v_i > 0.7 \text{ V} \tag{3-11} \end{cases}$$

(a)   (b) $v_i < 0.7$ V   (c) $v_i > 0.7$ V

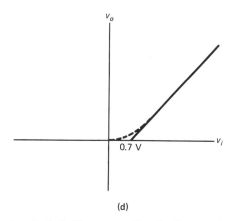

(d)

Figure 3-8 Series diode half-wave rectifier for first-quadrant operation, two equivalent circuits, and transfer characteristic.

The transfer characteristic curve is of the form shown in Figure 3-8(d). The ideal transfer characteristic of (3-11) is represented by the solid line. In the vicinity of the knee voltage, the actual behavior is more like the dashed curve. Based on the assumed reference directions, the active region corresponds to positive voltage and positive current, so this circuit can be described as one with *first-quadrant* active operation.

The circuit can readily be converted to one in which the positive portion of the signal is rejected simply by reversing the diode, and this circuit is shown in Figure 3-9(a). For $v_i > -0.7$ V (more positive than -0.7 V), the diode is reverse biased, and the equivalent circuit is shown in Figure 3-9(b). When $v_i < -0.7$ V, the diode is forward biased, and the equivalent circuit is shown in Figure 3-9(c). The loop equation now reads

$$-v_i - 0.7 + v_o = 0 \tag{3-12}$$

or

$$v_o = v_i + 0.7 \tag{3-13}$$

The complete operation can be expressed in idealized form as

$$v_o = \begin{cases} 0 & \text{for } v_i > -0.7 \text{ V} \tag{3-14} \\ v_i + 0.7 & \text{for } v_i < -0.7 \text{ V} \tag{3-15} \end{cases}$$

(a) (b) $v_i > -0.7$ V (c) $v_i < -0.7$ V

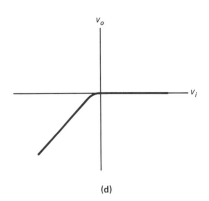

(d)

Figure 3-9 Series diode half-wave rectifier for third-quadrant operation, two equivalent circuits, and transfer characteristic.

The transfer characteristic curve is of the form shown in Figure 3-9(d). Active operation is now in the third quadrant (negative input and negative output).

The value $+0.7$ in (3-15) may bother some readers since we know that the output voltage magnitude is less than the input voltage magnitude. This is due to the fact that v_i is negative in the third quadrant, and this forces the magnitude of v_o to be less than the magnitude of v_i.

The operation and certain limitations of the preceding two rectifier circuits will be illustrated with an assumed signal chosen to illuminate certain points. Consider the signal shown in Figure 3-10(a) with the assumption that both positive and negative peaks are much larger in magnitude than 0.7 V. If the signal is applied as the input of the half-wave rectifier of Figure 3-8, the output would have a form similar to that of Figure 3-10(b). There is a cut-in effect near zero resulting from the fact that the input voltage must be close to 0.7 V to produce an output. In addition, there is a loss of about 0.7 V on the peaks. If the signal is applied to the rectifier circuit of Figure 3-9, the output is of the form shown in Figure 3-10(c), and similar effects are noted there.

(a)

(b)

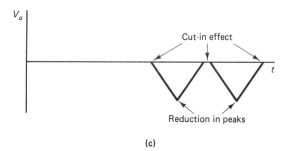

(c)

Figure 3-10   Input signal and outputs of the two rectifiers of Figures 3-8 and 3-9.

The relative effects of these degradations are much more significant at low signal levels. In some applications, they are unimportant, and in other applications, they may be unacceptable.

A different way to produce approximate rectification for some applications is the *shunt diode half-wave rectifier* shown in Figure 3-11(a). When $v_i < -0.7$ V, the diode is forward biased, and the equivalent circuit is shown in Figure 3-11(b). The output is $v_o \simeq -0.7$ V in this case. When $v_i > -0.7$ V, the diode is reverse biased, and the equivalent circuit is shown in Figure 3-11(c). The output without any load is $v_o = v_i$. As in previous circuits, it is assumed here that the reverse breakdown voltage of the diode is not reached.

The complete operation of the circuit can be expressed in idealized form as

$$v_o = \begin{cases} -0.7 \text{ V} & \text{for } v_i < -0.7 \text{ V} \qquad\qquad \textbf{(3-16)} \\ v_i & \text{for } v_i > -0.7 \text{ V} \qquad\qquad \textbf{(3-17)} \end{cases}$$

The transfer characteristic curve is shown in Figure 3-11(d), and first-quadrant active operation is observed. This circuit appears to have a very ideal relationship for positive voltage. However, the presence of the -0.7-V output for negative input voltages is a source of error and may not be acceptable for many applications. Further, any load on the output will affect the slope of the transfer characteristic (see Problems 3-11 and 3-12).

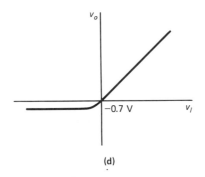

(d)

Figure 3-11 Shunt diode half-wave rectifier for first-quadrant operation. two equivalent circuits, and transfer characteristic.

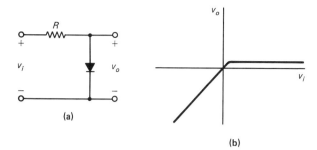

Figure 3-12 Shunt diode half-wave rectifier for third-quadrant operation and transfer characteristic.

(a)

(b)

The shunt diode converted to third-quadrant operation is shown in Figure 3-12(a) and the transfer characteristic is shown in Figure 3-12(b). Analysis of this circuit is left as an exercise for the reader, and the idealized results are

$$v_o = \begin{cases} 0.7 \text{ V} & \text{for } v_i > 0.7 \text{ V} \\ v_i & \text{for } v_i < 0.7 \text{ V} \end{cases}$$

(3-18)

(3-19)

All the basic half-wave rectifier circuits studied thus far are subject to errors, as we have noted. However, they are useful for noncritical applications. Highly linearized small-signal rectifier characteristics may be obtained with the use of an operational amplifier, and that is one of the topics of Chapter 16.

The examples that follow illustrate some of the types of calculations required in troubleshooting, analyzing, or designing these simple circuits.

Example 3-1

Consider the rectifier circuit of Figure 3-13(a). A source with time-varying voltage v_s and an internal resistance of 250 Ω is connected to the input.

(a)

(b)

(c)

(d)

Figure 3-13 Circuit of Example 3-1, equivalent circuits, and transfer characteristic.

(a) Determine a set of equations describing the relationship between the source voltage and output voltage, and sketch the transfer characteristic. **(b)** If v_s varies from -30 to 30 V, determine the peak values of the output voltage, the diode current, and the reverse voltage across the diode.

Solution

The rectifier circuit is exactly like that of Figure 3-8, and analysis would be exactly like that performed earlier if the source had no internal resistance. In fact, the relationship between v_i and v_o for the diode circuit alone would be of the form of (3-10) and (3-11). However, the source has an internal resistance, and we are interested in a relationship between the open-circuit source voltage v_s and the output voltage in this case. As we will see shortly, the internal source resistance has a significant effect on the output voltage.

(a) For $v_s > 0.7$ V, the equivalent circuit is shown in Figure 3-13(b). The circuit has the form of a single loop, and a current i has been chosen as the variable in the analysis. A clockwise loop equation reads

$$-v_s + 250i + 0.7 + 1000i = 0 \qquad \textbf{(3-20)}$$

Solution for i yields

$$i = -0.56 \times 10^{-3} + 0.8 \times 10^{-3} v_s \qquad \textbf{(3-21)}$$

The output voltage is

$$v_o = 1000i \qquad \textbf{(3-22)}$$

Substitution of (3-21) in (3-22) yields

$$v_o = -0.56 + 0.8v_s \qquad \text{for } v_s > 0.7 \text{ V} \qquad \textbf{(3-23)}$$

Of course, we can also state that

$$v_o = 0 \qquad \text{for } v_s < 0.7 \text{ V} \qquad \textbf{(3-24)}$$

The ideal transfer characteristic relating v_s to v_o is sketched in Figure 3-13(c). Observe that the independent variable is now v_s rather than v_i. From either (3-23) or the sketch, observe that the slope of the curve is now 0.8, a result of the fact that a fraction of the source voltage is dropped in the internal resistance. The value -0.56 in (3-23) *would be* the vertical intercept if the straight line in the transfer characteristic were extended downward.

(b) The peak values of the output voltage and diode current occur when the source voltage reaches its positive peak value. Substituting $v_s = 30$ V in (3-23) and (3-21), we obtain the following *peak* values:

$$v_o = -0.56 + 0.8 \times 30 = 23.44 \text{ V} \qquad \textbf{(3-25)}$$

$$i = -0.56 \times 10^{-3} + 0.8 \times 10^{-3} \times 30$$

$$= 23.44 \times 10^{-3} \text{ A} = 23.44 \text{ mA} \qquad \textbf{(3-26)}$$

The peak reverse diode voltage occurs at the negative peak of the input voltage. An equivalent circuit based on this condition is shown in Figure 3-13(d). For convenience, the source voltage is shown as an inverted positive voltage, and the corresponding reverse voltage V_{rp} across the diode has been labeled so that the value is positive. Since no current flows, we note that

$$V_{rp} = 30 \text{ V} \qquad \textbf{(3-27)}$$

An interesting point is that while the peak positive output voltage is reduced to 23.44 V, the peak diode reverse voltage is equal to the full negative peak value of the source voltage.

The peak diode current and the peak reverse diode voltage are important parameters in selecting a diode for a given application. These values must not exceed the corresponding maximum ratings for the diode.

3-4 LIMITERS

Diodes may be used to limit the allowable range of a variable signal. Many time-varying signals have a range of magnitudes that are generally predictable and for which the associated circuitry has been designed to accommodate. However, most signal sources may occasionally produce peaks or bursts outside the normal operating range. In addition, sudden noise spikes or stray pickup may result in overload conditions. Finally, the human element is always a factor, and an excessive level may accidentally be set during an adjustment or system connection.

Most electronic circuits are vulnerable to excessive signal levels and may accidentally be damaged or destroyed if safe levels are exceeded. In amplifiers designed to drive commercial broadcasting transmitters, a condition known as *overmodulation* can occur if the signal levels are excessive.

Circuits designed to limit the allowable range of a signal are known as *limiting circuits* or simply *limiters*. The limiting action may be gradual or abrupt. At this point we will study some of the simpler limiting circuits implemented with diodes, voltage sources, and possibly resistors. The particular circuits to be considered at this point are *nearly abrupt*. (They would be completely abrupt if the diodes were ideal.)

Abrupt limiting circuits are also referred to as *clamping circuits*, for reasons that will be clear shortly. However, there is another class of circuits called clamping circuits which perform a shifting level at the same time, and these circuits are studied in Chapter 16. To avoid confusion, we reserve the term "clamping circuits" for the group of circuits to be studied in Chapter 16.

In the analysis of the circuits here and in many other sections to follow, the schematic diagrams will often use the battery symbol in showing typical dc voltage for ease of illustration. In most applications, however, such voltages are obtained directly from available dc system voltages or indirectly through voltage reference circuits.

The basic form of a limiter circuit designed for positive voltage limiting is

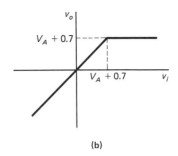

(a)

(b)

Figure 3-14 Limiter circuit with positive voltage limiting.

shown in Figure 3-14(a). The voltage v_i is the input signal voltage, and the voltage v_o is the output signal voltage.

The reader who has carefully studied the text up to this point should, hopefully, be able to infer forward- and reverse-bias conditions in this circuit without being "too formal" in the mathematical development. The algebraic sum of the sources around the loop tending to cause current flow in the FB direction is $v_i - V_A$, so when $v_i - V_A < 0.7$, or $v_i < V_A + 0.7$, the diode will be reverse biased. For this condition, there is no current flow around the loop, and $v_o = v_i$. However, when $v_i - V_A > 0.7$, or $v_i > V_A + 0.7$, the diode becomes forward biased. In this case the output voltage is the sum of the 0.7-V drop plus the dc voltage V_A, or $v_o = V_A + 0.7$.

A mathematical summary of the preceding conditions is

$$v_o = \begin{cases} v_i & \text{for } v_i < V_A + 0.7 \qquad \textbf{(3-28)} \\ V_A + 0.7 & \text{for } v_i > V_A + 0.7 \qquad \textbf{(3-29)} \end{cases}$$

The ideal transfer characteristic is shown in Figure 3-14(b).

As long as the positive peak of the input signal is less than about $V_A + 0.7$, the diode and dc voltage are isolated from the line and have no effect. However, when the signal level reaches about $V_A + 0.7$, the dc voltage and diode "take over" and prevent the output signal from increasing further.

Understand that the limiting action of this circuit, along with others to be considered in this section, is not precise because of the uncertainty of the forward diode drop. The limiting actually starts at a voltage slightly less than $V_A + 0.7$ and there is a slight increase in the output as the input increases. However, many applications of limiters do not require precise levels of the limiting value, and this type of circuit may be perfectly suitable. More accurate limiting levels may be established with operational amplifiers.

The corresponding form of a limiter circuit designed for negative voltage limiting is shown in Figure 3-15(a). The algebraic sum of the sources around the loop tending to cause forward bias (counterclockwise current flow) is $-v_i - V_B$, so when $-v_i - V_B < 0.7$ or $v_i > -V_B - 0.7$, the diode will be reverse biased. In this case, $v_o = v_i$. However, when $-v_i - V_B > 0.7$ or $v_i < -V_B - 0.7$, the diode becomes forward biased. In this case the output voltage is $v_o = -V_B - 0.7$.

A summary of the preceding conditions is

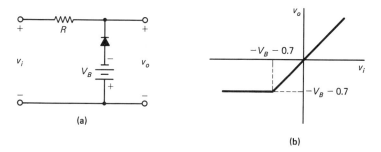

(a)

(b)

Figure 3-15   Limiter circuit with negative voltage limiting.

$$v_o = \begin{cases} v_i & \text{for } v_i > -V_B - 0.7 & \text{(3-30)} \\ -V_B - 0.7 & \text{for } v_i < -V_B - 0.7 & \text{(3-31)} \end{cases}$$

The ideal transfer characteristic is shown in Figure 3-15(b).

The preceding two circuits may be effectively combined to produce a limiter circuit capable of limiting on both positive and negative levels, and the resulting circuit is shown in Figure 3-16(a). Although the circuit now contains two diodes, it is easy to deduce from the earlier simpler circuits that the left-hand shunt path will conduct only for positive voltages, and the right-hand shunt path will conduct only for negative voltages. Thus, when either shunt path is conducting, the other is necessarily reverse biased and has no effect (unless, of course, the reverse breakdown voltage of the diode is inadvertently exceeded). A summary for this circuit is

$$v_o = \begin{cases} v_i & \text{for } -V_B - 0.7 < v_i < V_A + 0.7 & \text{(3-32)} \\ -V_B - 0.7 & \text{for } v_i < -V_B - 0.7 & \text{(3-33)} \\ V_A + 0.7 & \text{for } v_i > V_A + 0.7 & \text{(3-34)} \end{cases}$$

The ideal transfer characteristic is shown in Figure 3-16(b).

If the signal level is very low, and if limiting in the range of ± 0.7 V or so is acceptable, it is possible to limit directly with a pair of parallel diodes, as shown in Figure 3-17(a). In this case it is necessary to scrutinize the constant voltage diode

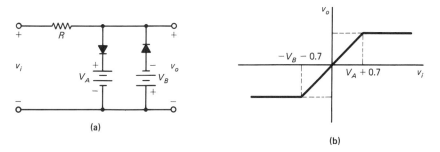

(a)

(b)

Figure 3-16   Limiter circuit with both positive and negative voltage limiting.

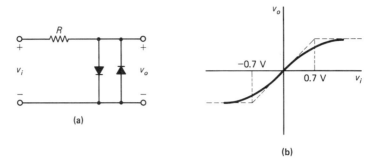

Figure 3-17 Limiter circuit with positive and negative voltage limiting at very low signal levels.

model carefully since operation will be in the vicinity of the knee voltage. The transfer characteristic is shown in Figure 3-17(b). The dashed line represents the characteristic based on the constant voltage drop assumption, and the solid line represents a typical realistic characteristic. In practice, the peak value of the normal signal should be well below 0.7 V in order not to be affected by the diodes. This type of circuit is often used on integrated-circuit chips for protection.

A few additional comments about all the circuits considered in this section are in order. First, the resistance R in either of the circuits may represent an actual circuit resistance, a source internal resistance, or a combination of both. However, there must be a minimum resistance for the circuit to function properly. Imagine, for example, that one of these limiting circuits were placed directly across a nearly ideal voltage source. As the input voltage started to increase above the level required to forward bias the diode, there would be no resistance to separate v_i from the output, and the diode would attempt to absorb the voltage difference. The result would be a sharp increase in current, and the diode could be destroyed quickly. The resistance required is a function of the maximum loop current permitted, and this is limited by either the diode or the reference voltage.

If the Thévenin equivalent resistance of the source is sufficiently high, it may serve as the resistance R. Thus it may be necessary to know something about the nature of the source before implementing these limiting circuits.

Frequently, a schematic diagram will be encountered in which one or more diodes (together with possible series voltage references) appear to be connected directly across a source. In such a case, the source is assumed to contain sufficient internal resistance to limit the current, as we have discussed.

The analysis of the circuits has assumed no loading at the output. In effect, this means that any additional circuitry following the limiting circuit will have sufficiently high input resistance that the open-circuit voltage v_o will remain the same with the load added. If this condition is not met, some attenuation of the signal level will occur, and the limiting levels will be reduced.

Clippers

Most of the limiter circuits considered here may also be used as *clipping circuits* or *clippers*. A clipping circuit is one in which one or both peaks of the input signal are intentionally operated above the limiting level so that the peaks are deliberately suppressed or "clipped." Thus the major difference between a *limiter* and a *clipper* is intent rather than circuit function.

Clipping circuits can be used to remove undesirable amplitude noise from certain types of signals. One must be cautious, however, to ensure that the signal itself is not severely altered in the process. Examples of signals that can be clipped for reducing noise levels are frequency-modulated signals and many types of digital data signals.

Clippers can also be used to reshape certain types of waveforms. For example, consider the sine wave of Figure 3-18(a), and assume that it is applied as the input to the circuit of Figure 3-16(a). Assume that $V_A = V_B$ and that the peak sine wave voltage is much larger than the limiting voltage. In this case the output would appear in the form shown in Figure 3-18(b). The result is a flat-top signal having some qualities of a square wave. (There are much better ways of generating a good square wave from a sine wave, as we will see later. This example was chosen simply to illustrate wave shaping.)

Zener Limiters

Zener diodes may be used as limiters in certain configurations. In fact, the use of a properly selected zener diode may eliminate the need for a separate reference

(a)

(b)

Figure 3-18 Application of circuit of Figure 3-16 as a clipper.

voltage. However, since a zener diode acts like a regular junction diode when forward biased, it is necessary to consider this fact when analyzing a circuit containing a zener.

A zener circuit useful for limiting positive voltages (and very small negative voltages) is shown in Figure 3-19(a). For $0 < v_i < V_Z$, the diode is reversed biased and less than the zener level, so the diode appears as an open circuit. There is no current flow and the output voltage without load is simply $v_o = v_i$. For $v_i > V_Z$, the zener conducts and holds the output voltage at V_Z for any subsequent increase of the input.

When $v_i < 0$, the diode behavior is that of a conventional diode. Thus very little conduction occurs until v_i reaches a level of about -0.7 V, and the diode then becomes forward biased and holds the output to that level.

A complete summary of the operation of this circuit is

$$v_o = \begin{cases} v_i & \text{for } -0.7 \text{ V} < v_i < V_Z & \textbf{(3-35a)} \\ -0.7 \text{ V} & \text{for } v_i < -0.7 \text{ V} & \textbf{(3-35b)} \\ V_Z & \text{for } v_i > V_Z & \textbf{(3-35c)} \end{cases}$$

The transfer characteristic is shown in Figure 3-19(b).

This circuit is primarily useful for limiting a voltage that is generally expected to be always positive or zero but for which a negative output up to about -0.7 V should be acceptable.

The circuit is readily modified to accommodate a signal that is generally expected to be negative simply by reversing the diode, and this circuit is shown in Figure 3-20(a). Without going through all the details, the output can be expressed as

$$v_o = \begin{cases} v_i & \text{for } -V_Z < v_i < 0.7 \text{ V} & \textbf{(3-36a)} \\ -V_Z & \text{for } v_i < -V_Z & \textbf{(3-36b)} \\ 0.7 \text{ V} & \text{for } v_i > 0.7 \text{ V} & \textbf{(3-36c)} \end{cases}$$

(a)

(b)

Figure 3-19 Zener diode used to limit positive voltage.

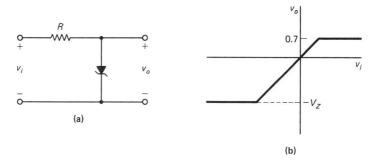

Figure 3-20 Zener diode used to limit negative voltage.

Back-to-Back Zener Limiter

A zener configuration that can be used to provide limiting for both positive and negative voltages is shown in Figure 3-21. The two diodes are connected in what is called a *back-to-back* connection; that is, they are in series and oriented in opposite directions. To provide some generality here, diode D1 is assumed to have a zener voltage V_{Z1}, and diode D2 is assumed to have a zener voltage V_{Z2}. For symmetrical signals, the zener levels are usually chosen to be the same.

A key point in the analysis of the circuit is that for either polarity of input signal, one diode will be forward biased and one will be reverse biased. Thus the critical transition voltage magnitudes are $V_{Z1} + 0.7$ and $V_{Z2} + 0.7$. The reader may wish to draw more detailed equivalent circuits to illuminate the development that follows.

For $0 < v_i < V_{Z1} + 0.7$, D1 is reverse biased and acts as an open circuit. Similarly, for $-V_{Z2} - 0.7 < v_i < 0$, D2 is reverse biased. For $v_i > V_{Z1} + 0.7$, zener reverse breakdown occurs in D1, and the output voltage is the sum of V_{Z1} and the forward biased drop of D2 (i.e., $v_o = V_{Z1} + 0.7$). Similarly, for $v_i < -V_{Z2} - 0.7$, zener reverse breakdown occurs in D2 and $v_o = -V_{Z2} - 0.7$.

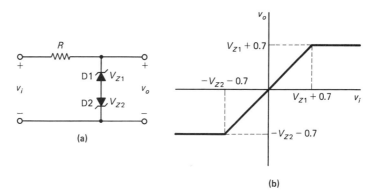

Figure 3-21 Back-to-back zener diode limiter circuit.

A summary of the preceding results follows.

$$v_o = \begin{cases} -V_{Z2} - 0.7 & \text{for } v_i < -V_{Z2} - 0.7 & \text{(3-37a)} \\ v_i & \text{for } -V_{Z2} - 0.7 < v_i < V_{Z1} + 0.7 & \text{(3-37b)} \\ V_{Z1} + 0.7 & \text{for } v_i > V_{Z1} + 0.7 & \text{(3-37c)} \end{cases}$$

The transfer characteristic is shown in Figure 3-21(b).

Example 3-2

A certain signal source is connected to the input of an amplifier, and to protect the amplifier from possible excess signal levels, the dual limiting circuit of Figure 3-22(a) is connected across the line. Two reference voltages of ±6 V are used to establish the limiter levels. The Thévenin output resistance of the source is 500 Ω, and no additional series resistance is added. The input to the amplifier is essentially an open circuit. (a) Determine the approximate levels of the input voltage at which limiting action occurs. (b) Write equations describing the transfer characteristic. (c) Assume that a positive "spike" of 15 V appears at the source. Determine the conducting diode current based on this condition. (d) For the same condition as (c), determine the reverse voltage across the nonconducting diode.

Figure 3-22 Limiter circuit of Example 3-2 shown two ways and equivalent circuit when $v_i = 15$ V.

Solution

(a) The circuit as shown in Figure 3-22(a) is representative of the way it might appear on a circuit diagram. However, to analyze it, a circuit model showing the effective series resistance can be drawn as in Figure 3-22(b). The two voltage references have also been drawn as "batteries" so that the equivalence of this circuit to that of Figure 3-16(a) is clearly demonstrated.

The level of limiting for a positive input corresponds to the +6-V reference plus the forward-biased drop of 0.7 V for D1 or 6.7 V. By symmetry, the corresponding negative limiting value is −6.7 V. Thus, for −6.7 V < v_i < 6.7 V, the signal is coupled from the source to the amplifier, and the diodes ideally have no effect.

(b) The equations may be expressed as

$$v_o = \begin{cases} v_i & \text{for } -6.7 < v_i < 6.7 \quad (\text{or } |v_i| < 6.7 \text{ V}) \quad \textbf{(3-38)} \\ -6.7 \text{ V} & \text{for } v_i < -6.7 \text{ V} \quad \textbf{(3-39)} \\ 6.7 \text{ V} & \text{for } v_i > 6.7 \text{ V} \quad \textbf{(3-40)} \end{cases}$$

The transfer characteristic will be of the form shown in Figure 3-16(b), with the two voltages labeled as +6.7 V and −6.7 V.

(c) With an assumed input voltage of 15 V, the equivalent circuit is shown in Figure 3-22(c). Note that the constant voltage model for D1 is assumed, but D2 is assumed to be an open circuit.

The current in diode D1 is i_{D1}, and since D2 is open, this current also flows through the 500-Ω resistance. A KVL equation for the left-hand loop is

$$-15 + 500i_{D1} + 0.7 + 6 = 0 \quad \textbf{(3-41)}$$

or

$$i_{D1} = \frac{8.3}{500} = 16.6 \text{ mA} \quad \textbf{(3-42)}$$

(d) Referring again to Figure 3-22(c), note that the reverse voltage v_{R2} across D2 has been shown in the direction to yield a positive value. This voltage can readily be determined by a loop encompassing the two 6-V sources, the 0.7-V drop and the desired voltage v_{R2}. Starting below the left-hand 6-V source and moving clockwise, we have

$$-6 - 0.7 + V_{R2} - 6 = 0 \quad \textbf{(3-43)}$$

or

$$V_{R2} = 12.7 \text{ V} \quad \textbf{(3-44)}$$

By symmetry, if the circuit were analyzed for $v_i = -15$ V, the magnitudes of diode current and reverse voltage would be the same, but the roles of the two diodes would be interchanged.

The use of diodes and resistors to implement simple combinational logic gates is illustrated in this section. A logic gate is a circuit having two or more possible output states (levels and/or conditions), and in which the existing state is a function of the possible input states. Combinational logic circuits are among the most basic of digital circuits.

Integrated-circuit logic functions are widely available at low cost, so the discrete circuits discussed here have limited utility in modern technology. However, they are included for two purposes: (1) they provide good drill in analyzing diode circuits having several states, and (2) they orient the reader to think in terms of logic levels and conditions.

For digital logic circuits, binary variables are employed. A binary variable is one having only two levels. The two levels may be denoted as "0" and "1," *low* and *high*, or *off* and *on*. Since all voltage levels are subject to variation due to losses, power-line variations, component tolerances, and so on, a range is usually specified for each binary variable, and a "forbidden" region is provided as a buffer zone. For example, a voltage level between 0 and 0.8 V may be defined as a binary *0* or *low* level, and a voltage level between 2 V and 5 V may be defined as a binary *1* or *high* level. The range between 0.8 and 2 V would then be undefined.

The logic gates discussed here assume a voltage level equal to or near zero for a logical 0 and a positive voltage (well above 0.7 V) for a logical 1.

OR Gate

A two-input *OR gate* implemented with diodes and a resistor is shown in Figure 3-23. The two inputs are v_1 and v_2, and the output is v_o. The output v_o is high if *either* v_1 or v_2 is high. This circuit will be analyzed in detail in Example 3-3.

Figure 3-23 Simple two-input diode logic OR gate.

AND Gate

A two-input *AND gate* implemented with diodes and a resistor is shown in Figure 3-24. The two inputs are v_1 and v_2, and the output is v_o. The output v_o is high only if v_1 *and* v_2 are high. This circuit will be analyzed in detail in Example 3-4.

Figure 3-24 Simple two-input diode AND gate.

Example 3-3

Consider the OR gate of Figure 3-25(a), in which logic levels of 0 and 5 V are employed. Determine the output voltage for each of the four combinations of the two input voltages, and verify that the OR function is performed.

Figure 3-25 OR gate of Example 3-3 and equivalent circuits of two states.

Solution

As stated in the problem, there are four combinations of input variables. These correspond to (1) $v_1 = v_2 = 0$; (2) $v_1 = 5$ V, $v_2 = 0$; (3) $v_1 = 0$ V, $v_2 = 5$ V; and (4) $v_1 = v_2 = 5$ V. In general, for n inputs, there are 2^n different combinations of input variables.

Since the circuit contains two diodes, for any combination of input variables, it might be necessary to guess at the four possible sets of diode conditions to determine the correct one that could exist. However, we use our "intuition" to infer the correct one in each case and verify that it is correct. Let's consider each of the four input combinations.

(1) $v_1 = v_2 = 0$. This one can be analyzed by inspection. This circuit is completely passive with both input voltages zero, so obviously $v_o = 0$.

(2) $v_1 = 5$ V, $v_2 = 0$. An equivalent circuit to be verified is shown in Figure 3-25(b). It is assumed that D1 is forward biased and D2 is reverse biased. The first condition is verified by noting that the net potential around the loop would result in clockwise current. The second condition is verified by noting that the net potential across D2 is more positive at the cathode than the anode.

A loop equation reads

$$-5 + 0.7 + 1000i_R = 0 \tag{3-45}$$

This yields

$$i_R = \frac{4.3}{1000} = 4.3 \text{ mA} = i_{D1} \tag{3-46}$$

The output voltage and reverse diode voltage v_{R2} (for D2) are the same.

$$v_o = v_{R2} = 1000i_R = 4.3 \text{ V} \tag{3-47}$$

Note that the output is high when one of the variables is high. Do not be disturbed by the fact that this voltage is 4.3 V rather than 5 V. The levels of digital variables frequently change through a circuit, but as long as they remain within a defined range, they can be interpreted properly.

(3) $v_1 = 0$, $v_2 = 5$ V. By the symmetry of the circuit, the condition is the same as part (b), but with the roles of D1 and D2 interchanged. Thus the output is high again.

(4) $v_1 = v_2 = 5$ V. An equivalent circuit to be verified is shown in Figure 3-25(c). Both diodes are forward biased in this case. This is readily verified by noting that current will flow upward through both of these branches, combine, and flow down through the load.

A loop equation around either loop that includes the 1-kΩ resistor is

$$-5 + 0.7 + 1000i_R = 0 \tag{3-48}$$

This yields

$$i_R = \frac{4.3}{1000} = 4.3 \text{ mA} \tag{3-49}$$

and

$$v_o = 4.3 \text{ V} \tag{3-50}$$

By symmetry, each of the two source branches should supply half of this current.

$$i_{D1} = i_{D2} = \frac{4.3}{1000} = 2.15 \text{ mA} \qquad (3\text{-}51)$$

The results of this problem can be summarized in the table that follows:

v_2	v_1	v_0
Low	Low	Low
Low	High	High
High	Low	High
High	High	High

The output is thus high if *either* v_1 *or* v_2 is high. The preceding table is called a *truth table*.

Example 3-4

Consider the AND gate of Figure 3-26, in which logic levels of 0 and 5 V are employed. Determine the output voltage for each of the four combinations of the two input voltages, and verify that the AND function is performed.

Figure 3-26   AND gate of Example 3-4 and equivalent circuits of two states.

Solution

The solution proceeds in the same manner as that of the OR gate in Example 3-3, and the same comments concerning possible combinations of input variables apply here. Each of the four combinations will be considered.

(1) $v_1 = v_2 = 0$. The equivalent circuit for this condition is shown in Figure 3-26(b). The circuit has been redrawn for ease of analysis, and the 5-V supply has been shown as a battery. The net voltage around either loop containing this voltage and a diode is such that current flows in the direction of i_{D1} and i_{D2}, so both diodes are forward biased. A loop equation including the voltage source and either diode is

$$-5 + 1000 i_R + 0.7 = 0 \qquad (3\text{-}52)$$

and

$$i_R = \frac{4.3}{1000} = 4.3 \text{ mA} \qquad (3\text{-}53)$$

By symmetry, this current will divide evenly between the diodes and

$$i_{D1} = i_{D2} = \frac{4.3}{2} = 2.15 \text{ mA} \qquad (3\text{-}54)$$

By direct inspection,

$$v_o = 0.7 \text{ V} \qquad (3\text{-}55)$$

This voltage must be interpreted as a *low* condition, so the output is low when both inputs are low.

(2) $v_1 = 5$ V, $v_2 = 0$. The equivalent circuit is shown in Figure 3-26(c). In this case D2 is assumed to be forward biased and D1 is assumed to be reverse biased. The first condition is verified by noting that the net voltage in the loop encompassing the 5-V voltage source and D2 is such that current must flow in the direction of i_{D2}. A loop equation in this case reads

$$-5 + 1000 i_R + 0.7 = 0 \qquad (3\text{-}56)$$

Solution for i_R yields

$$i_R = \frac{4.3}{1000} = 4.3 \text{ mA} = i_{D2} \qquad (3\text{-}57)$$

To verify the reverse-biased condition for v_{D1}, a loop equation is written around the loop containing the 5-V source and the 0.7-V drop across D2. We have

$$-5 - v_{D1} + 0.7 = 0 \qquad (3\text{-}58)$$

This yields

$$v_{D1} = -4.3 \text{ V} \qquad (3\text{-}59)$$

Since the assumed reference was positive from anode to cathode, but the value turns out to be negative, this diode is reverse biased.

The output voltage is simply

$$v_o = 0.7 \text{ V} \tag{3-60}$$

which is a *low* condition, as in the previous case.

(3) $v_1 = 0$, $v_2 = 5$ V. By the symmetry of the circuit, the condition is the same as part (b), but with D1 and D2 reversed. Thus the output is low again.

(4) $v_1 = v_2 = 5$ V. The equivalent circuit is shown in Figure 3-26(d). Both diodes are reverse biased in this case. This is verified by writing a loop equation around either loop containing the 5-V source and one of the inputs. No current is assumed to flow, so there is no voltage drop across the 1-kΩ resistor. For the loop containing v_{D1} and the 5-V source, we have

$$-5 + v_{D1} + 5 = 0 \tag{3-61}$$

which results in

$$v_{D1} = 0 \tag{3-62}$$

By symmetry, we deduce that

$$v_{D2} = 0 \tag{3-63}$$

Since these voltages are below the knee voltages, both diodes are reverse biased. The output voltage in this case is

$$v_o = 5 \text{ V} \tag{3-64}$$

The results of this problem can be summarized in the following truth table:

v_2	v_1	v_0
Low	Low	Low
Low	High	Low
High	Low	Low
High	High	High

The output is thus high only if v_1 *and* v_2 are high.

3-6 ZENER DIODE REGULATOR CIRCUITS

A general treatment of power supplies, including the process of ac-to-dc conversion, filtering and electronic regulation, will be given in Chapter 20 after the reader has developed sufficient background for full comprehension. However, enough material has been covered already to permit the analysis and design of dc zener regulator circuits. As a means of providing some common applications of zener diodes, these regulator circuits are discussed in this section.

Zener regulator circuits are used to obtain a reasonably regulated dc voltage from a larger available dc voltage. Since electronic devices require different voltage levels for operation, it is possible to employ a few fixed power supply voltages and to obtain the required fixed voltages by employing various zener circuits. A fundamental requirement at the outset is that the available dc voltage must be *larger* in magnitude than the desired voltage to be derived.

Although the analysis and design computations for zener circuits involve simple dc calculations and equally simple circuits, many beginners find the process a bit tricky, particularly when input and/or load conditions are subject to variations. The difficulty is in understanding what happens to certain circuit variables when a change occurs in a different variable.

The basic form of a zener regulator circuit arranged for positive voltages is shown in Figure 3-27. A dc voltage V_1 is available, and it is desired to derive a stable smaller dc voltage V_L by means of a zener diode. The diode is placed across the output in reverse-biased form, so the zener voltage V_Z is the desired load voltage (i.e., $V_Z = V_L$). The resistance R_L represents some load of interest and the load current is I_L. Since V_Z appears across R_L, Ohm's law indicates that

$$\boxed{I_L = \frac{V_L}{R_L} = \frac{V_Z}{R_L}}$$

(3-65)

The current through the diode is I_Z, and the current flowing from the source is I_1. From KCL applied to the node above the zener, we have

$$-I_1 + I_Z + I_L = 0$$

(3-66)

or

$$\boxed{I_1 = I_Z + I_L}$$

(3-67)

Stated in words, the source must supply a current equal to the sum of the zener current and the load current.

A KVL equation for the left-hand loop reads

$$-V_1 + R_s I_1 + V_Z = 0$$

(3-68)

Figure 3-27 Basic form of zener regulator circuits.

or

$$V_1 = R_s I_1 + V_Z \qquad \textbf{(3-69)}$$

The input voltage must thus equal the sum of the voltage drop across R_s plus the zener (or output) voltage.

The preceding equations are all straightforward based on the conditions given. Reflecting on the assumptions, suppose that the variables in the preceding equations were all fixed at constant values. If this were true, the situation would be quite simple. The utility of the zener diode is that it can maintain a reasonably constant voltage over a range of circuit parameter variations.

It was stated in Section 2-7 that the voltage V_Z does vary somewhat with bias current, and the effect can be modeled by a dynamic zener impedance Z_{ZT}. Let ΔI_Z represent a change in zener current and let ΔV_Z represent the corresponding change in voltage. Recall that those quantities are related by

$$\Delta V_Z = Z_{ZT} \Delta I_Z \qquad \textbf{(3-70)}$$

The impedance is positive, so an increase in current results in an increase in voltage.

In the development that follows, a reasonable approximation will be made that might bother some readers, but it is typical of the types of assumptions made in "real-life" electronics. In studying the effects of various changes for input and load, we assume initially that V_Z *remains fixed.* Once these computations are made, we can then observe the net change in current and employ (3-70) to estimate the net change in zener voltage. This process is justified because ΔV_Z is typically much smaller than V_Z and usually can be ignored for most computations outside of the zener itself.

Based on the preceding assumption, a basic concept will be emphasized at the outset. *As long as the zener diode is biased in the breakdown region, it will automatically adjust the current I_Z that it absorbs in order to attempt to maintain the voltage V_Z at a nearly constant level.* This process is fundamental to the analysis of a circuit and will appear in various forms in the discussion that follows.

Changes in Input Voltage

First, assume that the input voltage increases from V_1 to a new value V_1' as shown in Figure 3-28(a). If V_Z is to remain constant, the voltage across R_s must increase since the resistance must absorb the difference between V_1 and V_Z. Since R_s is fixed, the only way the resistive drop $R_s I_1$ can increase is for I_1 to increase. I_1 is composed of two components: I_Z and I_L. I_L cannot increase since it is fixed by V_Z and R_L. Thus I_Z must increase; that is, the zener absorbs more current when the input voltage increases. The opposite pattern occurs when V_1 decreases (i.e., I_Z decreases).

To summarize: *For fixed load current, the zener current increases as the input voltage increases, and vice versa.*

(a) Input voltage increases: If V_1 increases to V'_1, I_Z increases to I'_Z, I_1 increases to I'_1, and V_{R_s} increases to V'_{R_s} so as to keep V_Z and I_L constant.

(b) Load current increases: If I_L increases to I'_L, I_Z decreases to I'_Z so as to keep I_1, V_{R1}, and V_Z constant.

Figure 3-28 Separate effects of increased input voltage and increased load current.

Changes in Load Current

Next, assume that the load current increases from a value I_L to a new value I'_L as shown in Figure 3-28(b). (Note that this corresponds to a decrease in R_L.) If V_Z is to remain constant, the voltage across R_s must remain constant since V_1 has not changed. Since the resistive voltage drop is $R_s I_1$ and R_s is fixed, I_1 cannot change. However, $I_1 = I_Z + I_L$, and since I_L has increased I_Z must decrease to maintain a constant sum. The opposite pattern occurs when I_L decreases (i.e., I_Z increases).

To summarize: *For fixed input voltage, the zener current decreases as the load current increases, and vice versa.*

Changes in Input Voltage and Load Current

When both input voltage and load current are subject to variations, the variation in zener current is a combination of the two separate effects discussed previously. Look at each combination of changes, and determine what adjustment in zener current must occur to keep the voltage across the diode constant.

Worst-Case Effects

Refer to the zener diode characteristic shown in Figure 3-29. For proper circuit operation, the zener current must always be in the range from I_{ZK} to I_{ZM}. If the current magnitude ever drops below I_{ZK}, the circuit loses its ability to regulate, and the voltage magnitude drops drastically. Conversely, if the current magnitude exceeds I_{ZM}, the diode will be destroyed.

In any practical design, it is important that some investigation be made to ensure that the expected range of diode current always remain in the allowable range. For this purpose, estimated values of the minimum zener current $I_{Z(min)}$ and the maximum zener current $I_{Z(max)}$ should be made.

Assume first that the minimum and maximum expected values of V_1 and I_L are somehow estimated. Denote these values as $V_{1(min)}$, $V_{1(max)}$, $I_{L(min)}$, and $I_{L(max)}$, respectively. Even when the load current is expected to remain nearly constant, it could be disconnected accidentally, or an open circuit in the load could occur. Since this represents an increase in zener current, a good worst-case choice for the minimum-load current is $I_{L(min)} = 0$ for many situations.

In general, from (3-67),

$$I_Z = I_1 - I_L \tag{3-71}$$

From (3-68), I_1 is

$$I_1 = \frac{V_1 - V_Z}{R_s} \tag{3-72}$$

Substituting (3-72) in (3-71) results in

$$I_Z = \frac{V_1 - V_Z}{R_s} - I_L \tag{3-73}$$

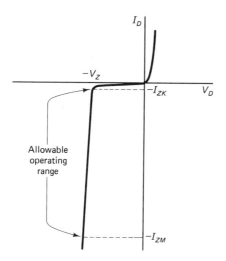

Figure 3-29 Allowable operating range for zener regulator circuits.

The minimum value of I_Z corresponds to the minimum value of V_1 and the maximum value of I_L.

$$I_{Z(min)} = \frac{V_{1(min)} - V_Z}{R_1} - I_{L(max)} \qquad (3\text{-}74)$$

The maximum value of I_Z corresponds to the maximum value of V_1 and the minimum value of I_L.

$$I_{Z(max)} = \frac{V_{1(max)} - V_Z}{R_1} - I_{L(min)} \qquad (3\text{-}75)$$

Although these formulas will allow us to "plug in" the different values and obtain the results directly, the reader is encouraged to think through the process in each case rather than memorize formulas.

Design of a Regulator Circuit

In designing a regulator circuit, a diode having a value of V_Z equal to the desired load voltage is used. Assume that the nominal values of I_L and V_1 are given. A nominal zener operating current I_Z must be selected, and the design then reduces to determining R_s. From (3-72), R_s is selected as

$$R_s = \frac{V_1 - V_Z}{I_1} = \frac{V_1 - V_Z}{I_Z + I_L} \qquad (3\text{-}76)$$

The value of I_Z is selected from available zener diode data. If both I_L and V_1 are expected to remain nearly constant, the value of data test current I_{ZT} may be a good choice.

Worst-case variations of V_1 and I_L should then be estimated. As indicated earlier, the value $I_{L(min)} = 0$ may be a good choice for the lower worst-case value of the load current. The worst-case range of I_Z should then be calculated as discussed earlier. If $I_{Z(max)}$ exceeds I_{ZM} or if $I_{Z(min)}$ is lower than I_{ZK}, the design cannot be achieved with the conditions assumed.

Example 3-5

Consider the zener regulator circuit Figure 3-30. Assume initially that $V_1 = 15$ V, $R_L = 100\ \Omega$, and $V_Z = 10$ V for these conditions. (a) Determine I_L, I_1, and I_Z. (b) Assume that V_1 changes to 16 V but that all other parameters remain the same. Repeat the computations of (a). (c) With $V_1 = 16$ V, assume that R_L is changed to 125 Ω. Repeat the computations of (a). (d) If the zener impedance is $Z_{ZT} = 5\ \Omega$, estimate the final zener voltage in (c).

Figure 3-30 Circuits of Example 3-5.

Solution

(a) For $R_L = 100 \ \Omega$ and $V_L = V_Z = 10$ V, the load current is

$$I_L = \frac{10}{100} = 0.1 \text{ A} = 100 \text{ mA} \tag{3-77}$$

The current is

$$I_1 = \frac{V_1 - V_Z}{40} = \frac{15 - 10}{40} = 0.125 \text{ A} = 125 \text{ mA} \tag{3-78}$$

The zener current I_Z is then

$$I_Z = I_1 - I_L = 125 - 100 = 25 \text{ mA} \tag{3-79}$$

(b) When V_1 changes to 16 V, the voltage across R_s will increase to maintain V_Z close to 10 V. Actually, V_Z changes slightly, but that change will be ignored for the present analysis. The new value of I_1 is

$$I_1 = \frac{16 - 10}{40} = 0.15 \text{ A} = 150 \text{ mA} \tag{3-80}$$

Since V_L and R_L are both assumed as constants, we have

$$I_L = 100 \text{ mA} \tag{3-81}$$

The zener current changes to

$$I_Z = 150 - 100 = 50 \text{ mA} \tag{3-82}$$

(c) When R_L is changed to 125 Ω, the load current is

$$I_L = \frac{10}{125} = 0.08 \text{ A} = 80 \text{ mA} \tag{3-83}$$

As in (b), we have assumed that V_Z remains at 10 V. Since $V_1 = 16$ V, the current I_1 is the same as in (b); that is,

$$I_1 = 150 \text{ mA} \tag{3-84}$$

The zener current is now

$$I_Z = 150 - 80 = 70 \text{ mA} \tag{3-85}$$

Reviewing the problem thus far, we see that the zener current increased when the input voltage increased and again when the load current decreased.

(d) Over the range of this problem, the zener current changed from 25 mA initially to 70 mA in (c). Thus $\Delta I_Z = 70 - 25 = 45$ mA. The change ΔV_Z in zener voltage can be estimated as

$$\Delta V_Z = Z_{ZT} \Delta I_Z = 5 \times 45 \times 10^{-3} = 0.225 \text{ V} \tag{3-86}$$

The final voltage V_Z' is then

$$V_Z' = 10 + 0.225 = 10.225 \text{ V} \tag{3-87}$$

One could correctly argue that this value when applied in (c) would result in a different value of current than was obtained in that part. Similarly, an adjustment in (b) would result in a different value. Such adjustments can be made when it is necessary to be as precise as possible. However, the approximate nature of the data involved usually makes the approach here sufficient for most purposes.

Example 3-6

A zener regulator circuit of the form shown in Figure 3-27 is to be designed to establish a 6.8-V reference from a fixed 10-V supply. The load current requirement is fixed at 100 mA. The load current requirement is fixed at 100 mA. The zener is to be biased at 20 mA. **(a)** Determine the value of R_s required. **(b)** If the load is accidentally removed from the zener, determine the maximum diode current.

Solution

(a) The net current I_1 through R_s is the load current plus the zener current, that is

$$I_1 = I_L + I_Z = 100 + 20 = 120 \text{ mA} \tag{3-88}$$

The voltage V_{R1} across R_s must be

$$V_{R1} = V_1 - V_Z = 10 - 6.8 = 3.2 \text{ V} \tag{3-89}$$

The value of R_s is then

$$R_s = \frac{V_1}{I_1} = \frac{3.2}{120 \times 10^{-3}} = 26.7 \ \Omega \tag{3-90}$$

(b) If the load is accidentally removed, the zener will absorb the entire current. Thus

$$I_{Z(\text{max})} = 120 \text{ mA} \tag{3-91}$$

The maximum diode current rating should be based on this possible worst-case effect.

DRILL PROBLEMS

For all problems except those involving zener diodes, assume that reverse breakdown does not occur within the range of signal voltages given. Assume a 0.7-V drop for forward-biased diodes unless otherwise indicated.

3-1. Consider the rectifier circuit of Figure 3-8(a) with $R = 1 \text{ k}\Omega$. An ideal time-varying voltage source is applied across the input terminals. If the source voltage varies from -20 to 20 V, determine the peak values of **(a)** the output voltage, **(b)** the diode current, and **(c)** the reverse voltage across the diode.

3-2. Consider the rectifier circuit of Figure 3-9(a) with $R = 2$ kΩ. An ideal time-varying voltage source is applied across the input terminals. If the source voltage varies from -15 to 15 V, determine the peak values of **(a)** the output voltage, **(b)** the diode current, and **(c)** the reverse voltage across the diode.

3-3. Repeat Problem 3-1 if the source voltage is nonsymmetrical and varies from -30 to 25 V.

3-4. Repeat Problem 3-2 if the source voltage is nonsymmetrical and varies from -15 to 20 V.

3-5. Consider the rectifier circuit of Figure 3-11(a) with $R = 1$ kΩ. An ideal time-varying voltage source is applied across the input terminals. If the source voltage varies from -20 to 20 V, determine the peak values of **(a)** the output voltage, **(b)** the diode current, and **(c)** the reverse voltage across the diode.

3-6. Consider the rectifier circuit of Figure 3-12(a) with $R = 2$ kΩ. An ideal time-varying voltage source is applied across the input terminals. If the source voltage varies from -15 to 15 V, determine the peak values of **(a)** the output voltage, **(b)** the diode current, and **(c)** the reverse voltage across the diode.

3-7. Repeat Problem 3-5 if the source voltage is nonsymmetrical and varies from -30 to 25 V.

3-8. Repeat Problem 3-6 if the source voltage is nonsymmetrical and varies from -15 to 20 V.

3-9. Consider the rectifier circuit of Figure P3-9. A source with time-varying voltage v_s and an internal resistance of 500 Ω is connected to the input. **(a)** Determine a set of equations describing the relationship between the source voltage and output voltage, and sketch the transfer characteristic. **(b)** If v_s varies from -30 to 30 V, determine the peak values of the output voltage, the diode current, and the reverse voltage across the diode.

Figure P3-9

3-10. Consider the rectifier circuit of Figure P3-10. A source with time-varying voltage v_s and an internal resistance of 300 Ω is connected to the input. **(a)** Determine a set of equations describing the relationship between the source voltage and output voltage, and sketch the transfer characteristic. **(b)** If v_s varies from -18 to 18 V, determine the peak values of the output voltage, the diode current, and the reverse voltage across the diode.

Figure P3-10

3-11. Consider the rectifier circuit of Figure P3-11. Determine a set of equations describing the relationship between v_i and v_o (*Hint:* Redraw the parallel combination of the diode and resistor and place the resistor on the left. Then, apply Thévenin's theorem looking back from the diode.)

Figure P3-11

3-12. Consider the rectifier circuit of Figure P3-12. Determine a set of equations describing the relationship between v_i and v_o. (See hint for Problem 3-11.)

Figure P3-12

3-13. A certain circuit is designed to limit positive voltages and is of the form shown in Figure 3-14(a) with $R = 1$ kΩ and $V_A = 6$ V. **(a)** Write equations for the transfer characteristics and sketch the form. **(b)** Determine the output voltage for each of the following values of input voltage: $v_i = -12, -5, 5,$ and 12 V. **(c)** If a "spike" of 30 V appears on the input, determine the diode current i_D.

3-14. A certain circuit is designed to limit negative voltages and is of the form shown in Figure 3-15 with $R = 2$ kΩ and $V_B = 12$ V. **(a)** Write equations for the transfer characteristic and sketch the form. **(b)** Determine the output voltage for each of the following values of input voltage: $v_i = -20, -6, 6,$ and 20 V. **(c)** If a "spike" of -50 V appears on the input, determine the diode current i_D.

3-15. A certain circuit is designed to limit both positive and negative voltages and is of the form shown in Figure P3-15. The Thévenin output resistance of the source is 1000 Ω, and no additional series resistance is added. The load is an amplifier input open circuit. **(a)** Determine the approximate levels of the input voltage at which limiting action occurs. **(b)** Write equations describing the transfer characteristic and sketch it. **(c)** Assume that a positive spike of 30 V appears at the source. Determine the conducting diode current based on this condition. **(d)** For the same condition as (c), determine the reverse voltage across the nonconducting diode.

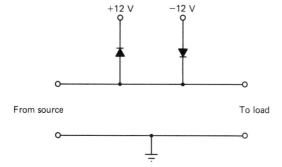

+12 V   −12 V

From source   To load

Figure P3-15

3-16. Repeat the analysis of Problem 3-15 for the circuit of Figure P3-16. All conditions not shown in the figure are the same as in Problem 3-15.

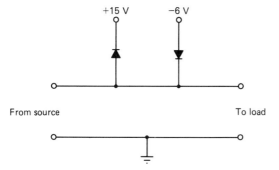

+15 V   −6 V

From source   To load

Figure P3-16

3-17. A zener limiting circuit of the form shown in Figure 3-19 has $V_Z = 6.2$ V and $R = 200$ Ω. **(a)** Write equations for the transfer characteristic and sketch it. **(b)** If $v_i = 20$ V, determine v_o and the diode current. Assume that any load connected across the output draws negligible current.

3-18. A zener limiting circuit of the form shown in Figure 3-20 has $V_Z = 5.1$ V and $R = 300$ Ω. **(a)** Write equations for the transfer characteristic and sketch it. **(b)** If $v_i = -25$ V, determine v_o and the diode current. Assume that any load connected across the output draws negligible current.

3-19. A zener limiting circuit of the form shown in Figure 3-21 has $V_{Z1} = V_{Z2} = 6.2$ V and $R = 200$ Ω. **(a)** Write equations for the transfer characteristic and sketch it. **(b)** If $v_i = 20$ V, determine v_o and the diode current. Assume that any load connected across the output draws negligible current.

3-20. A zener limiting circuit of the form shown in Figure 3-21 has $V_{Z1} = 5.1$ V, $V_{Z2} = 3.3$ V, and $R = 300$ Ω. **(a)** Write equations for the transfer characteristic and sketch it. **(b)** If $v_i = 25$ V, determine v_o and the diode current. **(c)** If $v_i = -25$ V, determine v_o and the diode current.

3-21. In the circuit of Example 3-5 (Figure 3-30), assume that V_1 changes to 17 V

Drill Problems

but that all other parameters remain the same. **(a)** Determine I_L, I_1, and I_Z. **(b)** With $V_1 = 17$ V, assume that R_L is changed to 80 Ω. Repeat the computations of (a).

3-22. In the circuit of Example 3-5 (Figure 3-30), assume that V_1 changes to 14.6 V but that all other parameters remain the same. **(a)** Determine I_L, I_1, and I_Z. **(b)** With $V_1 = 14.6$ V, assume that R_L is changed to 125 Ω. Repeat the computations of (a).

3-23. Consider the zener regulator circuit of Figure P3-23. Assume initially that $V_1 = 18$ V, $R_L = 150$ Ω, and $V_Z = 15$ V. **(a)** Determine I_L, I_1, and I_Z. **(b)** Assume that V_1 changes to 18.5 V but that all other parameters remain the same. Repeat the computations of (a). **(c)** With $V_1 = 18.5$ V, assume that R_L is changed to 170 Ω. Repeat the computations of (a). **(d)** If the zener impedance is 12 Ω, estimate the final zener voltage in (c).

Figure P3-23

3-24. Consider the negative voltage zener regulator circuit of Figure P3-24. Assume initially that $V_1 = 5$ V, $R_L = 40$ Ω, and $V_Z = 3.0$ V. **(a)** Determine I_L, I_1, and I_Z. **(b)** Assume that V_1 changes to 5.5 V but that all other parameters remain the same. Repeat the computations of (a). **(c)** With $V_1 = 5.5$ V, assume that R_L is changed to 60 Ω. Repeat the computations of (a). **(d)** If the zener impedance is 6 Ω, estimate the final zener voltage in (c).

Figure P3-24

DESIGN PROBLEMS

3-25. Design a simple half-wave rectifier circuit to meet the following specifications:

Applications of Diode-Resistor Circuits   Chap. 3

Positive output for positive input

Open-circuit signal voltage range from -30 to 30 V

Peak diode current ≤ 20 mA

Ideal voltage source input

Negligible loading on output

Draw a schematic diagram, and specify the *minimum* value of the resistance R and the diode peak reverse voltage rating. (In practice, one should select ratings well above the minimum values.)

3-26. In the design of Problem 3-25, assume that the voltage source has an internal resistance of 600 Ω. Specify again the *minimum* values of the *external* resistance R and the diode peak reverse voltage rating.

3-27. Design a limiter circuit employing standard junction diodes to meet the following specifications:

Symmetrical limiting at levels of about ± 15.7 V

Worst-case calculations to be based on assumed peak values of ± 50 V

Peak diode current ≤ 40 mA

Ideal voltage source input

Negligible loading on output

Assume that the system contains $+15$-V and -15-V power supplies. Draw a schematic diagram and specify the *minimum* values of the resistance R and the diode peak reverse voltage rating.

3-28. In the design of Problem 3-27, assume that the voltage source has an internal resistance of 600 Ω. Specify again the *minimum* values of the *external* resistance R and the diode peak reverse voltage rating.

3-29. Design a limiter circuit utilizing two zener diodes to meet the following specifications:

Symmetrical limiting at levels of about ± 5.8 V

Worst-case calculations to be based on assumed peak values of ± 50 V

Peak diode current ≤ 40 mA

Ideal voltage source input

Negligible loading on output

Draw a schematic diagram and specify the values of the two zener voltages. Specify the *minimum* value of the resistance R.

3-30. Assume in Problem 3-29 that nonsymmetrical limiting at levels of $+12.7$ V and -4 V is desired. All other conditions and assumptions are unchanged. Repeat the design and the design calculations of Problem 3-29.

3-31. Design a three-input circuit to satisfy the following requirements:

> Output is low if all three inputs are low.
> Output is high if either input or any input combination is high.

Assume at the input that a high level is 5 V and a low level is 0. However, an output high level can be from 4 to 5 V. Diode current is not to exceed 2.5 mA. Assume ideal sources at input and no loading at output.

3-32. Design a three-input circuit to satisfy the following requirements:

> Output is high if all three inputs are high.
> Output is low if either input is low.

Assume at the input that a high level is 5 V and a low level is 0. However, an output low level can be from 0 to 1 V. Diode current is not to exceed 2.5 mA. Assume ideal sources at input and no loading at output.

3-33. Design a zener regulator circuit to establish a 5.1-V reference from a 15-V supply. The load current requirement is 50 mA and the zener is to be biased at 20 mA. If the load is accidentally removed from the zener, determine the maximum diode current.

3-34. Design a zener regulator circuit to establish a -12-V reference from a -15-V supply. The load resistance is 500 Ω, and the zener is to be biased at 20 mA. If the load is accidentally removed from the zener, determine the maximum diode current.

TROUBLESHOOTING PROBLEMS

3-35. A circuit of the form shown in Figure P3-9 is not working properly. The following dc voltages are measured:

$v_s = 10$ V (open-circuit voltage with load temporarily disconnected)

$v_i = 10$ V

$v_o = 0$

Indicate the likely trouble.

3-36. A circuit of the form shown in Figure P3-10 is not working properly. The following dc voltages are measured:

$v_s = 12$ V (open-circuit voltage with load temporarily disconnected)

$v_i = 9$ V

$v_o = 9$ V

Indicate the likely trouble.

3-37. A circuit of the form shown in Figure P3-11 is not working properly. The following dc voltages are measured:

$$v_i = 12 \text{ V}$$

$$v_o = 12 \text{ V}$$

Identify two *possible* circuit faults.

3-38. A circuit of the form shown in Figure P3-11 is implemented without checking diode specifications and worst-case conditions. When the circuit fails to function properly, the following dc conditions are noted:

When $v_i = 16$ V, $v_o = 12$ V.
When $v_i = 32$ V, $v_o = 24$ V.
When $v_i = 40$ V, $v_o = 24$ V.

Indicate the likely cause of the problem.

COMPUTER PROBLEMS

3-39. Consider the diode rectifier circuit of Figure 3-8(a). Write a computer program to calculate v_o and the current i. The program should be capable of detecting forward or reverse bias and changing the form of the calculations accordingly.

Input data: R, v_i
Output data: v_o, i

3-40. Consider the circuit form of Figure 3-13(a) but with an arbitrary source resistance R_s and an arbitrary load resistance R_L. Write a computer program to calculate i, v_o, and v_i. The program should be capable of detecting forward or reverse bias and changing the form of the calculations accordingly.

Input data: R_s, R_L, v_s
Output data: i, v_o, v_i

3-41. Consider the basic form of the zener regulator circuit as given in Figure 3-27. Write a program to analyze the worst-case effects in the circuit as a result of parameter variations.

Input data: R, V_Z, $V_{1(\text{min})}$, $V_{1(\text{max})}$, $I_{L(\text{min})}$, $I_{L(\text{max})}$, Z_{ZT}
Output data: $I_{Z(\text{min})}$, $I_{Z(\text{max})}$, V_Z'

3-42. Write a program to design a zener regulator circuit of the form shown in Figure 3-27.

Input data: V_1, V_Z, I_L, I_Z
Output data: R_s

CHAPTER 4

Bipolar Junction Transistors

OBJECTIVES

After completing this chapter, the reader should be able to:

1. Identify the two major classes of transistors.
2. Draw a simplified layout representing the construction of a bipolar junction transistor (BJT).
3. Identify the two types of BJTs according to the polarities of the semiconductor materials.
4. Draw the schematic of and identify the three terminals of a BJT.
5. State the three operating regions of a BJT, and identify the bias conditions for each.
6. Draw equivalent circuit models for each of the three BJT operating regions.
7. Sketch the form of the collector characteristics of a typical BJT, and label all variables involved.
8. Define the dc current gain of a BJT.
9. State the relationship between base current, emitter current, and collector current.
10. Discuss the difference between a small-signal and a power transistor.
11. Inspect the specification sheets of typical BJTs to determine pertinent data.
12. Apply the derating factor to determine the power rating of a BJT at a specified temperature.

4-1 OVERVIEW

The development of the bipolar junction transistor in the late 1940s had a most significant impact on the future of electronics. Prior to that time, virtually all electronic control and amplification functions were performed with vacuum tubes. Over the next two decades, a major transition took place, with transistors and related semiconductor devices replacing vacuum tubes. Transistors had the advantages of reliability, smaller size, lower power dissipation, and lower costs.

136

The transition from vacuum tubes to transistors was only the beginning of electronics as we know it today. In the 1960s, semiconductor manufacturers developed the *integrated circuit* (IC), in which the equivalent of a number of transistors was contained on a single chip. This process continued through the levels of *large-scale integration* (*LSI*) and *very-large-scale integration* (*VLSI*). At the time of this writing, integrated circuits containing the equivalent of hundreds of thousands of transistors are available, and this trend will doubtless continue.

Despite these major developments, the serious electronics specialist must still understand how to deal with an individual (or *discrete*) transistor. The state of the art has never developed and probably never will develop to the point where the user can blindly connect integrated-circuit chips without regard to interactions and interface problems. In dealing with such problems, the need for considering transistor action at the input and output terminals of the chip is essential. In many situations, the voltage, current, and power levels are different, and discrete transistors may be required to match properly the levels between chips. The control of external devices from IC chips often requires transistors to serve as buffers or drivers. Finally, integrated circuits are not available to perform many functions at higher frequencies, and there is still heavy utilization of discrete transistors for such applications.

Transistors and integrated circuits are available in a variety of packages, depending on the power level and other factors. Some of the different types of transistor and integrated-circuit packages are illustrated in Figure 4-1. The transistors in

Figure 4-1 Typical packages for small-signal transistors along with
some integrated circuits. (Courtesy of Motorola, Inc.)

this photograph are of the *small-signal* type (to be defined later). Integrated circuits contain the equivalent of many different transistors and may have many external connections (referred to as *pins*).

There are two major classes of transistors: (1) the *bipolar junction transistor* (BJT) and (2) the *field-effect transistor* (FET). There are various categories within each individual class. Integrated circuits utilize both classes of transistors.

When the term "transistor" is used without a modifier in this book, it will be interpreted as either the major type being discussed at that point in the text, or the point of discussion could apply to either type. Where specific properties of individual types are discussed, the acronyms BJT and FET will often be used.

This chapter deals with the *bipolar junction transistor* and its operating characteristics. The major focus is on the external characteristics relating voltage and current at specific sets of terminals.

4-2 BIPOLAR JUNCTION TRANSISTOR

The *bipolar junction transistor* (BJT) in its basic form consists of three sections of semiconductor material. The three sections are denoted as the *emitter, base,* and *collector*. There are two possible configurations of this basic form and they are denoted as the (1) *npn* type and the (2) *pnp* type. The choice of these designations will be clear shortly.

A simplified layout form of an *npn* BJT is shown in Figure 4-2(a). The emitter and collector constitute the outer sections of the BJT, and they are both composed of *n*-type semiconductor material. The base is a thin layer of *p*-type material sandwiched between the emitter and collector, and it is very lightly doped. The emitter is heavily doped and the collector has an intermediate level of doping. Another distinction between the emitter and collector is that the collector region is larger and can dissipate more heat.

The schematic symbol for the *npn* transistor is shown in Figure 4-2(b). As far

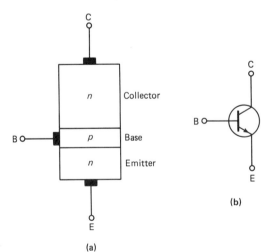

(a)

(b)

Figure 4-2 Physical layout of *npn* bipolar junction transistor and schematic symbol.

as a user is concerned, the contact with the "outside world" is through the three terminals, with *E* denoting *emitter, B* denoting *base,* and *C* denoting *collector*. Later we will see that the arrow in the emitter of the schematic symbol is pointing in the direction of positive conventional current flow for that terminal.

A simplified layout form of a *pnp* BJT is shown in Figure 4-3(a). The various doping properties discussed earlier apply to the *pnp* transistor, but its emitter and collector consist of *p*-type semiconductor material, and its base is composed of a thin layer of *n*-type material.

For either transistor type, the semiconductor junction between the base and emitter is called the *base–emitter junction* or *base–emitter diode*. The junction between the base and collector is called the *base–collector junction* or *base–collector diode*.

In the sense of the definitions given earlier and bias conditions to be discussed later, it is helpful to visualize a BJT as two diodes connected in series, with the base connection at the midpoint. Such forms for both the *npn* and *pnp* transistors are shown in Figure 4-4. These diagrams are particularly useful in deducing forward-and/or reverse-bias conditions for the two junctions. However, it must be stressed that *the BJT is much more than the simple connection of two diodes back to back.* The connection to the base terminal is shown as a dashed line to remind us that this is not a simple connection of two diodes.

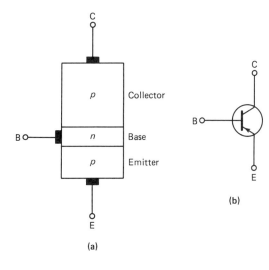

(a)

(b)

Figure 4-3 Physical layout of *pnp* bipolar junction transistor and schematic symbol.

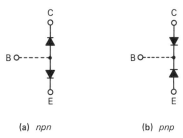

(a) *npn*

(b) *pnp*

Figure 4-4 Visualization of bipolar junction transistor as two diodes.

As a result of some complex semiconductor phenomena, a small current flow in the lightly doped base can have a very large effect on current flow from collector to emitter (*npn*) or from emitter to collector (*pnp*). Since our objective in this book is circuit analysis, troubleshooting, and design, we deal with this phenomenon through the external characteristics rather than the internal physics. The remainder of this chapter is devoted to a development of the properties of such external characteristics.

4-3 BJT OPERATING REGIONS

Since the BJT contains a base–emitter junction and a base–collector junction, each conceivably could be forward biased or reverse biased for a given set of circuit conditions. With two junctions, there are four possible combinations of bias conditions: (1) both junctions reverse biased, (2) both junctions forward biased, (3) base–emitter junction forward biased and base–collector junction reverse biased, and (4) base–emitter junction reverse biased and base–collector junction forward biased. Of these four conditions, the last one is not used often. When condition 4 does arise, it can be viewed in a manner similar to condition 3, but with the collector effectively becoming the emitter and the emitter becoming the collector. The characteristics are usually not optimum in that form since the emitter and collector are constructed differently. We will eliminate condition 4 from consideration and focus on the other three important cases. A summary of these three cases and the designations of the regions of operation are given in Table 4-1. Each condition will now be discussed in detail.

TABLE 4-1 TRANSISTOR OPERATING REGIONS[a]

Base emitter	Base collector	Region
RB	RB	Cutoff
FB	FB	Saturation
FB	RB	Active

[a]RB, reverse bias; FB, forward bias.

Cutoff

This condition corresponds to *reverse bias* for *both base–emitter* and *base–collector* junctions. In effect, both diodes act nearly like open circuits under these conditions. An ideal equivalent circuit is shown in Figure 4-5.

Figure 4-5 Ideal model of BJT in *cutoff* region.

Actually, there are some small temperature-dependent reverse leakage currents that will exist under reverse-bias conditions. For normal temperature conditions and stable circuit designs, these currents are usually negligible and will be ignored at the present time. However, these currents may be significant in some applications.

Saturation

This condition corresponds to *forward bias* for *both base–emitter* and *base–collector* junctions. As was the case for an individual junction diode in Chapter 3, various models may be applied for the two effective diodes in the saturated transistor. The most ideal model is to assume a perfect short circuit for both the base–emitter and base–collector diodes. This ideal model, shown in Figure 4-6, corresponds to the ideal diode model of Chapter 2. In effect, this ideal model indicates that all three transistor terminals would "come together" and assume the same potential under conditions of ideal saturation. This is in contrast to ideal cutoff, where all terminals are uncoupled from each other. The assumption of ideal saturation is used where rough checks of saturation conditions are made and where high accuracy is not required.

Recall that the constant voltage source model is the most widely used form for representing a single diode. In the same sense, a constant voltage source model could be used to represent both the base–emitter and base–collector junctions. Such a model would have the form shown in Figure 4-7, where V_{BE} and V_{BC} are the respective forward-biased constant voltage drops for the diodes. The polarities shown here represent an *npn* transistor, and both would be reversed for a *pnp* unit. If V_{BE} and V_{BC} each had a value of 0.7 V, the net voltage drop between collector and emitter would be zero since the two voltages are opposite in direction.

In practice, the model of Figure 4-7 as given is somewhat clumsy to apply and is not used often. Under most conditions of saturation, the collector current is much greater than the base current, and the two sources in Figure 4-7 would have to be made slightly different to account for the small collector–emitter saturation voltage drop. We will drop this model from further consideration.

Figure 4-6 Ideal model of BJT in *saturation* region.

Figure 4-7 Possible form for a constant voltage source saturation model of an *npn* transistor.

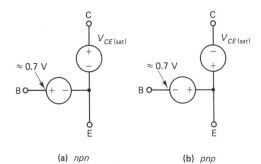

(a) *npn* (b) *pnp*

Figure 4-8 Best constant voltage source models of BJT in saturation region.

A better model for saturation is shown in Figure 4-8, with (a) representing an *npn* transistor and (b) representing a *pnp* transistor. In this model the standard silicon forward-biased voltage drop of 0.7 V for the base–emitter junction is assumed, with the polarity a function of the forward-biased direction of that junction. The base current is generally small compared with the collector current, and the constant voltage drop assumption is reasonably valid. However, we need to remind ourselves that the value 0.7 V is an approximation and may vary with different transistors and operating conditions.

To account for the fact that the voltage between collector and emitter saturation is not exactly zero, a small voltage source, denoted as $V_{CE(sat)}$, is placed between these terminals. The value of $V_{CE(sat)}$ typically varies from less than 0.1 V at small saturation currents to as much as 1 V or more at very large saturation currents.

In many applications involving saturation, the value of $V_{CE(sat)}$ is not known and cannot easily be determined. In such applications it may be necessary to assume that $V_{CE(sat)} = 0$. For such an assumption, the $V_{CE(sat)}$ sources in Figure 4-8 are replaced by short circuits. If, however, the application requires that the exact collector-to-emitter saturation voltage be determined, these simplified models predict invalid results.

Active Region

This condition corresponds to *forward bias* for the *base–emitter junction* and *reverse bias* for the *base–collector junction*. This means that for an *npn* transistor, the base is more positive than the emitter, and the collector is more positive than the base. The sense of polarities is reversed for a *pnp* transistor.

As transistors are connected in various circuits for specific applications, any one of the three terminals may be considered as a *common* terminal. The other two terminals may then be considered as input and output terminals, respectively.

The most widely employed common terminal is the emitter terminal, and any circuit connected in such a fashion is referred to as a *common-emitter* configuration. The most widely available transistor data are obtained through common-emitter measurement circuits and references.

An *npn* transistor, with three possible voltages labeled, is shown in Figure 4-9.

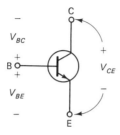

Figure 4-9 Three possible voltages for an *npn* transistor.

Using the reference directions shown, the following quantities are defined:

$$V_{BE} = \text{base–emitter voltage}$$
$$V_{BC} = \text{base–collector voltage}$$
$$V_{CE} = \text{collector–emitter voltage}$$

In all cases, the first subscript refers to the assumed positive reference. Thus, if a given voltage is positive, the terminal represented by the first subscript is more positive than the other terminal. On the other hand, if a given voltage is negative, the terminal represented by the first subscript is more negative than the terminal represented by the second subscript. In that case it is often more convenient to rewrite the voltage as a positive value and reverse the order of the subscript. The reference + and − signs are also reversed in the process. For example, if $V_{CE} = -10$ V, the assumed positive reference terminal (collector) is actually more negative than the emitter terminal. This could be rewritten as $V_{EC} = 10$ V, where the reference positive terminal is now identified as the emitter terminal. In general, for any terminals A and B,

$$V_{AB} = -V_{BA} \tag{4-1}$$

For active-region operation, the base–emitter junction must be forward biased and the base–collector junction must be reverse biased. For an *npn* transistor, the first condition implies that $V_{BE} \approx 0.7$ V. The assumed reference direction for V_{BE} with an *npn* transistor is thus compatible with the actual polarity for forward bias.

The second requirement for active-region operation (reverse-biased base–collector junction) needs some further clarification. Although it is the base–collector junction that establishes the condition, the collector–emitter voltage is more convenient as a reference in the common-emitter configuration. A relationship between base–collector voltage and collector–emitter voltage can be determined from a loop equation.

Referring to Figure 4-9, a KVL equation can be written by forming a loop from the emitter to the base, on to the collector, and back to the emitter. This equation reads

$$-V_{BE} + V_{BC} + V_{CE} = 0 \tag{4-2}$$

Solving for V_{BC}, we obtain

$$\boxed{V_{BC} = V_{BE} - V_{CE}} \tag{4-3}$$

If the base–emitter junction in an *npn* transistor is to be reverse biased, we require that $V_{BC} < 0.7$ V. This constraint gives

$$V_{BE} - V_{CE} < 0.7 \qquad\qquad \textbf{(4-4)}$$

or

$$V_{CE} > V_{BE} - 0.7 \qquad\qquad \textbf{(4-5)}$$

If we assume that $V_{BE} = 0.7$ and substitute this value in (4-5), the condition for V_{CE} is determined as

$$\boxed{V_{CE} > 0} \qquad\qquad \textbf{(4-6)}$$

The assumed positive reference direction for V_{CE} in Figure 4-9 for an *npn* transistor is thus compatible with the actual polarity for reverse bias.

We have seen earlier that the saturation condition results in a small positive voltage (typically about 0.1 V), so there is a transition region of a few tenths of a volt between saturation and full active-region operation. This transition region will be evident on the characteristic curves to be presented later. For simplicity, the expression of (4-6) will be retained as given, but it should be understood that some realistic positive voltage for an *npn* transistor is required for full active-region operation.

The conditions for active-region operation, with V_{CE} used as a criterion, are summarized for both *npn* and *pnp* transistors in Table 4-2. Note that for *pnp* transistors, the voltages are stated two ways. In the first way, the assumed positive reference terminals are the same as for the *npn* transistor, which results in negative voltages for both V_{BE} and V_{CE}. However, by reversing the subscripts and assuming the emitter terminal as the positive reference, the values become positive.

A schematic summary of both *npn* and *pnp* transistors in the forms most appropriate for common-emitter references is given in Figure 4-10. As shown, all

TABLE 4-2 CONDITIONS FOR ACTIVE-REGION OPERATION

Condition	*NPN*	*PNP*
Base–emitter forward bias	$V_{BE} \approx 0.7$ V	$V_{BE} \approx -0.7$ V or $V_{EB} \approx 0.7$ V
Base–collector reverse bias	$V_{CE} > 0$	$V_{CE} < 0$ or $V_{EC} > 0$

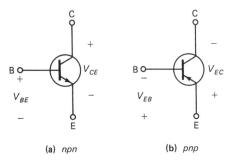

(a) *npn*   (b) *pnp*

Figure 4-10 Reference polarities of voltages for *npn* and *pnp* transistors.

voltages will be positive for active-region operation. In this book, the "upside-down" positive references will be assumed in most cases when working with *pnp* transistors.

Example 4-1

Classify the region of operation—*cutoff*, *saturation*, or *active*—for each of the following transistor types and conditions: (a) *npn*: $V_{BE} = 0.7$ V, $V_{CE} = 10$ V; (b) *npn*: $V_{BE} = 0.7$ V, $V_{CE} = 0$ V; (c) *npn*: $V_{BE} = -1$ V, $V_{CE} = 5$ V; (d) *pnp*: $V_{BE} = -0.7$ V, $V_{CE} = -10$ V; (e) *pnp*: $V_{EB} = 0.7$ V, $V_{EC} = 10$ V; (f) *pnp*: $V_{EB} = 0.7$ V, $V_{EC} = 0$ V; (g) *pnp*: $V_{BE} = 0.7$ V, $V_{EC} = 6$ V.

Solution

As the reader develops a "feel" for this process, it may be helpful to refer back to the cutoff and saturation models, Table 4-1 and particularly, Table 4-2. However, one should strive to develop the ability to predict the region of operation from the definition and the stated condition. Each part will now be considered and comments will be made as appropriate.

(a) For an *npn* transistor, $V_{BE} = 0.7$ V and $V_{CE} = 10$ V imply *active-region* operation.

(b) For an *npn* transistor, $V_{BE} = 0.7$ V and $V_{CE} = 0$ imply *saturation*.

(c) For an *npn* transistor, $V_{BE} = -1$ V (or $V_{EB} = 1$ V) implies reverse bias for the base–emitter junction. Coupled with this is $V_{CE} = 5$ V, so the transistor is in *cutoff*.

(d) For a *pnp* transistor, the conditions $V_{BE} = -0.7$ V and $V_{CE} = -10$ V imply the *active region*.

(e) For a *pnp* transistor, the conditions $V_{EB} = 0.7$ V and $V_{EC} = 10$ V imply the *active region*. Note that this case is actually the same as (d), but the intent was to illustrate different ways of specifying the voltages.

(f) For a *pnp* transistor, $V_{EB} = 0.7$ V and $V_{EC} = 0$ V imply *saturation*.

(g) For a *pnp* transistor, $V_{BE} = 0.7$ V implies *reverse bias* for the base–emitter junction. (Do not be misled by the value 0.7 V; it could have been any positive value below the reverse breakdown level. This value was selected to test the reader's alertness!) Coupled with this is $V_{EC} = 6$ V, so the transistor is in *cutoff*.

The operating characteristics of a transistor will now be explained by discussing a laboratory experiment that can be used to obtain certain characteristic curves for the transistor. Although the transistor to be discussed is hypothetical and its characteristics are somewhat idealized for clarity, the procedure can be readily verified in an electronics laboratory.

The circuit to be used in the discussion is shown in Figure 4-11. This circuit contains an *npn* transistor, two adjustable dc voltage sources, a resistance to ease adjustment of and limit the maximum current in the base circuit, two voltmeters, and two ammeters. The voltage source on the left will be referred to as the *base bias supply*, and the source on the right will be referred to as the *collector bias supply*.

Our major focus will be on the active region since many more data are required to characterize that condition, but cutoff and saturation conditions will also be evident. Note that the base and collector bias voltage sources have the proper polarities for active-region operation with an *npn* transistor. If the transistor were a *pnp* type, the polarities of both sources (along with voltmeter and ammeter terminals) would have to be reversed. Observe that the transistor is connected in a common-emitter configuration since the emitter is common between the input (base–emitter circuit) and output (collector–emitter circuit).

Figure 4-11 Measurement circuit for obtaining transistor characteristics.

Base Characteristics

First, we will concentrate on a set of measurements to be made on the base side of the circuit of Figure 4-11. The curves obtained are referred to as the *base* or *input characteristics* of the BJT.

Assume that the collector–emitter voltage V_{CE} is set at a fixed nominal positive voltage (e.g., 10 V). Assume then that the base bias voltage source is varied from zero to some positive value. At a number of points, the base–emitter voltage V_{BE} and the base current I_B are measured.

At very low values of V_{BE} (well under 0.1 V), there is no measurable base current. However, as V_{BE} approaches 0.7 V, the base current increases rapidly, and the resulting curve would have the form shown in Figure 4-12.

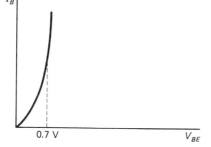

Figure 4-12 Form of base characteristic of BJT with V_{CE} set to a constant value.

We recognize immediately that this curve has the same form as the current versus voltage curve for a forward-biased junction diode. Indeed, the knee voltage is at about the same level (i.e., about 0.7 V for silicon). There is, however, a major difference. *For a given forward voltage, the base current in the transistor is much smaller than the current for a junction diode.* On the hypothetical transistor curve shown, the current values are likely in the range from about 20 to 100 μA. A typical junction diode might have current values 100 times as large for the same forward voltage.

Recall that the collector–emitter voltage V_{CE} is being held constant. Suppose that we change V_{CE} to a new value and obtain a new curve. The new curve will have the same shape but will be shifted slightly. The shift is so slight that there is very little need to obtain more than about two curves. Instead, Figure 4-13 shows the form of two extreme cases. The one on the left corresponds to setting the collector bias voltage supply to zero (i.e., $V_{CE} = 0$). The right curve is for some value $V_{CE} \gg 0$ (but below the collector–base junction breakdown voltage, of course).

The shift of the input characteristic with the change in collector–emitter voltage is called the *Early effect*. The direction of the shift in the curve is to the right as V_{CE} increases. If V_{CE} increases and V_{BE} is fixed, I_B decreases. Conversely, if V_{CE} increases and I_B is fixed, V_{BE} increases.

For many applications, the shift in the input characteristics as V_{CE} is changed is insignificant. Unless otherwise indicated, it will be ignored in most of our subsequent work.

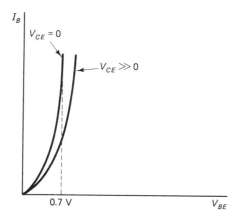

Figure 4-13 Forms of base characteristics of BJT for two extreme cases of V_{CE}.

As in the case of a diode, much analysis with transistors will utilize the constant voltage drop assumption for a forward-biased base–emitter junction; that is, $V_{BE} = 0.7$ V for an *npn* transistor and $V_{EB} = 0.7$ V for a *pnp* transistor.

Collector Characteristics

We are now ready to discuss some very important measurements performed on the collector side of the circuit of Figure 4-11. The set of curves to be obtained are referred to as the *collector* or *output characteristics*.

To enhance the discussion, we assume some numerical values, which have been chosen as representative of many small-signal transistors. However, these values simply illustrate the process and are not intended to suggest any general results. Actual values vary widely from one transistor type to another.

Basically, the goal is to obtain a plot of collector current I_C as a function of collector–emitter voltage V_{CE}. However, since base current I_B has a large controlling effect on collector current, all three variables must be considered. This could suggest a three-dimensional plot, which is hardly desirable.

In a situation involving three variables, the standard approach is to set one of the variables at a fixed value and obtain a two-dimensional curve involving the other two variables. The first variable is then set to a new fixed value and a second curve is obtained. This process is repeated as many times as desired until a so-called *family of curves* is obtained. The variable that is fixed for a given curve is referred to as a *parameter* for this purpose.

The *collector characteristics* of a BJT are a family of curves displaying collector current versus collector–emitter voltage with base current as the parameter. The collector characteristics for our hypothetical transistor are shown in Figure 4-14. These curves will be discussed in the context of the measurement system of Figure 4-11.

First, $I_B = 0$ is established. This is achieved by either setting the dc voltage on the base side to zero or simply opening the base circuit. In this case $I_C = 0$ for most

Figure 4-14 Collector characteristics of hypothetical BJT.

practical purposes as V_{CE} is varied. (There is a very small leakage current, as mentioned previously, but its effect will be ignored at this time.) This limiting case corresponds to the cutoff region, and it is a starting point for transition into the active region.

To move into the active region, the dc voltage on the base side is increased until forward bias is achieved in the base–emitter diode. The particular base current that is established is referred to as the *base bias current*. For our hypothetical transistor, a base bias current $I_B = 20$ μA is set. The collector voltage is then varied and the collector current is measured.

At very small values of collector–emitter voltage, there is a sharp increase in collector current. Recall that there is a transition of a few tenths of a volt between saturation and active-region operation. Once full active-region operation is attained, the collector current of this transistor levels off at about 2 mA for $I_B = 20$ μA. As V_{CE} increases over a fairly wide range, I_C increases slightly but not significantly. A good approximation for many transistors is to assume that it remains constant as V_{CE} varies.

We pause here to marvel at the process that is taking place. Earlier, the analogy of the transistor as two diodes connected in series was noted. However, if that model exactly described the transistor, there would be no collector current since the base–collector junction is reverse biased. The special construction of the transistor permits a small base current flowing through the forward-biased base–emitter junction to control a much larger collector current through the reverse-biased collector–base junction.

Two assumptions need to be pointed out. First, V_{CE} cannot be increased indefinitely. Like all *pn* junctions, there is a reverse breakdown voltage for the collector–base junction at which the collector current would increase sharply, and this could destroy the transistor. We assume that V_{CE} is always kept below that level. Second, we have seen earlier that the input characteristic is affected slightly by variations in V_{CE}. We will assume in our experiment that the base bias voltage is adjusted as necessary to keep I_B constant as V_{CE} is varied.

Next, assume that $I_B = 40$ μA is established by increasing the base bias voltage source. As V_{CE} is varied, the initial behavior is similar to that obtained in the previous curve. In this case, however, the collector current reaches a value of 4 mA at the leveling point.

This process is repeated here for several other values of base current, and the family of collector characteristics shown in Figure 4-14 is obtained. Observe that there is a slight shift to the right of the voltage at which the leveling occurs as higher values of bias current are established.

Look ahead to Figure 4-15 and note the three regions of operation for a transistor. It is helpful to visual these as transistor operation is discussed. Returning to Figure 4-14, some important properties of the active region can be deduced. In the active region, the collector voltage has very little effect on the collector current. In contrast, however, the *base current has a very large effect on the collector current*. In effect, the small base current controls the much larger collector current. It is this property that makes the transistor a very useful device. As we will see throughout the text, this type of control mechanism is fundamental to all useful electronic

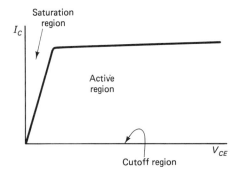

Saturation region

I_C

Active region

Cutoff region

V_{CE}

Figure 4-15 Three regions of operation for BJT shown on collector characteristics.

devices; that is, a small variable at the input controls a much larger variable at the output.

DC Current Gain

An important transistor parameter, referred to as the *dc current gain*, will now be defined. This quantity has traditionally been denoted as β_{dc}, particularly in textbooks. However, for certain reasons irrelevant to our present train of thought, it is denoted on specifications sheets as h_{FE}. Except where we are specifically using specifications sheets, we will use the symbol β_{dc}.

The definition of dc current gain is

$$\beta_{dc} = \frac{I_C}{I_B} \qquad (4\text{-}7)$$

Stated in words, the *dc current gain is the ratio of total dc collector current to total dc base current*. Both numerator and denominator of (4-7) must be expressed in the same units, but the resulting quantity has no units (i.e., it is *dimensionless*).

Let's see what β_{dc} is for our hypothetical transistor. Suppose that we first pick the curve for $I_B = 20\ \mu A$ and note that $I_C = 2$ mA. Expressing both quantities in milliamperes, we obtain

$$\beta_{dc} = \frac{2\ mA}{0.02\ mA} = 100 \qquad (4\text{-}8)$$

If we used instead the curve for $I_B = 40\ \mu A$, in which case $I_C = 4$ mA, we would obtain

$$\beta_{dc} = \frac{4\ mA}{0.04\ mA} = 100 \qquad (4\text{-}9)$$

which is obviously the same result as before. In effect, the collector current has doubled because the base current has doubled, but the ratio remains the same.

As indicated at the beginning, these curves have been idealized for clarity. Most transistor curves are not quite as "nice." In general, the value of β_{dc} will vary somewhat with different values of base current at which it is calculated. Further, as we move to the right, I_C usually increases slightly for a given I_B, so β_{dc} will also vary with V_{CE}.

The variation of β_{dc} with different conditions is discussed later, so let's not confuse the situation too much now. We will learn how to analyze and even design circuits under conditions where β_{dc} can vary widely. The main point now is to understand the definition and the all-important implications of this quantity.

β_{dc} provides a measure of how large the control property is between base and collector. The small base current has a very pronounced effect on the much larger collector current when β_{dc} is large. The value of β_{dc} for transistors can vary from less than 50 to more than 500. The value of 100 for our hypothetical transistor is typical for general-purpose transistors.

The characteristics that we have investigated have been for an *npn* transistor. A similar experiment could have been performed for a *pnp* transistor if all polarities of sources and instruments are reversed. In the strict sense of polarities and references as defined for the *npn* transistor, the various voltages and currents for a *pnp* transistor would appear in the third quadrant (negative current and negative voltage). Indeed, transistor curve tracers, which are instruments providing an oscilloscopic display of the characteristic curves, show "upside-down" or third-quadrant displays for *pnp* transistors. However, for analysis purposes, positive values may be obtained for a *pnp* transistor by reversing the defined directions of currents and reversing subscripts and reference directions for voltages, as noted earlier.

A summary of the assumed directions and polarity conventions to be employed in this text for both *npn* and *pnp* transistors is shown in Figure 4-16. The directions for the *npn* transistor of (a) are the same as for the hypothetical experiment discussed previously (although emitter current was not discussed). For the *pnp* transistor of (b), note that I_B is *out* of the base, I_E is *into* the emitter, and I_C is *out* of the collector. The pertinent voltages are now V_{EB} and V_{EC}. *With these directions and polarities, all quantities are positive for active-region operation.*

One more helpful point to remember is that *the arrow in the schematic symbol of a transistor has the direction of positive current flow in the emitter.*

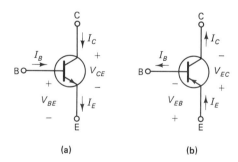

(a)   (b)

Figure 4-16 Recommended reference directions for *npn* and *pnp* transistors.

Example 4-2

Some measurements with a nonideal *npn* BJT have been made, and the results for three different conditions are summarized below:

	I_B (μA)	I_C (mA)	V_{CE} (V)
(a)	20	1.8	10
(b)	100	11	10
(c)	100	11.2	20

Determine the value of β_{dc} at each point.

Solution

For each point, the value of β_{dc} is

$$\beta_{dc} = \frac{I_C}{I_B} \tag{4-10}$$

The value of V_{CE} obviously does not appear in the calculation. However, the different values are given to substantiate the fact that operation is in the active region and to show that β_{dc} does typically vary with V_{CE}.

(a) For the first point, we have

$$\beta_{dc} = \frac{1.8 \times 10^{-3}}{20 \times 10^{-6}} = 90 \tag{4-11}$$

(b) For the second point, I_B is five times as great as for the first point but V_{CE} is the same. We have

$$\beta_{dc} = \frac{11 \times 10^{-3}}{100 \times 10^{-6}} = 110 \tag{4-12}$$

(c) For the third point, I_B is the same as for the second point but V_{CE} has been doubled. The value of β_{dc} is

$$\beta_{dc} = \frac{11.2 \times 10^{-3}}{100 \times 10^{-6}} = 112 \tag{4-13}$$

These results should be interpreted simply as illustrations of typical variations. For a given transistor *type*, the variation β_{dc} from one unit to another is even greater, as we will see later.

Example 4-3

Suppose that the same transistor for which measurements were performed in Example 4-2 is used in a circuit application in which $I_B = 90$ μA and $V_{CE} = 12$ V. Estimate the approximate value of I_C expected.

Solution

It is immediately noted that the operating conditions do not correspond exactly to the conditions under which the test was performed. However, since only an estimate of the approximate value is desired, a reasonable approach is to use the value of β_{dc} closest to the operating point.

The given conditions correspond closely to those of (b), for which it was determined that $\beta_{dc} = 110$. Employing that assumption, I_C is estimated as

$$I_C = \beta_{dc} I_B = 110 \times 90 \times 10^{-6} = 9.9 \text{ mA} \qquad \textbf{(4-14)}$$

This problem illustrates a typical estimation process of the type so commonly encountered in working with practical electronic circuits. It is possible with detailed equivalent circuits to extrapolate between points and, perhaps, obtain more accurate results. However, in many applications, the approach taken here is sufficient considering the uncertainty of the transistor characteristics.

4-5 ACTIVE-REGION DC MODEL

We consider next a dc model that can be used for predicting transistor operation in the active region. The adjective *dc* may be misleading and needs to be qualified. Much later in the text, some so-called *ac* or *small-signal* models are developed. Those models deal with *small changes* about some nominal operating point and, in general, have different forms than those discussed here. The terms *dc model* and *ac model* are widely used.

The dc model to be discussed here can be applied to certain signal-type situations as long as it is the *total* voltage and/or current that is of interest and as long as the transistor high-frequency and switching effects are negligible. Thus, in a number of early chapters, the dc model is used for a number of applications where time-varying signals are involved, but in which total voltage and current levels are used.

The basic dc models for the BJT are shown in Figure 4-17, with (a) representing the form for a *npn* transistor and (b) representing the form for a *pnp* transistor. The directions of all currents are labeled so that they will be positive quantities as given. If the direction of a given variable is reversed in working a problem, a negative sign should be used. The assumed knee voltage for silicon (i.e., 0.7 V) has been assumed.

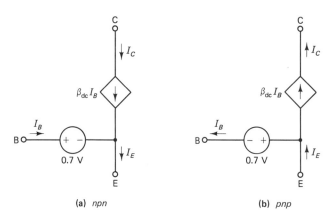

Figure 4-17 Active-region dc models for *npn* and *pnp* transistors. (Reverse leakage current is neglected.)

(a) *npn*

(b) *pnp*

The base–emitter junction is represented by a constant voltage source model since it is forward biased. For an *npn* transistor, the base is, therefore, about 0.7 V higher than the emitter terminal. For a *pnp* transistor, however, the emitter is about 0.7 V higher than the base voltage. The positive direction of base current is *into* the base for an *npn* transistor and *out* of the base for a *pnp* transistor.

To account for the effect of the base control, a current source with value $\beta_{dc} I_B$ is placed between collector and base terminals. This source is a *dependent* or *controlled* source since it is a function of a variable in a different part of the circuit. In effect, the collector current I_C is equal to this current source value since there is no shunting resistance (i.e., $I_C = \beta_{dc} I_B$).

At the junction between collector branch, base branch, and emitter branch, an application of KCL yields

$$I_E = I_B + I_C \qquad (4\text{-}15)$$

Stated in words, *the emitter current is the sum of base current plus collector current*. Note that this property applies to both *npn* and *pnp* transistors as long as all quantities are expressed as positive numbers.

In many situations, the base current is much smaller than the collector current. As long as I_B is known, one might as well use the exact form of (4-15) when required. However, as we will see later, in many circuits the base current may not be known accurately. In that event, the following rule is convenient: *In many active-region transistor circuits, the emitter current is only slightly larger than the collector current, and for many applications, the approximation $I_E \approx I_C$ is valid.*

It must be understood that the models of Figure 4-17 apply only if the transistor is based in the active region, which means that the base–emitter junction is forward biased and $V_{CE} > 0$ for an *npn* unit (or $V_{EC} > 0$ for a *pnp* unit). Suppose that one is analyzing a circuit problem and either of the two following conditions result: (1) the calculated base current *appears* to be negative and (2) the calculated collector–emitter voltage for an *npn* unit *appears* to be negative (or the opposite situation for a *pnp* unit). The models given here do not provide reverse leakage effects, so both of the preceding possibilities are *impossible*. The model would, therefore, be invalid under those circumstances, and further analysis would be required to determine the exact region of operation. The conclusion is that this model is valid only for the active region.

Example 4-4

Construct the active-region dc model for the hypothetical *npn* transistor whose characteristics were given in Figure 4-14.

Solution

It was determined earlier that $\beta_{dc} = 100$ for the transistor. The equivalent circuit will be of the form of Figure 4-17(a), and it is shown in Figure 4-18 for this particular transistor.

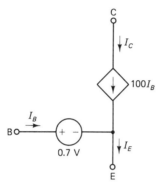

Figure 4-18 Active-region dc model for *npn* transistor of Example 4-4.

Example 4-5

A certain *pnp* transistor has $\beta_{dc} = 120$. Construct the active-region dc equivalent circuit for the transistor.

Solution

The equivalent circuit will be of the form of Figure 4-17(b), and it is shown in Figure 4-19 for this particular transistor. Compare this equivalent circuit of the *pnp* transistor with that of the *npn* transistor of Example 4-4 (Figure 4-18) so that you are clear about the directions of the current sources and the base–emitter junction voltage drops.

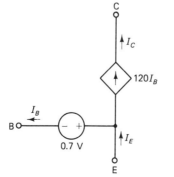

Figure 4-19 Active-region dc model for *pnp* transistor of Example 4-5.

Example 4-6

Some conditions concerning one or more of the three transistor currents I_B, I_C, and I_E for different transistors are given in the different parts of this problem. Active-region operation is assumed in all cases. In each case, *determine exactly*, if possible, or *estimate*, if necessary, the values of the other currents. **(a)** $I_B = 50$ μA, $I_C = 3$ mA; **(b)** $I_B = 80$ μA, $I_E = 5$ mA; **(c)** $I_C = 12$ mA, $I_E = 12.15$ mA; **(d)** $I_E = 7$ mA.

Solution

The basic relationship to be employed is that of (4-15). However, in certain parts, we may be required to assume that $I_E \approx I_C$.

(a) $I_B = 50$ μA, $I_C = 3$ mA. In this case the value of I_E may be determined exactly as

$$I_E = 50 \times 10^{-6} + 3 \times 10^{-3} = 3.05 \times 10^{-3} \text{ A} = 3.05 \text{ mA} \qquad \textbf{(4-16)}$$

(b) $I_B = 80$ μA, $I_E = 5$ mA. In this case the value of I_C may be determined exactly as

$$I_C = I_E - I_B = 5 \times 10^{-3} - 8 \times 10^{-6} = 4.92 \times 10^{-3} \text{ A} = 4.92 \text{ mA}$$
$$\textbf{(4-17)}$$

(c) $I_C = 12$ mA, $I_E = 12.15$ mA. In this case the value of I_B may be determined exactly as

$$I_B = I_E - I_C = 12.15 \times 10^{-3} - 12 \times 10^{-3} = 0.15 \times 10^{-3} \text{ mA}$$
$$= 150 \text{ } \mu A \qquad \textbf{(4-18)}$$

(d) $I_E = 7$ mA. All that can be done in this case is simply to estimate that

$$I_C \approx 7 \text{ mA} \qquad \textbf{(4-19)}$$

The value of I_B cannot be determined or even estimated closely. If one wanted to make a "ballpark" guess, an assumption of $\beta_{dc} = 100$ might be reasonable. This would lead to a guess of $I_B = 7 \times 10^{-3}/100 = 70$ μA.

In some design procedures studied later, an attempt is made to estimate the *worst-case* or *maximum* possible value of I_B. For active-region operation, the minimum expected value of β_{dc} for the transistor would then be used, since that value would predict the maximum possible base current. If, for example, the minimum value of β_{dc} for the particular transistor were 35, the maximum expected value of base current would be $I_B = 7 \times 10^{-3}/35 = 200$ μA. Although this type of process may bother some readers accustomed to exact numerical results, it is often the best that can be done with "real-life" electronic devices.

As the next step in our study of bipolar junction transistors, we will observe some actual transistor specifications to see how they relate to the developments in the text. Any actual real-life circuit design and development must be performed in conjunction with a detailed study of the specifications of the devices used.

At the peak of discrete transistor utilization, there were several thousand different transistor types available. Many of these have been discontinued and will be found only in older equipment. Because of the close similarity between many different types, replacement charts have been developed to assist in replacing unavailable types with the available types. The major transistor manufacturers have transistor manuals that provide data on individual transistor characteristics.

Transistors are often loosely classified as either (1) *small-signal transistors* or (2) *power transistors*. The lines between these categories are not rigidly defined, and some transistors could be classified as either type, depending on their application. As a reasonable guide, small-signal transistors are those in which the transistor power level is generally less that 0.5 W, and power transistors usually have power levels exceeding 0.5 W.

A major factor in the classification is the physical size and construction. Small-signal transistors usually have small physical sizes and no special provisions for heat transfer. In contrast, power transistors often have relatively large structures and provisions for *heat sinks,* which are devices used to facilitate heat transfer out of the device.

Power transistors are considered as a separate topic in Chapter 19. At this point we focus on some representative specifications of the so-called small-signal general-purpose transistors.

There are a number of small-signal transistors that are quite popular due to their wide availability, low cost, and versatility. We focus on one of these for the purpose of studying typical characteristics. The one selected is the 2N3904, for which a complete set of specifications is reproduced in Appendix D. We are prepared at this time to investigate only the dc characteristics. Later portions of the specifications will be studied in subsequent chapters.

Refer now to the specifications sheets of the 2N3903/2N3904 transistors in Appendix D. Manufacturers frequently group several similar semiconductor types together, as we noted for diodes in Chapter 2. We focus on the data for the 2N3904.

The first group of specifications we study are the *absolute maximum ratings*. First, the maximum values of three voltages, V_{CEO}, V_{CBO}, and V_{EBO}, are listed. The first two subscripts represent transistor terminals in accordance with our standard notation, but the third subscript O is new to us. It represents an *open* condition for the terminal not listed in the first two subscripts. Thus V_{CEO} represents collector–emitter voltage with the base open, V_{CBO} represents collector–base voltage with the emitter open, and V_{EBO} represents emitter–base voltage with the collector open. The maximum values of these three voltages are 40 V, 60 V, and 6 V, respectively.

It should be noted that all three of these voltages represent reverse-bias conditions. If the emitter is open, the maximum collector–base voltage is 60 V. However, in the more common situation where the emitter is not open, the maximum collector–emitter voltage (with base open) is 40 V. The maximum emitter–base voltage is only 6 V. In the active and saturation regions, the base–emitter junction is forward biased, and this need not be a concern. However, in applications where cut-off is encountered, care must be taken not to exceed this relatively low reverse-bias rating for the base–emitter junction. The maximum rating of the collector current is specified as $I_C = 200$ mA.

The total power absorbed by a transistor is denoted on some specifications sheets as P_D and on others by P_T. The latter symbol is used on the particular specifications sheets considered here.

The total power absorbed by a transistor consists of the power absorbed by the collector–emitter side of the circuit plus the power absorbed by the base–emitter side of the circuit. In many (but not all) applications, the second power level is considerably smaller than the first power level and may be neglected. For dc power analysis, let V_{CE} represent the dc collector–emitter voltage, and let I_C represent the dc collector current. Under conditions in which the base power is negligible, the total dc power may be approximated as

$$P_T = V_{CE}I_C \qquad (4\text{-}20)$$

For ambient temperatures of 25°C or less, the maximum value of P_T for the 2N3904 is specified as 350 mW.

Observe that a *derating factor* follows the maximum power specification. If the *ambient temperature* T_A exceeds 25°C, the amount of heat that can be transferred out of the transistor decreases. The change in power is a linear function of the increase in temperature.

Let D represent the derating factor. The basic unit is watts/°C (W/°C), but for small-signal transistors, the value is usually specified in mW/°C. The derating factor specified for the 2N3904 is 2.8 mW/°C.

Let $P_T(T_A)$ represent the maximum power dissipation at ambient temperature T_A, where $T_A > 25$°C, and let $P_T(25)$ represent the specified maximum power dissipation at $T_A = 25$°C. The following relationship is used to derate the power dissipation:

$$P_T(T_A) = P_T(25) - (T_A - 25)D \qquad (4\text{-}21)$$

It should be noted from the preceding data that there are separate maximum ratings for *voltage, current,* and *power,* and *all three* variables must be maintained below these ratings. For example, there are combinations of voltage and current that fall within the allowable range of each, but whose product exceeds the maximum power level. In the same fashion, some voltage and current combinations result in power levels below the maximum power level, but in which either the voltage or the current maximum is exceeded. For any design application, it is essential that *all three conditions be checked* before further consideration of the transistor is made.

The last set of maximum ratings are the *temperature* ratings. The symbol T_J represents the transistor junction temperature, and it is required to be in the range −55 to 135°C. Storage and lead temperature ratings are also provided.

The next data to be considered are the *electrical characteristics*. The first three quantities listed are the reverse-biased *breakdown* voltages. The breakdown condition is denoted by (BR). The remaining subscripts follow the pattern defined previously. As a matter of fact, the *minimum* values of the three voltages given represent the same voltages listed under *absolute maximum ratings* considered previously. This may seem confusing, but it is indicating that if the respective voltages are kept below the maximum ratings considered earlier, breakdown will not occur.

The next two data values represent *leakage* currents under *cutoff* conditions. The collector cutoff current I_{CEV} represents collector leakage current when the base–emitter junction is reverse biased. The base cutoff current I_{BEV} represents base current under the same condition. Both of the values are given as 50 nA.

The next complete block of data provides important information on the dc current gain. The symbol h_{FE} is used and, as noted earlier, this is the same as β_{dc}. However, do not confuse this with another symbol, h_{fe}, which appears at a different point in the specifications sheets and relates to the small-signal or ac data.

Values of dc current gain are specified at a number of different operating points. At one operating point ($V_{CE} = 1$ V, $I_C = 10$ mA), the minimum value is listed as 100 and the maximum value is listed as 300. This does not mean that h_{fe} for a *particular* 2N3904 will vary from 100 to 300. Rather, it means that out of a large stock of 2N3904 transistors, the values of β_{dc} could be expected to range from 100 to 300.

At other points, only minimum values are given. The wide variations shown here are typical and represent the kind of uncertainty associated with transistor specifications.

The next block of data is entitled *collector–emitter saturation voltage*. The maximum values of $V_{CE(sat)}$ are provided at two operating points. These values are 0.2 V and 0.3 V, respectively.

The final block of static data is entitled *base–emitter saturation voltage*, and this quantity is denoted as $V_{BE(sat)}$. This value represents the usual forward biased base–emitter junction drop. At one point, the minimum value is listed as 0.65 V, and the maximum value is listed as 0.85 V. At the second point, only the maximum value of 0.95 V is listed. We have been assuming and will continue to assume the common value of 0.7 V as representative for moderate bias conditions.

The large section of data on the third page entitled *dynamic characteristics* relates to small-signal, switching, and delay characteristics of the transistor and is discussed later in the book.

Some graphical data are provided and are useful in studying the dc properties of the transistor. At the top of the fifth page of the specifications, a family of curves of h_{FE} (or β_{dc}) for the 2N3904 as a function of I_C with T_A as a parameter is given. For a given temperature, h_{FE} shows a slight increase as I_C varies from 0.1 mA to slightly more than 10 mA, but then it begins to decrease with increased collector current.

Note the temperature dependency of h_{FE} as indicated by the different curves. The value of h_{FE} increases with increasing temperature. Finally, note that $V_{CE} = 1$ V for three of the curves, but $V_{CE} = 5$ V for the fourth. There is little difference in the two curves for the different values of V_{CE} at $T_A = 25°C$, indicating only minimal variation of h_{FE} with V_{CE}.

The variations of h_{FE} with I_C, T_A, and V_{CE} shown here are reasonably typical, but exact data for a given transistor should be consulted when such information is critical. For a reasonable range of operating conditions for many transistors, h_{FE} usually displays only a moderate change with I_C and V_{CE}, but the change with temperature is more pronounced. For a stable temperature, therefore, it is usually reasonable to use as an estimate the value of h_{FE} at the closest operating voltage and current for which data are available.

The bottom figure on the fifth page shows the base–emitter voltage as a function of collector current with temperature as a parameter. Note that V_{BE} *decreases* with *increasing* temperature. One rule of thumb is that V_{BE} *decreases* by about 2.2 mV for each *increase* of 1°C in the temperature.

The top figure on the sixth page of the specifications shows the collector–emitter saturation voltage as a function of the base current with collector current as a parameter. The steep vertical lines represent the values of base current at which saturation starts to occur. As the curves flatten out, the transistor is driven further into saturation, and the value of $V_{CE(\text{sat})}$ is reduced.

The bottom figure on the sixth page shows the collector–emitter saturation voltage as a function of collector current with temperature as a parameter. Note that $V_{CE(\text{sat})}$ increases with increasing temperature.

Example 4-7

Determine the maximum power rating of a 2N3904 at an ambient temperature $T_A = 40°C$.

Solution

It was noted in the text that the value of power dissipation of 25°C is $P_T(25) = 350$ mW and the derating factor is $D = 2.8$ mW/°C. Since both quantities are given in milliwatts, the analysis can be performed directly with such units. The desired quantity is $P_T(40)$, and we have

$$P_T(40) = P_T(25) - (40 - 25) \times 2.8$$
$$= 350 - 15 \times 2.8 = 308 \text{ mW} \qquad \textbf{(4-22)}$$

Note the significant reduction in power dissipation at the higher temperature.

4-7 TESTING TRANSISTORS

A number of different procedures can be used for testing transistors. This includes the use of general-purpose instruments such as an ohmmeter as well as specialized instruments designed specifically for transistors.

Ohmmeter Tests

An ohmmeter can provide some quick checks on a BJT in much the same way as for a diode. The power should be off and the transistor should be removed from the circuit in which it is connected.

Since a BJT can be considered in a sense as two diodes in series, check the forward and reverse resistances of the base–emitter diode and the base–collector diode. The discussion of Section 2-10 concerning diodes is applicable here. Depending on whether the unit is an *npn* or a *pnp* type, the appropriate forward- and reverse-bias directions can be inferred.

In addition to the diode tests, check the resistance between collector and emitter with the ohmmeter leads in both combinations. The resistance should be high in both directions for this test since there are two diodes in series with opposite directions. A common transistor trouble is a collector–emitter short, so if you read zero to a few thousand ohms between emitter and collector in either direction, the transistor is shorted.

Transistor Testers

If the transistor fails any of these quick checks, it is bad. However, it may pass the preceding tests and still have some faults. The ohmmeter only provides dc tests and does not indicate problems with gain or signal conditions. For example, the value of β_{dc} may be too low. Active transistor checks may be made with special transistor testers. These testers usually provide a measure of the value of β_{dc} for the transistor as well as other checks.

The most elaborate means for checking a transistor is with a *curve tracer*. These oscilloscope-based units actually display the collector characteristic curves on the screen. Curve tracers are used primarily in electronic circuit design and testing.

Tests Performed in a Circuit

When a transistor is energized within a circuit, various dc voltages may be measured as a check on operation. It is usually more convenient to measure all voltages with respect to ground. Voltages between terminals may then be determined by taking the differences between ground-referred voltages.

Based on your understanding of voltage levels for different transistor terminals, check the various voltages to ensure proper conditions. For example, if the base voltage with respect to ground is $V_B = 3.7$ V and the emitter voltage with respect to ground is $V_E = 3$ V, the value of $V_{BE} = 3.7 - 3 = 0.7$ V, and this indicates the forward-bias condition for the base–emitter junction of an *npn* transistor.

DRILL PROBLEMS

4-1. Classify the region of operation—*cutoff, saturation,* or *active*—for each of the following transistor types and conditions: **(a)** *npn*: $V_{BE} = -2$ V,

$V_{CE} = 12$ V; **(b)** *npn*: $V_{BE} = 0.7$ V, $V_{CE} = 15$ V; **(c)** *npn*: $V_{BE} = 0.7$ V, $V_{CE} = 0$; **(d)** *pnp*: $V_{EB} = 0.7$ V, $V_{CE} = -15$ V; **(e)** *pnp*: $V_{BE} = 3$ V, $V_{EC} = 10$ V; **(f)** *pnp*: $V_{BE} = -0.7$ V, $V_{CE} = 0$.

4-2. Based on the dc voltages shown, classify the region of operation—*cutoff, saturation,* or *active*—for each of the transistors shown in Figure P4-2.

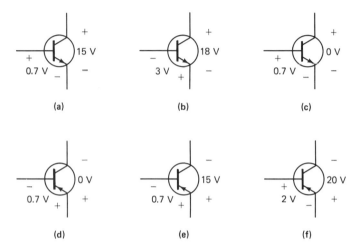

Figure P4-2

4-3. Determine V_{BC}, the voltage from base to collector, for each of the conditions in Problem 4-1.

4-4. Determine V_{BC}, the voltage from base to collector, for each of the circuits of Problem 4-2 (Figure P4-2).

4-5. For a certain transistor and operating conditions, $I_B = 40$ μA and $I_C = 10$ mA. Determine β_{dc} and I_E.

4-6. For a certain transistor and operating conditions, $I_B = 100$ μA and $I_E = 2$ mA. Determine β_{dc} and I_C.

4-7. For a certain transistor and operating conditions, $I_B = 30$ μA and $\beta_{dc} = 120$. Determine I_C and I_E.

4-8. For a certain transistor and operating conditions, $I_C = 50$ mA and $\beta_{dc} = 200$. Determine I_B and I_E.

4-9. The collector characteristics of a certain hypothetical *npn* BJT are shown in Figure P4-9. Determine the value of β_{dc}.

I_C (mA)

$I_B = 100\ \mu A$

20

$I_B = 75\ \mu A$

15

$I_B = 50\ \mu A$

10

$I_B = 25\ \mu A$

5

0

V_{CE} **Figure P4-9**

4-10. A certain idealized *npn* BJT has $\beta_{dc} = 250$ over a wide range of conditions. Sketch the collector characteristic curves corresponding to $I_B = 10\ \mu A$, $20\ \mu A$, and $30\ \mu A$. Label the collector current axis properly.

4-11. Construct the active-region dc model for the transistor of Problem 4-9.

4-12. A certain *pnp* transistor has $\beta_{dc} = 150$. Construct the active-region dc model.

4-13. A certain transistor has $I_B = 100\ \mu A$ and $I_E = 2$ mA. Determine I_C.

4-14. A certain transistor has $I_B = 100\ \mu A$ and $I_C = 2$ mA. Determine I_E.

4-15. A certain transistor has $I_C = 1.51$ mA and $I_E = 1.53$ mA. Determine I_B.

4-16. A certain transistor has $I_C = 10$ mA. The value of β_{dc} is specified to be in the range 50 to 200. Estimate I_E and determine the maximum and minimum possible values of I_B.

4-17. Determine the maximum power rating of a 2N3904 at an ambient temperature $T_A = 30°C$.

4-18. Determine the maximum power rating of a 2N3904 at an ambient temperature $T_A = 100°C$.

4-19. A rule of thumb stated in the text is that V_{BE} decreases by about 2.2 mV for each increase of 1°C in the temperature. Refer to the bottom figure on the fifth page of the 2N3904 specifications. At $I_C = 1$ mA, note the value of $V_{BE(sat)}$ at 25°C. Using the rule of thumb, calculate $V_{BE(sat)}$ at +125°C and compare with the value given on the figure.

4-20. Repeat the analysis of Problem 4-19 at $I_C = 10$ mA.

4-21. Look ahead to Problem 4-23 and note the definition and expression for α_{dc}. Calculate α_{dc} for the following values of β_{dc}: **(a)** 50; **(b)** 100; **(c)** 200; **(d)** 5000. What can you conclude about α_{dc}?

Drill Problems   **163**

4-22. Look ahead to Problem 4-24 and note the expression for β_{dc} in terms of α_{dc}. Calculate β_{dc} for each of the following values of α_{dc}: **(a)** 0.9; **(b)** 0.99; **(c)** 0.999.

DERIVATION PROBLEMS

4-23. Early transistor references utilized an α_{dc} gain parameter, which is defined as

$$\alpha_{dc} = \frac{I_C}{I_E}$$

Show that α_{dc} can be expressed in terms of β_{dc} as

$$\alpha_{dc} = \frac{\beta_{dc}}{1 + \beta_{dc}}$$

4-24. From the results of Problem 4-23, show that β_{dc} can be expressed in terms of α_{dc} as

$$\beta_{dc} = \frac{\alpha_{dc}}{1 - \alpha_{dc}}$$

4-25. Show that emitter current can be expressed in terms of base current as

$$I_E = (1 + \beta_{dc})I_B$$

4-26. The transistor power dissipation P_T is $P_T = V_{CE}I_C$. For a given value of P_T, sketch the form of the curve of I_C versus V_{CE}. Since I_C and V_{CE} are limited to $I_{C(max)}$ and $V_{CE(max)}$, show how the curve and these limits define a safe operating region for the transistor.

TROUBLESHOOTING PROBLEMS

4-27. You have been assigned the task of checking circuit conditions for a number of BJTs in a complex circuit. Most of the BJTs have none of their terminals connected to the circuit common ground. However, the common terminal of your voltmeter is connected to the circuit ground, and all voltages indicated are referred to this ground. The instrument is a digital voltmeter, and both positive and negative voltages are indicated. For each of the following measurements and transistor types, identify the region of operation: *cutoff*, *saturation*, or *active*. **(a)** *npn*: $V_B = 6$ V, $V_E = 8$ V, $V_C = 24$ V; **(b)** *npn*: $V_E = -3$ V, $V_C = 10$ V, $V_B = -2.3$ V; **(c)** *npn*: $V_C = -9$ V, $V_E = -9$ V, $V_B = -8.3$ V; **(d)** *pnp*: $V_C = 2$ V, $V_E = 20$ V, $V_B = 20.7$ V; **(e)** *pnp*: $V_E = 0$ V, $V_C = 0$ V, $V_B = -0.7$ V; **(f)** *pnp*: $V_E = 12$ V, $V_B = -12.7$ V, $V_C = -24$ V.

4-28. Repeat Problem 4-27 for the following transistor types and measurements:

(a) *npn*: $V_E = 6$ V, $V_C = 12$ V, $V_B = 6.7$ V; (b) *npn*: $V_C = 0$ V, $V_E = 0$ V, $V_B = 0.7$ V; (c) *npn*: $V_E = 10$ V, $V_C = 22$ V, $V_B = 9.3$ V; (d) *pnp*: $V_C = 12$ V, $V_E = 12$ V, $V_B = 11.3$ V; (e) *pnp*: $V_C = -8$ V, $V_E = -3$ V, $V_B = -2$ V; (f) *pnp*: $V_E = -10$ V, $V_C = -18$ V, $V_B = -10.7$ V.

4-29. While troubleshooting a circuit containing an *npn* transistor that *should* be operating in the active region, the following ground-referred voltages are measured: $V_B = 12$ V, $V_E = 11.3$ V, $V_C = 11.3$ V. What can you conclude about the transistor?

4-30. While troubleshooting a circuit containing a *pnp* transistor that *should* be operating in the active region, the following ground-referred voltages are measured: $V_B = 2$ V, $V_E = 0$ V, $V_C = -12$ V. What can you conclude about the transistor?

CHAPTER 5

BJT-Resistor Circuit Applications

OBJECTIVES

After completing this chapter, the reader should be able to:

1. Demonstrate the notational conventions for dc variables, total instantaneous variables, and time-varying or ac variables.
2. Analyze various dc circuit forms containing a transistor, resistors, and a power supply.
3. Analyze or design a circuit in which a BJT is used as a switch.
4. Analyze or design a circuit in which a BJT is used as a current source.
5. Analyze or design a basic emitter-follower circuit.
6. Analyze a stable bias circuit.
7. Design a stable bias current from operating point specifications.
8. Construct a dc load line for either the base circuit or the collector circuit of a BJT.
9. Demonstrate how changes in different transistor circuit parameters and variables affect the load line.

5-1 OVERVIEW

The bipolar junction transistor was introduced in Chapter 4, and the terminal characteristics were discussed. We are now ready to start exploring some of the various circuit applications that utilize the BJT.

At this point in the book, all circuits considered will utilize only transistors and resistors (plus power supplies and, possibly, diodes). Primary focus will be on dc analysis and operating characteristics. However, many of these circuits can be used with time-varying inputs and outputs provided that instantaneous voltages and currents are considered.

Among the applications to be considered are BJT electronic switches, current sources, and emitter-follower circuits. These circuits find numerous applications in interfacing between integrated circuits, for buffers and drivers, and in switching loads.

The establishment of a stable dc bias circuit for a BJT is considered in detail. Both the analysis of existing circuits and the design of such circuits are discussed.

5-2 NOTATION AND SYMBOLS

The beginning electronics student is often bewildered by the many different symbols and subscripts encountered in the study of electronic devices and circuits. This author is convinced that the learning process is facilitated by an orderly process of notation, and this section is devoted to developing a basis for much of the remainder of the book. Although all possible situations cannot be covered this early, the *general pattern* is discussed here, and this should enhance all developments that follow.

The circuit variables that appear most often in electronic applications are voltages and currents (although power may also be considered as a circuit variable in some cases). Depending on the circuit and the desired purpose of analysis, these variables may appear in at least *three forms*: (1) *dc form*, (2) *total instantaneous form*, and (3) *time-varying signal (or ac) form*.

The *dc form* for variables is used when the primary objective in the analysis or design is to determine various dc voltages, currents, and power in a circuit or to characterize a fixed quantity. No variation of these quantities for the given set of conditions is normally assumed in a dc analysis. All voltages and currents could be measured with dc instruments, and oscilloscopic presentations of dc voltages would simply be horizontal lines. As a general rule, *dc or fixed variables are represented by uppercase symbols*. The subscripts may be either uppercase or lowercase. Examples of dc variables used in this chapter are I_C, V_{CE}, and V_{R_C}. The convention for constant values associated with a waveform (e.g., peak value) follows the same pattern as for dc values.

The *total instantaneous form* for variables is used when the primary objective in the analysis or design is to observe the combined effect of both a dc level and a variation about that level as a result of a signal. Oscilloscopic presentations of total instantaneous voltages must be done with the vertical channel set for dc (direct coupled) input. Measurements of total instantaneous variables made with instruments must be done with some care since different types of instruments respond in different ways to complex waveforms. As a general rule, *total instantaneous variables are represented by lowercase symbols with uppercase subscripts*. Examples of total instantaneous variables are i_C, v_{CE}, and v_{BE}.

The *time-varying signal or ac* form for variables is used when the primary objective in the analysis or design is to study the effect of the circuit on signals. In this case, the dc levels are ignored, and only time-varying variables are shown. Oscilloscopic presentations of time-varying voltages can be facilitated by setting the vertical channel for ac input. In this manner, the dc offsets are eliminated, and the viewer can optimize the display of the time-varying component. As a general rule, *time-varying signal variables are represented by lowercase symbols with lowercase subscripts*. Examples of time-varying instantaneous variables are i_c, v_{ce}, and v_{be}. The term *ac variable* is often used as a more casual and abbreviated description of a *time-varying signal variable*, and we will use that term at many points in the text.

The difference between a total instantaneous variable and a time-varying signal variable is illustrated in Figure 5-1. The variable of interest here is the voltage between collector and emitter. The total instantaneous voltage is denoted as v_{CE}, and it is shown in Figure 5-1(a). Observe that this voltage varies up and down around a dc level. If the time axis were the baseline of an oscilloscope, the presentation would have the form shown with *dc coupling* on the input.

The time-varying signal form of the voltage is shown in Figure 5-1(b), and it is denoted as v_{ce} in accordance with the notation discussed previously. This form would result on an oscilloscopic display if the input were changed to *ac coupling*. Note that the peak value of this waveform is denoted as V_p. Although the signal itself is time varying and is denoted by a lowercase symbol, its peak value is a constant property of the waveform and is denoted by an uppercase symbol. This particular signal is a sine wave, but time-varying signals assume many different forms.

The pattern on subscripts was discussed in some detail in Section 1-10. A review of this notation, along with some additional points relative to electronic circuit conventions, will be given here.

The vast majority of electronic circuits contain a common circuit ground. *Voltage variables with a single subscript represent voltages referred to the common*

(a)

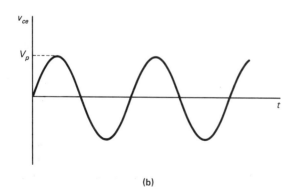

(b)

Figure 5-1 Typical instantaneous voltage and the time-varying (ac) portion of the voltage.

circuit ground. For example, V_C represents the dc collector voltage with respect to ground, v_C represents the total instantaneous collector voltage with respect to ground, and v_c represents the signal portion of the collector voltage with respect to ground.

Voltages with respect to circuit ground are the most useful in testing and troubleshooting circuits. In many test procedures, the ground terminal of the instrument is connected to the circuit ground, and the other instrument terminal is moved around from point to point in the circuit. In fact, through power supply safety grounding, it may be difficult and even hazardous with some instruments to measure directly voltages other than those with respect to ground. Thus there is a very practical association in the electrical circuit analysis of voltages referred to common ground.

Voltage variables with two parallel subscripts represent voltages between two points in a circuit, with the first subscript being the assumed positive reference point. For example, V_{CE} represents the dc voltage from collector to emitter, v_{CE} represents the total instantaneous voltage from collector to emitter, and v_{ce} represents the time-varying (or ac) part of the voltage from collector to emitter.

Voltages between two points in a circuit can always be expressed in terms of the differences between the ground-referred voltages. For example, if the ground-referred total instantaneous collector and emitter voltages are denoted as v_C and v_E, respectively, the collector-to-emitter total instantaneous voltage v_{CE} is

$$v_{CE} = v_C - v_E \qquad\qquad \textbf{(5-1)}$$

In some circuits it will be necessary to identify voltages across specific circuit components. In such cases the subscript will often assume the form of the circuit component for convenience in notation. In some cases, this may involve a "three-level" symbol, but the resulting variable will be more obvious in the process. For example, the dc voltage across a resistance R_C will be denoted as V_{R_C}.

It is customary to identify power supply voltages used to establish bias for transistors by a double subscript containing the letter twice representing the transistor terminal closest to the voltage. Examples are V_{CC} for a collector dc supply, V_{EE} for an emitter dc supply, and V_{BB} for a base dc supply.

The double-subscript notation can be used for currents when necessary, and this was explained in Section 1-10. As it turns out, however, the currents involved with most basic semiconductor devices can be adequately defined by a single subscript or by a subscript indicating the component through which the current is flowing. For example, I_C represents the dc collector current, and I_{R_1} represents the dc current flowing through a resistance R_1.

Let us now illustrate some of these conventions for a portion of some hypothetical transistor circuit. The reader need not be concerned about how the particular circuit works at this point since the circuit is not even complete. Rather, the intent is simply to illustrate the terminology.

Consider the portion of some *npn* transistor circuit shown in Figure 5-2, which has been labeled with a number of dc voltages and currents. First, we note that the collector terminal is connected through a collector resistance R_C to a positive bias

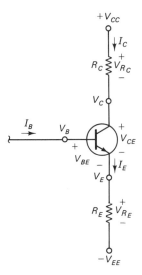

Figure 5-2 Illustration of a number of possible transistor dc voltages and currents.

supply voltage V_{CC}. (The negative terminal of this bias voltage is connected to the reference ground.) The emitter terminal is connected through an emitter resistance R_E to a negative supply voltage $-V_{EE}$. (The positive terminal of this supply voltage is also connected to the reference ground.)

The base dc current, collector dc current, and emitter dc current are denoted as I_B, I_C, and I_E, respectively, and are shown in their normal positive directions. The dc voltages at the three transistor terminals, as referred to the circuit common ground, are denoted as V_B, V_C, and V_E. In addition, the dc voltage V_{BE} between base and emitter and the dc voltage V_{CE} between collector and emitter are also identified. Finally, the dc voltage V_{R_C} across R_C, and the dc voltage V_{R_E} across R_E are identified. Both of the latter voltages are defined in the directions that will result in positive values.

Assume next that some external signal is applied to this circuit and it is desired to study the total instantaneous voltages and currents. The notation for this condition is shown in Figure 5-3. Not all the variables used in the dc analysis are considered in this case.

First, the power supplies $+V_{CC}$ and $-V_{EE}$ remain the same since these are fixed quantities. The total instantaneous currents for the three terminals are now denoted as i_B, i_C, and i_E, respectively. The total instantaneous voltages at the three terminals are denoted as v_B, v_C, and v_E, respectively. Finally, the total instantaneous base–emitter and collector–emitter voltages are denoted as v_{BE} and v_{CE}, respectively.

Later in the text we will learn how to construct special circuit models for dealing only with the time-varying signal portions of the various voltages and currents. In such cases, the notation introduced here for such variables (lowercase symbols and lowercase subscripts) will be used extensively.

The reader may find it instructive to refer back to this section, along with Sections 1-10 and 1-11, and review these conventions as new devices and circuits are introduced throughout the book.

Figure 5-3 Circuit of Figure 5-2 with assumed signal and a number of possible transistor instantaneous voltages and currents.

5-3 ACTIVE-REGION BJT CIRCUIT DRILL

A dc equivalent circuit for the BJT in the active region was developed in Chapter 4. In this section some relatively simple circuit configurations utilizing this equivalent circuit are studied and analyzed in detail. The circuits considered at this point are not necessarily intended to represent complete circuits or particular applications. Rather, the intent is to provide drill in constructing circuit models and in computing various voltages and currents so that the reader may develop a "feel" for the process. However, the concepts established from these drill circuits will arise extensively in later sections as more complete circuit applications are considered.

First, consider the circuit shown in Figure 5-4(a) containing an *npn* BJT. The

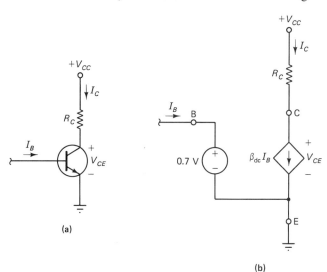

Figure 5-4 Basic *npn* BJT common-emitter circuit and active region dc model used for drill circuit.

base current I_B is assumed to be supplied from a circuit whose details are not yet specified. The emitter is connected to the circuit ground, and the collector is connected through a resistance R_C to a positive dc supply voltage V_{CC}. These "vertical" connections to $+V_{CC}$ above and ground below are used extensively in circuit diagrams. Remember that the negative terminal of the V_{CC} supply is also grounded.

The equivalent circuit model in the linear active region is shown in Figure 5-4(b). The model for an *npn* transistor as given in Chapter 4 replaces the transistor between the three terminals, but the remainder of the circuit is not changed.

Before analyzing this circuit, we wish to emphasize that the portion of the circuit consisting of the collector and emitter terminals, the resistance, and the power supply constitutes a closed loop. The circuit is redrawn for this purpose in Figure 5-5(a), and the model oriented toward this new layout is shown in Figure 5-5(b). Be sure that you understand the equivalence of the forms of Figures 5-4 and 5-5. In most subsequent work, the more standard forms of Figure 5-4 will be used.

Let us now write a KVL equation for the collector loop. Returning to the forms of Figure 5-4, the following process will be used: Start at ground and move up through the power supply to V_{CC}. Next, move down through R_C to the collector, and then on down through the transistor back to ground. The KVL equation reads

$$-V_{CC} + R_C I_C + V_{CE} = 0 \qquad (5\text{-}2)$$

If Figure 5-5 had been used in formulating this equation, the KVL loop would be counterclockwise in direction. It turns out that counterclockwise loops are often more convenient for the collector circuits of *npn* transistors. Solving for V_{CE} in (5-2), we obtain

$$\boxed{V_{CE} = V_{CC} - R_C I_C} \qquad (5\text{-}3)$$

Before commenting on this result, we will consider the comparable situation with a *pnp* transistor. This circuit is shown in Figure 5-6(a), and the dc active-region model is shown in Figure 5-6(b). In this case, the power supply is a negative voltage

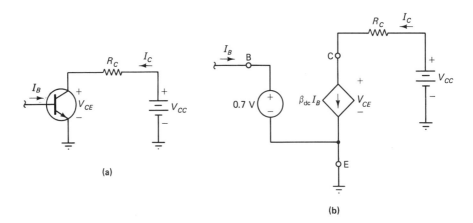

(a)

(b)

Figure 5-5 Alternative way to draw circuit of Figure 5-4 to illustrate collector loop.

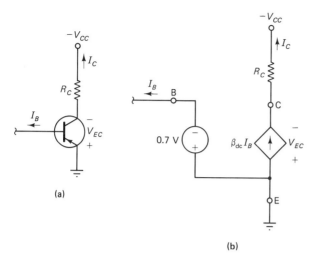

(a)

(b)

Figure 5-6 Basic *pnp* BJT common-emitter circuit and active-region dc model used for drill calculations.

$-V_{CC}$ with respect to ground. Note that conventions of voltage and currents are reversed in accordance with the suggestions of Chapter 4.

Before analyzing this circuit, an alternative circuit layout is shown in Figure 5-7(a), and the equivalent-circuit model having a similar form is shown in Figure 5-7(b). This circuit and model are *exactly the same* as those of Figure 5-6. This "upside-down" form with emitter on the top and collector on the bottom is often used for *pnp* transistors. Many beginners are confused by changing circuit layouts to different forms, so learn to recognize this form when it occurs.

For the *pnp* transistor, the following KVL process is used: Start at ground and move through the transistor to the collector and on through R_C to $-V_{CC}$. Finally, move through $-V_{CC}$ to ground (a voltage rise). The loop equation reads

$$V_{EC} + R_C I_C - V_{CC} = 0 \qquad \text{(5-4)}$$

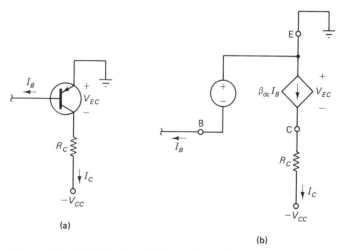

(a)

(b)

Figure 5-7 "Upside-down" form of *pnp* circuit of Figure 5-6.

Solving for V_{EC}, we obtain

$$\boxed{V_{EC} = V_{CC} - R_C I_C} \tag{5-5}$$

Comparing (5-5) with (5-3), we note that the collector loop equations for *npn* and *pnp* transistors have the same form, but with V_{EC} for the *pnp* unit replacing V_{CE} for the *npn* unit. All variables and parameters in both equations have positive values, so the equations behave in the same manner.

Although one could and should be able to "walk through" this process when required, an intuitive point to remember is this: *The voltage across the transistor on the collector side is the power supply voltage minus the collector current times the collector resistance.*

The collector current I_C in the *npn* circuit of Figure 5-4 or the *pnp* current of Figure 5-6 is given by

$$\boxed{I_C = \beta_{dc} I_B} \tag{5-6}$$

Curves displaying idealized variations of I_C and V_{CE} (or V_{EC}) with I_B are shown in Figure 5-8. In the ideal case, I_C varies linearly with base current until saturation is reached. However, V_{CE} (or V_{EC}) decreases as I_B increases according to (5-3) or (5-5), which is deduced by substituting I_C in terms of I_B from (5-6). This decrease in voltage across the transistor results from an increasing portion of the supply voltage being dropped across R_C as I_C increases.

We will now relate the equations and Figure 5-8 to some properties of significance at cutoff and saturation.

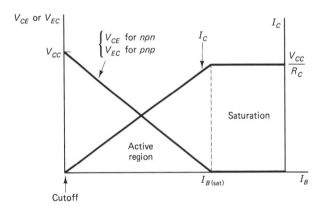

Figure 5-8 Ideal variation of collector–emitter (or emitter–collector) voltage and collector current with base current.

Cutoff

In the cutoff region, $I_B = 0$ and $I_C = 0$. Since $I_C = 0$, there is no drop across R_C, and the full power supply voltage appears across the transistor. These results for cutoff are summarized as follows:

$$I_B = 0 \qquad \text{(5-7)}$$

$$I_C = 0 \qquad \text{(5-8)}$$

$$V_{CE} = V_{CC} \quad npn \qquad \text{(5-9a)}$$

$$V_{EC} = V_{CC} \quad pnp \qquad \text{(5-9b)}$$

This situation corresponds to the left-hand side of Figure 5-8.

Saturation

For this purpose we will assume at this time that $V_{CE(sat)} = 0$ in the saturation model. As I_B increases, I_C increases, and the drop across R_C increases. The value of V_{CE} (or V_{EC}) decreases as shown in Figure 5-8 until it reaches zero. At that point, all the supply voltage V_{CC} appears across R_C and there is none left for the transistor. The value of I_C at that point will be denoted as $I_{C(sat)}$ and it is determined by setting $V_{CE} = 0$ in (5-3) or $V_{EC} = 0$ in (5-5). The result is

$$I_{C(sat)} = \frac{V_{CC}}{R_C} \qquad \text{(5-10)}$$

This value of I_C is the maximum collector current that can flow in the circuit. For $I_C < I_{C(sat)}$, the active-region model is applicable and $I_C = \beta_{dc} I_B$. However, once $I_{C(sat)}$ is reached, further increases in I_B result in no additional increase in collector current. In effect, the active-region model is no longer valid, and the saturation-region model is applicable.

Let $I_{B(sat)}$ represent the value of base current that will result in collector saturation. This value is determined as

$$I_{B(sat)} = \frac{I_{C(sat)}}{\beta_{dc}} \qquad \text{(5-11)}$$

The value of I_B can be made larger than $I_{B(sat)}$, but I_C cannot increase beyond $I_{C(sat)}$ for the given conditions. When $I_B > I_{B(sat)}$, the control process between base and collector is lost, and I_C remains at $I_{C(sat)}$. When $I_B > I_{B(sat)}$, the base is said to be *overdriven*.

In actual practice, the value of V_{CE} at saturation is $V_{CE(sat)}$. When it is necessary to consider this voltage, a more accurate value of the collector saturation current is

$$I_{C(sat)} = \frac{V_{CC} - V_{CE(sat)}}{R_C} \qquad \text{(5-12)}$$

The problem with using this more exact form is that the value of $V_{CE(sat)}$ is a function of the collector current. In turn, the collector current depends on $V_{CE(sat)}$. Thus some estimating is involved in attempting an exact solution. For many noncritical applications, the assumption $V_{CE(sat)} = 0$ is acceptable.

Along with the analysis procedures, the reader should strive to develop the following visualization process from this section: As the base current increases, the collector current increases, but the voltage magnitude between collector and emitter decreases. This process will arise frequently in later applications.

Example 5-1

Consider the circuit of Figure 5-9 containing an *npn* transistor. For $\beta_{dc} = 50$, determine the values of I_C, I_B, and v_i that result in saturation. Assume that $V_{BE} = 0.7$ V and $V_{CE(sat)} = 0$.

+15 V

$R_C = 12$ kΩ

I_C

I_B

R_B
200 kΩ

$+$
V_{BE}
$-$

$+$
V_{CE}
$-$

v_i

I_E

Figure 5-9 Circuit of Example 5-1.

Solution

At saturation with $V_{CE(sat)} = 0$, the full power supply voltage appears across R_C. Thus

$$I_{C(sat)} = \frac{V_{CC}}{R_C} = \frac{15 \text{ V}}{12 \times 10^3 \text{ }\Omega} = 1.25 \text{ mA} \tag{5-13}$$

The value of I_B that causes saturation is

$$I_{B(sat)} = \frac{I_{C(sat)}}{\beta_{dc}} = \frac{1.25 \times 10^{-3} \text{ A}}{50} = 25 \text{ }\mu\text{A} \tag{5-14}$$

Next, a clockwise loop equation is formed for the base circuit.

$$-v_i + R_B I_B + V_{BE} = 0 \tag{5-15}$$

Solving for v_i and substituting $I_B = 25$ μA, $V_{BE} = 0.7$ V, and $R_B = 200$ kΩ, we obtain

$$v_i = R_B I_B + V_{BE} = 2 \times 10^5 \times 25 \times 10^{-6} + 0.7 = 5.7 \text{ V} \tag{5-16}$$

Example 5-2

Consider the circuit of Example 5-1 (Figure 5-9) again. With the same assumptions as in Example 5-1, determine I_B, I_C, I_E, and V_{CE} for each of the following values of v_i: **(a)** 0; **(b)** 2 V; **(c)** 4 V; **(d)** 10 V.

Solution

At the outset, it must be stressed that there is a questionable assumption in this problem. The value of β_{dc} in a typical transistor does vary with different operating points, and the assumption of a constant value $\beta_{dc} = 50$ at different operating points is not completely valid. Despite that, the results with such an assumption are often as best as can be done, so we will proceed on that basis.

The active-region model is shown in Figure 5-10. The analysis that follows may be applied to this active-region model or to the original circuit.

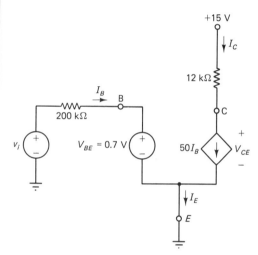

Figure 5-10 Active-region model of circuit of Figure 5-9 as used in Example 5-2.

(a) $v_i = 0$. In this case the base–emitter junction is reverse biased and cutoff exists. We thus have

$$I_B = I_C = I_E = 0 \qquad \text{(5-17)}$$

and

$$V_{CE} = 15 \text{ V} \qquad \text{(5-18)}$$

For the next several steps, a clockwise loop equation for the base circuit will be written as was done in Example 5-1. However, the intent in this case is to determine I_B as a function of v_i. We have

$$-v_i + R_B I_B + V_{BE} = 0 \qquad \text{(5-19)}$$

Solving for I_B and substituting pertinent values, we obtain

$$I_B = \frac{v_i - V_{BE}}{R_B} = \frac{v_i - 0.7}{2 \times 10^5} \qquad (5\text{-}20)$$

Note the form of (5-20) for the base loop. The voltage $v_i - 0.7$ is the net voltage across R_B. This voltage divided by R_B yields the base current.

Once I_B is known, the collector current in the active region is

$$I_C = 50\,I_B \qquad (5\text{-}21)$$

The emitter current is

$$I_E = I_B + I_C \qquad (5\text{-}22)$$

Finally, the collector–emitter voltage in the active region is

$$V_{CE} = V_{CC} - R_C I_C = 15 - 12 \times 10^3 I_C \qquad (5\text{-}23)$$

(b) $v_i = 2$ V. If this value is substituted in (5-20), we obtain

$$I_B = \frac{2 - 0.7}{2 \times 10^5} = 6.5\ \mu\text{A} \qquad (5\text{-}24)$$

The collector current from (5-21) is

$$I_C = 50 \times 6.5 \times 10^{-6} = 0.325\ \text{mA} \qquad (5\text{-}25)$$

The emitter current from (5-22) is

$$I_E = 0.325\ \text{mA} + 0.0065\ \text{mA} = 0.3315\ \text{mA} \qquad (5\text{-}26)$$

Note that I_E differs from I_C by only about 2%. The collector–emitter voltage from (5-23) is

$$V_{CE} = 15 - 12 \times 10^3 \times 0.325 \times 10^{-3} = 11.1\ \text{V} \qquad (5\text{-}27)$$

(c) $v_i = 4$ V. This step proceeds in the same manner as (b), and only the calculations will be listed.

$$I_B = \frac{4 - 0.7}{2 \times 10^5} = 16.5\ \mu\text{A} \qquad (5\text{-}28)$$

$$I_C = 50 \times 16.5 \times 10^{-6} = 0.825\ \text{mA} \qquad (5\text{-}29)$$

$$I_E = 0.825\ \text{mA} + 0.0165\ \text{mA} = 0.8415\ \text{mA} \qquad (5\text{-}30)$$

$$V_{CE} = 15 - 12 \times 10^3 \times 0.825 \times 10^{-3} = 5.1\ \text{V} \qquad (5\text{-}31)$$

(d) $v_i = 10$ V. Assume that we proceed as in parts (b) and (c), in which case I_B is calculated as

$$I_B = \frac{10 - 0.7}{2 \times 10^5} = 46.5\ \mu\text{A} \qquad (5\text{-}32)$$

If we were not alert, this value of I_B might lead us to calculate

$I_C = 50 \times 46.5 \times 10^{-6} = 2.325$ mA, which might further lead us to calculate $V_{CE} = 15 - 12 \times 10^3 \times 2.325 \times 10^{-3} = -12.9$ V! Something is wrong because V_{CE} cannot be negative in this circuit!

The answer, of course, is that we have reached the saturation point and the active-region model is no longer valid. In this circuit we can quickly refer back to the results of Example 5-1 and note that $I_{B(sat)} = 25$ μA and $I_{C(sat)} = 1.25$ mA. Thus I_B from (5-32) exceeds the value of $I_{B(sat)}$, and further calculations of I_C and V_{CE} based on the active-region model are meaningless. Thus we conclude that

$$I_C = I_{C(sat)} = 1.25 \text{ mA} \qquad \text{(5-33)}$$

$$V_{CE} = 0 \qquad \text{(5-34)}$$

$$I_E = 1.25 \text{ mA} + 0.0465 \text{ mA} = 1.2965 \text{ mA} \qquad \text{(5-35)}$$

Some comments are in order. First, while I_C cannot exceed $I_{C(sat)}$, I_B can continue to increase beyond the value of base current that causes saturation. As we will see in Section 5-2, a good design approach for circuits in which saturation is desired is to ensure that I_B is large enough always to cause saturation. Second, since I_B continues to increase after saturation occurs but I_C remains constant, the assumption that $I_E \simeq I_C$ becomes less valid for highly saturated circuits. The difference between I_C and I_E increases from about 2% in previous steps to nearly 4% in this example. In highly saturated circuits, the difference may be much greater.

Example 5-3

The circuit of Figure 5-11(a) containing an *npn* transistor illustrates the simplest (but *not* the best) manner in which a single power supply may be used to

(a)

(b)

Figure 5-11   Circuit and active-region model for Example 5-3.

establish active-region bias (or possibly saturation). Determine the values of I_C and V_{CE} for each of the following values of β_{dc}: (a) 50; (b) 100; (c) 200. Assume that $V_{BE} = 0.7$ V and $V_{CE(sat)} = 0$.

Solution

The active-region model is shown in Figure 5-11(b). As in Example 5-2, both the circuit and its active-region model have been shown for clarity. In relatively simple circuit configurations, the complete analysis can often be performed directly on the original circuit configuration, and it may not be necessary to bother with the active-region model. However, the active-region model is another way of visualizing circuit operation, and it will be used as appropriate throughout the book.

Although there is only a single power supply, it is common for both the base and collector circuits. Thus the loop equations for both base and collector circuits contain the 15-V dc supply.

To form the KVL loop for the base circuit, we will start at ground, move up through the power supply (voltage rise), and then move down through R_B and the base–emitter junction drop to ground. We have

$$-15 + 220 \times 10^3 I_B + 0.7 = 0 \qquad \text{(5-36)}$$

Solving for I_B, we obtain

$$I_B = \frac{15 - 0.7}{220 \times 10^3} = 65 \ \mu\text{A} \qquad \text{(5-37)}$$

The collector current in the active region is

$$I_C = \beta_{dc} I_B = 65 \times 10^{-6} \beta_{dc} \qquad \text{(5-38)}$$

The collector loop equation is formed in a manner similar to that of the base circuit, and we have

$$-15 + 2 \times 10^3 I_C + V_{CE} = 0 \qquad \text{(5-39)}$$

Solving for V_{CE}, we obtain

$$V_{CE} = 15 - 2 \times 10^3 I_C \qquad \text{(5-40)}$$

The results for each value of β_{dc} will be considered next.

(a) $\beta_{dc} = 50$. When this value is substituted in (5-38) and (5-40), we obtain

$$I_C = 65 \times 10^{-6} \times 50 = 3.25 \text{ mA} \qquad \text{(5-41)}$$

$$V_{CE} = 15 - 2 \times 10^3 \times 3.25 \times 10^{-3} = 8.5 \text{ V} \qquad \text{(5-42)}$$

(b) $\beta_{dc} = 100$. Proceeding as in part (a), we obtain

$$I_C = 65 \times 10^{-6} \times 100 = 6.5 \text{ mA} \qquad \text{(5-43)}$$

$$V_{CE} = 15 - 2 \times 10^3 \times 6.5 \times 10^{-3} = 2 \text{ V} \qquad \text{(5-44)}$$

(c) $\beta_{dc} = 200$. If the procedure of the preceding two steps were followed without further consideration, the values of I_C and V_{CE} would be calculated as 13 mA and -11 V, respectively! However, these impossible results quickly tell us that saturation has been reached, and the active-region model is no longer valid.

At saturation, and with $V_{CE(sat)} = 0$, the collector current is

$$I_C = I_{C(sat)} = \frac{V_{CC}}{R_C} = \frac{15}{2000} = 7.5 \text{ mA} \qquad \textbf{(5-45)}$$

which is the maximum collector current for the given circuit. The value of V_{CE} is, of course,

$$V_{CE} = V_{CE(sat)} = 0 \qquad \textbf{(5-46)}$$

This particular circuit configuration had a fixed value of base current in all three parts. From the results it can be concluded that as β_{dc} increases, I_C increases, and V_{CE} decreases for a fixed base current. There is some value of β_{dc} that causes saturation—can the reader determine this value?—and the transistor would remain saturated for subsequent increases in β_{dc}.

This problem illustrates the difficulty of establishing fixed values of I_C and V_{CE} in the active region when β_{dc} of a given transistor type varies widely. We deal with this process in Sections 5-7 and 5-8 and learn how a fixed operating point is established through proper stable circuit design.

Example 5-4

Consider the circuit of Figure 5-12(a) containing a *pnp* transistor with $\beta_{dc} = 50$. Determine I_B, I_C, I_E, and V_{EC}. Assume that $V_{EB} = 0.7$ V.

(a)

(b)

Figure 5-12 Circuit of Example 5-4 drawn two ways.

Solution

The circuit as given in Figure 5-12(a) is an "upside-down" form which, in certain ways, is more convenient for *pnp* transistors than the standard form since most voltages have the more positive terminal above the negative terminal. However, to provide maximum insight, the circuit is redrawn with emitter at the bottom in Figure 5-12(b). Comparing this circuit with that of Example 5-3 as shown in Figure 5-11(a), it is noted that both circuits have identical structures (but with opposite polarities and directions, of course).

In this case we work directly with the actual circuit diagram rather than the model. To form the base loop, we start at ground and move on through the base–emitter junction and R_B to the power supply negative terminal. Moving from that point back to ground (a voltage rise) completes the loop. We have

$$V_{EB} + 200 \times 10^3 I_B - 12 = 0 \tag{5-47}$$

Solving for I_B and substituting $V_{EB} = 0.7$ V results in

$$I_B = \frac{12 - V_{EB}}{200 \times 10^3} = \frac{12 - 0.7}{200 \times 10^3} = 56.5 \ \mu\text{A} \tag{5-48}$$

The collector current is

$$I_C = \beta_{dc} I_B = 50 \times 56.5 \times 10^{-6} = 2.825 \text{ mA} \tag{5-49}$$

To form the collector loop, we start again at ground, move to the collector, on to the power supply, and back to ground. The KVL equation is

$$V_{EC} + 1.8 \times 10^3 I_C - 12 = 0 \tag{5-50}$$

Solving for V_{EC} and substituting I_C from (5-49), we obtain

$$V_{EC} = 12 - 1.8 \times 10^3 I_C = 12 - 1.8 \times 10^3 \times 2.825 \times 10^{-3}$$

$$= 6.915 \text{ V} \tag{5-51}$$

The emitter current is

$$I_E = I_B + I_C = 56.5 \times 10^{-6} \text{ A} + 2.825 \times 10^{-3} \text{ A}$$

$$= 2.8815 \text{ mA} \tag{5-52}$$

5-4 BJT SWITCHES

One of the most common applications of the bipolar junction transistor is that of an electronic switch. This particular application of discrete transistors remains quite important even with the widespread existence of integrated circuits.

A typical application of a switch involves some load that is to be turned on or off by a control signal. The control signal might be a signal appearing at the output of a digital logic circuit or a microprocessor. The power level of the control signal will be very small and incapable of switching a load. However, the control signal may be capable of providing enough base drive to switch a transistor between two extreme levels, and the transistor may then serve to switch the load.

Most applications of the BJT as switches involve two levels of a control signal. When one level of the control signal is applied to the transistor, it operates in the cutoff region and acts as an open circuit. When the second level of the control signal is applied to the transistor, the state changes all the way to the saturation region and acts essentially as a short circuit.

One of the most common switching configurations is a *series* switch utilizing an *npn* transistor, a positive dc supply voltage, and a control voltage having the levels zero and a positive voltage. The form of this circuit is shown in Figure 5-13(a).

When $v_i = 0$, the base–emitter junction of the transistor is reverse biased, and the transistor is open. The equivalent circuit as viewed by the load is shown in Figure 5-13(b). Neglecting reverse leakage current, the switch is open.

Next assume that $v_i = V_i$, where V_i is a positive voltage of sufficient magnitude to produce saturation. In this case, the ideal equivalent circuit is shown in Figure 5-13(c). In the ideal case, the entire supply voltage appears across the load, and the switch is closed. In practice, there is a small saturation voltage drop across the transistor switch, so the voltage across the load is slightly less than V_{CC}. This factor must be considered in the design if the load is very voltage sensitive.

In the design of the *npn* BJT switch, it is necessary that the positive input level of the control signal drive the transistor into saturation. If one knew exactly the value of β_{dc}, the base circuit could theoretically be designed to reach the exact value of saturation when the positive control voltage is applied. However, this minimum design approach is unsatisfactory for several reasons. First, the value of β_{dc} for a given transistor is subject to change with collector voltage, temperature, and other factors. Second, if one is designing for production, it is impractical to produce a "custom design" for each individual transistor. As pointed out earlier, the value of β_{dc} for a given transistor type in a large stock may vary by a $5:1$ range or more.

The key to this dilemma is to design around a *worst-case* assumption. It has been noted in several examples and the text that once saturation is reached, this condition remains as base current is increased further. The approach is to select the most pessimistic conditions concerning β_{dc} (and any other circuit variables) and to ensure that base current is sufficient to produce saturation for those conditions. It follows that saturation will also occur for more favorable conditions.

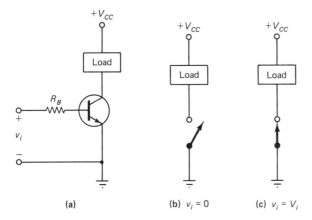

(a)   (b) $v_i = 0$ (c) $v_i = V_i$

Figure 5-13 Transistor switch using an *npn* transistor to close switch with positive voltage.

As far as the base circuit is concerned, it is the *minimum* value of β_{dc} that must be used in the worst-case design. If there is sufficient base drive to ensure saturation with the minimum value of β_{dc}, all other values of β_{dc} must necessarily produce saturation.

There are several rules of thumb that have been used in this endeavor. One is to use the specified minimum value of β_{dc} and design the base circuit to produce a base drive current perhaps twice the value required (to provide additional leeway) for this value of β_{dc}.

A second rule of thumb is to select an arbitrarily low β_{dc}, well below a realistic minimum value of β_{dc}, and design around that value. The value $\beta_{dc} = 10$, for example, has been used in many designs. If the positive value of the control voltage v_i is subject to uncertainty, the worst-case design approach would be to select the *minimum* expected value of that voltage as the basis for the design.

Another factor that must be considered in the design is the additional effect of the output resistance of the control voltage on the base drive current. In the basic circuit of Figure 5-13, a signal voltage v_i and resistance R_B are shown. In practice, the Thévenin equivalent circuit of the control voltage will be in series with R_B. In some cases, this resistance will be very small in comparison with R_B and may be neglected. However, its value should always be estimated, and its possible effect in series with R_B must be considered.

A different form of a *series* switch utilizing a *pnp* transistor, a negative dc supply voltage, and a control voltage having the levels zero and a negative voltage is shown in Figure 5-14(a). The resulting conditions for this circuit are shown in Figure 5-14(b) and (c). All conditions in this circuit are simply the negative mirror images of those in Figure 5-13.

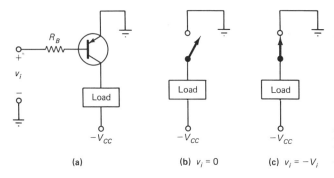

Figure 5-14 Transistor switch using a *pnp* transistor to close switch with negative voltage.

(a)   (b) $v_i = 0$ (c) $v_i = -V_i$

Example 5-5

A common relay switching circuit is shown in Figure 5-15. The switching operation is actually a two-step process in the following sense. First, the low power control signal turns on the transistor switch, which can *sink* (absorb to ground) sufficient current to turn on the relay. The relay switch contacts then close, and these mechanical contacts may be designed to handle a very large

+12 V   Relay

Assumed to
be negligible

To load

R_B

$100 \leqslant \beta_{dc} \leqslant 300$

v_i

Figure 5-15 Circuit of Example 5-5.

current (such as an industrial load or motor control circuit, etc.) The relay coil requires more power than the small control signal is capable of providing. The transistor in turn is incapable of handling the large external load current. However, the transistor can handle the relay control current, so it serves as a *buffer* or *interface* between the control signal and the relay coil.

The diode across the coil is reverse biased during the period when the relay coil is excited, so its effect will be neglected in the analysis that follows. However, when current has been flowing through the coil and the transistor is turned off, the effect is to terminate current through an inductor. When an attempt is made abruptly to terminate current through an inductance, a large negative voltage is generated. Such a voltage may have destructive effects unless some means are provided for suppressing it. The diode serves this purpose since it becomes forward biased under the conditions discussed previously and provides an inductive discharge path.

Assume that design for production is being performed and that the β_{dc} range for the transistor is from 100 to 300, as noted on the diagram. Assume that the control voltage has two levels: (1) $v_i = 0$ and (2) $v_i = 6$ V. Assume that the output resistance of the control circuit is small so that it can likely be neglected in comparison with R_B. Assume that the relay coil operates at about 12 V and has a coil resistance of 100 Ω (not shown). Finally, assume that $V_{BE} = 0.7$ V and $V_{CE(sat)} = 0$. Using the criterion of twice the base current for worst-case conditions, complete the design by specifying the value of R_B.

Solution

Since the low value of the control voltage is $v_i = 0$, the transistor will be at cutoff and the relay will be open for that condition. Thus the design calculations are based on $v_i = 6$ V.

First, the value of $I_{C(sat)}$ is determined. When the transistor is saturated, the 12-V supply appears across the relay coil, and since the relay coil has a dc resistance of 100 Ω, we have

$$I_{C(sat)} = \frac{12 \text{ V}}{100 \text{ }\Omega} = 0.12 \text{ A} \tag{5-53}$$

The base drive required to create this collector current is first determined under the worst-case condition of $\beta_{dc} = 100$ (minimum value). This value, denoted as $I_{B(sat)}$, is

$$I_{B(sat)} = \frac{0.12}{100} = 0.0012 \text{ A} = 1.2 \text{ mA} \tag{5-54}$$

Based on the criterion specified, we will choose

$$I_B = 2I_{B(sat)} = 2 \times 1.2 \text{ mA} = 2.4 \text{ mA} \tag{5-55}$$

From the base circuit with $v_i = 6$ V, we can calculate R_B as follows:

$$R_B = \frac{6 - 0.7}{2.4 \times 10^{-3}} = 2208 \text{ }\Omega \tag{5-56}$$

If the output resistance of the control source were an appreciable fraction of the required value of R_B, it should be subtracted from R_B. For example, assume that the control source has an output resistance of 600 Ω. The external resistance required would then be $R_B = 2206 - 600 = 1606$ Ω.

5-5 BJT CURRENT SOURCES

Bipolar junction transistors may be connected in a manner that will result in a nearly constant value of current for a variable load resistance. A circuit that supplies a current which is independent of the load resistance is, of course, a current source. The development here will be based on a single transistor current source.

A current source utilizing an *npn* BJT is shown in Figure 5-16. At first glance, this circuit seems somewhat similar to a transistor switch. However, there are two major differences in the circuit layout and operation as compared with the switch. First, the current source contains a resistance R_E between emitter and ground. This resistance will play a major role in establishing a different mode of operation, as we will see shortly. Second, the current source is operated in the active region, while the switch utilizes the cutoff and saturation regions.

We will now analyze the current source. Assume at this time that an ideal voltage source v_i appears between base and ground. If the emitter were grounded, as has been the case with most circuits considered thus far, it would be obvious that the transistor would start conducting for $v_i \simeq 0.7$ V. However, the presence of R_E may puzzle the reader in determining the point where active-region operation begins.

To assist in developing a "feel" for this circuit, momentarily assume that $v_i < 0.7$ V, so that cutoff exists. In this case the emitter current is zero, and there is no voltage across R_E. This means that the emitter is at ground potential, and the initial value of v_i required to produce current is the same as if R_E were not there (i.e., $v_i = 0.7$ V).

Figure 5-16 Single-transistor
constant current source.

When v_i reaches about 0.7 V, current starts to flow through the transistor and a small drop appears across R_E. This drop is in opposition to v_i, so the net base–emitter voltage is decreased very slightly, but active-region operation remains. This is a case where we must alter our thinking to realize that the base–emitter voltage is not exactly 0.7 V, but it varies slightly on either side. If v_i is increased further, collector and emitter current increase further, but the drop across R_E also increases, again tending to stabilize the base–emitter voltage to a nearly constant value. Thus, while the knee voltage in this circuit is the same as if the emitter were grounded, the process of conduction is much more gradual as the input voltage increases.

A brief pause will be made here for an important comment. This circuit represents a form of *negative feedback*, a subject that will constitute a major topic later in the text. The emitter resistance is the significant factor and it can result in a highly stabilized circuit.

Let us now write a KVL clockwise loop equation for the base circuit. We have

$$-v_i + V_{BE} + R_E i_E = 0 \qquad (5\text{-}57)$$

Solving for i_E, we obtain

$$i_E = \frac{v_i - V_{BE}}{R_E} \qquad (5\text{-}58)$$

Although we must accept "slight" variations in V_{BE} for the circuit to function properly, from an overall circuit point of view we will assume that $V_{BE} = 0.7$ V. Further, we will assume that β_{dc} is relatively large for the transistor, which means that $i_E \simeq i_C$. These assumptions result in

$$\boxed{i_C \simeq i_E = \frac{v_i - 0.7}{R_E}} \qquad (5\text{-}59)$$

The preceding equation gives us an exciting result! It says that the collector current is a function only of the input voltage and the emitter resistance, two quantities that are easily controlled (assuming, of course, that we disregard slight variations in the 0.7-V value). The very elusive parameter β_{dc} does not even appear in the equation! In effect, we have circumvented the problem with the wide β_{dc} variation through the use of a stabilizing emitter resistance.

This circuit forms the basis of bias stabilizing networks and is considered from that point of view in Sections 5-7 and 5-8. The focus at this point is on the implementation of a current source, and the discussion here addresses that application.

So far, most of the analysis has centered on the base–emitter part of the circuit. In effect, the collector current is determined from conditions established in the base–emitter circuit and the emitter resistance. However, the collector–emitter part of the circuit must be studied to determine the range of conditions that ensure active-region operation.

Let v_L represent the voltage across the load. If this were a simple resistive load, v_L would be related to i_C by Ohm's law. In general, however, the load could represent a nonlinear device (e.g., a diode). A KVL equation for the collector–emitter loop reads

$$-V_{CC} + v_L + v_{CE} + R_E i_E = 0 \qquad \text{(5-60)}$$

Solution for v_{CE} yields

$$v_{CE} = V_{CC} - v_L - R_E i_E \qquad \text{(5-61a)}$$

$$= V_{CC} - v_L - R_E i_C \qquad \text{(5-61b)}$$

where $i_C \simeq i_E$ has been substituted in the last form. Stated in words, the preceding equation indicates that the voltage across the transistor is the supply voltage minus the voltage drops of the load and the emitter resistance.

For active-region operation, $v_{CE} > 0$, and the circuit will function as a current source for that condition. In practice, since $V_{CE(\text{sat})}$ is a few tenths of a volt, some additional allowance should be made for that condition.

Returning to the expression for collector current of (5-59), the design of a current source will depend on the objective and given conditions. If one particular fixed value of load current is desired, v_i and/or R_E are selected to produce the required i_C. In some applications it is desired to provide an adjustable current source. This could easily be accommodated by providing an adjustment for v_i. This situation results in a *voltage-controlled current source*, which will be studied in more general terms later. Examples 5-6 and 5-7 will illustrate some of the typical situations encountered in this type of circuit.

Thus far we have assumed that v_i is an ideal voltage source, which simplified the base–emitter circuit analysis. In practice, however, the source will usually have some output resistance in its Thévenin model. A circuit diagram containing an effective source resistance R_S is shown in Figure 5-17. A loop equation is

$$-v_i + R_S i_B + V_{BE} + R_E i_E = 0 \qquad \text{(5-62)}$$

Solving for i_E, letting $V_{BE} = 0.7$ V, and assuming that $i_C \simeq i_E$, there results

$$i_C = \frac{v_i - 0.7 - R_S i_B}{R_E} \qquad \text{(5-63)}$$

When this result is compared with (5-59), it is noted that the numerator of (5-63)

Figure 5-17 Single-transistor constant current source with effect of source resistance considered.

contains the additional term $R_S i_B$. Since $i_C = \beta_{dc} i_B$, $i_B = i_C / \beta_{dc}$. If this relationship were substituted in (5-63), a more exact equation for i_C could be determined, with β_{dc} appearing as a parameter.

In practice, the more exact procedure suggested by the preceding paragraph may not be warranted. If β_{dc} is reasonably large and if R_S is not too large, the product $R_S i_B$ is usually quite small and may be neglected. The term "reasonably large" and "not too large" probably puzzle the reader, so the best way to resolve the dilemma is to make a *worst-case* calculation of $R_S i_B$ based on the worst-case value of base current $i_{B(max)}$. The quantity $i_{B(max)}$ is that base current that would be required to produce i_C if the value of β_{dc} were at the minimum value (i.e., $\beta_{dc(min)}$). This base current is

$$i_{B(max)} = \frac{i_C}{\beta_{dc(min)}} \tag{5-64}$$

If the product $R_S i_{B(max)}$ is small compared with 0.7 V, the term $R_S i_B$ in (5-63) may be ignored, and the simplified equation of (5-59) may be assumed.

Example 5-6

The circuit of Figure 5-18 is designed to produce a nearly constant current through a variable resistive load R_L. An ideal 5-V source is used to establish the current. **(a)** Determine the value of the current I_C. **(b)** Determine the range of R_L over which the circuit will function properly. It is assumed that β_{dc} is sufficiently large to justify the approximations used in the text.

Figure 5-18 Circuit of Example 5-6.

Solution

(a) Uppercase symbols are used for I_C and the various voltages because the input is no longer considered as an instantaneous voltage, but is a fixed 5-V source. The load current I_C is determined from (5-59) as

$$I_C = \frac{v_i - 0.7}{R_E} = \frac{5 - 0.7}{430} = 10 \text{ mA} \tag{5-65}$$

The voltage across the emitter resistance V_{R_E} is

$$V_{R_E} = 4.3 \text{ V} \tag{5-66}$$

This voltage remains constant as long as active-region operation occurs.

(b) When $R_L = 0$, the value of V_{CE} is simply

$$V_{CE} = 12 - 4.3 = 7.7 \text{ V} \tag{5-67}$$

which is certainly well within the active region. As R_L increases, however, its voltage increases, and V_{CE} decreases. Thus there will be some maximum value of R_L at which active-region operation ceases.

In general,

$$V_{CE} = 12 - 4.3 - R_L I_C = 7.7 - R_L I_C \tag{5-68}$$

The value of $R_{L(\text{max})}$ is determined by setting $V_{CE} = 0$. Thus

$$0 = 7.7 - R_{L(\text{max})} I_C$$

or

$$R_{L(\text{max})} = \frac{7.7}{I_C} = \frac{7.7}{0.01} = 770 \ \Omega \tag{5-69}$$

Thus the circuit functions as a current source provided that R_L is within the range $0 < R_L < 770 \ \Omega$. When $R_L \geq 770 \ \Omega$, the BJT is saturated.

Example 5-7

The circuit of Figure 5-19 is to be used in a production design as a current source driver for the LED shown. The advantage of using a current source is that the light intensity of an LED is a function of the current, and if the current

Figure 5-19 LED driver circuit of Example 5-7.

is constant, the light intensity will be more uniform even though the LED voltage will vary with different diodes. **(a)** For a desired LED current of 20 mA, determine the value of R_E required. **(b)** If the LED voltage drop can vary from 1.5 to 2.5 V, determine the expected range of V_{CE} in order to check active-region operation for all conditions.

Solution

(a) The value of R_E is determined from (5-59) as

$$R_E = \frac{2 - 0.7}{0.02} = 65 \ \Omega \tag{5-70}$$

(b) The collector–emitter voltage V_{CE} is

$$V_{CE} = 5 - V_{R_E} - V_D = 5 - 1.3 - V_D = 3.7 - V_D \tag{5-71}$$

The maximum value of V_{CE}, denoted as $V_{CE(max)}$, occurs for the minimum value of V_D.

$$V_{CE(max)} = 3.7 - 1.5 = 2.2 \ V \tag{5-72}$$

The minimum value of V_{CE}, denoted as $V_{CE(min)}$, occurs for the maximum value of V_D.

$$V_{CE(min)} = 3.7 - 2.5 = 1.2 \ V \tag{5-73}$$

Thus the range of V_{CE} is from 1.2 to 2.2 V and active-region operation should be ensured under all conditions.

5-6 EMITTER FOLLOWER

A BJT circuit having application both in discrete form and in integrated circuits is the *emitter follower*. The simplest dc coupled form of such a circuit utilizing an *npn* transistor is shown in Figure 5-20(a). The collector terminal is connected directly to the positive power supply terminal, and the load is connected between the emitter

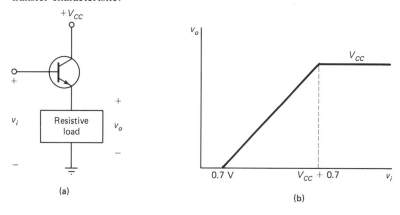

Figure 5-20 Basic *npn* dc-coupled emitter-follower circuit and ideal transfer characteristic.

and ground. In this simple form of the emitter follower, *it is assumed that both the input v_i and the load have dc paths to ground.*

The circuit is almost identical in form to the single transistor current source of Section 5-5. What distinguishes the emitter follower, however, is the fact that the desired output variable is the voltage v_o across the load as a function of the input voltage variable v_i.

For $v_i \geq 0.7$ V, the base–emitter junction is forward biased, and current flows through the transistor to the load. Assuming a base–emitter voltage drop of 0.7 V, a loop equation for the base circuit yields

$$-v_i + 0.7 + v_o = 0 \qquad \text{(5-74)}$$

or

$$\boxed{v_o = v_i - 0.7} \qquad \text{(5-75)}$$

From (5-75) we see that as the input voltage increases, the output voltage also increases, but differs from the input by 0.7 V. In effect, the output voltage at the emitter *follows* the input as suggested by the name of the circuit.

The transfer characteristic of the emitter follower is shown in Figure 5-20(b). The output is zero for $v_i < 0.7$ V since the base–emitter junction will be reverse biased. For $v_i < 0.7$ V, the straight-line characteristic of (5-75) applies as long as active-region operation occurs. However, when v_i becomes sufficiently positive that the voltage between collector and emitter approaches zero, saturation occurs, and the output can increase no further. Ideally, this would occur when $v_i = V_{CC} + 0.7$ and $v_o = V_{CC}$ as shown. In practice, saturation occurs a few tenths of a volt below this level.

Since the output voltage is close to, but always smaller than, the input voltage, a beginner might logically question the value of this circuit. The value lies in the *isolation* that the circuit provides. A typical application might be as follows: A voltage close to a level v_i is required to furnish a moderate or large amount of current to a load. However, the source itself is incapable of supplying the amount of current required. A transistor having the necessary current capability is selected, and the emitter follower is used to separate the source from the load. All the source need do is provide the much smaller base current drive, while the transistor supplies the required larger load current. The circuit thus serves as a *buffer*.

The actual base current drive is a function of the required load current and the value of β_{dc} for the transistor. Let i_o represent the required load current, which is, of course, the emitter current (i.e., $i_o = i_E$). Since $i_E = i_B + i_C$ and $i_C = \beta_{dc} i_B$, the following relationship is established between load current and base current:

$$i_o = i_B + i_C = i_B + \beta_{dc} i_B = (1 + \beta_{dc}) i_B \qquad \text{(5-76)}$$

Solving for i_B, we obtain

$$i_B = \frac{i_o}{1 + \beta_{dc}} \qquad \text{(5-77)}$$

The last result indicates how much smaller i_B is than the required load current, which provides the isolation inherent in the circuit.

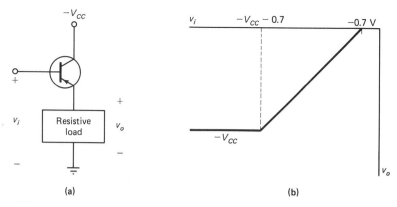

(a)   (b)

Figure 5-21 Basic *pnp* emitter-follower circuit and ideal transfer characteristic.

Understand that the output voltage differs from the input voltage by the 0.7-V drop, so any application of this circuit in this simple form must be one in which this difference poses no problem. Later we will see how additional feedback can be used with integrated circuits to overcome this difference.

A *pnp* transistor can be used to implement an emitter follower for a negative input voltage, and this circuit is shown in Figure 5-21(a). Since the same standard positive input and output references are used here as for the previous circuit, the active region of operation is now for $v_i \leq -0.7$ V. In this region the output is given by

$$v_o = v_i + 0.7 \tag{5-78}$$

Note that since $v_i < 0$, the magnitude of v_o will be less than the magnitude of v_i. The transfer characteristic for this circuit is shown in Figure 5-21(b). Saturation in this circuit occurs at an output level of $-V_{CC}$ in the ideal case.

Example 5-8

The emitter-follower circuit of Figure 5-22 is used to isolate a source v_i from a 100-Ω load. The small difference between v_i and v_o is not critical for the par-

Figure 5-22 Emitter-follower circuit of Example 5-8.

ticular application. **(a)** Assuming that $V_{BE} = 0.7$ V, determine v_o in terms of v_i. **(b)** Assuming that $V_{CE(sat)} = 0$, determine the range of input voltage for which the circuit functions. **(c)** For $v_i = 12$ V and $\beta_{dc} = 50$, determine the value of base current i_B.

Solution

(a) From a base KVL loop equation or from (5-75), the output voltage is simply

$$v_o = v_i - 0.7 \tag{5-79}$$

(b) The smallest value of the input at which the circuit starts to function is $v_i = 0.7$ V. The largest value corresponds to saturation. At saturation $v_o = 15$ V and $v_i = 15.7$ V. Thus the ideal range is 0.7 V $< v_i < 15.7$ V. In practice, the upper value would be reduced slightly due to the realistic collector–emitter saturation voltage.

(c) For $v_i = 12$ V, the output voltage is

$$v_o = 12 - 0.7 = 11.3 \text{ V} \tag{5-80}$$

The output load current is

$$i_o = \frac{11.3 \text{ V}}{100 \text{ } \Omega} = 113 \text{ mA} \tag{5-81}$$

The value of i_B is

$$i_B = \frac{i_o}{1 + \beta_{dc}} = \frac{113 \text{ mA}}{51} = 2.216 \text{ mA} \tag{5-82}$$

The advantage of this circuit is clear from these results. While the required load current is 113 mA, the current supplied by the input source need only be 2.216 mA.

5-7 STABLE BIAS CIRCUITS WITH SINGLE POWER SUPPLY

Many applications of bipolar junction transistors require that a stable set of conditions be established at a specific point in the active region. A stable operating point is one in which the collector current and collector–emitter voltage should remain nearly constant and predictable even though transistor characteristics of the given type vary widely from one unit to another.

If a given individual transistor were being used for a specific circuit, the operating point could be adjusted experimentally around the actual β_{dc}, and this might suffice for a limited special application. This approach is unsatisfactory for production, however, since transistor characteristics vary widely within a given type. As we have seen earlier, the variation of β_{dc} for a given transistor type may swing the operating point all the way from well in the active region to saturation with a simple bias circuit.

In this section a very popular stable bias circuit utilizing a single power supply will be analyzed and discussed. The circuit is the most widely used for discrete BJT active-region applications. This circuit is a prelude for dealing with basic BJT amplifier configurations to be discussed in later chapters.

Our approach in this section is that of analyzing a given circuit to determine properties of the operating point established. The design of such circuits is considered in the next section, so all examples considered in this section are analyzed as they are given.

The key to the single power supply stable bias circuit is the concept of the constant current source as developed in Section 5-5. In this case, however, both the collector–emitter circuit and base–emitter circuits will be biased from a single power supply.

The basic stable bias circuit as applied to an *npn* transistor using a positive bias power supply (with respect to circuit ground) is shown in Figure 5-23(a). Generally,

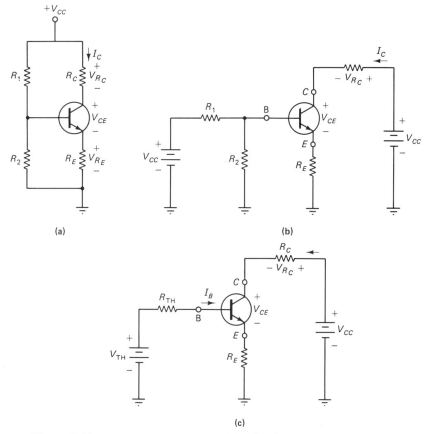

Figure 5-23 Single power supply stable bias for an *npn* transistor and equivalent circuits.

the circuit includes the transistor, an emitter resistance R_E, a collector resistance R_C, and a voltage-divider network consisting of R_1 and R_2. In all circuit configurations that follow, R_1 will represent the resistance adjacent to the collector resistance, and R_2 will represent the resistance adjacent to the emitter resistance. This arrangement should be noted because in some of the upside-down or *pnp* configurations, R_2 may be above R_1 on the schematic.

An intuitive approach may be formulated as follows: The voltage divider on the base side results in a positive voltage well below V_{CC} appearing at the base. This voltage in consort with R_E establishes an emitter current in the same manner as discussed for the current source concept of Section 5-5. Since emitter and collector currents are nearly the same, the collector current turns out to be dependent primarily on the dc voltage across R_2 and the emitter resistor R_E.

To illustrate the analysis, the circuit is first redrawn as shown in Figure 5-23(b). Although there is only one power supply, the circuit may be studied *as if* there were two supplies of equal value as shown, and this allows the base–emitter circuit and collector–emitter circuits to be separated.

Next, a Thévenin equivalent circuit looking back from the base (point B) will be determined. The value of the Thévenin voltage V_{TH} is determined from the voltage-divider rule as

$$V_{TH} = \frac{R_2}{R_1 + R_2} V_{CC} \tag{5-83}$$

The Thévenin equivalent resistance is

$$R_{TH} = R_1 \| R_2 = \frac{R_1 R_2}{R_1 + R_2} \tag{5-84}$$

Using this Thévenin equivalent circuit, a simplified circuit model is shown in Figure 5-23(c). The base current I_B may be considered as flowing from a source V_{TH} in series with R_{TH}.

From the base equivalent circuit, a clockwise KVL loop equation is written as follows:

$$-V_{TH} + R_{TH}I_B + V_{BE} + R_E I_E = 0 \tag{5-85}$$

The current I_B is given by

$$I_B = \frac{I_C}{\beta_{dc}} \tag{5-86}$$

The largest value of this term corresponds to the minimum value of β_{dc}. In the practical design of the bias circuit (to be discussed in Section 5-8), the resistances R_1 and R_2 are chosen such that the largest value of this term is very small compared with the other terms in (5-85). It is convenient to relate this term specifically to the term $R_E I_E$, so the following inequality is required:

$$\frac{R_{\text{TH}} I_C}{\beta_{\text{dc}}} \ll R_E I_E \qquad \text{for a stable design} \qquad \textbf{(5-87)}$$

Since $I_C \simeq I_E$, this inequality may be restated as

$$\frac{R_{\text{TH}}}{\beta_{\text{dc}}} \ll R_E \qquad \text{for a stable design} \qquad \textbf{(5-88)}$$

or

$$R_{\text{TH}} \ll \beta_{\text{dc}} R_E \qquad \text{for a stable design} \qquad \textbf{(5-89)}$$

Based on the assumed inequality, the second term in (5-85) may be neglected. The value of I_E (and I_C) may then be closely approximated as

$$\boxed{I_C \approx I_E \approx \frac{V_{\text{TH}} - V_{BE}}{R_E} = \frac{V_{\text{TH}} - 0.7}{R_E}} \qquad \textbf{(5-90)}$$

The result is in agreement with the development of the current source in Section 5-5. In this case the voltage establishing the current is the open-circuit voltage V_{TH}. Once again, we see that the current is virtually independent of the transistor β_{dc} provided, of course, that β_{dc} is large and that the inequality discussed earlier is met.

On the collector side, the voltages across the individual portions of the circuit may now be determined. The voltage across R_C, denoted as V_{R_C}, is determined as

$$\boxed{V_{R_C} = R_C I_C} \qquad \textbf{(5-91)}$$

The voltage across R_E, denoted as V_{R_E}, is

$$\boxed{V_{R_E} = R_E I_E \approx R_E I_C} \qquad \textbf{(5-92)}$$

The voltage V_{CE} from collector to emitter is

$$\boxed{V_{CE} = V_{CC} - V_{R_E} - V_{R_C}} \qquad \textbf{(5-93)}$$

In effect, the dc supply voltage V_{CC} is divided into three parts: V_{R_E}, V_{R_C}, and V_{CE}.

The single supply bias circuit for a *pnp* circuit using a negative power supply with respect to ground is shown in Figure 5-24(a). This circuit layout is, as shown, in the same form as for the *npn* transistor. The corresponding "upside-down" layout with ground at the top for the same circuit is shown in Figure 5-24(b). Note that when the circuit is shown in this form, the resistors R_2 and R_E are at the top, and R_1 and R_C are at the bottom.

The expressions for V_{TH}, R_{TH}, and I_E or I_C are exactly the same as for the *npn* transistor. However, I_E and I_C are defined in the direction from emitter to collector so that they are positive quantities. Similarly, the voltage across the transistor is defined as V_{EC}, which will be a positive quantity. The two resistive voltage drops retain the same notation as before, but the reference directions are reversed. The ex-

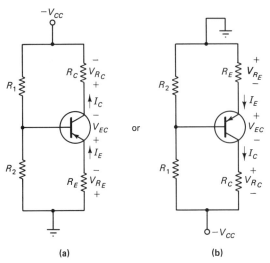

(a)   or   (b)

Figure 5-24 Single negative power supply stable bias for a *pnp* transistor shown two ways.

pressions for V_{R_E} and V_{R_C} will be the same as for the *npn* transistors [i.e., (5-91) and (5-92)]. The expression for V_{EC} will be

$$V_{EC} = V_{CC} - V_{R_C} - V_{R_E} \qquad (5\text{-}94)$$

In some applications, there may be available only a positive dc power supply voltage with respect to ground. This is ideal if only *npn* transistors are available. However, suppose that it is desired to use a *pnp* transistor in such a case. Although the basic approach would be to bias the collector to a negative level with respect to the emitter, *the same effect can be achieved by biasing the emitter to a positive level with respect to the collector*. A stable bias arrangement to achieve this result is shown in Figure 5-25. Basically, the voltage across the complete circuit has the same layout as for the *pnp* circuit of Figure 5-24(b). The two circuit forms differ only with respect to the ground point. Thus the preceding discussion for the *pnp* circuits of Figure 5-24 applies to this circuit with respect to the currents and voltages

Figure 5-25 Method for establishing stable bias for a *pnp* transistor with a single positive power supply.

across components. The voltages with respect to ground in the two circuits are, of course, different. As in the "upside-down" form of Figure 5-24(b), note that R_2 and R_E are at the top, and R_1 and R_C are at the bottom.

Example 5-9

(a) For the circuit of Figure 5-26, determine I_E, I_C, V_{RC}, V_{RE}, and V_{CE}. **(b)** If the minimum value of β_{dc} for the transistor is 50, determine the maximum possible value of I_B. **(c)** Check the inequality of (5-89) as a means of assessing the relative stability of the bias point.

Figure 5-26 Circuit of Example 5-9.

Solution

(a) The value of V_{TH} is

$$V_{TH} = \frac{4700}{4700 + 1 \times 10^4} \times 15 = 4.80 \text{ V} \tag{5-95}$$

Assuming that $V_{BE} = 0.7$ V, we have

$$I_C \simeq I_E = \frac{4.8 - 0.7}{3900} = 1.05 \text{ mA} \tag{5-96}$$

The three voltage drops around the transistor are

$$V_{RE} = 3.9 \times 10^3 \times 1.05 \times 10^{-3} = 4.1 \text{ V} \tag{5-97}$$

$$V_{RC} = 5.1 \times 10^3 \times 1.05 \times 10^{-3} = 5.36 \text{ V} \tag{5-98}$$

$$V_{CE} = 15 - 5.36 - 4.1 = 5.54 \text{ V} \tag{5-99}$$

(b) The maximum base current $I_{B(max)}$ corresponds to $\beta_{dc(min)}$ and is

$$I_{B(max)} = \frac{1.05 \times 10^{-3}}{50} = 21 \text{ } \mu\text{A} \tag{5-100}$$

(c) The value of R_{TH} is

$$R_{TH} = 4.7 \times 10^3 \| 10^4 = 3197 \ \Omega \qquad \text{(5-101)}$$

The value of $\beta_{dc} R_E$ is $50 \times 3.9 \times 10^3 = 195,000$. From the inequality of (5-89), we see that $3197 \ll 195,000$, and the circuit should be stable.

5-8 DESIGN FOR A STABLE BIAS POINT

In the preceding section, the analysis of a stable biasing circuit was performed, and the relationships for predicting the approximate circuit bias conditions were developed. In this section, an approach for actually designing such a circuit will be given. The design of a stable bias operating point is a necessary part of an overall amplifier design using BJTs.

First, it must be understood that there is no one single procedure for a design such as this. Various restrictions may result in modifications of any design procedure. Further, there is not necessarily any one final design that is "best"; a number of different designs could all be acceptable. Some individuals develop an intuitive approach with practice that may be superior to a formal approach. Nevertheless, there is a strong motivation for an approach that will guide the inexperienced person to an acceptable result.

The approach here is based on one provided by Hayt and Neudick* and is judged by this author to be one of the easiest methods to apply. It will be given initially as a step-by-step procedure. However, with some practice, you will understand the logic behind the steps, and it then becomes less "cookbook" in form.

Assumed Conditions

Refer to the circuit diagram of Figure 5-27 in the development that follows. Assume that the value of V_{CC} is given and that a desired value of I_C is to be established for a particular transistor type. The exact value of β_{dc} need not be known since the intent of the circuit is to establish a specified value of I_C for any transistor of the given type. However, the minimum value of β_{dc}, denoted as $\beta_{dc(min)}$, at the given operated point should be estimated from specifications if possible. If this is not practical, even a pessimistic reasonable guess, such as $\beta_{dc(min)} = 25$, may be acceptable in some cases.

The power supply voltage V_{CC} is divided into three parts: V_{R_C}, V_{CE}, and V_{R_E}, in which

$$\boxed{V_{CC} = V_{R_C} + V_{CE} + V_{R_E}} \qquad \text{(5-102)}$$

The choice of the three voltages will depend on the ultimate application of the circuit. For some amplifier designs, the two voltages V_{R_C} and V_{CE} are chosen to be the

---

*W. H. Hayt, Jr., and G. W. Neudick, *Electronic Circuit Analysis and Design,* 2nd ed. (Boston: Houghton Mifflin, 1983).

Figure 5-27 Circuit form of stable bias circuit with variables used in design.

same (i.e., $V_{R_C} = V_{CE}$). In some cases, all three voltages are chosen to be equal; that is,

$$V_{R_C} = V_{CE} = V_{R_E} = \frac{V_{CC}}{3} \qquad \textbf{(5-103)}$$

The latter choice is quite good when the peak expected output signal voltage is well below $V_{CC}/3$, and when there is no additional loading on the output. However, it tends to be somewhat "wasteful" of the available dc voltage, as it is not necessary to use as much of the available voltage as $V_{CC}/3$ for V_{R_E} to achieve a stable design.

One approach is to use a 40–40–20 split, which amounts to

$$V_{R_C} = V_{CE} = 0.4 V_{CC} \qquad \textbf{(5-104)}$$

and

$$V_{R_E} = 0.2 V_{CC} \qquad \textbf{(5-105)}$$

Other possibilities may be used, and the reader will be guided on the approach at this point in the text.

As a general rule, the larger the value of V_{R_E}, the more stable the circuit, but the less dynamic range of the output signal.

Once the assumptions of I_C and the three voltages are made, the design procedure is given as follows:

Design steps

1. Determine R_C as

$$R_C = \frac{V_{R_C}}{I_C} \qquad \textbf{(5-106)}$$

2. Assuming that $I_E \approx I_C$, determine R_E as

$$R_E = \frac{V_{R_E}}{I_E} \approx \frac{V_{R_E}}{I_C}$$ (5-107)

3. Calculate

$$V_{R_2} = V_{R_E} + V_{BE} = V_{R_E} + 0.7$$ (5-108)

where $V_{BE} = 0.7$ V has been assumed.

4. Calculate V_{R_1} as

$$V_{R_1} = V_{CC} - V_{R_2}$$ (5-109)

5. Compute the maximum value of I_B, indicated as $I_{B(\max)}$, using the minimum value of β_{dc}, indicated as $\beta_{dc(\min)}$.

$$I_{B(\max)} = \frac{I_C}{\beta_{dc(\min)}}$$ (5-110)

6. Select a current I_{R_1} through the resistance R_1 to be much larger than the maximum value of I_B. This can be stated as

$$I_{R_1} = K I_{B(\max)}$$ (5-111)

where K is a constant to be selected. The choice of this constant is arbitrary, but a value between 10 and 100 is usually satisfactory. As a general rule, larger values of K result in a more stable circuit but increase loading on any signal source at the input (to be considered in amplifier configurations) and increased circuit power dissipation.

7. Calculate R_1 as

$$R_1 = \frac{V_{R_1}}{I_{R_1}}$$ (5-112)

8. Calculate I_{R_2} as

$$I_{R_2} = I_{R_1} - I_{B(\max)}$$ (5-113)

9. Calculate R_2 as

$$R_2 = \frac{V_{R_2}}{I_{R_2}}$$ (5-114)

This completes the preliminary design calculations.

In general, the four resistance values obtained from this process will usually not turn out to be standard values. One can then substitute the nearest standard values as a trial approach. In most cases it is best to round R_1 and R_2 in the same direction; that is, use the next *larger* standard values for *both*, or use the next *smaller standard* values for *both*. However, rounding in the same direction may not be the best approach for R_E and R_C since the changes in both values accentuate the variation of V_{CE} from the design goal.

Once a tentative design is determined, return to the basic equation for I_C as given by (5-90) and determine I_C for the actual component values selected. Next, compute V_{R_C}, V_{R_E}, and V_{CE} based on this I_C. If all values are acceptable, the design may be considered complete as far as the work of this chapter is concerned. (Other considerations for amplifier applications may have to be studied.)

If this design seems unsatisfactory due to either an unacceptable I_C or to unacceptable voltages in the collector loop, some additional work is required. Sometimes, the change of one or two of the four resistances may resolve the problem. In other cases it may be necessary to choose a different value of K and start all over.

Example 5-10

Design a stable bias circuit for an *npn* transistor based on the following specifications and conditions:

Power supply voltage = 15 V
Desired collector current = 2 mA
Desired $V_{RC} = V_{CE} = 6$ V
Desired $V_{R_E} = 3$ V
$\beta_{dc(min)} = 50$

Select $K = 30$. In addition to the idealized design, investigate the results of using standard values on I_C and the three voltages.

Solution

The procedure delineated in the text will be followed exactly in the steps that follow.

$$R_C = \frac{V_{R_C}}{I_C} = \frac{6}{2 \times 10^{-3}} = 3 \text{ k}\Omega \tag{5-115}$$

$$R_E = \frac{V_{R_E}}{I_C} = \frac{3}{2 \times 10^{-3}} = 1.5 \text{ k}\Omega \tag{5-116}$$

$$V_{R_2} = V_{R_E} + 0.7 = 3 + 0.7 = 3.7 \text{ V} \tag{5-117}$$

$$V_{R_1} = V_{CC} - V_{R_2} = 15 - 3.7 = 11.3 \text{ V} \tag{5-118}$$

$$I_{B(\text{max})} = \frac{I_C}{\beta_{\text{dc(min)}}} = \frac{2 \times 10^{-3}}{50} = 40 \ \mu\text{A} \tag{5-119}$$

$$I_{R_1} = K I_{B(\text{max})} = 30 \times 40 \ \mu\text{A} = 1.2 \ \text{mA} \tag{5-120}$$

$$R_1 = \frac{V_{R_1}}{I_{R_1}} = \frac{11.3}{1.2 \times 10^{-3}} = 9417 \ \Omega \tag{5-121}$$

$$I_{R_2} = I_{R_1} - I_{B(\text{max})} = 1.2 \times 10^{-3} - 40 \times 10^{-6} = 1.16 \ \text{mA} \tag{5-122}$$

$$R_2 = \frac{V_{R_2}}{I_2} = \frac{3.7}{1.16 \times 10^{-3}} = 3190 \ \Omega \tag{5-123}$$

The circuit based on these idealized values is shown in Figure 5-28. (Momentarily ignore the values in parentheses.) Let us now calculate the value of I_C based on these values and the idealized relationship of (5-90). First V_{TH} is calculated as

$$V_{\text{TH}} = \frac{3190}{3190 + 9417} \times 15 = 3.796 \ \text{V} \tag{5-124}$$

The value of I_C is then

$$I_C = \frac{3.796 - 0.7}{1.5 \times 10^{-3}} = 2.064 \ \text{mA} \tag{5-125}$$

The reader may be puzzled by the fact that even with the ideal resistance values, I_C differs somewhat (about 3%) from the design goal. This difference is a result of the approximate nature of the design procedure in assuming the worst-case value of I_B. In fact, this procedure compensates somewhat for the idealized nature of the result of (5-90) in the following manner: When (5-90) is employed, the drop resulting from $R_{\text{TH}} I_B$ is ignored. Said differently, (5-90) assumes that $\beta_{\text{dc}} = \infty$. Any finite value of β_{dc} will result in a smaller value of I_C than (5-90) predicts. Thus the actual value of I_C will be somewhat smaller than the value predicted by (5-125).

Now, let's explore the possibility of using standard values (see Appendix

+15 V

9417 Ω
(9.1 kΩ)

3 kΩ

3190 Ω
(3 kΩ)

1.5 kΩ

Figure 5-28 Design of Example 5-10 (rounded standard values are in parentheses).

BJT-Resistor Circuit Applications   Chap. 5

A). The closest standard 5% value to R_1 is 9.1 kΩ, and the closest standard 5% value to R_2 is 3 kΩ. Both of these are less than the ideal values, so we are rounding in the same direction as suggested. The rounded values for R_1 and R_2 are shown in parentheses on Figure 5-28.

It turns out that the ideal values of R_E (1.5 kΩ) and R_C (3 kΩ) are standard values, so we are in good shape there. Based on using the preceding standard values of R_1 and R_2, the value of I_C will now be calculated. First V_{TH} is calculated as

$$V_{TH} = \frac{3 \times 10^3}{3 \times 10^3 + 9.1 \times 10^3} \times 15 = 3.719 \text{ V} \qquad \textbf{(5-126)}$$

The value of I_C is

$$I_C = \frac{3.719 - 0.7}{1.5 \times 10^3} = 2.013 \text{ mA} \qquad \textbf{(5-127)}$$

This value of I_C turns out to be closer to the design goal with standard rounded values than with the values obtained from the design process. However, this result is somewhat misleading since, as pointed out earlier, the actual value of I_C will be somewhat smaller than the result of (5-127) suggests.

Next, the voltages are calculated based on the value of I_C of (5-127).

$$V_{R_E} = 2.013 \times 10^{-3} \times 1.5 \times 10^3 = 3.02 \text{ V} \qquad \textbf{(5-128)}$$

$$V_{R_C} = 2.013 \times 10^{-3} \times 3 \times 10^3 = 6.04 \text{ V} \qquad \textbf{(5-129)}$$

$$V_{CE} = 15 - 3.02 - 6.04 = 5.94 \text{ V} \qquad \textbf{(5-130)}$$

Although V_{R_C} and V_{CE} do not turn out to be exactly the same (they rarely do with real-life values), they differ by only about 1.7%. Considering the variations and tolerances involved, this is about as good as can usually be achieved with this type of process.

5-9 LOAD-LINE ANALYSIS

For most transistor dc analysis and design procedures, the various equivalent circuits considered throughout the chapter are the most direct means for solving problems. There is, however, a graphical procedure for determining transistor variables in certain circuits from the voltage–current characteristics. This procedure involves superimposing a liner circuit equation on the voltage–current characteristics and determining the point of intersection of the two functions. The linear circuit equation is called a *load line*.

Circuits that are directly amenable to load-line analysis are usually limited to those that can be reduced to simple series configurations for both collector and base circuits. It is also quite difficult to employ load-line analysis when there is a resistance between emitter and ground.

Before getting involved with the details, it should be stressed that the most powerful outcome of load-line analysis is the *visualization process* of operating conditions that it promotes. It is somewhat limited as an exact analysis and design tool, but one can gain tremendous insight into the effects of various circuit parameters and their variations by visualizing the effects on the load line.

The process of load-line analysis will be developed with the dc circuit of Figure 5-29(a). For this purpose, both base and collector circuits have been constructed as loops with power supplies assumed for each. Two load lines can be constructed corresponding to the respective sides of the circuit, and these will be designated as the *base circuit load line* and the *collector circuit load line*.

Base Circuit Load Line

Assume that the BJT base characteristic providing a graph of I_B versus V_{BE} is given as shown in Figure 5-29(b). (Recall that there is some variation of this characteristic

(a) Circuit

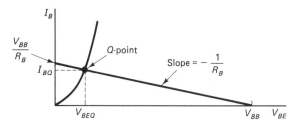

(b) Base circuit load line

(c) Collector circuit load line

Figure 5-29 BJT circuit and input and output load lines.

BJT-Resistor Circuit Applications   Chap. 5

with V_{CE} but we are ignoring that for this purpose.) A KVL equation for the base–emitter loop of Figure 5-29(a) reads

$$-V_{BB} + R_B I_B + V_{BE} = 0 \qquad \text{(5-131)}$$

Solving for I_B, we obtain

$$\boxed{I_B = \left(-\frac{1}{R_B}\right) V_{BE} + \frac{V_{BB}}{R_B}} \qquad \text{(5-132)}$$

This result has been expressed in a different form than earlier in the chapter for a special reason. The variables are assumed to be V_{BE} and I_B, with V_{BE} considered as the independent variable and I_B considered as the dependent variable.

When (5-132) is compared with the standard form of a linear equation $y = mx + b$, the following associations are readily noted:

$$m = \text{slope} = -\frac{1}{R_B} \qquad \text{(5-133)}$$

$$b = \text{vertical intercept} = \frac{V_{BB}}{R_B} \qquad \text{(5-134)}$$

Alternatively, the vertical intercept could be obtained by simply setting $V_{BE} = 0$ in (5-132) and noting that $I_B = V_{BB}/R_B$ for this case.

The horizontal intercept is readily obtained by setting $I_B = 0$ in (5-131) and solving for V_{BE}. The result is

$$\text{horizontal intercept} = V_{BB} \qquad \text{(5-135)}$$

The straight-line equation corresponding to (5-132) is the *base circuit load line*, and it can readily be superimposed on the base–emitter characteristics as shown in Figure 5-29(b). *The intersection of the base–emitter load line with the input characteristic determines the input operating conditions.*

One slight notational problem needs to be addressed before proceeding further. Throughout the chapter, we have used notation such as I_B, V_{BE}, I_C, V_{CE}, and so on, to represent dc transistor currents and voltages for a given set of conditions, and this has not created any problems. For this graphical analysis, however, the characteristic curves are labeled with these same quantities for *any possible operating conditions* for the transistor. Yet for a given set of circuit conditions, there will be only *one possible set of values* of the transistor variables that can exist. We need a separate notational form that distinguishes the *particular* set of values that actually exist from the *possible* range of all values that could exist from the characteristics.

We resolve this potential dilemma by first defining the widely used term *Q-point* as a reference to the particular set of variables that actually satisfy all circuit relationships. (The term *Q-point* is a contraction for *quiescent point*, which is significant in amplifier analysis, as we will see later.) For the moment, we will simply refer to the Q-point as the actual operating point in which all circuit constraints are satisfied. From a notational point of view, the symbol Q will be added to the end of the subscript. Thus the values of I_B, V_{BE}, I_C, and V_{CE} at the Q-point will be denoted as I_{BQ}, V_{BEQ}, I_{CQ}, and V_{CEQ}, respectively.

At various points in the text, the Q subscript will be omitted when it does not contribute to the clarity of any analysis or when it could be misinterpreted. It will be used primarily where the concept of a designated operating point is to be delineated and where ambiguity in notation would result without it.

Returning now to the base–emitter load line of Figure 5-29(b), we see that the intersection of the load line with the base characteristic results in a base-to-emitter voltage V_{BEQ} and a base current I_{BQ}. While no actual values are shown, we would expect V_{BEQ} to be close to 0.7 V. The load line, however, indicates the actual value of V_{BEQ} that would result.

Collector Circuit Load Line

Assume that the BJT collector characteristics are of the form shown in Figure 5-29(c). A KVL equation for the collector–emitter loop of Figure 5-29(a) reads

$$-V_{CC} + R_C I_C + V_{CE} = 0 \qquad (5\text{-}136)$$

Solving for I_C, we have

$$I_C = \left(-\frac{1}{R_C}\right)V_{CE} + \frac{V_{CC}}{R_C} \qquad (5\text{-}137)$$

in which, again, the straight-line equation form $y = mx + b$ has been used. The independent variable in this case is V_{CE}, and the dependent variable is I_C.

From the basic straight-line equation properties, we have

$$\text{slope} = -\frac{1}{R_C} \qquad (5\text{-}138)$$

$$\text{vertical intercept} = \frac{V_{CC}}{R_C} \qquad (5\text{-}139)$$

By setting $I_C = 0$ in (5-136), the horizontal intercept is determined as

$$\text{horizontal intercept} = V_{CC} \qquad (5\text{-}140)$$

The straight-line equation corresponding to (5-137) is the *collector circuit load line*, and it is superimposed on the collector characteristics of Figure 5-29(c). In this case there is a whole family of curves with which it could intersect. *The correct solution is the point on the particular curve $I_B = I_{BQ}$ with which the load line intersects.* Thus the base current has pronounced influence on the collector circuit by determining which characteristic curve represents the applicable circuit condition.

For the characteristics of Figure 5-29(c), the Q-point is observed at the intersection of the curve corresponding to $I_B = I_{BQ}$ and the load line. The values of V_{CE} and I_C at this point are designated as V_{CEQ} and I_{CQ} as shown.

Although load lines can be drawn for both base–emitter and collector–emitter circuits as we have shown, most references to the term *load line* without a modifier refer to the collector–emitter load line. The reason is that the collector–emitter load line usually provides much more information about a variety of circuit conditions. In the remainder of this section, the use of the collector–emitter load line for this purpose is explored.

In the remainder of this book, the term *load line* without a modifier will be interpreted to mean the collector–emitter load line for a BJT or an analogous output circuit load line for an FET (drain–source load line, as we will see later).

To observe the visualization process that the load line provides, a series of circuit element and parameter variations will be considered. No quantitative results will be considered, but representative loads will be shown and discussed qualitatively.

Operating Regions

First, the three BJT operating regions are illustrated by the three load lines of Figure 5-30. For the cutoff region in (a), the Q-point is all the way to the right at the horizontal intercept. This condition corresponds to $I_B = 0$.

(a)

(b)

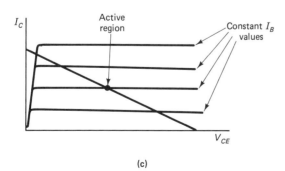

(c)

Figure 5-30 Load lines showing the three BJT operating regions.

The saturation region is illustrated in (b). The Q-point in this case corresponds to the intersection of the load line with the saturation region of the characteristic curves.

The active region is illustrated in (c). In this case, the Q-point is established at some intersection between cutoff and saturation. In many applications, an intersection at or near the midpoint is desired.

Variation of Base Current

Consider the load line of Figure 5-31. The direction of the arrows indicate the manner in which base current variations affect the Q-point if all other circuit conditions are fixed. From any point on the load line, a decrease in base current causes the Q-point to move toward the cutoff point. This results in an increase in collector–emitter voltage and a decrease in collector current, as is readily noted from the projections of the Q-point to the two axes. Conversely, an increase in base current causes the Q-point to move toward saturation. This results in a decrease in collector–emitter voltage and an increase in base current.

We will see in Chapter 10 that the process of movement along the load line is the basis of amplifier actions in a transistor. Such movement would be caused by the signal to be amplified.

Variation of Collector Bias Voltage

When the collector power supply bias voltage is changed and all other circuit parameters are fixed, the result is a completely new load line. Assume first a collector power supply voltage V_{CC1}, and then assume that this voltage is changed to a larger value V_{CC2}. Two possible load lines are shown in Figure 5-32. As long as R_C remains constant, the two load lines have the same slope. However, the respective horizontal intercepts (V_{CC1} and V_{CC2}) and the vertical intercepts (V_{CC1}/R_C and V_{CC2}/R_C) are different.

If the base current were fixed, the two Q-points would be determined from the intersections of the curve corresponding to $I_B = I_{BQ}$ with the two load lines. For the hypothetical curves of Figure 5-32, note that the first Q-point (denoted as Q_1) is almost in the saturation region, while the second Q-point (denoted as Q_2) is much

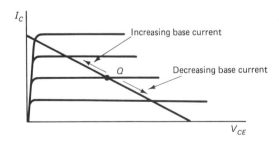

Figure 5-31 Load line showing variation of base current.

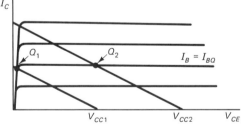

Figure 5-32 Load lines showing variation of collector power supply voltage.

closer to the middle of the active region. This illustrates the type of visualization that is very worthwhile with the load line.

Variation of Collector Resistance

When the collector resistance R_C is varied and all other circuit parameters are fixed, the result is a completely new load line. In this case, however, the slope varies.

Assume first a resistance R_{C1}, and then assume that this resistance is changed to a larger value R_{C2}. Two possible load lines are shown in Figure 5-33. As long as V_{CC} remains constant, the horizontal intercept remains constant. However, both the respective slopes $(-1/R_{C1}$ and $-1/R_{C2})$ and the vertical intercepts $(V_{CC}/R_1$ and $V_{CC}/R_2)$ are different.

If the base current were fixed, the two Q-points would be determined from the intersections of the curve corresponding to $I_B = I_{BQ}$ with the load lines. For the hypothetical curves of Figure 5-33, note that the first Q-point (Q_1) is close to the middle of the active region, while the second Q-point is at saturation.

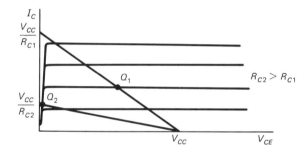

Figure 5-33 Load lines showing variation of collector resistance.

Example 5-11

A certain idealized *npn* transistor has the collector characteristics shown in Figure 5-34. The transistor is connected in the circuit of Figure 5-29(a) with

Figure 5-34 Collector characteristics for Example 5-11.

Solution

(a) The load line is most easily constructed by first determining the two intercepts. The horizontal intercept is simply $V_{CC} = 20$ V. The vertical intercept is $V_{CC}/R_C = 20/2000 = 0.01$ A $= 10$ mA. A straight line is drawn between these points as shown in Figure 5-34.

Before determining the Q-point, the value of I_{BQ} must be calculated. This current is

$$I_{BQ} = \frac{V_{BB} - V_{BE}}{R_B} = \frac{6 - 0.7}{132.5 \times 10^3} = 40 \ \mu\text{A} \qquad \textbf{(5-141)}$$

The Q-point is then located at the intersection of the load line with the curve corresponding to $I_B = 40 \ \mu$A as shown in Figure 5-34. The values of collector–emitter voltage and collector current at this point are $V_{CEQ} = 10$ V and $I_{CQ} = 5$ mA.

(b) The saturation point is located on the left side of the load line where the characteristic curves intersect the load line. The value of collector current at this point is $I_{C(\text{sat})} = 10$ mA. The value of $I_{B(\text{sat})}$ is the *smallest* value of I_B that results in saturation and this value is $I_{B(\text{sat})} = 80 \ \mu$A. Note that there is a small value of collector–emitter voltage at this point [i.e., $V_{CE(\text{sat})}$], but it is very small for the idealized characteristics.

5-10 TTL INTEGRATED LOGIC CIRCUITS

In this section an introduction to logic operations performed with BJTs will be provided. The purpose is to demonstrate a very important application of BJTs as well as to relate the study of electronic devices in this book to the study of digital circuits in texts emphasizing digital electronics.

One of the most widely employed concepts in digital electronics is the so-called *transistor-transistor logic* (TTL) circuit family. A wide variety of integrated-circuit chips are available in TTL form, and a large percentage of digital application circuits employ TTL devices.

You will learn about designing, analyzing, and troubleshooting TTL circuits from a separate course and/or book devoted to the field of digital electronics. Our purpose here is to show how the basic circuit functions of such chips are directly related to the material covered in the past two chapters. Very few users of integrated-circuit modules understand all the details of the chip operation. However, it is helpful to visualize certain important input/output functions so that appropriate drive, load, and interfacing conditions can be met.

NAND Circuit

The TTL operation will be illustrated with a discussion of a two-input NAND gate, whose typical integrated circuit form is shown in Figure 5-35. The NAND designation is a contraction of the terms *NOT* and *AND*. A *NOT* operation is an inversion of the logical state (i.e., a logical 1 is changed to a logical 0, and vice versa).

The two-input circuit of Figure 5-35 contains four transistors and three diodes. The two input diodes on the left are limiting diodes used to prevent excessive voltage on the input transistor, and they do not contribute directly to circuit operation. The input transistor may seem strange in that it has two emitter terminals, but they may be considered essentially as parallel paths.

Above various points on the circuit diagrams are two numbers, one of which is in parentheses ( ), and the other of which is in brackets [ ]. The first number represents a *typical* value of voltage in volts when *either* or *both* input voltages are at a logical 0 level. The second number represents a *typical* value of voltage in volts when *both* input voltages are at a logical 1 level.

Logic 0 at Either Input

First, assume that either or both inputs have a logical 0 value. TTL specifications at the input stipulate that this is any voltage level between zero and 0.8 V. A typical value of 0.3 V is assumed in the analysis. For this condition, the base–emitter junction of T1 will be forward biased, and current will flow from the 5-V source, through the 4-kΩ resistor out through emitter (or emitters) of T1, and through the input (not shown) to ground. The input is said to be *sinking* the current, and for that

Values in parentheses and brackets are typical voltage levels
(in volts) for different input conditions (see text)

Figure 5-35   Form of a transistor-transistor logic (TTL) NAND gate.

reason, the TTL is referred to as *current sinking* logic. The base of T1 is about 0.7 V higher than the emitter or about 1 V to ground, as shown.

To forward bias the base–collector junction of T1, a base voltage of about 2.1 V would be required. This particular voltage represents the sum of the T1 0.7-V base–collector junction, the 0.7-V base–emitter drop of T2 and the 0.7-V base–emitter drop of T4. Since the base voltage is only 1 V, the base–collector junction of T1 is reverse biased, and no base drive is applied to either T2 or T4. Consequently, T2 and T4 are in the cutoff state.

Since no collector current flows in T2, the voltage at the bottom of the 1.6-kΩ resistor is high enough to forward bias T3, and this stage acts essentially as an emitter follower. Because of the presence of both base–emitter drop and a diode drop, the typical output shown is about 1.4 V below the base of T3 or about 3 V. This voltage level is within the range of a logical 1.

Logic 1 at Both Inputs

Next assume that *both* inputs are at a logical 1 level. TTL specifications at the input stipulate that this is any voltage level between 2 and 5 V, as indicated. For this condition the base–emitter junction of T1 is reverse biased. The voltage at the base of T2 is then sufficient to forward bias the combination of the base–collector junction of T1 and the base–emitter junctions of T2 and T4. The typical value of 2.1 V at the base of T1 is the sum of the base–collector forward drop of T1 plus the two base–emitter drops of T2 and T3 (i.e., 3×0.7 V = 2.1 V).

Since T2 is now forward biased, its collector voltage (1 V) is too low to forward bias the series combination of the T3 base–emitter voltage plus the diode drop. Thus T3 is uncoupled from the output. However, T4 is driven into saturation by the emitter of T2, and the output voltage of T4 is the collector–emitter saturation voltage. A typical value of 0.3 V is shown. TTL specifications at the output stipulate a maximum level of 0.4 V for a logical 0.

A summary of TTL specifications is given in Table 5-1. A truth table depicting logic conditions for the NAND gate is given in Table 5-2.

TABLE 5-1 TTL LOGIC-LEVEL SPECIFICATIONS

Logical 0 level at input:	0 to 0.8 V
Logical 1 level at input:	2 to 5 V
Logical 0 level at output:	0 to 0.4 V
Logical 1 level at output:	2.4 to 5 V

TABLE 5-2 TRUTH TABLE FOR NAND GATE

Input 1	Input 2	Output
0	0	1
0	1	1
1	0	1
1	1	0

5-11 DARLINGTON COMPOUND CONNECTION

A circuit that may be utilized to achieve a very high value of β is the *Darlington compound connection*. This circuit is used both in discrete form and in integrated-circuit packages.

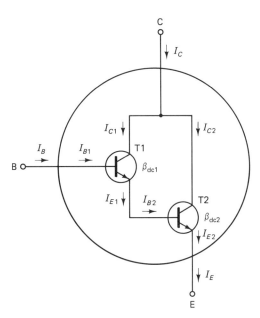

Figure 5-36 Darlington compound connection with different variables labeled.

The basic form of the Darlington compound pair using *npn* transistors is shown in Figure 5-36. As far as the three *external* connections shown outside the circle are concerned, the pair acts like a single transistor, so those variables are labeled without subscripts. However, the individual transistor variables are labeled with subscripts 1 and 2 as noted.

The collector current I_{C1} is related to the base current I_{B1} by

$$I_{C1} = \beta_{dc1} I_{B1} \tag{5-142}$$

The emitter current I_{E1} is

$$I_{E1} = I_{B1} + I_{C1} = I_{B1} + \beta_{dc1} I_{B1} = (1 + \beta_{dc1}) I_{B1} \tag{5-143}$$

This current acts as the base current for the second transistor (i.e., $I_{B2} = I_{E1}$).

The collector current I_{C2} is

$$I_{C2} = \beta_{dc2} I_{B2} = \beta_{dc2} I_{E1} \tag{5-144}$$

Substitution of (5-143) in (5-144) yields

$$I_{C2} = \beta_{dc2}(1 + \beta_{dc1}) I_{B1} \tag{5-145}$$

The total collector current I_C is

$$I_C = I_{C1} + I_{C2} \tag{5-146}$$

Substitution of (5-142) and (5-145) in (5-146) yields

$$I_C = \beta_{dc1} I_{B1} + \beta_{dc2}(1 + \beta_{dc1}) I_{B1} \tag{5-147}$$

$$= (\beta_{dc1} + \beta_{dc2} + \beta_{dc1}\beta_{dc2}) I_B \tag{5-148}$$

where the substitution $I_{B1} = I_B$ was made.

The net value of β, denoted simply as β_{dc}, is

$$\beta_{dc} = \frac{I_C}{I_B} = \beta_{dc1} + \beta_{dc2} + \beta_{dc1}\beta_{dc2} \qquad \text{(5-149)}$$

Under relatively large individual values of β, this result is often estimated simply as

$$\beta_{dc} \approx \beta_{dc1}\beta_{dc2} \qquad \text{(5-150)}$$

Example 5-12

A Darlington compound connection of the form shown in Figure 5-36 has $\beta_{dc1} = 100$ and $\beta_{dc2} = 80$. **(a)** Determine the exact value of β_{dc} for the compound pair. **(b)** Determine an estimate of β_{dc} based on (5-150).

Solution

(a) The exact value of β_{dc} is

$$\beta_{dc} = \beta_{dc1} + \beta_{dc2} + \beta_{dc1}\beta_{dc2}$$
$$= 100 + 80 + 100 \times 80 = 8180 \qquad \text{(5-151)}$$

(b) The estimated value of β_{dc} based on (5-150) is

$$\beta_{dc} \approx \beta_{dc1}\beta_{dc2} = 100 \times 80 = 8000 \qquad \text{(5-152)}$$

This value differs from the exact value by about 2.2%. This estimate is often adequate for quick checks considering the wide variation and uncertainty in parameter values.

Note the very large value of β_{dc} obtained through this process. Darlington compound pairs, with appropriately matched transistors, are available as stock items.

DRILL PROBLEMS

Note: Unless indicated otherwise, the following assumptions can be made in all problems:

1. $V_{BE} = 0.7$ V for *npn* transistors and $V_{EB} = 0.7$ V for *pnp* transistors with forward-biased base–emitter junctions.

2. $V_{CE(sat)} = 0$ for *npn* transistors and $V_{EC(sat)} = 0$ for *pnp* transistors.

5-1. Consider the circuit of Figure P5-1. If $\beta_{dc} = 80$, determine $I_{C(sat)}$, $I_{B(sat)}$, and the value of v_i that results in saturation.

Figure P5-1

216

5-2. Consider the circuit of Figure P5-2. If $\beta_{dc} = 125$, determine $I_{C(sat)}$, $I_{B(sat)}$, and the value of v_i that results in saturation.

Figure P5-2

5-3. Consider the circuit of Problem 5-1 (Figure P5-1) again. With the same assumptions as in Problem 5-1, determine I_B, I_C, I_E, and V_{CE} for each of the following values of v_i: **(a)** 0; **(b)** 3 V; **(c)** 6 V; **(d)** 9 V.

5-4. Consider the circuit of Problem 5-2 (Figure P5-2) again. With the same assumptions as in Problem 5-2, determine I_B, I_C, I_E, and V_{EC} for each of the following values of v_i: **(a)** 0; **(b)** -2 V; **(c)** -4 V; **(d)** -6 V.

5-5. For the circuit of Figure P5-5, determine the values of I_C and V_{CE} for each of the following values of β_{dc}: **(a)** 50; **(b)** 100; **(c)** 150.

Figure P5-5

5-6. For the circuit of Figure P5-6, determine the values of I_C and V_{EC} for each of the following values of β_{dc}: **(a)** 50; **(b)** 100; **(c)** 150.

Figure P5-6

217

5-7. For the circuit of Figure P5-7, determine the values of I_C and V_{CE} for each of the following values of R_B: **(a)** 200 kΩ; **(b)** 390 kΩ; **(c)** 560 kΩ.

Figure P5-7

5-8. For the circuit of Figure P5-8, determine the values of I_C and V_{EC} for each of the following values of R_B: **(a)** 100 kΩ; **(b)** 200 kΩ; **(c)** 270 kΩ.

Figure P5-8

5-9. For the circuit of Figure P5-9, determine the values of I_C and V_{CE} for each of the following values of R_C: **(a)** 1 kΩ; **(b)** 2 kΩ; **(c)** 3 kΩ.

Figure P5-9

5-10. For the circuit of Figure P5-10, determine the values of I_C and V_{EC} for each of the following values of R_C: **(a)** 1 kΩ; **(b)** 2 kΩ; **(c)** 3 kΩ.

−15 V

I_C

180 kΩ

R_C

V_{EC}

$\beta_{dc} = 80$

Figure P5-10

5-11. The circuit of Figure P5-11 is used to establish a constant current through a resistive load R_L. Determine **(a)** the current I_C and **(b)** the range of R_L over which the circuit will function properly.

+15 V

I_C

R_L Load

6 V

2650 Ω

Figure P5-11

5-12. The circuit of Figure P5-12 is used to establish a constant current through a resistive load R_L. Determine **(a)** the current I_C and **(b)** the range of R_L over which the current will function properly.

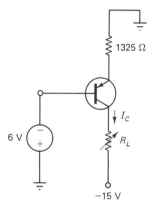

1325 Ω

I_C

6 V

R_L

−15 V

Figure P5-12

5-13. The emitter–follower circuit of Figure P5-13 is used to isolate a source v_i from a 200-Ω load. **(a)** Determine an expression for v_o in terms of v_i. **(b)** Determine the range of input voltage for which the circuit functions. **(c)** For $v_i = 20$ V and $\beta_{dc} = 60$, determine the value of base current i_B.

+24 V

v_i

200 Ω v_o

Figure P5-13

5-14. The emitter-follower circuit of Figure P5-14 is used to isolate a source v_i from a 120-Ω load. **(a)** Determine an expression for v_o in terms of v_i. **(b)** Determine the range of input voltage for which the circuit functions. **(c)** For $v_i = -12$ V and $\beta_{dc} = 40$, determine the value of the base current i_B.

−15 V

v_i

120 Ω v_o

Figure P5-14

5-15. In the circuit of Figure P5-15, $V_{CC} = 24$ V, $R_1 = 20$ kΩ, $R_2 = 8.2$ kΩ, $R_C = 4.3$ kΩ, and $R_E = 3.3$ kΩ. **(a)** Determine I_C, I_E, V_{RC}, V_{RE}, and V_{CE}.

+V_{CC}

R_1

R_2

I_C

R_C V_{RC}

V_{CE}

I_E

R_E V_{RE}

Figure P5-15

(b) If the minimum value of β_{dc} for the transistor is 50, determine the maximum possible value of I_B. **(c)** Check the inequality (5-89) as a means of assessing the relative stability of the circuit.

5-16. Repeat the analysis of Problem 5-15 if $V_{CC} = 30$ V, $R_1 = 27$ kΩ, $R_2 = 10$ kΩ, $R_C = 2$ kΩ, and $R_E = 1.5$ kΩ.

5-17. For the circuit of Figure P5-17, determine I_C, I_E, V_{RC}, V_{RE}, and V_{EC}.

Figure P5-17

5-18. For the circuit of Figure P5-18, determine I_C, I_E, V_{RC}, V_{RE}, and V_{EC}.

Figure P5-18

5-19. A certain idealized *npn* transistor has the collector characteristics shown in Figure P5-19. The transistor is connected in the circuit of Figure 5-29(a)

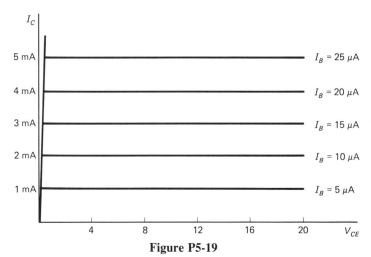

Figure P5-19

with the following parameter values: $V_{BB} = 5$ V, $V_{CC} = 20$ V, $R_B = 430$ kΩ, $R_C = 4$ kΩ. **(a)** Draw the dc load line and determine V_{CEQ} and I_{CQ}. **(b)** From the load line determine the values of $I_{C(\text{sat})}$ and $I_{B(\text{sat})}$.

5-20. Repeat the analysis of Problem 5-19 if the parameters are changed to the following values: $V_{BB} = 6$ V, $V_{CC} = 16$ V, $R_B = 1.06$ MΩ, $R_C = 8$ kΩ.

5-21. The circuit of Figure P5-21 is referred to as a *complementary common-emitter switch*. Assume that $V_{CE(\text{sat})} = 0.1$ V for the *npn* transistor and $V_{EC(\text{sat})} = 0.1$ V for the *pnp* transistor. Assume that $\beta_{\text{dc(min)}} = 25$. Determine v_o for **(a)** $v_i = 0$ V and **(b)** $v_i = 5$ V.

Figure P5-21

5-22. The circuit of Figure P5-22 can be used as an emitter-follower switch. Determine v_o for **(a)** $v_i = -0.7$ V and **(b)** $v_i = 5.7$ V.

Figure P5-22

5-23. The form of a certain *resistor-transistor logic (RTL)* gate is shown in Figure P5-23. For your analysis, assume idealized input values of 0 V for logical 0, 5 V for logical 1, and $V_{CE(sat)} = 0$. **(a)** Assume that both inputs are at a logical 0 level. Determine v_o. **(b)** Assume that either input is at a logical 1 level. Determine v_o. **(c)** Construct a truth table for the four possible inputs. Identify the type of circuit.

Figure P5-23 Form of resistor-transistor logic (RTL) NOR gate.

5-24. The form of a certain *diode transistor logic (DTL)* gate is shown in Figure P5-24. For your analysis, assume *idealized* input values of 0 V for logical 0, 5 V for logical 1, and $V_{CE(sat)} = 0$. **(a)** Assume that either input is at a logical 0 level. Determine v_o. **(b)** Assume that both inputs are at a logical 1 level. Determine v_o. **(c)** What type of function is performed by this gate? (You may assume that the output transistor saturates when its base–emitter junction is forward biased.)

Figure P5-24 Form of a diode-transistor logic (DTL) NAND gate.

5-25. One relatively stable circuit sometimes used to bias a BJT is referred to as *collector-feedback bias* and is shown in Figure P5-25. Show that the collector current can be expressed as

$$I_C = \frac{V_{CC} - 0.7}{R_C + (R_B + R_C)/\beta_{dc}}$$

Figure P5-25

5-26. In this exercise, an *exact* expression for I_C for the stable bias circuit of Figure 5-23(a) will be derived with β_{dc} retained in the equation. **(a)** In the equivalent circuit of Figure 5-23(c), write a loop equation for the base–emitter loop recognizing that the current through R_E is $I_E = I_B + I_C$. **(b)** By substituting $I_B = I_C/\beta_{dc}$, show that I_C can be expressed as

$$I_C = \frac{\beta_{dc}(V_{TH} - V_{BE})}{R_{TH} + (\beta_{dc} + 1)R_E}$$

DESIGN PROBLEMS

5-27. The circuit of Figure P5-27 is to be used to switch the 24-V source across the load. Assume that the control voltage has two levels: (1) $v_i = 0$ and (2) $v_i = 12$ V. The load is resistive and has a value of 400 Ω. Assume that design for production is being performed, and the β_{dc} variation for the transistor type is indicated. Using the criterion of twice the base current for worst-case conditions, complete the design by determining the value of R_B.

Figure P5-27

5-28. The circuit of Figure P5-28 is to be used to switch the -24-V source across the load. Assume that the control voltage has two levels: (1) $v_i = 0$ and (2) $v_i = -6$ V. The load is resistive and has a value of 600 Ω. Assume that design for production is being performed, and the β_{dc} variation of the transistor type is indicated. Using the criterion of twice the base current for worst-case conditions, complete the design by determining the value of R_B.

Figure P5-28

5-29. Design a current source using an *npn* transistor to provide a current of 1 mA through a variable load. The available power supply voltage is $+15$ V, and this voltage will be used for biasing both the collector and the base circuits.

5-30. Design a current source using a *pnp* transistor to provide a current of 3 mA through a variable load. The available power supply is -15 V, and this voltage will be used for biasing both the collector and the base circuits.

5-31. Design a stable bias circuit for an *npn* transistor based on the following specifications and conditions:

> Power supply voltage = 24 V
> Desired collector current = 1 mA
> Desired $V_{RC} = V_{CE} = 10$ V
> Desired $V_{RE} = 4$ V
> $\beta_{dc(min)} = 50$

Select $K = 20$.

5-32. Design a stable bias circuit for an *npn* transistor based on the following specifications and conditions:

> Power supply voltage = 15 V
> Desired collector current = 2 mA
> Desired $V_{RC} = V_{CE} = V_{RE} = 5$ V
> $\beta_{dc(min)} = 40$

Select $K = 20$.

5-33. Repeat the design of Problem 5-31 with $K = 80$.

5-34. Repeat the design of Problem 5-32 with $K = 80$.

5-35. You are troubleshooting a circuit of the form shown in Figure P5-15, which should be operating in the active region, and you measure the following voltages with respect to ground: $V_{CC} = 15$ V, $V_C = 7.7$ V, $V_E (= V_{R_E}) = 7.7$ V. Which of the following *could be* the source of the problem? **(a)** R_1 open; **(b)** R_2 open; **(c)** R_C open; **(d)** R_E open.

5-36. You are troubleshooting a circuit of the form shown in Figure P5-15, which should be operating in the active region, and you measure the following voltages with respect to ground: $V_{CC} = 15$ V, $V_C = 15$ V, $V_E (= V_{R_E}) = 0$. All but one of the following *could be* the source of the problem. Identify the one condition that could *not* cause the problem. **(a)** R_1 open; **(b)** R_2 shorted; **(c)** R_C open; **(d)** R_E open.

5-37. If the transistor in Figure P5-1 is initially biased in the active region, indicate whether V_{CE} would *increase*, *decrease*, or *remain nearly the same* for each of the following troubles. **(a)** The 91-kΩ resistor is open; **(b)** the 91-kΩ resistor is shorted; **(c)** the 2-kΩ resistor is open; **(d)** the 2-kΩ resistor is shorted; **(e)** the transistor develops a short between collector and emitter; **(f)** temperature changes cause β_{dc} to increase.

5-38. For the current source of Figure P5-11, indicate whether I_C would *increase*, *decrease*, or *remain nearly the same* for each of the following changes. **(a)** The 6-V source decreases to 4 V; **(b)** the 2650-Ω resistance changes in resistance to 3.5 kΩ; **(c)** the 15-V power supply increases to 17 V; **(d)** the load resistance R_C increases by 20%, but the transistor remains in the active region; **(e)** the value of β_{dc} increases by 20%.

COMPUTER PROBLEMS

5-39. Write a program to determine all voltages and currents in a circuit of the *form* of Figure 5-9, but with all circuit parameters entered as input data. The program should be capable of testing to see if saturation occurs, and the output should be adjusted accordingly.

Input data: V_{CC}, R_C, R_B, v_i, β_{dc}
Output data: I_B, I_C, V_{CE}

5-40. Write a program to analyze the stable bias circuit of Figure 5-23(a) and to determine all voltages and currents. In this problem the computation of I_C will be based on the widely employed approximation of (5-90).

Input data: V_{CC}, R_C, R_E, R_1, R_2
Output data: I_C, V_{R_C}, V_{R_E}, V_{CE}

5-41. Write a program based on the same objectives as in Problem 5-40, but use the exact expression for I_C given in Problem 5-26.

> Input data: All data of Problem 5-40 and β_{dc}
> Output data: Same as Problem 5-40

5-42. Write a program to design a stable bias circuit of the form of Figure 5-23(a).

> Input data: I_C, V_{CE}, V_{R_C}, V_{R_E}, $\beta_{dc(min)}$, K
> Output data: R_C, R_E, R_1, R_2

CHAPTER 6
Field-Effect Transistors

OBJECTIVES

After completing this chapter, the reader should be able to:

1. Draw a simplified layout representing the construction of a field-effect transistor (FET).
2. Identify the major classifications of field-effect transistors (JFET versus IGFET, depletion mode versus enhancement mode, etc.)
3. Identify the two types of FETs according to the polarities of the semiconductor materials.
4. Draw the schematic symbols of the different types of FETs and identify the three terminals.
5. State the three operating regions of an FET and identify conditions for each.
6. Draw equivalent-circuit models for each of the three FET operating regions.
7. Define the following terms for a depletion-mode FET: gate–source cutoff voltage, pinchoff voltage, zero-bias drain current, transition voltage.
8. Define the gate–source threshold voltage for an enhancement-mode IGFET.
9. Sketch the form of the drain characteristics of a typical FET, and label all variables involved.
10. Sketch the form of a transfer characteristic, and label the variables involved.
11. State the relationships between drain current and gate–source voltage for both depletion-mode FETs and enhancement-mode FETs in the beyond-pinchoff region.
12. State the relationships for conductance and resistance of an FET in the ohmic region.
13. Discuss the similarities and differences between the operating characteristics of BJTs and FETs.
14. Inspect the specifications sheets of typical FETs to determine pertinent data.

6-1 CLASSIFICATIONS

The second major class of transistor is the *field-effect transistor* (*FET*). Like the BJT, the FET is composed of semiconductor materials such as silicon. Many modern integrated circuits utilize FET technology, and discrete FETs are available as individual components.

The field-effect transistor is often referred to as a *unipolar* transistor because conduction occurs as a result of a single type of charge carrier (either holes *or* electrons) in a given unit. This is in contrast to the *bipolar* junction transistor, where conduction occurs as a result of both types of charge carriers in a given unit.

Although there are certain similarities between the BJT and the FET that will become evident as we study FET operation, there are significant differences. In its basic form, the BJT is considered as a *current-controlled* device, while the FET is a *voltage-controlled* device. The base control terminal of a BJT draws a moderate current from the signal source connected to it, so the effective input resistance is relatively low. In contrast, the corresponding control terminal of an FET (the *gate*, as we will see shortly) requires virtually no current, so the effective input resistance is very high. For the insulated gate type of FET, the input resistance may be as high as hundreds of megohms.

The classification scheme for FETs is quite complex and requires some study for mastery of the pertinent terminology. A brief overview of this structure will be made at the beginning, although the significance of the various terms will not be clear until the different types of units are studied in the chapter.

A "family tree" of the FET classification structure is shown in Figure 6-1. One major class of FET is the *junction field-effect transistor*, referred to as the *JFET*. All JFETs operate in what is called the *depletion mode*, in which conduction in the semiconductor is controlled by a *depletion layer*.

A second major class of FET is the *insulated gate field-effect transistor*, referred to as the *IGFET*. The IGFET is further classified as either a *depletion-mode* device or an *enhancement-mode* device. A depletion-mode IGFET functions in the

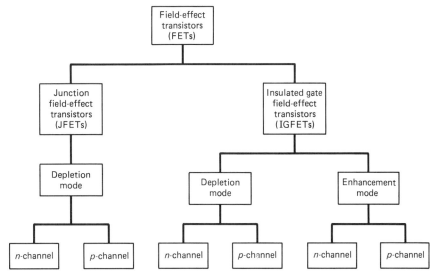

Figure 6-1 Family tree showing different classifications of field-effect transistors.

same manner as a JFET. An enhancement-mode IGFET, however, requires a process of channel enhancement in order to allow current flow.

Depletion-mode devices (either JFETs or IGFETs) are referred to as "normally on" devices since they are capable of some conductivity in their basic forms with no control voltage. In contrast, enhancement-mode devices are referred to as "normally off" devices since they will not conduct in their basic forms without the assistance of an "enhancement voltage."

All the different FETs are available as either *n-channel* types or *p-channel* types. This last classification is similar to *npn* and *pnp* BJTs and relates to the polarities of the semiconductor material used in the device construction. To simplify the categorization somewhat, the developments in the next several sections assume only *n*-channel devices. Once the operation and general properties of these *n*-channel units are thoroughly understood, it is a straightforward process to infer the properties of the corresponding *p*-channel units simply by reversing all current directions and voltage polarities.

6-2 JUNCTION FIELD-EFFECT TRANSISTORS

The coverage of field-effect transistors will begin with a discussion of the *junction field-effect transistor*, hereafter referred to as the JFET. As indicated in the preceding section, we focus on the *n*-channel JFET.

N-Channel JFET

A simplified construction of an *n*-channel JFET is indicated by the sketch of Figure 6-2(a). A lightly doped bar of *n*-type semiconductor material (usually silicon) forms the *channel*, and contacts are placed at both ends. These ends are designated as the *source* and the *drain*, respectively. Two segments of heavily doped *p*-type semicon-

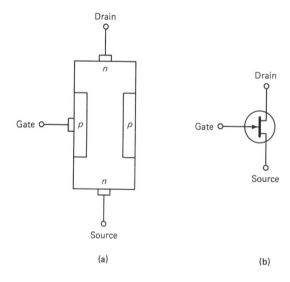

(a)

(b)

Figure 6-2 Basic construction of an *n*-channel JFET and the schematic symbol.

ductor are embedded in the sides. The two *p*-type segments are internally connected and constitute the *gate*. Current can flow along the channel, and it is heavily influenced by the voltage at the gate through the creation of a *depletion layer*.

The interface between the *p*-type gate and the *n*-type channel constitutes a *pn* junction diode. In the normal operation of the *n*-channel JFET, however, the gate is always biased *negatively* (or possibly at zero) with respect to the source, so this diode remains reverse biased. Consequently, the only current flow in the gate will be the reverse current of the gate–channel diode.

The schematic symbol of the *n*-channel JFET is shown in Figure 6-2(b). The direction of the arrow can be remembered from the concept of the *pn* junction between the *p*-type gate and the *n*-type channel.

6-3 FET OPERATING REGIONS

There are three regions of operation for a JFET. These are referred to as the (1) *cutoff region*, (2) *ohmic region*, and (3) *beyond pinchoff region*. There are both similarities and differences between these three regions and the three regions for a BJT. A comparison will be made after we have studied in detail the various FET properties.

Test Circuit

The circuit properties of the JFET will be introduced by showing the results of some possible experiments performed on a hypothetical *n*-channel JFET. The characteristics of the hypothetical JFET have been idealized somewhat for illustration, but they are representative of real-life devices.

Consider the test circuit of Figure 6-3. The circuit is connected in the *common-source configuration*, for which most data and operating parameters are determined. Adjustable dc voltage supplies are connected to both gate and drain terminals with respect to the source terminal. These bias supplies will be referred to as the *gate bias supply* and the *drain bias supply*, respectively. Note that the *gate* voltage is

Figure 6-3   Test circuit for hypothetical *n*-channel JFET.

biased *negatively* with respect to the source, while the *drain* voltage is biased *positively* with respect to the source.

Ohmic Region

The first experiment to be discussed involves varying the drain–source voltage V_{DS} over a very small positive range from zero up to, perhaps, 0.1 V or so. Refer to Figure 6-4 in the discussion that follows.

First, assume that the gate–source voltage is set to $V_{GS} = 0$. This could be achieved by setting the gate bias supply to zero or simply removing that bias supply and connecting the gate directly to the source. Assume now that the drain bias supply is adjusted in very small positive steps from zero and that I_D is measured as a function of V_{DS}. The current I_D increases approximately linearly with V_{DS}, as shown by the top curve in Figure 6-4. The resulting straight line, which starts at the origin, has the same form as for a passive resistor. The slope has the units of amperes/volt or siemens, so the slope represents the conductance. Thus *the JFET is acting like a resistor in this region*; hence the term *ohmic region* is quite appropriate.

Next, assume that we set $V_{GS} = -1$ V. (Note that the positive reference for V_{GS} is at the gate terminal, but the negative value takes care of the reverse voltage.) The curve of I_D versus V_{DS} is still a straight line, but it assumes a smaller slope as shown by the line labeled $V_{GS} = -1$ V in Figure 6-4.

In Example 6-1 we will calculate the conductance and resistance for each line of our hypothetical JFET. For the moment, however, we will simply observe the trends.

The experiment is continued for $V_{GS} = -2$ V and $V_{GS} = -3$ V, and the slopes of the two lines continue to decrease as shown in Figure 6-4. In the ohmic region, the JFET assumes the role of a *voltage-controlled resistor*. For a fixed value of V_{GS},

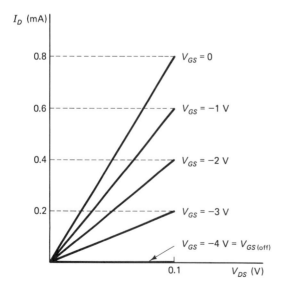

Figure 6-4   Drain characteristics of hypothetical *n*-channel JFET in linear portion of ohmic region.

the resistance is nearly constant for *small positive values* of V_{DS}. However, by varying V_{GS}, the resistance can be made to change.

The ohmic region actually extends farther to the right (more positive) than the range of V_{DS}, as we will see shortly. It also extends somewhat into the negative range of V_{DS}. However, the curves become nonlinear for increasing positive values or for negative values of V_{DS}, and the effective resistance is no longer a constant.

In the basic form shown, the assumption of a nearly constant resistance is generally valid only for small positive values of V_{DS}. By special feedback circuits, it is possible to extend the constant resistance both to negative values and more positive values of V_{DS}.

It might seem logical to introduce a mathematical model and an equivalent-circuit model for the ohmic region at the point. However, some of the terminology required in these models is best defined by first discussing the cutoff region and the beyond-pinchoff region. Consequently, models for the ohmic region will be delayed until Section 6-5. For the moment, remember that the JFET acts like a resistance in the ohmic region.

Cutoff Region

At some negative value of V_{GS}, the effective conduction width of the channel will be reduced completely to zero by depletion, and no further drain current can flow. The value of gate–source voltage at which drain current stops flowing is called the *gate–source cutoff voltage*, and it is denoted by $V_{GS(off)}$. For the characteristics of Figure 6-4, the value of the gate–source cutoff voltage is observed to be $V_{GS(off)} = -4$ V. This condition also corresponds to the beginning of the *cutoff region*. Further increases in the negative voltage applied to the gate have no effect since the cutoff region has been reached, and no current flows in the channel. (This statement assumes that the magnitude of the voltage is not large enough to cause reverse junction breakdown.)

An ideal equivalent circuit for a JFET in the cutoff region is shown in Figure 6-5. This equivalent circuit has the same form as the model for a BJT in the cutoff region.

Figure 6-5   Ideal model of FET in *cutoff* region.

Example 6-1

Determine the conductance and resistance values associated with each line for the hypothetical JFET characteristics of Figure 6-4.

Solutions

Let g_{DS} represent the conductance in siemens from drain to source and let $r_{DS} = 1/g_{DS}$ represent the resistance in ohms. The slope g_{DS} in each case is determined by dividing the change in current along the line by the corresponding change in drain–source voltage. For the $V_{GS} = 0$ line, we have

$$g_{DS} = \frac{0.8 \times 10^{-3} \text{ A}}{0.1 \text{ V}} = 8 \times 10^{-3} \text{ S} = 8 \text{ mS} \qquad \text{(6-1)}$$

The resistance r_{DS} is

$$r_{DS} = \frac{1}{8 \times 10^{-3}} = 125 \text{ } \Omega \qquad \text{(6-2)}$$

For the $V_{GS} = -1$ V line, we have

$$g_{DS} = \frac{0.6 \times 10^{-3} \text{ A}}{0.1 \text{ V}} = 6 \times 10^{-3} \text{ S} = 6 \text{ mS} \qquad \text{(6-3)}$$

and

$$r_{DS} = \frac{1}{6 \times 10^{-3}} = 166.7 \text{ } \Omega \qquad \text{(6-4)}$$

The reader is invited to repeat this process for the other lines, and a short table summarizing these values follows:

V_{GS} (V)	g_{DS} (mS)	r_{DS} (Ω)
0	8	125
-1	6	166.7
-2	4	250
-3	2	500
-4	0	∞

Observe that as V_{GS} becomes more negative, the conductance decreases and the resistance increases. For $V_{GS} = V_{GS(\text{off})}$ (-4 V in this case), the conductance becomes zero and the resistance becomes infinite. This represents an open circuit between drain and source since the cutoff region has been reached.

Beyond-Pinchoff Region

The next experiment to be discussed utilizes the same circuit as before (Figure 6-3) and follows the same procedure as the preceding one except for one major difference. The drain–source voltage will now be increased to a much larger value for each curve than for the ohmic region experiment. The complete set of curves so obtained are called the *output characteristics* or the *drain characteristics*. The latter term is used most often in this book.

The form of the drain characteristics of the hypothetical transistor are shown in Figure 6-6. The scale is quite different here than for Figure 6-4. The ohmic region of Figure 6-4 corresponds to a tiny portion of the curves near the origin in Figure 6-6. Observe that the overall drain characteristics do not increase linearly as they did for the small value of V_{DS}. Instead, their slopes decrease and eventually approach zero.

After the curves flatten out, there is very little additional change in I_D for increasing V_{DS}. This region is referred to as the *beyond-pinchoff region*. It is also called the *saturation region*, for reasons discussed shortly. However, the latter designation could be confusing when the FET is compared with the BJT. Recall that saturation for a BJT corresponds to a collector–emitter voltage near zero, so the meaning of saturation in that case is quite different than for an FET. We will use the term *beyond-pinchoff region* in this text.

A brief explanation of the phenomenon follows. As V_{DS} increases, the voltage between gate and drain (i.e., V_{GD}) becomes increasingly negative. This negative voltage results in a depletion region along the upper portion of the channel. When a depletion region was formed along the bottom of the channel by a negative value of V_{GS}, current eventually stopped flowing. In the present case, however, current has already been established since V_{GS} is assumed to be more positive than $V_{GS(off)}$. The result is that the current flow becomes saturated, and very little further increase can occur. The channel is said to be "pinched off," and a nearly constant value of current results for a fixed value of V_{GS}.

From Figure 6-6 it is noted that the curve corresponding to $V_{GS} = 0$ flattens out at about $V_{DS} = 4$ V. Recall, however, that $V_{GS(off)} = -4$ V. The equality between

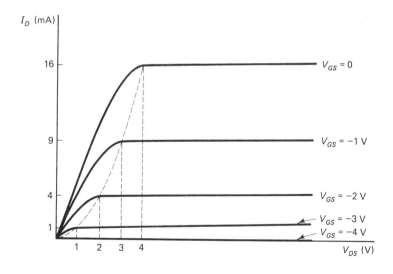

Figure 6-6 Drain characteristics of hypothetical *n*-channel JFET for full range of V_{DS}.

these magnitudes is more than a coincidence. The beyond-pinchoff region for an ideal JFET is reached when $V_{DS} = -V_{GS(off)}$ for $V_{GS} = 0$.

It is convenient for this purpose to define a quantity called the *pinchoff voltage*, and it is denoted by V_p. The pinchoff voltage is related to the cutoff voltage by

$$V_p = -V_{GS(off)} \qquad (6\text{-}5)$$

Thus, for $V_{GS} = -4$ V, $V_p = 4$ V.

The transition from the ohmic region to the beyond-pinchoff region occurs for $V_{DS} = V_p$ along the curve corresponding to $V_{GS} = 0$. For other values of V_{GS}, however, the transition point is shifted back to the left as noted.

Let $V_{DS(tran)}$ represent the value of drain–source voltage for an ideal JFET at which this transition occurs. It is related to the pinchoff voltage V_p and the gate–source voltage by

$$\boxed{V_{DS(tran)} = V_p + V_{GS}} \qquad (6\text{-}6)$$

When $V_{GS} = 0$, (6-6) reduces to $V_{DS(tran)} = V_p$, as already observed. Note that $V_p > 0$ and $V_{GS} < 0$ in (6-6) for an n-channel JFET. Note also that V_p is a constant for a given transistor, while the other two quantities vary with the operating curve.

Example 6-2

Calculate the drain–source transition voltages for each characteristic curve of the hypothetical JFET whose output characteristics were given in Figure 6-6.

Solution

The gate–source cutoff voltage was determined earlier to be $V_{GS(off)} = -4$ V, and the pinchoff voltage was then noted to be $V_p = 4$ V. The values of the transition voltage $V_{DS(tran)}$ are determined from (6-6) as

$$V_{DS(tran)} = V_p + V_{GS} = 4 + V_{GS} \qquad (6\text{-}7)$$

For $V_{GS} = 0$ we have

$$V_{DS(tran)} = 4 + 0 = 4 \text{ V} \qquad (6\text{-}8)$$

as expected. For $V_{GS} = -1$ V we have

$$V_{DS(tran)} = -1 + 4 = 3 \text{ V} \qquad (6\text{-}9)$$

The other results are tabulated as follows:

V_{GS} (V)	$V_{DS(tran)}$ (V)
0	4
−1	3
−2	2
−3	1
−4	0

The last case (cutoff) could be considered meaningless, but it is included to demonstrate that the transition point approaches $V_{DS(tran)} = 0$ as V_{GS} approaches $V_{GS(off)}$. The transition voltage is farthest to the right for $V_{GS} = 0$. These trends are readily observed in Figure 6-6.

The dashed curve superimposed on the characteristic curves of Figure 6-6 represents the locus of transition voltages. The region to the left of this curve is the ohmic region, and the region to the right is the beyond-pinchoff region.

6-4 CHARACTERISTICS IN BEYOND-PINCHOFF REGION

The forms of a few idealized n-channel JFET drain characteristic curves were shown in Figure 6-6. The reader may have observed the apparent nonlinear spacing between the curves; that is, the spacing between successive curves increases as V_{GS} is varied from cutoff to zero. This pattern is characteristic of field-effect transistors and is to be expected.

In this section a mathematical model for predicting the drain current as a function of the gate–source voltage applicable to the *beyond-pinchoff region* will be presented. The reader must understand that this equation is approximate and must be used with some caution in working with actual JFET characteristics. However, its form is sufficiently close to the actual behavior of typical JFET characteristics that it is widely used in analysis and design.

First, we define a parameter I_{DSS} known as the *zero-bias drain current*. Specifically, it is the value of I_D in the beyond-pinchoff region corresponding to $V_{GS} = 0$. For the characteristics of Figure 6-6, $I_{DSS} = 16$ mA.

The drain current in the beyond-pinchoff region follows a square-law functional relationship and can be closely approximated for many JFETs as

$$I_D = I_{DSS} \left[1 - \frac{V_{GS}}{V_{GS(off)}} \right]^2 \qquad \textbf{(6-10)}$$

For an n-channel JFET, $V_{GS(off)}$ is negative and V_{GS} is either negative or zero. Because of potential sign difficulties, some persons prefer to express (6-10) in the alternative form

$$I_D = I_{DSS}\left[1 - \frac{|V_{GS}|}{|V_{GS(off)}|}\right]^2 \tag{6-11}$$

in which the magnitudes of V_{GS} and $V_{GS(off)}$ are used. I_D and I_{DSS} are both positive for an n-channel JFET, so magnitude bars are not required for them.

Observe that V_{DS} does not appear in (6-10) and (6-11). The reason is that this model assumes no variation of I_D with V_{DS} in the beyond-pinchoff region. In practice, there is a slight increase in I_D as V_{DS} is increased, but this will be ignored in our idealized analysis.

Transfer Characteristic

The drain characteristics of a hypothetical n-channel JFET have been displayed and discussed. An alternative presentation, which is often more useful in designing circuits, is the *transfer characteristic*. The *transfer characteristic* is a plot of I_D versus V_{GS} in the beyond-cutoff region. The transfer characteristic of the hypothetical n-channel JFET considered earlier is shown in Figure 6-7. By comparing Figure 6-7 with Figure 6-6, the reader may readily verify that the values of V_{GS} and I_D at the circled points of Figure 6-7 correspond to the flat portions of the drain characteristics of Figure 6-6. Since a large number of individual drain characteristic curves could have been obtained in Figure 6-6 by assuming more values of V_{GS}, the continuous curve of Figure 6-7 represents the locus of all such values of V_{GS} and I_D. The square-law form of the drain current is rather evident from the curvature in Figure 6-7.

Reverse Form of Equation

The form of either (6-10) or (6-11) expresses I_D as a dependent function of the independent variable V_{GS}. However, suppose it is necessary to determine the value of V_{GS}

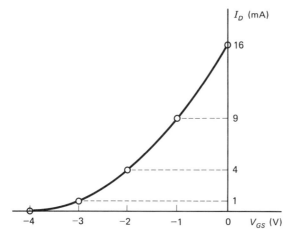

Figure 6-7 Transfer characteristic of hypothetical n-channel JFET considered in Figure 6-6.

that would produce a desired I_D. For this purpose it is necessary to solve for V_{GS} as a function of I_D.

To accomplish this task, the right-hand side of (6-10) is first expanded. The variable V_{GS} will appear both as a linear term and a second-degree term. The resulting equation is a quadratic equation in the variable V_{GS}. The steps leading to the solution of V_{GS} will be left as an exercise for the reader (Problem 6-39). All quadratic equations have two solutions, but one solution in this case is simply a mathematical value that does not correspond to the physical limitations of the problem. The proper solution for I_D as a function of V_{GS} is the following equation:

$$V_{GS} = V_{GS(\text{off})}\left(1 - \sqrt{\frac{I_D}{I_{DSS}}}\right)$$

(6-12)

Graphical Analysis

A normalized form of the beyond-pinchoff equation may readily be formulated, which permits a graphical approach to the analysis. First, both sides of (6-10) are divided by I_{DSS}, which results in

$$\frac{I_D}{I_{DSS}} = \left[1 - \frac{V_{GS}}{V_{GS(\text{off})}}\right]^2$$

(6-13)

For convenience, momentarily define the following normalized variables:

$$X = \frac{V_{GS}}{V_{GS(\text{off})}}$$

(6-14)

$$Y = \frac{I_D}{I_{DSS}}$$

(6-15)

The result of (6-13) may then be expressed in the simpler "pure" mathematical form

$$Y = (1 - X)^2$$

(6-16)

The result of (6-16) is shown in graphical form in Figure 6-8. The abscissa is labeled in terms of both $V_{GS}/V_{GS(\text{off})}$ and in terms of $|V_{GS}|/|V_{GS(\text{off})}|$ since both ratios are the same.

This curve applies to both n-channel and p-channel devices as long as the quantities are appropriately interpreted. As we will see later, it also applies to depletion-mode IGFETs (but not to enhancement-mode IGFETs). Because of the normalized nature of the variables, it is applicable for any values of $V_{GS(\text{off})}$ and I_{DSS}. Do not be confused by the fact that the abscissa is reversed in direction from the earlier form of the transfer characteristic curve. The right-hand side corresponds to cutoff and the left-hand side corresponds to $V_{GS} = 0$.

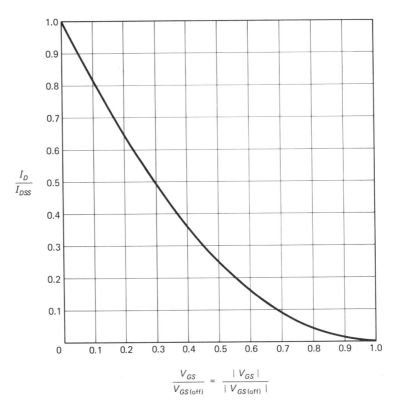

$$\frac{V_{GS}}{V_{GS\,(off)}} = \frac{|\,V_{GS}\,|}{|\,V_{GS\,(off)}\,|}$$

Figure 6-8   Normalized form of transfer characteristic.

Example 6-3

Using the mathematical model for the drain current in the beyond-pinchoff region, verify that the circled values of I_D versus V_{GS} in Figure 6-7 (or the final values of I_D versus V_{DS} in Figure 6-6) represent the idealized values.

Solution

First, we note that $I_{DSS} = 16$ mA and $V_{GS(off)} = -4$ V, as determined previously. We will illustrate both forms of (6-10) and (6-11). The current I_D can be expressed in either of the following forms:

$$I_D = 16 \times 10^{-3}\left(1 - \frac{V_{GS}}{-4}\right)^2 \qquad \textbf{(6-17)}$$

or

$$I_D = 16 \times 10^{-3}\left(1 - \frac{|V_{GS}|}{4}\right)^2 \qquad \textbf{(6-18)}$$

where $|V_{GS(\text{off})}| = 4$ V was used in (6-18).

From either (6-17) or (6-18), it is immediately noted that when $V_{GS} = 0$, $I_D = 16 \times 10^{-3}$ A $= 16$ mA, as expected.

Consider $V_{GS} = -1$ V and equation (6-17). The expression for I_D is

$$I_D = 16 \times 10^{-3}\left(1 - \frac{-1}{-4}\right)^2 = 16 \times 10^{-3}\left(1 - \frac{1}{4}\right)^2$$

$$= 9 \times 10^{-3} \text{ A} = 9 \text{ mA} \qquad \textbf{(6-19)}$$

Let's verify that (6-18) produces the same result.

$$I_D = 16 \times 10^{-3}\left(1 - \frac{1}{4}\right)^2 = 9 \times 10^{-3} \text{ A} = 9 \text{ mA} \qquad \textbf{(6-20)}$$

where $|V_{GS}| = 1$ V was used in (6-20).

Comparing the use of (6-17) and (6-18) in (6-19) and (6-20), respectively, we can see the possible sign difficulty in (6-17) and (6-19). Despite that, however, the form of (6-17) is usually more desirable when several separate algebraic manipulations are required. The point to remember is that the quantity in parentheses will always represent a numerical *difference* in the final analysis. In the remainder of the example, we use the form of (6-17). For $V_{GS} = -2$ V, we have

$$I_D = 16 \times 10^{-3}\left(1 - \frac{-2}{-4}\right)^2 = 16 \times 10^{-3}\left(1 - \frac{1}{2}\right)^2$$

$$= 4 \times 10^{-3} \text{ A} = 4 \text{ mA} \qquad \textbf{(6-21)}$$

For $V_{GS} = -3$ V,

$$I_D = 16 \times 10^{-3}\left(1 - \frac{-3}{-4}\right)^2 = 16 \times 10^{-3}\left(1 - \frac{3}{4}\right)^2$$

$$= 1 \times 10^{-3} \text{ A} = 1 \text{ mA} \qquad \textbf{(6-22)}$$

Finally, for $V_{GS} = -4$ V,

$$I_D = 16 \times 10^{-3}\left(1 - \frac{-4}{-4}\right)^2 = 16 \times 10^{-3}(1 - 1)^2 = 0 \qquad \textbf{(6-23)}$$

as we should expect. The preceding values are readily verified in either Figure 6-7 or 6-6.

Example 6-4

Tests on a certain n-channel JFET in the beyond-pinchoff region are performed. When gate and source are connected together, $I_D = 10$ mA. When an adjustable dc source is connected between gate and source, the drain current ceases to flow when $V_{GS} = -2$ V. **(a)** Determine an equation for predicting I_D as a function of V_{GS} in the beyond-pinchoff region. **(b)** Determine the value of I_D when $V_{GS} = -1$ V.

Solution

(a) From the statement of the test, we surmise that $I_{DSS} = 10$ mA and $V_{GS(\text{off})} = -2$ V. The equation for predicting I_D in the beyond-pinchoff region is

$$I_D = I_{DSS}\left[1 - \frac{V_{GS}}{V_{GS(\text{off})}}\right]^2 = 0.01\left(1 - \frac{V_{GS}}{-2}\right)^2 = 0.01\left(1 + \frac{V_{GS}}{2}\right)^2 \quad \textbf{(6-24)}$$

(b) For $V_{GS} = -1$ V, we have

$$I_D = 0.01\left(1 - \frac{-1}{-2}\right)^2 = 0.0025 \text{ A} = 2.5 \text{ mA} \quad \textbf{(6-25)}$$

Example 6-5

For the n-channel JFET of Example 6-4, determine the value of V_{GS} that produces $I_D = 5$ mA three ways: **(a)** by direct solution of the quadratic equation; **(b)** by use of the reverse equation as given by (6-12); **(c)** by use of the graph of Figure 6-8.

Solution

(a) The equation for I_D was stated in (6-24). We set $I_D = 5$ mA $= 5 \times 10^{-3}$ A and this results in

$$5 \times 10^{-3} = 0.01\left(1 - \frac{V_{GS}}{-2}\right)^2 \quad \textbf{(6-26)}$$

This equation is simplified by dividing both sides by 0.01, and the right-hand side is expanded. This results in

$$0.5 = 1 + V_{GS} + \frac{V_{GS}^2}{4} \quad \textbf{(6-27)}$$

This equation is rearranged in standard quadratic form as

$$V_{GS}^2 + 4V_{GS} + 2 = 0 \quad \textbf{(6-28)}$$

Application of the quadratic formula yields

$$V_{GS} = \frac{-4 \pm \sqrt{16 - 8}}{2} = \frac{-4 \pm 2.828}{2} \qquad \text{(6-29)}$$

When the two values are computed, one turns out to be $V_{GS} = -3.414$ V. This is an *impossible* physical solution since this voltage is more negative than the pinchoff voltage and represents the cutoff region. This superfluous value is a mathematical solution but not a physically possible solution. The other solution is the correct one and is

$$V_{GS} = -0.586 \text{ V} \qquad \text{(6-30)}$$

When the preceding mathematical process is followed, the superfluous solution can always be determined by observing the physical limits of the problem. The correct solution is always the one involving *opposite* signs in the numerator of the quadratic solution, as in (6-29).

(b) Substitution of the pertinent values in (6-12) results in

$$V_{GS} = -2\left(1 - \sqrt{\frac{5 \times 10^{-3}}{10 \times 10^{-3}}}\right) = -2(1 - \sqrt{0.5})$$

$$= -2(0.293) = -0.586 \text{ V} \qquad \text{(6-31)}$$

(c) To utilize Figure 6-8, we first normalize the value of I_D by forming I_D/I_{DSS}. This value is $5 \times 10^{-3}/10 \times 10^{-3} = 0.5$. For an ordinate of 0.5, the corresponding abscissa is about $V_{GS}/V_{GS(\text{off})} = 0.29$. We then calculate V_{GS} as

$$V_{GS} = 0.29 \times (-2) = -0.58 \text{ V} \qquad \text{(6-32)}$$

Comparing the results of (a), (b), and (c), we note that (a) and (b) produce exactly the same values, while (c) produces approximately the same result to the expected accuracy of the graph.

6-5 OHMIC-REGION CHARACTERISTICS

Early in the discussion of the JFET, we observed an idealized drain voltage–current characteristic in the ohmic region, and the reader may wish to refer back to Figure 6-4. It was noted that for very small values of V_{DS}, the curves of I_D versus V_{DS} were straight lines whose slopes were a function of V_{GS}. In this region the FET behaves like a voltage-controlled resistor.

A mathematical model for the ideal characteristics of these straight lines will now be given. Let g_{DS} represent the drain-to-source conductance in siemens corresponding to any one of these curves, and let g_{DSO} represent the conductance of the line representing $V_{GS} = 0$. The value of conductance is simply the slope of a particular line. It turns out that g_{DS} can be closely approximated for most JFETs as

$$g_{DS} = g_{DSO}\left[1 - \frac{V_{GS}}{V_{GS(\text{off})}}\right]$$ (6-33)

The idealized value of g_{DSO} is

$$g_{DSO} = -\frac{2I_{DSS}}{V_{GS(\text{off})}}$$ (6-34)

At first glance, g_{DSO} may appear to have a negative value, but this is an illusion. The quantity I_{DSS} is positive and $V_{GS(\text{off})}$ is negative, so the latter negative value in consort with the negative sign in front results in a positive value of g_{DSO}. To delineate this point, an alternative form of g_{DSO} for an n-channel JFET is

$$g_{DSO} = \frac{2I_{DSS}}{|V_{GS(\text{off})}|}$$ (6-35)

Let $r_{DS} = 1/g_{DS}$ represent the resistance associated with a given line. This value may be expressed as

$$r_{DS} = \frac{r_{DSO}}{1 - V_{GS}/V_{GS(\text{off})}}$$ (6-36)

where

$$r_{DSO} = \frac{1}{g_{DSO}} = -\frac{V_{GS(\text{off})}}{2I_{DSS}} = \frac{|V_{GS(\text{off})}|}{2I_{DSS}}$$ (6-37)

The equivalent circuit of the n-channel JFET in the *linear* portion of the ohmic region is shown in Figure 6-9. For the resistance to be assumed as a constant value, the value of V_{DS} must be limited to a small range, typically from 0 to about 0.1 V.

From the preceding relationships, it can be inferred that the maximum conductance and minimum resistance correspond to $V_{GS} = 0$. As the magnitude of V_{GS} increases (i.e., V_{GS} becomes more negative), the conductance decreases and the resistance increases.

One particular limiting case of the ohmic region corresponds to the use of an FET as an *analog switch*. In this type of circuit, the FET is alternately switched from

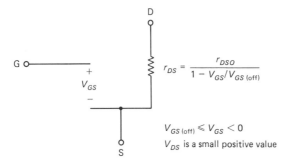

$$r_{DS} = \frac{r_{DSO}}{1 - V_{GS}/V_{GS(\text{off})}}$$

$V_{GS(\text{off})} \leqslant V_{GS} < 0$

V_{DS} is a small positive value

Figure 6-9 Ohmic-region model for n-channel JFET for small positive values of V_{DS}.

the lowest resistance condition to the cutoff region. Some FETs are manufactured specifically for this purpose. More will be said about this application in Chapter 7.

Example 6-6

Using the mathematical model for the conductance in the lower portion of the ohmic region, verify that the values of conductance computed in Example 6-1 (corresponding to the lines of Figure 6-4) represent the idealized values.

Solution

We note again that $I_{DSS} = 16$ mA and $V_{GS(off)} = -4$ V for the idealized JFET considered earlier. The value of g_{DSO} from (6-34) is

$$g_{DSO} = \frac{-2(16 \times 10^{-3})}{-4} = 8 \times 10^{-3} \text{ S} = 8 \text{ mS} \qquad \textbf{(6-38)}$$

The conductance g_{DS} from (6-33) is

$$g_{DS} = 8 \times 10^{-3}\left(1 - \frac{V_{GS}}{-4}\right) \qquad \textbf{(6-39)}$$

For $V_{GS} = 0$, it is readily noted that

$$g_{DS} = 8 \times 10^{-3} \text{ S} = 8 \text{ mS} \qquad \textbf{(6-40)}$$

For $V_{GS} = -1$ V, the conductance is

$$g_{DS} = 8 \times 10^{-3}\left(1 - \frac{-1}{-4}\right) = 8 \times 10^{-3}\left(1 - \frac{1}{4}\right)$$

$$= 6 \times 10^{-3} \text{ S} = 6 \text{ mS} \qquad \textbf{(6-41)}$$

The reader is invited to repeat this process at $V_{GS} = -2$ V, -3 V, and -4 V, and the conductance values are 4 mS, 2 mS, and 0, respectively. These values are readily verified from the results of Example 6-1. Alternatively, we could have used (6-36) and (6-37) to compute the resistance instead of the conductance.

Before leaving this example, we need to bring the reader back to the "real world" again. In this example and several others earlier in the chapter, we have verified various relationships for the JFET with the hypothetical JFET characteristics provided. These characteristics were "rigged" to produce these characteristics exactly in the ideal forms. Seldom (if ever) do any actual characteristics work out so nicely. Most actual FET characteristics can be approximated by the models provided, but they should be interpreted as reasonable approximations rather than exact results. As we shall see later, the actual parameters of most FETs of a given type vary widely from one device to another.

We will now consider a circuit model that can be used to show the dc operation of an n-channel JFET in the beyond-pinchoff region in a complete circuit. The basic form of this model is shown in Figure 6-10.

The drain side of this model consists of an ideal dependent current source whose value represents the drain current. This model assumes that the drain current remains constant as V_{DS} changes with the applicable region.

We immediately see that the drain circuit of the n-channel JFET has a striking similarity to the collector circuit active-region dc model of an *npn* bipolar junction transistor; that is, both devices act nearly like ideal current sources, and the current sources have the same direction. However, there are differences in the operation of the devices, as we will see.

The dc current source for the *BJT* is a function of the *base current*, while the dc current source for the *FET* is a function of the *gate–source* voltage. For the dc model, *we thus consider the BJT to be a current-controlled device, while we consider the FET to be a voltage-controlled device.* (Later we will see that the BJT can also be considered to be voltage controlled in amplifier applications.)

The value of the current source for the BJT was simply $\beta_{dc}I_B$, while the value of the current source in the FET is a somewhat more complex function of V_{GS}. Some significant differences between the BJT and the FET are evident on the input sides. The base side of a BJT looks into a forward-biased *pn* junction, in which a nearly constant voltage drop appears. Current must flow into the base circuit of a BJT, and the loading effect on the external circuit must be considered in applications. In contrast, *the input of an FET is a very large resistance, which for many applications can be assumed to be infinite.* This assumption has been made in Figure 6-10 and is noted by the open circuit between gate and source terminals. This means that in this model no current would flow into the gate terminal.

Actually, there is a small current (typically less than 1 nA) that flows in the gate of a JFET, and this could be represented by a very large resistance between gate and source. However, for most applications, this resistance can be ignored. For other classes of FETs to be considered later, the resistance can be as high as hundreds of megohms.

Figure 6-10 Beyond-pinchoff dc model for n-channel JFET.

6-7 INSULATED GATE FIELD-EFFECT TRANSISTORS

The second broad category of field-effect transistors is the *insulated gate field-effect transistor*, widely referred to as the *IGFET*. The most common type of IGFET is the *metal-oxide semiconductor field-effect transistor* or *MOSFET*. In practice, the terms IGFET and MOSFET are both used casually to refer to all transistors of this broad category. We use the term IGFET in most places in this book since not all IGFETs are MOSFETs.

IGFETs can be classified in two ways according to their conducting status: (1) *depletion mode* or "normally on," and (2) *enhancement mode* or "normally off." Each category will now be discussed.

Depletion-Mode IGFET

The construction of an *n*-channel depletion mode IGFET is suggested by the sketch of Figure 6-11. As in the case of an *n*-channel JFET, there is a channel with *source* and *drain* contacts at the respective ends. A *p*-region called the *substrate* physically reduces the conducting path to a narrow channel. A thin layer of *silicon dioxide* (SiO_2) insulates the metallic gate from the channel in the common case of a MOSFET. Thus, the *pn* junction diode that exists in the JFET does not appear in the IGFET. The result of this insulation layer is that the gate current of the IGFET is even smaller than that of the JFET, which has already been indicated to be quite small.

As in the case of the JFET, the width of the conduction channel is controlled by the gate–source voltage. If the gate–source voltage in a depletion-mode IGFET is sufficiently negative, the channel is completely blocked and no current flows. This corresponds to the channel cutoff voltage.

As a general rule, the drain characteristics of an *n*-channel depletion-mode IGFET are very similar to those of the *n*-channel JFET except for one difference. In

Figure 6-11 Basic construction of an *n*-channel depletion-mode IGFET.

the n-channel JFET, V_{GS} is usually not permitted to assume positive values since this would forward bias the gate–channel junction. Because of the insulation layer, V_{GS} in an n-channel depletion-mode IGFET may assume positive values without affecting the high input resistance. Strictly speaking, this corresponds to enhancement-mode operation of the depletion-mode device. A separate discussion of devices specifically designed for enhancement-mode operation will be made shortly.

Although some IGFETs have the substrate terminal available as a fourth terminal, many devices are manufactured with this terminal connected internally to the source. The schematic diagram of an n-channel depletion-mode IGFET with substrate connected to source internally is shown in Figure 6-12.

Some typical drain characteristics of an n-channel depletion-mode IGFET are shown in Figure 6-13. Curves are shown for both negative and positive values of V_{GS} to illustrate that operation is possible for the IGFET with both polarities.

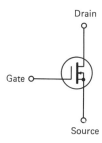

Figure 6-12 Schematic symbol of an n-channel depletion-mode IGFET (substrate is connected to source internally).

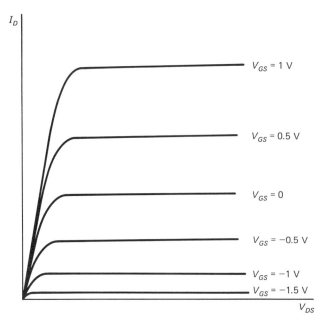

Figure 6-13 Typical drain characteristics of an n-channel depletion-mode IGFET showing operation for both negative and positive values of V_{GS}.

Enhancement-Mode IGFET

The construction of an n-channel enhancement-mode IGFET is suggested by the diagram of Figure 6-14. The p-type substrate completely separates the two n-type sections of the channel. The channel is thus blocked, and for that reason the enhancement-mode IGFET can be called a *normally off* IGFET.

When $V_{DS} > 0$, there is no conduction in an n-channel enhancement-mode IGFET for $V_{GS} \leq 0$. For conduction to occur, V_{GS} must assume some minimum *positive value*. A sufficient positive value of V_{GS} will attract free electrons from the n-type channel into the p-region in the vicinity of the gate. This results in an effective layer of n-type material which bridges across the substrate and permits conduction. Thus conduction has been *enhanced* by the presence of a positive voltage.

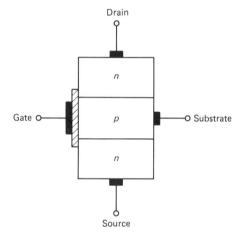

Figure 6-14 Basic construction of an n-channel enhancement-mode IGFET.

Threshold Voltage

The minimum positive value of V_{GS} that results in drain current flow for an n-channel enhancement-mode IGFET is called the *threshold voltage* and is denoted as $V_{GS(\text{th})}$. Thus $I_D = 0$ for $V_{GS} < V_{GS(\text{th})}$. The threshold voltage for an n-channel enhancement-mode IGFET is analogous to the gate–source cutoff voltage for the n-channel depletion-mode IGFET and the JFET. However, note that for n-channel devices, $V_{GS(\text{th})} > 0$ and $V_{GS(\text{off})} < 0$.

The schematic symbol of an n-channel enhancement-mode IGFET is shown in Figure 6-15. The substrate is assumed to be connected to the source internally. The symbols for depletion-mode and enhancement-mode devices look somewhat similar and the reader is encouraged to compare Figures 6-12 and 6-15 to observe the difference.

Figure 6-15 Schematic symbol of an n-channel enhancement-mode IGFET (substrate is connected to source internally).

The results of a possible experiment performed on a hypothetical *n*-channel enhancement-mode IGFET is discussed next. As in the case of the JFET considered earlier in the chapter, the characteristics have been idealized somewhat.

Consider the test circuit shown in Figure 6-16 containing an *n*-channel enhancement-mode IGFET. If this circuit is compared with the JFET circuit of Figure 6-3, the main difference is that the gate voltage bias supply for the *n*-channel enhancement-mode device has its positive terminal connected to the gate rather than to ground.

Refer to the drain characteristics of Figure 6-17 in the discussion that follows. If $V_{GS} = 0$ (or a small positive voltage), $I_D = 0$ as V_{DS} is varied over a reasonable

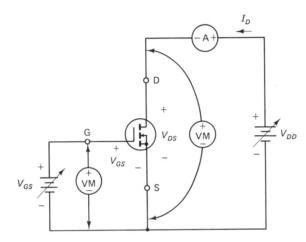

Figure 6-16 Test circuit for hypothetical *n*-channel enhancement-mode IGFET.

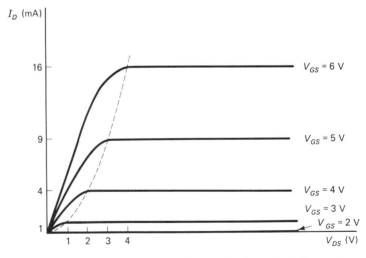

Figure 6-17 Drain characteristics of hypothetical *n*-channel enhancement-mode IGFET.

positive range because of the "normally-off" property. The IGFET is biased in the cutoff region until V_{DS} is increased beyond the threshold voltage $V_{GS(th)}$.

For this particular device, it is assumed that $V_{GS(th)} = 2$ V. Thus, for $V_{GS} > 2$ V, drain current is measured as V_{DS} is changed in positive increments.

Beyond-Pinchoff Region

Observe in Figure 6-17 that the transition point from the ohmic region to the beyond-pinchoff region shifts farther to the right on the drain characteristics as V_{GS} increases. For depletion-mode IGFETs and JFETs, a convenient quantity called the pinchoff voltage was used in characterizing the transition point. However, there is no particular need to use such a definition for enhancement-mode IGFETs as the threshold voltage may be used directly in the formulation.

Let $V_{DS(tran)}$ represent the transition voltage from the ohmic region to the beyond-pinchoff region. For an ideal n-channel enhancement-mode IGFET, this voltage is

$$V_{DS(tran)} = V_{GS} - V_{GS(th)} \qquad \text{(6-42)}$$

All the variables in (6-42) are positive quantities for an n-channel unit.

The curves of I_D versus V_{DS} for various values of V_{GS} are shown in Figure 6-17. The curves are similar in shape to those of the JFET and depletion-mode IGFET, but they correspond to a different range of V_{GS}. Like the JFET and the depletion-mode IGFET, the enhancement-mode IGFET possesses an ohmic region in which the device behaves like a voltage-controlled resistance. At small values of V_{DS}, the resistance is approximately constant. More will be said about the ohmic region later.

6-9 ENHANCEMENT-MODE IGFET MODELS

The drain current I_D as a function of the gate–source voltage V_{GS} for an n-channel enhancement-mode IGFET can be described very closely in the beyond-pinchoff region by the equation

$$I_D = K\left[\frac{V_{GS}}{V_{GS(th)}} - 1\right]^2 \qquad \text{(6-43)}$$

where $V_{GS(th)}$ is the threshold voltage and K is a constant. The constant K is determined by matching the actual characteristics of a given unit to the mathematical model. In effect, it is a curve-fitting constant chosen to match the physical conditions. (Should we say a "fudge factor"?)

Transfer Characteristic

As for the JFET and depletion-mode IGFET, a *transfer characteristic* for the IGFET may also be displayed. The transfer characteristic of the hypothetical n-channel IGFET is shown in Figure 6-18. By comparing Figure 6-18 with Figure 6-17, the

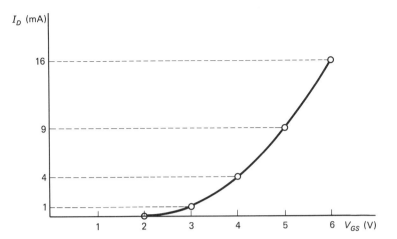

Figure 6-18 Transfer characteristic of hypothetical n-channel enhancement-mode IGFET.

reader may readily verify that the values of V_{GS} and I_D at the circled points of Figure 6-18 correspond to the flat portions of the drain characteristics of Figure 6-17. Once again, the square-law form of the drain current is evident.

Beyond-Pinchoff DC Model

A dc circuit model of the n-channel enhancement-mode IGFET in the beyond-pinch-off region is shown in Figure 6-19. The form is the same as that of the n-channel JFET given in Figure 6-10 except for the expression for I_D. In both models, the gate has been considered as an open circuit. In actual practice, this assumption is even more valid for the IGFET than for the JFET because of the insulated gate structure.

Reverse Equation

Determination of the value of V_{GS} that produces a given value of I_D in (6-43) is similar to the process considered in Section 6-4 for depletion-mode devices. It will be

Figure 6-19 Beyond pinchoff model for n-channel enhancement-mode EGFET.

left as an exercise for the reader (Problem 6-40) to show that

$$V_{GS} = V_{GS(\text{th})}\left(\sqrt{\frac{I_D}{K}} + 1\right)$$

(6-44)

Ohmic-Region Characteristics

The ohmic-region mathematical model of the enhancement-mode IGFET will now be considered. For $V_{GS} > V_{GS(\text{th})}$ and very small values of V_{DS} (typically ≤ 0.1 V), the IGFET behaves approximately like a constant value of resistance. The conductance g_{DS} is of the form

$$g_{DS} = g_{DSO}\left[\frac{V_{GS}}{V_{GS(\text{th})}} - 1\right]$$

(6-45)

The idealized value of g_{DSO} in this case is

$$g_{DSO} = \frac{2K}{V_{GS(\text{th})}}$$

(6-46)

where K is the same constant appearing in (6-43).

An expression for the resistance r_{DS} is

$$r_{DS} = \frac{1}{g_{DS}} = \frac{r_{DSO}}{\dfrac{V_{GS}}{V_{GS(\text{th})}} - 1}$$

(6-47)

where

$$r_{DSO} = \frac{1}{g_{DSO}} = \frac{V_{GS(\text{th})}}{2K}$$

(6-48)

A circuit model for the enhancement-mode IGFET in the ohmic region is shown in Figure 6-20.

$$r_{DS} = \frac{r_{DSO}}{V_{GS}/V_{GS\,(\text{th})} - 1}$$

$V_{GS} > V_{GS\,(\text{th})}$

V_{DS} is a small positive value

Figure 6-20 Ohmic region model for n-channel enhancement-mode IGFET for small positive values of V_{DS}.

Example 6-7

Calculate the drain–source transition voltage for each characteristic curve of the hypothetical enhancement-mode IGFET of Figure 6-17.

Solution

The threshold voltage was observed to be $V_{GS(th)} = 2$ V. An expression for transition voltage $V_{GS(tran)}$ is determined from (6-42) as

$$V_{DS(tran)} = V_{GS} - 2 \qquad (6-49)$$

For each value of V_{GS} on Figure 6-17, the transition voltage is easily calculated, and a short table of values follows.

V_{GS} (V)	$V_{DS(tran)}$ (V)
2	0
3	1
4	2
5	3
6	4

These trends are readily observed in Figure 6-17. The dashed curve superimposed on the characteristic curves represents the locus of transition voltages. The ohmic region is located to the left of this curve, and the beyond-pinchoff region is located to the right.

Example 6-8

Using the mathematical model for the drain current in the beyond-pinchoff region, verify that the circled values of I_D versus V_{GS} in Figure 6-18 (or the final values of I_D versus V_{DS} in Figure 6-17) represent the idealized values.

Solution

Since $V_{GS(th)} = 2$ V, the form of the equation for I_D from (6-43) is

$$I_D = K\left(\frac{V_{GS}}{2} - 1\right)^2 \qquad (6-50)$$

This equation clearly predicts $I_D = 0$ when $V_{GS} = 2$ V, as it should. However, K must be determined before any subsequent values of I_D can be predicted.

In an actual application of the equation, one could select a known condition for I_D and V_{GS} to determine K. In our application, we will use the fact that $I_D = 1$ mA when $V_{GS} = 3$ V. When these values are substituted in (6-50), there results

$$1 \times 10^{-3} = K\left(\frac{3}{2} - 1\right)^2 = K(0.25) \tag{6-51}$$

Solving for K, we obtain

$$K = 4 \times 10^{-3}\,\text{A} \tag{6-52}$$

Note that K has the units of amperes.

The equation for I_D can be expressed as

$$I_D = 4 \times 10^{-3}\left(\frac{V_{GS}}{2} - 1\right)^2 \tag{6-53}$$

For $V_{GS} = 4$ V, we can determine I_D as

$$I_D = 4 \times 10^{-3}\left(\frac{4}{2} - 1\right)^2 = 4 \times 10^{-3}\,\text{A} = 4\,\text{mA} \tag{6-54}$$

which is readily verified.

The reader is invited to show that the values for I_D of 9 mA and 16 mA are readily determined when $V_{GS} = 5$ V and 6 V, respectively. As was the case of the JFET considered earlier, these characteristics were idealized to fit the exact mathematical form.

In an actual IGFET, the value of K might depend somewhat on the point initially used to establish the curve-fitting process. In such a case, the best fit would probably be in the region in which operation of the IGFET is expected.

6-10 GENERAL OVERVIEW OF FETS

To simplify the development thus far in the chapter, only n-channel FETs have been considered. Thus we have considered the n-channel JFET, the n-channel depletion-mode IGFET, and the n-channel enhancement-mode IGFET. All of these devices are also available as p-channel components. The construction of a given p-channel device is similar to that of the corresponding n-channel device except that all n-type semiconductor sections are replaced by p-type sections, and vice versa. All current directions and voltage polarities are reversed.

The schematic symbols of all the n-channel FETs and the corresponding p-channel devices are summarized in Figure 6-21. The left-hand column represents the various n-channel FETs, which have all been introduced earlier in the chapter. The right-hand column represents the corresponding symbols for the complementary p-channel FETs. In the symbols, if a given arrow is directed *into* the device (irrespective of whether it is on the left or right), the device is an n-channel unit. If the arrow is directed *out* of the device, it is a p-channel unit.

The positive direction of I_D is *into the drain* (and, of course, *out* of the *source*)

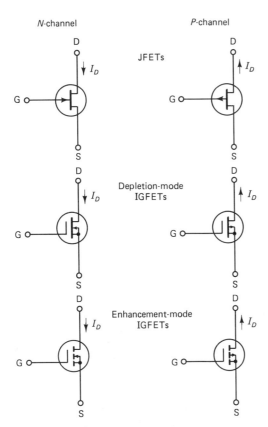

N-channel   P-channel

JFETs

I_D I_D

Depletion-mode
IGFETs

I_D I_D

Enhancement-mode
IGFETs

I_D I_D

Figure 6-21 Summary of schematic symbols for the various FETs.

for n-channel devices as we have seen throughout the chapter. By defining I_D as flowing *into the source and out of the drain* for p-channel devices, the value of I_D will be positive for normal operation. By performing this reversal in convention, the values of I_{DSS} and K will likewise be positive.

The reader will recall that when BJTs were studied, it was convenient to switch from V_{CE} and V_{BE} for *npn* transistors to V_{EC} and V_{EB} for *pnp* transistors, so that the resulting voltages were normally positive. Since we have just discussed reversing I_D for p-channel FETs, one might be led to conclude that we should switch references on the FET voltages in the same manner. Although this could be done, it is best to leave the pertinent FET voltages as V_{GS} and V_{DS} and let the signs assume whatever values are required.

The reason that we do not reverse the conventions on the FET voltages is that it creates some difficulties in working with the cutoff, pinchoff, and threshold voltage parameters and with identification of the beyond-pinchoff regions. Thus the subscripts of the voltages V_{GS} and V_{DS} will be maintained, but the signs will be changed as required.

A summary of the direction and sign patterns is provided in Table 6-1. For this purpose, it is convenient to classify FETs as either the (1) *depletion-mode group* or as the (2) *enhancement-mode group*. The depletion-mode group includes JFETs and depletion-mode IGFETs. As far as sign patterns and functional relationships are con-

TABLE 6-1 SUMMARY OF DIRECTIONS AND POLARITIES FOR DIFFERENT TYPES OF FETs IN BEYOND-PINCHOFF (OR POSSIBLY OHMIC) REGIONS

Depletion-Mode Group
(JFETs and Depletion-Mode IGFETs)

	N-Channel				P-Channel		
I_D	$V_{GS(off)}$	V_{GS}[a]	V_{DS}	I_D	$V_{GS(off)}$	V_{GS}[a]	V_{DS}
$\downarrow$	—	$V_{GS(off)} < V_{GS} \le 0$	$+$	$\uparrow$	$+$	$0 \le V_{GS} < V_{GS(off)}$	—

$$I_D = I_{DSS}\left[1 - \frac{V_{GS}}{V_{GS(off)}}\right]^2 = I_{DSS}\left[1 - \frac{|V_{GS}|}{|V_{GS(off)}|}\right]^2$$

[a]Depletion-mode IGFETs may also function with V_{GS} on other side of zero.

Enhancement-Mode Group

	N-Channel				P-Channel		
I_D	$V_{GS(th)}$	V_{GS}	V_{DS}	I_D	$V_{GS(th)}$	V_{GS}	V_{DS}
$\downarrow$	$+$	$V_{GS} > V_{GS(th)}$	$+$	$\uparrow$	—	$V_{GS} < V_{GS(th)}$	—

$$I_D = K\left[\frac{V_{GS}}{V_{GS(th)}} - 1\right]^2 = K\left[\frac{|V_{GS}|}{|V_{GS(th)}|} - 1\right]^2$$

cerned, these two types of FETS behave very much alike. The enhancement-mode group consists of the enhancement-mode IGFETs.

The left-hand column represents the n-channel devices and the right-hand column represents the p-channel devices. As noted previously, I_D is defined in a direction such that it is positive. The directions of the arrows remind us of these positive directions (assuming that the drain is at the top).

Refer first to the left-hand-side (n-channel) block for the depletion-mode group. We note that $V_{GS(off)}$ is negative (implying that V_p is positive) and V_{GS} normally assumes values in the range $V_{GS(off)} < V_{GS} < 0$. We note also that V_{DS} is positive.

Corresponding patterns for the p-channel block on the right-hand side are opposite. Thus $V_{GS(off)}$ is positive (implying that V_p is negative), the voltage V_{GS} normally assumes a value between 0 and the positive value $V_{GS(off)}$, and V_{DS} normally assumes a negative value.

The equation for I_D in the beyond-pinchoff region is listed below the depletion-mode blocks in two ways. Both equations as given apply to both n-channel and p-channel units. The current I_D and the parameter I_{DSS} are both defined as positive, as noted previously. The quantities V_{GS} and $V_{GS(off)}$ are both negative for an n-channel device and are both positive for a p-channel device. Thus the ratio $V_{GS}/V_{GS(off)}$ is positive for both cases. In the right-hand equation form, magnitudes are used, and the ratio is obviously positive.

Refer now to the left-hand-side block (n-channel) for the enhancement-mode group. We note that $V_{GS(th)}$ is positive and that V_{GS} normally assumes values in the range $V_{GS} > V_{GS(th)}$. It is also noted that V_{DS} is positive.

Corresponding inequalities for the p-channel block on the right-hand side are opposite. Thus $V_{GS(th)}$ is negative, the voltage V_{GS} normally assumes a lower value (more negative) than $V_{GS(th)}$, and V_{DS} assumes a negative value.

The equation for I_D in the beyond-pinchoff region is listed below the enhancement-mode blocks in two ways. Both equations as given apply to both n-channel and p-channel units. Since V_{GS} and $V_{GS(th)}$ both assume the same sign, the ratio $V_{GS}/V_{GS(th)}$ is always positive.

6-11 READING FET DATA SHEETS

A brief overview of some typical FET dc specifications is presented in this section. The particular FET group to be studied consists of the 2N4220 through the 2N4222 along with newer versions of these types (as indicated by the addition of A to the number). Refer to Appendix D for the data sheets on this group. These transistors are n-channel JFETs.

The absolute maximum ratings are listed near the top of the first data page. These data are similar in most ways to the corresponding BJT maximum ratings, and similar discussions of Section 4-6 are applicable here. The maximum values of the drain–source and drain–gate voltages are given as 30 V. The maximum gate–source voltage is indicated as -30 V. (Alternatively, this could have been stated as $+30$ V for the maximum source–gate voltage.) The maximum drain current is specified as 15 mA.

For a dc drain–source voltage V_{DS} and a dc drain current I_D, the power P_D dissipated in the FET is

$$P_D = V_{DS}I_D \tag{6-55}$$

The maximum power dissipation at 25°C is listed as 300 mW, and the derating factor is 2 mW/°C.

Various electrical characteristics at 25°C are listed in the table. The *gate–source breakdown voltage* is denoted as $V_{(BR)GSS}$ and the minimum value is listed as -30 V. Earlier it was indicated that the maximum gate–source voltage rating was -30 V, so the reader may be confused by the use of the terms "minimum" and "maximum."

This can be explained as follows: The *minimum* possible reverse breakdown voltage for the gate-source junction is 30 V. Therefore, if the maximum *gate–source* voltage is kept under 30 V, the junction should never encounter reverse breakdown. Thus, the absolute maximum ratings are our guide to safe design.

The normal mode of operation for the JFET is with the gate–channel reverse biased. The very small current that flows for this condition is called *gate reverse current*, and it is denoted as I_{GSS}. A typical value at 25°C is -0.1 nA. The negative sign indicates that the current is flowing out of the gate. The small value confirms our earlier practical assumption that the gate is nearly an open circuit. At a temperature of 100°C, however, the maximum value increases to -100 nA.

The next item is the *gate–source cutoff voltage*. Minimum and typical values are not listed, but the maximum values range from -4 V for the 2N4220 to -8 V for the 2N4222. The corresponding range for the pinchoff voltage is 4 to 8 V.

In practice, the exact gate–source cutoff is somewhat difficult to measure accurately. For that reason, an additional block of data indicating the gate–source voltage for a reference small drain current is given.

Refer now to the block entitled *zero-gate-voltage drain current*, which is the same as the *zero-bias drain current*. The range of I_{DSS} depends on the transistor type and varies from 0.5 to 3.0 mA for a 2N4220 to a range of 5.0 to 15 mA for a 2N4222.

The block entitled *static drain–source ON resistance* corresponds to $V_{GS} = 0$ and was denoted as r_{DSO} in this chapter. This typical value ranges from 500 to 300 Ω, depending on the transistor type. The remaining data on the first page refer to small-signal or ac parameters and are considered later in the text.

Refer now to the second and third pages of the data. Several sets of drain characteristics, corresponding to different cutoff voltages, have been provided. In each case, the corresponding transfer characteristic is shown. These curves should be interpreted as *representative* of the possible types of characteristics that might be obtained for transistors in the group.

DRILL PROBLEMS

6-1. The drain characteristics in the linear portion of the ohmic region for a certain idealized n-channel JFET are shown in Figure P6-1. Determine the values of conductance and resistance for each of the following values of V_{GS}: **(a)** 0; **(b)** -1 V; **(c)** -2 V.

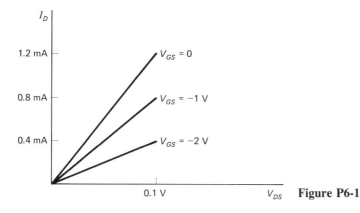

Figure P6-1

6-2. The drain characteristics in the linear portion of the ohmic region for a certain idealized n-channel JFET are shown in Figure P6-2. Determine the values of

Figure P6-2

conductance and resistance for each of the following values of V_{GS}: **(a)** 0; **(b)** -0.5 V; **(c)** -1 V; **(d)** -1.5 V.

6-3. The overall drain characteristics of the idealized JFET of Problem 6-1 are shown in Figure P6-3. **(a)** Determine the gate–source cutoff voltage and the pinchoff voltage. **(b)** Determine the transition voltage (from ohmic to beyond cutoff regions) for the following values of V_{GS}: 0, -1 V, -2 V.

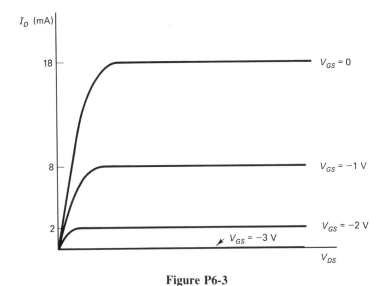

Figure P6-3

6-4. The overall drain characteristics of the idealized JFET of Problem 6-2 are shown in Figure P6-4. **(a)** Determine the gate–source cutoff voltage and the pinchoff voltage. **(b)** Determine the transition voltage (from ohmic to beyond pinchoff regions) for the following values of V_{GS}: 0, -0.5 V, -1 V, -1.5 V.

Field-Effect Transistors   Chap. 6

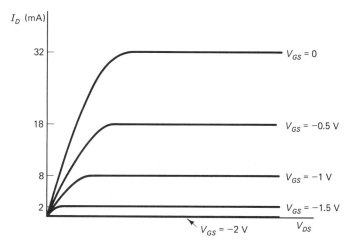

Figure P6-4

6-5. Refer to the idealized n-channel JFET characteristics of Figure P6-3. **(a)** Determine the zero-bias drain current. **(b)** Write an equation for I_D in the beyond-pinchoff region. **(c)** Verify that the result of (b) predicts the correct values of current for the curves of Figure P6-3.

6-6. Refer to the idealized n-channel JFET characteristics of Figure P6-4. **(a)** Determine the zero-bias drain current. **(b)** Write an equation for I_D in the beyond-pinchoff region. **(c)** Verify that the result of (b) predicts the correct values of current for the curves of Figure P6-4.

6-7. Using the results of Problem 6-5, determine for the given transistor the values of V_{GS} that produce the following drain current values (in beyond-pinchoff region): **(a)** 5 mA; **(b)** 10 mA; **(c)** 15 mA.

6-8. Using the results of Problem 6-6, determine for the given transistor the values of V_{GS} that produce the following drain current values (in beyond-pinchoff region): **(a)** 5 mA; **(b)** 12 mA; **(c)** 25 mA.

6-9. For the idealized JFET drain characteristics of Figure P6-3, construct the transfer characteristic.

6-10. For the idealized JFET drain characteristics of Figure P6-4, construct the transfer characteristic.

6-11. **(a)** Write an equation for the conductance in the linear portion of the ohmic region for the JFET of Problem 6-1. You may need to use some of the results of Problems 6-3 and 6-5. **(b)** Use the results of (a) to compute the conductance for each of the lines of Figure P6-1, and compare with the results of Problem 6-1.

6-12. **(a)** Write an equation for the conductance in the linear portion of the ohmic region for the JFET of Problem 6-2. You may need to use some of the results of Problems 6-4 and 6-6. **(b)** Use the results of (a) to compute the conductance for each of the lines of Figure P6-2 and compare with the results of Problem 6-2.

6-13. A certain n-channel JFET has a gate–source cutoff voltage of -5 V and a zero-bias drain current of 12 mA. Assume that the JFET is biased with a gate–source voltage of -2 V. Determine **(a)** the ideal drain–source resistance in the linear portion of the ohmic region and **(b)** the ideal drain current in the beyond-pinchoff region.

6-14. A certain p-channel JFET has a gate–source cutoff voltage of 4 V and a zero-bias drain current of 1 mA. Assume that the JFET is biased with a gate–source voltage of 1 V. Determine **(a)** the ideal drain–source resistance in the linear portion of the ohmic region and **(b)** the ideal drain current in the beyond-pinchoff region.

6-15. Draw a beyond-pinchoff dc model for the JFET of Problem 6-13.

6-16. Draw a beyond-pinchoff dc model for the JFET of Problem 6-14.

6-17. Assume that the drain characteristics of Figure P6-3 apply to some hypothetical n-channel depletion-mode IGFET. Calculate the idealized values of drain current in the beyond-pinchoff region for the following positive values of gate–source voltage: **(a)** 0.5 V; **(b)** 1 V.

6-18. Assume that the drain characteristics of Figure P6-4 apply to some hypothetical n-channel depletion-mode IGFET. Calculate the idealized values of drain current in the beyond-pinchoff region for the following positive values of gate–source voltage: **(a)** 0.5 V; **(b)** 1 V.

6-19. The drain characteristics of an idealized n-channel enhancement-mode IGFET are shown in Figure P6-19. **(a)** Determine the gate–source threshold voltage. **(b)** Determine the transition voltage (from ohmic to beyond-pinchoff region) for the following values of V_{GS}: 3 V, 4 V, 5 V, 6 V.

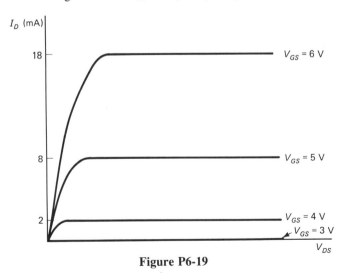

Figure P6-19

6-20. The drain characteristics of an idealized n-channel enhancement-mode IGFET are shown in Figure P6-20. **(a)** Determine the gate–source threshold voltage. **(b)** Determine the transition voltage (from ohmic to beyond-pinchoff region) for the following values of V_{GS}: 2 V, 2.5 V, 3.5 V, 4 V.

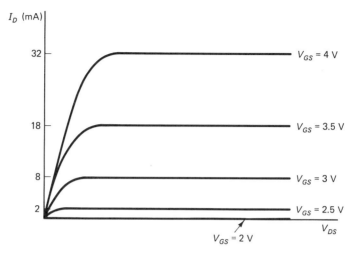

Figure P6-20

6-21. Refer to the idealized n-channel enhancement-mode characteristics of Figure P6-19. **(a)** Determine the value of K. **(b)** Write an equation for I_D in the beyond-pinchoff region. **(c)** Verify that the result of (b) predicts the correct values of current for the curves of Figure P6-19.

6-22. Refer to the idealized n-channel enhancement-mode characteristics of Figure P6-20. **(a)** Determine the value of K. **(b)** Write an equation for I_D in the beyond-pinchoff region. **(c)** Verify that the result of (b) predicts the correct values of current for the curves of Figure P6-20.

6-23. Using the results of Problem 6-21, determine for the given transistor the values of V_{GS} that produce the following values of drain current (in the beyond-pinchoff region): **(a)** 5 mA; **(b)** 12 mA; **(c)** 20 mA.

6-24. Using the results of Problem 6-22, determine for the given transistor the values of V_{GS} that produce the following values of drain current (in the beyond-pinchoff region): **(a)** 5 mA; **(b)** 12 mA; **(c)** 25 mA.

6-25. For the idealized IGFET drain characteristics of Figure P6-19, construct the transfer characteristic.

6-26. For the idealized IGFET drain characteristics of Figure P6-20, construct the transfer characteristic.

6-27. A certain n-channel enhancement-mode IGFET starts to conduct when the gate–source voltage is 3 V. When the gate–source voltage is 4 V, the drain current is 5 mA. Determine the drain current when the gate–source voltage is 5 V.

6-28. A certain p-channel enhancement-mode IGFET starts to conduct when the gate–source voltage is -2 V. When the gate–source voltage is -3 V, the drain current is 6 mA. Determine the drain current when the gate–source voltage is -4 V.

6-29. Draw a beyond-pinchoff dc model for the IGFET of Problem 6-27.

6-30. Draw a beyond-pinchoff dc model for the IGFET of Problem 6-28.

6-31. The drain current in a certain n-channel JFET can be expressed as

$$I_D = 6 \times 10^{-3}\left(1 + \frac{V_{GS}}{2.3}\right)^2$$

Determine **(a)** the gate–source cutoff voltage and **(b)** the zero-bias drain current.

6-32. The drain current in a certain p-channel JFET can be expressed as

$$I_D = 0.018\left(1 - \frac{V_{GS}}{4.5}\right)^2$$

Determine **(a)** the gate–source cutoff voltage and **(b)** the zero-bias drain current.

6-33. The drain current in a certain n-channel enhancement-mode IGFET can be expressed as

$$I_D = 0.012\left(\frac{V_{GS}}{2.8} - 1\right)^2$$

Determine the threshold voltage.

6-34. The drain current in a certain p-channel enhancement-mode IGFET can be expressed as

$$I_D = 0.002(V_{GS} + 4)^2$$

Determine the threshold voltage.

6-35. A certain n-channel JFET has $V_{GS(\text{off})} = -3$ V. Identify the region of operation (*cutoff*, *ohmic*, or *beyond-pinchoff*) for each of the following conditions: **(a)** $V_{GS} = -2$ V, $V_{DS} = 12$ V; **(b)** $V_{GS} = -1$ V, $V_{DS} = 1$ V; **(c)** $V_{GS} = -4$ V, $V_{DS} = 12$ V; **(d)** $I_D = 5$ mA, $V_{DS} = 8$ V.

6-36. A certain p-channel JFET has $V_{GS(\text{off})} = 4$ V. Identify the region of operation (*cutoff*, *ohmic*, or *beyond-pinchoff*) for each of the following conditions: **(a)** $V_{GS} = 5$ V, $V_{DS} = -15$ V; **(b)** $V_{GS} = 2$ V, $V_{DS} = -12$ V; **(c)** $V_{GS} = 1$ V, $V_{DS} = -2$ V; **(d)** $I_D = 8$ mA, $V_{DS} = -10$ V.

6-37. A certain *n-channel* enhancement-mode IGFET has $V_{GS(\text{th})} = 3$ V. Identify the region of operation (*cutoff*, *ohmic*, or *beyond-pinchoff*) for each of the following conditions: **(a)** $V_{GS} = 2$ V, $V_{DS} = 10$ V; **(b)** $V_{GS} = 5$ V; $V_{DS} = 3$ V; **(c)** $V_{GS} = 5$ V, $V_{DS} = 1$ V.

6-38. A certain p-channel enhancement-mode IGFET has $V_{GS(th)} = -4$ V. Identify the region of operation (*cutoff*, *ohmic*, or *beyond-pinchoff*) for each of the following conditions: **(a)** $V_{GS} = -7$ V, $V_{DS} = -2$ V; **(b)** $V_{GS} = -5$ V, $V_{DS} = -12$ V; **(c)** $V_{GS} = 0$ V, $V_{DS} = -10$ V.

DERIVATION PROBLEMS

6-39. Starting with (6-10) for I_D expressed in terms of V_{GS} for a depletion-mode FET, derive (6-12) for V_{GS} expressed in terms of I_D.

6-40. Starting with (6-43) for I_D expressed in terms of V_{GS} for an enhancement-mode FET, derive (6-44) for V_{GS} expressed in terms of I_D.

TROUBLESHOOTING PROBLEMS

6-41. While troubleshooting a circuit containing an n-channel JFET that *should* be operating in the beyond-pinchoff region, the following ground-referred voltages are measured: $V_S = 15$ V, $V_D = 36$ V, $V_G = 1$ V. What can you conclude about the *likely* region of operation for the transistor?

6-42. While troubleshooting a circuit containing a p-channel JFET that *should* be operating in the ohmic region, the following ground-referred voltages are measured: $V_D = 5$ V, $V_S = 30$ V, $V_G = 30$ V. What can you conclude about the *likely* region of operation for the transistor?

COMPUTER PROBLEMS

6-43. Write a program to determine the drain current for a JFET or depletion-mode IGFET (either n-channel or p-channel) in the beyond-pinchoff region.

   Input data: $V_{GS(off)}$, I_{DSS}, V_{GS}
   Output data: I_D

6-44. Write a program to determine the drain current for an enhancement-mode IGFET (either n-channel or p-channel) in the beyond-pinchoff region.

   Input data: $V_{GS(th)}$, K, V_{GS}
   Output data: I_D

6-45. Write a program to determine the gate–source voltage for a JFET or depletion-mode IGFET (either n-channel or p-channel) in the beyond-pinchoff region corresponding to a given drain current.

   Input data: $V_{GS(off)}$, I_{DSS}, I_D
   Output data: V_{GS}

6-46. Write a program to determine the gate–source voltage for an enhancement-mode IGFET (either n-channel or p-channel) in the beyond-pinchoff region corresponding to a given drain current.

Input data: $V_{GS(\text{th})}$, K, I_D
Output data: V_{GS}

CHAPTER 7

FET-Resistor Circuit Applications

OBJECTIVES

After completing this chapter, the reader should be able to:

1. Construct a dc load line for the drain circuit of an FET.
2. Demonstrate how changes in different transistor circuit parameters and variables affect the load line.
3. Analyze or design a circuit in which an FET is used as a switch.
4. Show how an FET biased in the ohmic region can be used as a voltage-controlled attenuator.
5. Discuss the different approaches for biasing an FET and their relative stability.
6. Analyze or design a self-bias circuit for an FET.
7. Construct a bias line on the transfer characteristic.
8. Discuss the operation of a self-plus-fixed bias circuit.
9. Analyze or design a source-follower circuit.

7-1 OVERVIEW

The field-effect transistor was introduced in Chapter 6, and the terminal characteristics were developed in detail. We are now ready to begin the consideration of various circuits that utilize the FET.

In this chapter all circuits considered will utilize only transistors and resistors (along with dc bias sources). Primary focus will be on dc analysis and operating characteristics. However, many of these circuits can be used with time-varying inputs and outputs provided that it is the instantaneous total voltages and currents that are of interest.

Among the applications to be considered are FET switches, voltage-controlled resistors, current sources, and source-follower circuits. Such circuits find applications in interfacing between integrated circuits, in buffers and drivers, for switching and multiplexing, and in electronically adjustable-gain control circuits.

The establishment of stable dc bias circuits for the different types of FETs is considered. Both the analysis of existing circuits and the design of these circuits are discussed.

7-2 FET DRAIN–SOURCE LOAD LINE

The concept of the load line for a BJT was introduced in Chapter 5. It is appropriate to begin the discussion of FET dc circuit analysis by showing a similar load line for a relatively simple circuit configuration. The transition from the linear ohmic region to the beyond-pinchoff region is more gradual for the FET than the transition from the saturation region to the active region for the BJT. The load line will help us visualize the process better.

Consider the circuit of Figure 7-1(a) containing an n-channel JFET, a resistance R_D in series with the drain, and two dc bias supplies. As it stands, the circuit is not necessarily intended to represent any useful application, but concepts inherent in this circuit will arise in practical application circuits later. Observe the directions of the polarities of the two bias sources for the n-channel JFET.

No current is assumed to flow in the gate. The equation for the gate–source portion of the circuit is simply

$$V_{GS} = -V_{GG} \tag{7-1}$$

Application of KVL for the drain–source loop in a counterclockwise direction yields

$$-V_{DD} + R_D I_D + V_{DS} = 0 \tag{7-2}$$

Solving for I_D, we obtain

$$\boxed{I_D = \left(-\frac{1}{R_D}\right)V_{DS} + \frac{V_{DD}}{R_D}} \tag{7-3}$$

This equation is of the form $y = mx + b$, and it is analogous to the collector–emitter loop equation of a BJT as given by (5-132).

Typical JFET drain characteristics are shown in Figure 7-1(b), and the linear equation of (7-3) may readily be superimposed on these curves. The vertical intercept is $-V_{DD}/R_D$, the horizontal intercept is V_{DD}, and the slope of the load line is $-1/R_D$.

Most of the properties discussed in Chapter 5 for the BJT load line apply here. The Q-point is determined from the intersection of the curve corresponding to $V_{GS} = -V_{GG}$ with the load line. If V_{GS} is varied, movement of the Q-point takes place along the load line. Once again, we see the *voltage control* property of the FET, which is in contrast to the *current control* property of the BJT. The subscript Q is added to any pertinent variables at the Q-point, as noted.

(a) Circuit

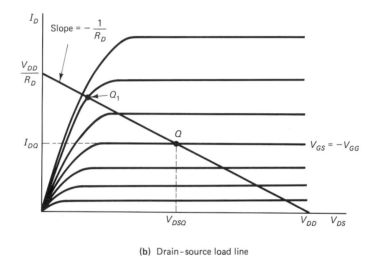

(b) Drain-source load line

Figure 7-1   JFET circuit and drain–source load line.

Suppose that V_{GS} were increased (made more positive) so that the Q-point moves to Q_1 in figure 7-1(b). We are now out of the pinchoff region and into the ohmic region. This may not always be apparent from the circuit model if a computation of V_{DS} is made from the mathematical relationship. A check of the transition voltage $V_{DS(tran)}$ would, of course, reveal that the beyond-pinchoff model would not be valid. However, the load-line interpretation helps us to visualize this process better.

If beyond-pinchoff operation is desired, make sure that the range of V_{GS} in the particular circuit is such that all applicable curves intersect with the load line in the flat region of the output characteristics. The calculated value of V_{DS} should always be greater than the pertinent transition voltage. If not, the beyond-pinchoff model is not valid, and some other approach must be used.

In applications involving the FET as a voltage-controlled resistor or as a switch, the load line should intersect the curves in the ohmic region. Furthermore, the range of V_{GS} (or other variables) must be such as to maintain operation in the ohmic region for that application.

A similar load line could be constructed for any of the FET types if the drain characteristics are known. The polarity of the gate–source voltage will depend, of course, on whether the device belongs to the depletion-mode group or the enhancement-mode group. For example, if the device is an n-channel enhancement-mode IGFET, the values of V_{GS} corresponding to the different output characteristics are positive and larger than $V_{GS(\text{th})}$.

Example 7-1

The circuit of Figure 7-2(a) contains an n-channel JFET. The JFET drain characteristics are assumed to be idealized and equal to those considered in Figure 6-6. Verify analytically that operation is in the beyond-pinchoff region, and determine I_D and V_{DS}.

(a)   **(b)**

Figure 7-2   Circuit of Example 7-1 and equivalent-circuit model.

Solution

A brief review of Figure 6-6 indicates that $I_{DSS} = 16$ mA and $V_{GS(\text{off})} = -4$ V. Since $V_{GS} = -1$ V in this circuit, it is obvious that the FET is not in the cutoff region. Strictly speaking, we do not yet know whether the FET is operating in the beyond-pinchoff region or in the ohmic region. We will proceed on the assumption that the circuit is operating in the beyond-pinchoff region and, subsequently, verify the validity of that assumption. An equivalent-circuit model is shown in Figure 7-2(b).

In the beyond-pinchoff region, the equation for I_D is

$$I_D = 16 \times 10^{-3}\left(1 - \frac{V_{GS}}{-4}\right)^2 \tag{7-4}$$

For $V_{GS} = -1$ V, we have

$$I_D = 16 \times 10^{-3}\left(1 - \frac{-1}{-4}\right)^2 = 9 \times 10^{-3} \text{ A} = 9 \text{ mA} \qquad \text{(7-5)}$$

For this particular value of V_{GS}, we could have determined I_D from Figure 6-6. The value of V_{DS} is

$$V_{DS} = 24 - (9 \times 10^{-3}) \times (1.2 \times 10^3) = 24 - 10.8 = 13.2 \text{ V} \quad \text{(7-6)}$$

The transition voltage for $V_{GS} = -1$ V is $V_{DS(\text{tran})} = 3$ V. Since the calculated value of V_{DS} is much greater than the transition voltage, assumption of beyond-pinchoff region operation is valid.

Example 7-2

For the circuit of Figure 7-2(a), assume that the drain resistance is changed to $R_D = 3$ kΩ. Attempt to repeat the analysis of Example 7-1.

Solution

The alert reader is probably aware that there is something different here because of the term "attempt." Proceeding with the questionable assumption of beyond-pinchoff operation, the calculated drain current would be the same as before (i.e., 9 mA). If a calculation of V_{DS} is attempted, the value obtained would be $V_{DS} = 24 - 9 \times 10^{-3} \times 3 \times 10^{-3} = 24 - 27 = -3$ V. This result is impossible! Not only is this value smaller (by 6 V) than the transition voltage of 3 V, but it is negative.

The difficulty lies in the fact that operation has shifted into the ohmic region, and the beyond-pinchoff model is no longer valid. If the calculated value of V_{DS} is less than the transition voltage (i.e., if $V_{DS} < V_{DS(\text{tran})}$), the circuit is operating in the ohmic region.

The situation here is analogous to that of the BJT when saturation occurs. For FET circuits not designed specifically for switching purposes, however, the ohmic region may represent a substantial range of V_{DS}, and a simple assumption such as $V_{DS} \simeq 0$ may represent a gross error.

If operation is shifted far enough into the ohmic region to justify the constant resistance model given in Chapter 6, a complete solution may be possible. This concept is considered in Section 7-4. However, the drain characteristics are somewhat nonlinear in the transition region between the linear part of the ohmic region and the beyond-pinchoff region, and a simple analysis is not possible. Although nonlinear models for this region have been developed, their lack of utility does not justify further consideration here.

Based on the material developed thus far, about all we can say is that V_{DS} is somewhere between 0 and 3 V and I_D is less than 9 mA. We will show in Example 7-3 how a better solution and better visualization can be achieved through the use of the load line.

Example 7-3

Construct load lines for the circuit conditions of **(a)** Example 7-1 and **(b)** Example 7-2.

Solution

The idealized output characteristics were originally given in Figure 6-6, and they are shown again in Figure 7-3. The horizontal scale of the latter figure has been extended somewhat because of the required range of V_{DS}.

(a) In Example 7-1, V_{DD} = 24 V and R_D = 1.2 kΩ. The horizontal intercept is 24 V, and the vertical intercept is 24 V/1.2 × 10³ Ω = 20 × 10⁻³ A = 20 mA. The resulting load line is L1 in Figure 7-3. The Q-point is determined from the intersection of L1 with the curve corresponding to V_{GS} = −1 V, and it is indicated as Q_1. The values of I_D and V_{DS} corresponding to the intersection are 9 mA and 13.2 V, respectively. These values are in agreement with the results of Example 7-1.

(b) In Example 7-2, V_{DD} = 24 V and R_D = 3 kΩ. The horizontal intercept is again 24 V, but the vertical intercept is 24 V/3 × 10³ Ω = 8 × 10⁻³ A = 8 mA. The resulting load line is L2 in Figure 7-3. The Q-point in this case is indicated as Q_2, and it is in the ohmic region. This portion of the ohmic region is somewhat nonlinear and we can only make graphical estimates of I_D and V_{DS}. Reasonable estimates are $I_D \simeq$ 7.2 mA and $V_{DS} \simeq$ 2.2 V.

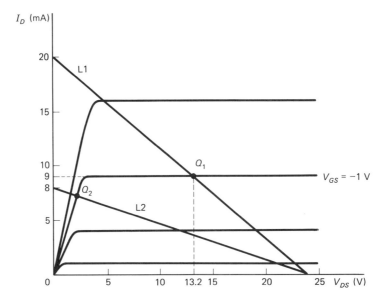

Figure 7-3   Drain characteristics and load lines of Example 7-2.

Field-effect transistors may be used as electronic switches in a manner quite similar to that of bipolar junction transistors. In fact, well-designed FET switches have certain advantages over BJT switches. Although it was largely ignored in Chapter 5, a saturated BJT always exhibits a small offset $V_{CE(\text{sat})}$. In contrast, an FET acts as a purely passive resistor near the origin in the lower portion of the ohmic region.

A rather extensive line of special-purpose FETs are manufactured and marketed for the specific application of electronic switching. These units are referred to as *analog switches*. Most of them are integrated-circuit modules containing one or more FETs along with necessary drive and level conversion circuitry. The input control levels are often chosen to be compatible with standard digital logic circuits.

Consideration of these special-purpose ICs is not within the intended objective of this book. Instead, we will investigate some of the concepts of switching utilizing basic FETs. These basic FETs can be used for limited noncritical switching functions. Most important for our purposes, they should help the reader to understand the general concept basic to more complex switching circuit configurations.

For any FET device used as a switch, the control signal must be capable of switching between two extremes: (1) cutoff region and (2) ohmic region. In the cutoff region, the FET will act like an open circuit, and the resulting switch will be open. In the ohmic region, the FET will act like a resistor. The design must be such that the resulting value of V_{DS} in the ohmic region is very small compared with the source voltage so that the FET approaches a short circuit. Said differently, the FET equivalent resistance in the ohmic region should be small compared with the net circuit series resistance.

Depending on whether the FET is a depletion-mode device or an enhancement-mode device and whether it is an n-channel or a p-channel device, a variety of possible control voltage levels and switched levels are possible. The cases that will be considered are intended as representative of the many possibilities.

N-Channel JFET Switch

Consider the n-channel JFET shown in Figure 7-4(a), in which a load is connected between the drain and a positive voltage V_{DD}. When the control input signal v_i is

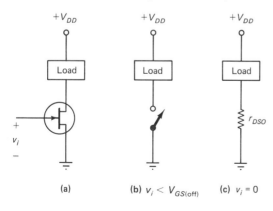

(a)   (b) $v_i < V_{GS(\text{off})}$ (c) $v_i = 0$

Figure 7-4 Electronic switch using an n-channel JFET to close switch with zero voltage.

more negative than $V_{GS(off)}$ (i.e., $v_i < V_{GS(off)}$), the FET is at cutoff, and the equivalent circuit is shown in Figure 7-4(b).

Assume next that $v_i = 0$. Operation now shifts to the $V_{GS} = 0$ curve. If the circuit has been designed properly, the intersection of the load line with the $V_{GS} = 0$ curve will be in the ohmic region, and the FET will act like a small resistance r_{DSO}. The equivalent circuit is shown in Figure 7-4(c). Most of the supply voltage V_{DD} will appear across the load, although a small portion will appear across the FET.

It was assumed in the preceding development that the load line intersected the $V_{GS} = 0$ curve in the ohmic region, and that assumption is fundamental to the operation of the FET as a switch. With a poor design, it is possible that the intersection will occur in the beyond-pinchoff region, and the desired switching operation will not be achieved. Once again, we see the utility of the load line in visualizing the process.

N-Channel Enhancement-Mode IFGET Switch

The same switching arrangement considered before, but with an n-channel enhancement-mode IGFET, is shown in Figure 7-5(a). The control input voltage v_i in this case consists of two nonnegative levels. If the control voltage is less than the threshold voltage (i.e., $v_i < V_{GS(th)}$), the FET is an open circuit as shown in (b). The value $v_i = 0$ would be a good choice in this case.

If $v_i > V_{GS(th)}$, and if the load line is such that operation shifts to the ohmic region, the equivalent circuit is of the form shown in Figure 7-5(c). In practice, this control level should be sufficiently larger than $V_{GS(th)}$ to shift the value of V_{DS} to a very small voltage level.

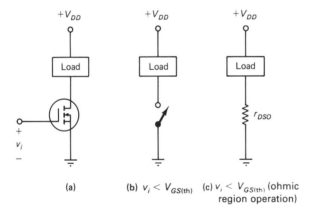

(a) (b) $v_i < V_{GS(th)}$ (c) $v_i < V_{GS(th)}$ (ohmic region operation)

Figure 7-5 Electronic switch using an n-channel enhancement-mode IGFET to close switch with positive voltage.

Example 7-4

It is desired to switch on and off the 6-kΩ load in the circuit of Figure 7-6. The FET characteristics are assumed to be those considered in Figure 6-6 and again in Figure 7-3. **(a)** Using a load line, show that the circuit is capable of

+24 V

6 kΩ v_L

v_i

Figure 7-6   Circuit of Example 7-4.

operating as a switch. **(b)** Determine the required control levels for v_i. **(c)** Estimate the voltage v_L across the load when the switch is on.

Solution

(a) The drain characteristics are shown again in Figure 7-7. To construct the load line, we first determine the horizontal and vertical intercepts. The horizontal intercept is 24 V, and the vertical intercept is 24 V$/6 \times 10^{-3}$ Ω $= 4 \times 10^{-3}$ A $= 4$ mA. The load line is then superimposed on the drain characteristics as shown in Figure 7-7. The intersection of the load line with the curves corresponding to the larger (less negative) values of V_{GS} is in the ohmic region, so operation as a switch is feasible.

(b) For $v_i = V_{GS(off)} = -4$ V, the FET is at cutoff, as indicated by Q_1 in Figure 7-7. In practical FETs, cutoff is not as abrupt as suggested by these idealized curves. To ensure cutoff, a voltage more negative than $V_{GS(off)}$ would probably be used. (Of course, it cannot be as negative as the gate–source reverse breakdown voltage.)

For $v_i = 0$, operation shifts over to Q_2, and the FET is well into the

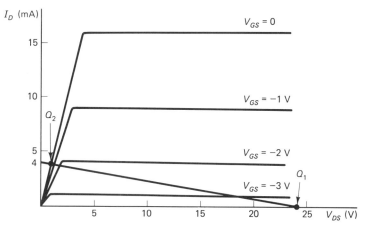

Figure 7-7   Drain characteristics of Example 7-4.

ohmic region. One might be tempted to increase the gate drive above this level, but this should *not* be done for the JFET. The reason is that the gate–source diode would then be forward biased, and this would drastically alter its operation. Thus $v_i = 0$ is the usual maximum positive limit for the *n*-channel JFET.

(c) A reasonable estimate for V_{DS} at point Q_2 is $V_{DS} \approx 1$ V. The load voltage is then $V_L \approx 24 - 1 = 23$ V. Thus we are not able to achieve perfect switching with this idealized "general-purpose" FET. Special-purpose analog switches could do much better.

7-4 VOLTAGE-CONTROLLED RESISTORS

We have observed that an FET acts as a voltage-controlled resistance in the ohmic region. This property may be used in a number of possible applications, such as voltage-controlled gain circuits, stabilization circuits, and tuning circuits. In this section, the concept is discussed in more detail and a representative example is shown.

In the same sense as for FET switches, we will choose to discuss the application of general-purpose FETs for this purpose. This should permit the reader to understand the concept of how the special-purpose devices work for this application.

Any circuit utilizing an FET for creating a voltage-controlled resistance must be designed so that operation is in the ohmic portion of the drain characteristics. Although the nonlinear portion may suffice for certain noncritical situations, the most exacting applications require operation in the linear portion of the ohmic region. In general, this means that the value of V_{DS} for the FET must be maintained at a relatively low lever. If the signal level is very low, this condition is usually easy to achieve. If the signal level is moderate or large, a large drain resistance would be required to drop the voltage across the FET to a small level.

Refer to the expanded drain characteristics of Figure 7-8. These curves repre-

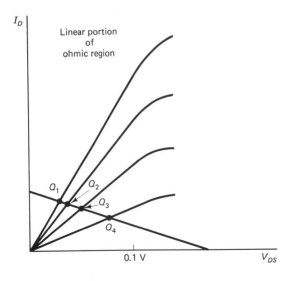

Figure 7-8 Expanded scale of drain characteristics in lower part of ohmic region and load line for voltage-controlled resistance.

sent those of a typical n-channel JFET in the lower portion of the ohmic region, in which each curve has a nearly constant slope over a small range. The value 0.1 V is typical and may vary with different units. The load line of an applicable circuit should intersect the characteristic curves in the linear region as indicated.

For the representative load line shown, the horizontal intercept is relatively small (< 0.1 V), indicating a low-voltage application. By varying V_{GS}, operation may be shifted between the different Q-points shown. At each point, the resistance is different. Higher input voltages are possible as long as the load line intersects in the pertinent portion of the output characteristics.

Example 7-5

The circuit of Figure 7-9 is to be used as a voltage-controlled *attenuator*. (*Attenuation* is the process of reducing a signal level, and an *attenuator* is a circuit that provides attenuation.) The FET characteristics in the ohmic region are assumed to be idealized and equal to those considered in Example 6-6. Let $A = v_o/v_i$ represent the signal transfer ratio. This transfer ratio is to be governed by the control voltage V_{GS}. The signal v_i is a small positive voltage with a peak value not exceeding 0.1 V. Compute the values of A corresponding to the following values of V_{GS}: **(a)** 0; **(b)** -1 V; **(c)** -2 V; **(d)** -3 V; **(e)** -4 V.

Figure 7-9 Voltage-controlled attenuator of Example 7-5.

Solution

The hypothetical JFET was assumed to have linear characteristics for a range of V_{DS} up to about 0.1 V, and since the signal level will not exceed a peak level of 0.1 V, operation in the linear portion of the ohmic region should be ensured.

In Example 6-6, the conductance g_{DS} of this JFET was determined to be

$$g_{DS} = 8 \times 10^{-3}\left(1 - \frac{V_{GS}}{-4}\right) \tag{7-7}$$

The corresponding resistance $r_{DS} = 1/g_{DS}$ is

$$r_{DS} = \frac{1}{g_{DS}} = \frac{125}{1 - V_{GS}/(-4)} \tag{7-8}$$

The transfer, or output/input, ratio of the circuit in Figure 7-9 is determined by a simple application of the voltage-divider rule with r_{DS} representing the resistance of the JFET. We have

$$A = \frac{v_o}{v_i} = \frac{r_{DS}}{r_{DS} + 2000} \qquad (7\text{-}9)$$

For each value of V_{GS}, the value of r_{DS} is computed from (7-8). This value is then substituted in (7-9) to determine A. A table of results follows.

	V_{GS} (V)	r_{DS} (Ω)	A
(a)	0	125	0.0588
(b)	-1	166.67	0.0769
(c)	-2	250	0.1111
(d)	-3	500	0.2
(e)	-4	∞	1

The voltage transfer ratio is varied form 0.0588 to 1 as V_{GS} is varied from 0 to -4 V. This illustrates one way in which the level of a signal can be controlled electronically by an external control signal using an FET as a voltage-controlled resistor.

By using a voltage-controlled attenuator in conjunction with a fixed amplifier gain, the net effect is that of a voltage-controlled gain circuit. Such circuits are very important for achieving automatic gain adjustment in systems using feedback techniques.

7-5 DEPLETION-MODE FET BIAS CIRCUITS

In this section, we investigate some of the most common circuit configurations for establishing dc bias for depletion-mode FETs. Bias circuits are required for providing appropriate operating conditions with various applications, such as amplifiers.

For bias analysis we lump together JFETs and depletion-mode IGFETs since the basic approaches in the circuit designs are virtually identical. (Strictly speaking, depletion-mode IGFETs can also be operated in the enhancement-mode, but that situation can be analyzed from the material given in Section 7-6.)

For ease in the development, n-channel units will be considered. The circuit configurations given can all be used for p-channel units if the polarities of all power supplies are reversed.

Fixed Bias

The most obvious way to establish an operating bias for a JFET is to use a separate fixed power supply. Consider the circuit of Figure 7-10 containing an n-channel JFET in a common-source configuration. The drain is connected through a drain re-

Figure 7-10 Fixed bias for an n-channel JFET using a separate power supply.

sistance R_D to a positive dc voltage $+V_{DD}$. Since the required value of V_{GS} for an n-channel JFET is negative, the gate is connected through a resistance R_G to a negative dc voltage $-V_{GG}$.

Until we study amplifier circuits, the reader may wonder why the negative dc voltage is not connected directly to the gate. The reason is that when an ac signal is applied to the input, there must be a resistance across which the input signal can be developed. Without R_G, the gate voltage can never vary and no signal amplification can occur. A similar property holds for the drain circuit with respect to R_D. We postpone full consideration to a later chapter, but realize that these resistors will eventually be required.

The actual gate current in an FET is extremely small and will be neglected for this analysis. Therefore, the dc voltage drop across R_G is essentially zero, and the value of V_{GS} is

$$V_{GS} = -V_{GG} \tag{7-10}$$

Thus the gate–source voltage is simply equal to the value of the negative bias supply. For beyond-pinchoff operation, which is normal for most amplifier applications, the drain current is simply

$$I_D = I_{DSS}\left[1 - \frac{V_{GS}}{V_{GS(\text{off})}}\right]^2 \tag{7-11}$$

While the process seems straightforward as long as I_{DSS} and $V_{GS(\text{off})}$ are known, this circuit has only limited application in production situations. The values of I_{DSS} and $V_{GS(\text{off})}$ of a given type vary widely from one transistor to another, and it is usually impractical to adjust V_{GS} to compensate.

To illustrate the potential difficulty, let us return to the concept of the transfer characteristic curve introduced in Section 6-4. Consider the two transconductance curves T1 and T2 shown in Figure 7-11. Curve T1 has parameters $I_{DSS}^{(1)}$ and $V_{GS(\text{off})}^{(1)}$, and curve T2 has parameters $I_{DSS}^{(2)}$ and $V_{GS(\text{off})}^{(2)}$. Assume that a certain drain current $I_D^{(1)}$ has been established with a fixed bias $V_{GS} = -V_{GG}$ for an FET with the characteristic T1. Now assume that for the same bias voltage, the first FET is replaced with one having the transfer characteristic T2. The drain current changes to a new value $I_D^{(2)}$,

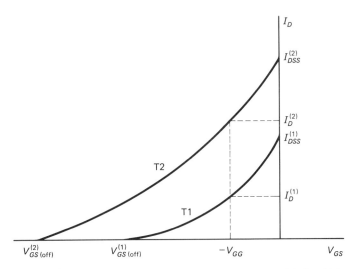

Figure 7-11 Illustration of shift in operating point with fixed bias.

which is more than double the first value in the illustration. The basic parameters for a given FET type can vary by 4:1 or more, so the potential stability of the operating point is very poor.

In summary, fixed bias circuits are practical only for limited applications where it is feasible to adjust the bias for individual transistors or, in some applications, where the actual bias level is not critical.

Self-Bias

We now investigate a circuit that can be used to provide a degree of stabilization of the operating point while eliminating the necessity of providing a separate power supply for the gate bias. A portion of the drain power supply voltage is used to establish bias, and because of the manner in which the bias is generated, it is referred to as *self-bias*.

Consider the circuit shown in Figure 7-12. As in the case of the fixed bias circuit, this circuit contains a drain resistance R_D and a gate resistance R_G. In addition, there is a source resistance R_S whose value is critical in establishing bias, as we will see shortly.

In the same fashion as for the fixed bias circuit, the resistance R_G serves as a load across which an input signal can be developed in an amplifier configuration. However, it also serves to provide a dc path between gate and ground, which is necessary in the self-bias circuit.

The value of R_G is usually quite large (typically, 100 kΩ or greater), and its exact value is usually not critical. However, R_S is much smaller (typically, < 1 kΩ), and its value has a critical effect on the Q-point.

We will start at the gate terminal and form a clockwise loop through the source resistor to ground and back to the gate. As this loop is formed, we will recognize

$+V_{DD}$

$\downarrow I_D$

R_D

$+$
V_{GS} $-$
$+$
$\downarrow I_D$

R_G

R_S

$-$

Figure 7-12 Self-bias circuit for an n-channel JFET.

that the voltage across R_S is $R_S I_D$. However, since we are assuming no gate current in the ideal JFET, the voltage drop across R_G is zero, as noted earlier. The loop equation thus reads

$$V_{GS} + R_S I_D + 0 = 0 \qquad (7\text{-}12)$$

where the last 0 on the left represents the assumed zero drop across R_G.

There is, of course, a very small gate current and, in order to neglect the drop across R_G, we are assuming that the product of R_G times this gate current is small compared with the other voltages. This condition is usually achieved by keeping R_G below a few megohms.

Solving for V_{GS} in (7-12), we have

$$\boxed{V_{GS} = -R_S I_D} \qquad (7\text{-}13)$$

Thus the gate-to-source voltage is simply the negative of the voltage drop across the source resistance. Since $R_S I_D$ is a positive value for an n-channel JFET, V_{GS} is a negative value, as required. We see why the circuit is called a self-bias circuit, since the FET generates its own bias through the voltage drop across R_S.

A different way of visualizing this process is as follows: The gate voltage with respect to ground is zero since the gate is connected through R_G to ground and there is negligible voltage drop across R_G. However, the FET source terminal voltage is now positive with respect to ground because of the drop across R_S. Lifting the source potential above ground while maintaining the gate potential at ground results in the same bias effect as maintaining the source terminal at ground and placing a negative bias voltage at the gate.

Assuming beyond-pinchoff-region operation, the drain current is given by

$$I_D = I_{DSS}\left[1 - \frac{V_{GS}}{V_{GS(\text{off})}}\right]^2 \qquad (7\text{-}14)$$

Let's inspect (7-13) and (7-14) to see what we have determined thus far. From (7-14), I_D is a function of V_{GS}. However, from (7-13), V_{GS} is, in turn, a function of

Sec. 7-5   Depletion-Mode FET Bias Circuits   **281**

I_D. It seems as if we are going around in circles! The key to the problem is that there are two equations, and as long as we have no more than two unknowns, we can solve the set of equations simultaneously.

Because of the uncertainty of the parameters and the fact that the expression for I_D itself is approximate, the resulting solution should be considered approximate at best. Nevertheless, there is value in the solution in predicting the approximate operating point when the circuit is given.

Assume for the moment that I_{DSS}, $V_{GS(\text{off})}$, and R_S are known, and that the two unknowns are V_{GS} and I_D. Simultaneous solution of (7-13) and (7-14) results in a quadratic equation, for which there are two possible solutions for each unknown. However, one each of the values represents a physically impossible situation and is rejected.

Self-Bias Solution

The solution that follows has been developed in a general manner that can be utilized for any depletion-mode device (i.e., JFETs and depletion-mode IGFETs). Let $|V_{GS(\text{off})}|$ represent the magnitude of the cutoff voltage and let I_{DSS} represent the zero-gate–drain current. (We are defining I_{DSS} as a positive value for both n-channel and p-channel devices, so we do not need magnitude bars for it.) As a convenient parameter that simplifies the equation that follows, we define B as

$$B = \frac{|V_{GS(\text{off})}|}{R_S I_{DSS}} \qquad (7\text{-}15)$$

The drain current I_D for the self-bias circuit is given by

$$I_D = I_{DSS}\left(B + \frac{B^2}{2} - \sqrt{B^3 + \frac{B^4}{4}}\right) \qquad (7\text{-}16)$$

for either an n-channel or a p-channel device. The magnitude of gate–source voltage is then

$$|V_{GS}| = R_S I_D \qquad (7\text{-}17)$$

The value of V_{GS} is actually negative for an n-channel device and positive for a p-channel device.

Example 7-6

The self-bias circuit shown in Figure 7-13 contains the hypothetical idealized n-channel JFET whose drain characteristics were given in Figures 6-6, 7-3, and 7-7. Determine analytically the values of I_D, V_{GS}, V_{R_S}, V_{R_D}, and V_{DS}.

Figure 7-13 Circuit of Example 7-6.

Solution

The values of I_{DSS} and $V_{GS(off)}$ for this JFET were determined earlier to be $I_{DSS} = 16$ mA and $V_{GS(off)} = -4$ V. The analytical procedure involving equations (7-16) and (7-17) will be used to determine I_D. First, B from (7-15) is calculated as

$$B = \frac{4}{150 \times 16 \times 10^{-3}} = 1.6667 \qquad \textbf{(7-18)}$$

This quantity is substituted in (7-16) and I_D is determined as

$$I_D = 16 \times 10^{-3}\left[1.6667 + \frac{(1.6667)^2}{2} - \sqrt{(1.6667)^3 + \frac{(1.6667)^4}{4}}\right]$$
$$= 7.913 \times 10^{-3} \text{ A} = 7.913 \text{ mA} \qquad \textbf{(7-19)}$$

The value of V_{GS} for an n-channel JFET is negative and is

$$V_{GS} = -R_S I_D = -150 \times 7.913 \times 10^{-3} = -1.187 \text{ V} \qquad \textbf{(7-20)}$$

The voltage V_{R_S} across the source resistance has the same magnitude as V_{GS}, but the positive terminal is at the top as shown in Figure 7-13. Thus

$$V_{R_S} = 1.187 \text{ V} \qquad \textbf{(7-21)}$$

As indicated earlier, it is assumed that the gate current is negligible and that there is no voltage drop across the 470-kΩ gate resistor. The voltage V_{R_D} across the drain resistance is

$$V_{R_D} = 1.5 \times 10^3 \times 7.913 \times 10^{-3} = 11.87 \text{ V} \qquad \textbf{(7-22)}$$

The voltage V_{DS} across the FET is the supply voltage minus the drops across R_D and R_S.

$$V_{DS} = 24 - 11.87 - 1.187 = 10.94 \text{ V} \qquad \textbf{(7-23)}$$

7-6 BIAS LINE

A great deal of insight on self-bias can be gained from a graphical approach. Assume for the moment that I_{DSS} and $V_{GS(\text{off})}$ are known. Consider the transfer characteristic displaying I_D as a function of V_{GS} in the beyond-pinchoff region, as shown in Figure 7-14. From (7-13), we can solve for I_D in terms of V_{GS}, and we have

$$I_D = -\frac{1}{R_S}V_{GS} \qquad \textbf{(7-24)}$$

This equation is a straight-line equation of the form $y = mx + b$, in which the vertical intercept is $b = 0$. The straight line thus intersects the origin, and the slope is $m = -1/R_S$.

The straight line given by (7-24) can be superimposed on the transfer characteristic as shown in Figure 7-14. This line is called the *bias line*, and it has some similarities to the load line.

The intersection of the bias line with the transfer characteristic establishes the operating or Q-point. The resulting values of V_{GS} and I_D at the Q-point are denoted as V_{GSQ} and I_{DQ} on Figure 7-14. When self-bias is employed, it is more convenient to work with a bias line on the transfer characteristic than with the load-line and drain characteristics.

The process that occurs in the circuit is that I_D settles at a value such that the value of $V_{GS} = -R_S I_D$ is the required voltage for causing I_D. This circuit represents a form of negative feedback, a major topic considered later in the book.

The self-bias circuit provides a moderate degree of stability to changes in transistor characteristics. To illustrate this concept, refer to Figure 7-15. Assume initially that a given FET has the transfer characteristic T_1 with parameters $V_{GS(\text{off})}^{(1)}$ and $I_{DSS}^{(1)}$. For a particular value of R_S, operation is at the point Q_1, corresponding to $V_{GS}^{(1)}$ and $I_D^{(1)}$, as shown.

Assume that the FET is replaced by one having a different transfer characteristic T_2 with parameters $V_{GS(\text{off})}^{(2)}$ and $I_{DSS}^{(2)}$. If the circuit had employed fixed bias with a value $V_{GS}^{(1)}$, operation would shift all the way up to Q_3. However, with self-bias and the same source resistance, operation shifts to Q_2 instead. The value of V_{GS} and I_D at Q_2 are $V_{GS}^{(2)}$ and $I_D^{(2)}$, respectively. This represents a more moderate shift in the value of I_D than with fixed bias. There is, therefore, a degree of adjustment in the self-bias circuit that maintains a moderate degree of control on the operating-point current.

Stabilization in the bias circuit usually requires that the drain current be kept within a reasonable range of variation. To achieve more stabilization with the basic self-bias circuit, it is necessary that the magnitude of the slope of the bias line be

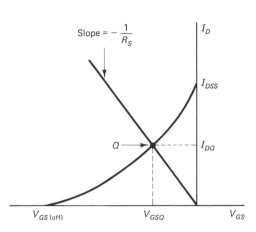

Figure 7-14 Bias line superimposed on transfer characteristic.

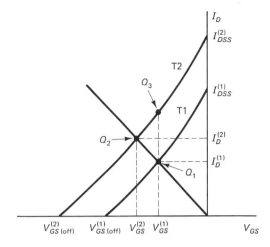

Figure 7-15 Change in operating point with self-bias circuit for different transistors.

smaller. In this fashion the change in I_D for two different Q-points would be less. However, there is a practical limit to this process, as demonstrated next.

Consider the two characteristics T_1 and T_2 in Figure 7-16. To minimize the variation in I_D, the bias line has been shown with a slope having a small magnitude. This corresponds to a relatively large value of R_S. Although I_D has a higher degree of stability in this case, there are two problems with this approach: (1) The operating point for I_D is forced to assume a relatively small value. Later we will see that the resulting gain in amplifier applications will be relatively small as a consequence. (2) A larger dc voltage across R_S will result with this approach, and less of the supply voltage is available for the transistor and the load.

In conclusion, the self-bias circuit is very useful for less critical circuits, where a degree of stability is sufficient. By combining the self-bias circuit with fixed bias, a greater amount of stability can be achieved without the undesirable results discussed previously. This is discussed in Section 7-8.

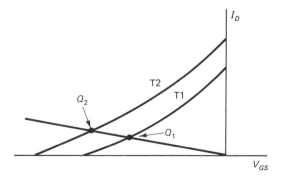

Figure 7-16 Illustration of how more stabilization with self-bias forces operation to small values of I_D.

Example 7-7

For the self-bias circuit of Example 7-6, determine the operating point using the bias line.

Solution

The transfer characteristic of the hypothetical JFET was given in Figure 6-7, and it is shown again in Figure 7-17. The bias line is a straight line through the origin with a slope $-1/R_S = -1/150 = -0.00667$ A/V. However, the easiest way to construct the bias line is to choose a convenient value of I_D, compute the corresponding voltage $V_{GS} = -150I_D$, and draw a straight line from the origin through this point. A convenient choice is $I_D = 10$ mA, in which case $V_{GS} = -150 \times 10^{-2} = -1.5$ V. Other points could have been chosen as well.

The intersection of this bias line with the transfer characteristic determines the operating point, and this is shown in Figure 7-17. It is obvious that the values of I_D and V_{GS} cannot be determined graphically to the degree of accuracy labeled. However, to show compatibility with the results of Example 7-6, the values obtained analytically have been labeled on the axes.

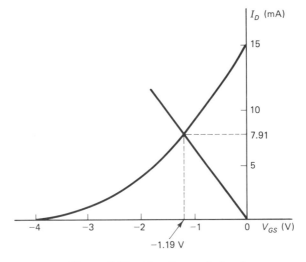

Figure 7-17 Bias-line analysis of Example 7-7.

7-7 DESIGN OF SELF-BIAS CIRCUIT

Consider the self-bias circuit of Figure 7-18 for an n-channel JFET. The design of this circuit involves the selection of a suitable operating point and the determination of the required source resistance R_S. Assume for the moment that the transconduc-

Figure 7-18 Self-bias circuit for JFET with variable for design identified.

tance characteristics are known *exactly* (i.e., the values of $V_{GS(off)}$ and I_{DSS} are known). For a desired operating-point drain current I_{DQ}, the value of V_{GSQ} that would produce this current is determined from either the transconductance graph or by solving the beyond-pinchoff equation.

Because of the nonlinear relationship between V_{GS} and I_D, values of voltage and current along the curve do not follow an intuitively simple form. Some normalized convenient values that aid in dealing with this relationship are summarized in Table 7-1. For example, $I_D = 0.5I_{DSS}$ when $V_{GS} = 0.2929V_{GS(off)}$, but $I_D = 0.25I_{DSS}$ when $V_{GS} = 0.5V_{GS(off)}$. The normalized curve of Figure 6-8 is also useful for this purpose.

TABLE 7-1 CONVENIENT GATE–SOURCE VOLTAGE AND DRAIN CURRENT BIAS COMBINATIONS

$V_{GS}/V_{GS(off)}$	I_D/I_{DSS}
0	1
0.1340	0.75
0.25	0.5625
0.2929	0.5
0.5	0.25
0.75	0.0625
1	0

Once V_{GSQ} is determined, R_S is then calculated as

$$R_S = \frac{|V_{GSQ}|}{I_{DQ}}$$

(7-25)

Referring to Figure 7-18, the power supply voltage V_{DD} is divided into three parts: (1) V_{R_D}, (2) V_{DS}, and (3) V_{R_S}. The value of V_{R_S} is, of course, the magnitude of V_{GSQ}; that is,

$$\boxed{V_{R_S} = |V_{GSQ}|} \qquad (7\text{-}26)$$

The remaining voltage is divided between V_{R_D} and V_{DSQ}, and R_D can be selected to achieve the required proportion. The required value of R_D is

$$\boxed{R_D = \frac{V_{R_D}}{I_D} = \frac{V_{DD} - V_{DSQ} - V_{R_S}}{I_D}} \qquad (7\text{-}27)$$

The preceding procedure is fine *if* we know the exact parameters for a given FET. Suppose, however, that the design is to be performed for a particular type of FET in which the range, but not the exact values, of the parameter is known. Because the self-bias circuit has only a moderate degree of stability, one must accept the fact that the operating point will vary somewhat with different transistors. The simplicity of the self-bias circuit may justify its use as long as there is sufficient leeway in the specifications to accommodate the expected variations.

Assume that for a given FET, the minimum and maximum values of I_{DSS} are $I_{DSS}^{(1)}$ and $I_{DSS}^{(2)}$, respectively. Similarly, assume that the values of $V_{GS(off)}$ corresponding to minimum and maximum magnitudes are $V_{GS(off)}^{(1)}$ and $V_{GS(off)}^{(2)}$, respectively. As a general rule, transistors with smaller values of I_{DSS} tend to correspond to smaller magnitudes of $V_{GS(off)}$, and vice versa. As a reasonable assumption, two extreme transconductance characteristic curves can be constructed, as shown in Figure 7-19. The curve T_1 is an assumed lower bound and T_2 is an assumed upper bound. All possible curves for the given FET are expected to fall in the bounded region, with some expected "tracking" of the characteristics.

Since R_S is a constant value, the bias line is unique and will intersect any applicable curve in the region. The range of possible drain currents will lie between $I_{DQ}^{(1)}$ and $I_{DQ}^{(2)}$, and the range of possible bias voltages will lie between $V_{GSQ}^{(1)}$ and $V_{GSQ}^{(2)}$.

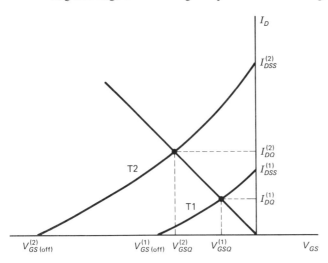

Figure 7-19 Range of operating points with self-bias circuit.

If the self-bias circuit is to be acceptable in a production sense, the extreme range of possible values must produce acceptable dc operating-point values. This usually means investigating the effects of the various circuit voltages and currents as well as amplifier gain parameters (to be studied later). For a given power supply voltage and a particular drain resistance R_D, the voltage drops V_{R_D} and V_{R_S} will be minimum, and V_{DSQ} will be maximum at $I_{DQ}^{(1)}$. Conversely, V_{R_D} and V_{R_S} will be maximum and V_{DSQ} will be minimum at $I_{DQ}^{(2)}$. Some detailed study along with a possible trial-and-error approach on the selection of R_S may be required to achieve an acceptable design.

For a p-channel JFET, the basic approach is the same, but the polarities of all voltages will be reversed. The preceding equations can therefore be modified to cover that situation if voltage magnitudes are used. For example, all voltages in (7-27) can be expressed as magnitudes, and the equation still applies. However, the two negative signs must be retained since the magnitude of voltage across the FET is the magnitude of the power supply voltage minus the other two voltage magnitudes in the loop.

Example 7-8

A certain n-channel JFET has the idealized characteristics given in Figures 6-6, 7-3, and 7-7. Design a self-bias circuit of the form of Figure 7-18 to achieve the following operating-point conditions: $I_{DQ} = 8$ mA and $V_{R_D} = V_{DSQ}$. The available power supply voltage is $V_{DD} = 15$ V.

Solution

Since $I_{DSS} = 16$ mA, $I_{DQ}/I_{DSS} = 8$ mA/16 mA $= 0.5$. The value of gate–source voltage required for this common operating-point condition is readily determined from Table 7-1. We note from the table that $V_{GS}/V_{GS(off)} = 0.2929$ for $I_D/I_{DSS} = 0.5$. Since $V_{GS(off)} = -4$ V, we have

$$V_{GSQ} = 0.2929 \times (-4) = -1.172 \text{ V} \tag{7-28}$$

The required value of R_S is

$$R_S = \frac{1.172 \text{ V}}{8 \times 10^{-3} \text{ A}} = 146.5 \ \Omega \tag{7-29}$$

The voltage V_{R_S} is

$$V_{R_S} = |V_{GSQ}| = 1.172 \ V \tag{7-30}$$

The dc voltage to be distributed between R_D and the drain–source drop is $15 - 1.172 = 13.828$ V. The specification calls for this voltage to be divided equally. Thus

$$V_{R_D} = V_{DSQ} = \frac{13.828}{2} = 6.914 \text{ V} \tag{7-31}$$

The drain resistance R_D is

$$R_D = \frac{V_{R_D}}{I_D} = \frac{6.914}{8 \times 10^{-3}} = 864.2\ \Omega \qquad (7\text{-}32)$$

The value of the gate resistance R_G is not critical, and a range of possible values is usually acceptable. The value selected would probably depend on specifications concerning the minimum input resistance, which usually relates to the circuit as an amplifier. Since this question will be addressed later, we will indicate simply a typical standard value of 470 kΩ for the present purpose.

The circuit diagram with the calculated values of R_S and R_D is shown in Figure 7-20. Understand that these calculated values are *not* standard resistance values. In a situation such as this, one approach is to use the nearest standard values and hope that the shift in operating point is minimal. Another approach is to try to select an operating point such that all component values are standard. Example 7-9 will deal with this particular circuit after the component values are rounded to standard values.

Figure 7-20 Circuit design of Example 7-8 with idealized values.

Example 7-9

Assume that R_D and R_S in the design of Example 7-8 are rounded to the nearest standard values, as shown in Figure 7-21. Assuming the ideal JFET characteristics, determine the actual values of I_{DQ}, V_{GSQ}, V_{R_S}, V_{R_D}, and V_{DSQ}.

Figure 7-21 Circuit of Example 7-9 representing modification of design of Example 7-8 to standard values.

Solution

The first step is to determine the values of I_{DQ} and V_{GSQ} for the given transistor characteristics and the value of R_S. It turns out that the same transistor characteristics and the value of R_S were employed in Example 7-6, so we can use some of the results of that analysis. Reviewing the results of that problem, we have

$$I_{DQ} = 7.913 \text{ mA} \tag{7-33}$$

$$V_{GSQ} = -1.187 \text{ V} \tag{7-34}$$

$$V_{R_S} = 1.187 \text{ V} \tag{7-35}$$

The remaining variables are different than those of Example 7-6 since the power supply voltage and drain resistance are different. We have

$$V_{R_D} = 7.913 \times 10^{-3} \times 820 = 6.489 \text{ V} \tag{7-36}$$

$$V_{DS} = 15 - 6.489 - 1.187 = 7.324 \text{ V} \tag{7-37}$$

The various calculated values can be compared with the target design goals of Example 7-8. The most significant changes are in V_{R_D} and V_{DS}. The design goal was to achieve equal voltages. The value of V_{R_D} is less than its design goal, and V_{DS} is greater than its design goal.

Example 7-10

Consider again the bias circuit design of Example 7-8. In this example we assume that the *exact* values of R_S and R_D calculated in Example 7-8 are employed. Obviously, this would produce the exact design goal operating point for the given exact transistor parameters. However, assume that the transistor production characteristics are subject to variation according to the following range:

$V_{GS(off)}$ variation: -6 to -2 V

I_{DSS} variation: 12 to 20 mA

The ideal values considered earlier of -4 V and 16 mA now represent the nominal midpoint values of the assumed range. Compute the range of operating-point conditions.

Solution

As discussed in the text, two extreme range transfer characteristic curves will be assumed. One corresponds to $V_{GS(off)} = -2$ V and $I_{DSS} = 12$ mA. The other corresponds to $V_{GS(off)} = -6$ V and $I_{DSS} = 20$ mA. All computations involving the first point will be denoted with a superscript (1), and similar computation at the second point can be denoted with a superscript (2).

The procedure of (7-15) and (7-16) can be employed to determine the operating-point current and gate–source voltage. The remaining calculations then proceed using basic circuit laws.

At the first point, the calculations are listed as follows:

$$B^{(1)} = \frac{2}{146.5 \times 12 \times 10^{-3}} = 1.1377 \tag{7-38}$$

$$I_{DQ}^{(1)} = 12 \times 10^{-3} \left[1.1377 + \frac{(1.1377)^2}{2} - \sqrt{(1.1377)^3 + \frac{(1.1377)^4}{4}} \right]$$

$$= 4.915 \text{ mA} \tag{7-39}$$

$$V_{GSQ}^{(1)} = -4.915 \times 10^{-3} \times 146.5 = -0.720 \text{ V} \tag{7-40}$$

$$V_{R_S}^{(1)} = 0.720 \text{ V} \tag{7-41}$$

$$V_{R_D}^{(1)} = 864.2 \times 4.915 \times 10^{-3} = 4.247 \text{ V} \tag{7-42}$$

$$V_{DS}^{(1)} = 15 - 0.720 - 4.247 = 10.033 \tag{7-43}$$

The reader is invited to perform similar computations for the other extreme condition. Such results, along with those computed previously, are summarized in the table that follows.

	Lower bound values[1]	Nominal values	Upper bound values[2]
$V_{GS(off)}$ (V)	−2	−4	−6
I_{DSS} (mA)	12	16	20
I_{DQ} (mA)	4.915	8	10.825
V_{GSQ} (V)	−0.720	−1.172	−1.586
V_{R_S} (V)	0.720	1.172	1.586
V_{R_D} (V)	4.247	6.914	9.356
V_{DSQ} (V)	10.033	6.914	4.058

While there is significant variation in the operating point, it is more stable than for fixed bias. For circuits in which this variation is acceptable, the self-bias circuit is a simple approach to the bias design.

7-8 FIXED-PLUS-SELF-BIAS CIRCUIT

An additional degree of stabilization in the drain current may be achieved by combining fixed bias with self-bias. If fixed bias alone were employed, a separate negative dc source would be required for an n-channel depletion-mode device, as noted earlier. However, when the combination of fixed and self-bias is used, the fixed voltage required with an n-channel device is positive, and this voltage can be obtained from the drain supply through a voltage divider. This approach is referred to in many books as voltage-divider bias.

+V_DD

R_1 R_D

I_D

+
V_{DS}
−

+
V_{GS} −

+
R_2 V_G
−

R_S

Figure 7-22   Self-plus-fixed-bias circuit for n-channel JFET.

The basic form of the circuit is shown in Figure 7-22. Assuming negligible gate current, the dc voltage V_G at the gate with respect to ground is

$$V_G = \frac{R_2}{R_1 + R_2} V_{DD} \qquad (7\text{-}44)$$

Since this voltage is positive, it is opposite in direction to the required gate–source voltage. However, the drain current flowing through R_S produces a self-bias voltage $R_S I_D$, which turns out to be larger in magnitude than V_G. If the circuit is designed properly, the net value of V_{GS} will turn out to be negative, as we will see shortly.

Starting at ground and moving clockwise around the gate–source loop, the KVL equation is

$$-V_G + V_{GS} + I_D R_S = 0 \qquad (7\text{-}45)$$

Solving for I_D, we obtain

$$I_D = -\frac{1}{R_S} V_{GS} + \frac{V_G}{R_S} \qquad (7\text{-}46)$$

Consider that V_{GS} is the independent variable and I_D is the dependent variable. The result is a straight-line equation, and it represents the bias line for the fixed-plus-self-bias circuit.

To see how the higher degree of stabilization is achieved, refer to Figure 7-23. The two transfer characteristic curves T_1 and T_2 represent the extreme range situations, as considered earlier. The slope of the bias line is $-1/R_S$, and the horizontal intercept is determined as V_G. Because of the relatively large value of V_G and the fact that it is located to the right of the origin, the slope of the bias line may be reduced dramatically (by increasing R_S) in comparison to that of the fixed-bias circuit. The two extreme intersection points, corresponding to respective drain current values of $I_{DSS}^{(1)}$ and $I_{DSS}^{(2)}$, are much closer together than for the case of self-bias alone.

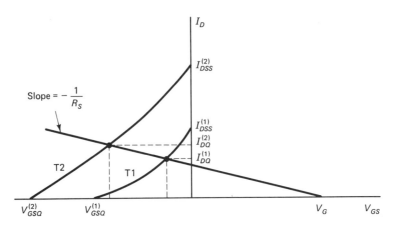

Figure 7-23 Range of operating points with self-plus-fixed-bias circuit.

7-9 ENHANCEMENT-MODE FET BIAS CIRCUITS

In the preceding section, some of the common methods for establishing an operating point for JFETs and depletion-mode IGFETs were discussed. In this section, similar circuits for enhancement-mode IGFETs are discussed. While the basic approach is largely the same, the polarity of the required bias voltage changes certain conditions for the possible circuit configurations.

Fixed Bias

As in the case of depletion-mode devices, fixed bias may be employed with an enhancement-mode device. Consider the circuit of Figure 7-24(a), containing an n-

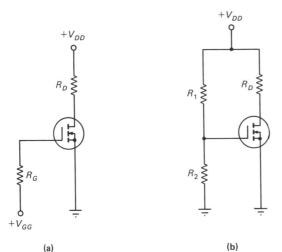

(a)

(b)

Figure 7-24 Fixed-bias circuits for n-channel enhancement-mode IGFET.

channel IGFET in a common-source configuration. The drain is connected through a drain resistance R_D to a positive dc voltage $+V_{DD}$. In contrast to an n-channel depletion-mode device, which requires a negative value of V_{GS} for bias, the n-channel enhancement-mode device requires a positive bias voltage. In this circuit, the gate is connected through a resistance R_G to a positive bias voltage V_{GG}.

Since the drain and gate bias voltages have the same polarity, the drain power supply may be used to establish bias for both terminals. A circuit using this approach to achieve fixed bias for the n-channel IGFET is shown in Figure 7-24(b). Since the required value of V_{GS} is much smaller than V_{DD}, the voltage divider, consisting of R_1 and R_2, reduces the voltage to the appropriate level. Assuming no gate current, the Q-point gate–source voltage V_{GSQ} is

$$\boxed{V_{GSQ} = \frac{R_2}{R_1 + R_2} V_{DD}} \qquad (7\text{-}47)$$

The fixed-bias circuit for enhancement-mode IGFETs is thus simpler than for depletion-mode devices since only one dc bias supply is required. However, it suffers from the same uncertainty in operating point due to the wide variation in transistor parameters.

Self-Plus-Fixed-Bias Circuit

The self-bias circuit utilizing a source resistance does not work for enhancement-mode devices because the voltage drop across the source resistor has the wrong polarity to bias the gate. However, by combining the fixed-bias voltage-divider circuit with self-bias, a stable biasing circuit can be achieved.

The form of the biasing circuit is shown in Figure 7-25. Operation is similar to the corresponding circuit for depletion-mode devices, with one difference. For depletion-mode devices, the voltage drop across R_S is larger than the drop across R_2, so that the net value of V_{GS} is negative. For enhancement-mode devices, the voltage drop across R_2 is larger than the drop across R_S, and the net value of V_{GS} is positive.

Figure 7-25 Self-plus-fixed-bias circuit for n-channel enhancement-mode IGFET.

The dc voltage V_G at the gate with respect to ground is

$$\boxed{V_G = \frac{R_2}{R_1 + R_2} V_{DD}}$$ (7-48)

The current I_D produces a self-bias voltage $R_S I_D$ across R_S. A clockwise loop around the gate–source circuit results in

$$-V_G + V_{GS} + I_D R_S = 0$$

Solution for I_D yields

$$\boxed{I_D = -\frac{1}{R_S}V_{GS} + \frac{V_G}{R_S}}$$ (7-49)

This result is the same as (7-46) for depletion-mode devices. However, the applicable region of operation will be different for enhancement-mode devices, as we will see shortly.

Refer to Figure 7-26 for the discussion that follows. The two transfer characteristic curves T_1 and T_2 represent extreme range situations for the IGFET under consideration. Curve T_1 has a threshold voltage $V_{GS(\text{th})}^{(1)}$ and curve T_2 has a threshold voltage $V_{GS(\text{th})}^{(2)}$. The bias line intersects the two curves at $I_{DQ}^{(1)}$ and $I_{DQ}^{(2)}$, respectively. The variation in these operating-point currents is much less than the variation in characteristic curve parameters.

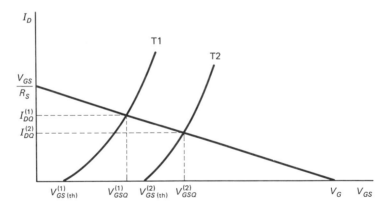

Figure 7-26 Range of operating points with self-plus-fixed-bias circuit.

7-10 FET CONSTANT CURRENT SOURCES

The FET closely approximates an ideal current source in the beyond-pinchoff region. It is therefore capable of being utilized for applications in which a constant current source is required.

We saw in Chapter 5 that when a BJT was utilized as a constant current source, the value of the current could be established to within a reasonably close value by appropriate circuit design. This is possible because the base–emitter voltage drop does not vary considerably from one unit to another. For the FET, however, the wide variation in gate–source voltage for a given drain current makes it difficult to specify what the actual current will be for a general-purpose FET. The general-purpose FET is best suited for an application in which a current source is needed but for which the exact value is either not critical or can be adjusted as required.

Constant Current Diodes

Some FET circuits are specifically manufactured for the purpose of establishing constant current sources. The schematic diagram of such a device is shown in Figure 7-27. A bias resistor appears in the source lead, and the gate terminal is connected directly to one of the input leads. There is no provision for an external gate terminal, so the gate resistance is not required. Instead, the bias voltage developed across the source resistor is applied directly to the gate. Since there are only two external terminals, the device is a diode as far as external connections are concerned.

In a particular application, it is required that the gate–source voltage exceed a certain minimum positive voltage to ensure beyond-pinchoff operation. However, over a wide range of gate–source voltage, a constant current is established.

In a sense, the *constant current diode* provides the opposite type of regulation achieved by the *constant voltage zener diode* introduced in Chapter 2. The value of current for a given diode type is subject to some variation between different units.

Figure 7-27   FET constant current source "diode."

7-11 DC SOURCE FOLLOWER

A basic BJT circuit considered in Chapter 5 was the dc *emitter follower*. A similar type of circuit exists for the FET, and it is called a *source follower*.

The basic form of a source follower intended for dc input and output is shown in Figure 7-28. For a given input v_i, a KVL equation around the gate–source loop reads

$$-v_i + v_{GS} + i_D R_S = 0 \qquad (7\text{-}50)$$

or

$$i_D = -\frac{1}{R_S} v_{GS} + \frac{v_i}{R_S} \qquad (7\text{-}51)$$

Figure 7-28 Basic form of an
n-channel JFET dc source follower.

The output voltage is

$$v_o = R_S i_D \qquad (7\text{-}52)$$

An alternative interpretation is simply

$$\boxed{v_o = v_i - v_{GS} = v_i + |v_{GS}|} \qquad (7\text{-}53)$$

The form of (7-51) is a type of bias equation relating i_D and v_{GS}, which has been investigated earlier. However, it differs from earlier forms in that v_i is now a variable. For a given value of v_i, however, a bias line can be constructed. We consider next two possible values of v_i and the resulting bias lines.

Refer to Figure 7-29 in the discussion that follows. First consider a relatively small possible value of v_i, denoted as $v_i^{(1)}$. The bias line for this voltage is denoted as B1 on Figure 7-29, and the corresponding gate–source voltage is $v_{GS}^{(1)}$. From (7-53), we can deduce that the dc value of v_o will be larger than v_i by $|v_{GS}^{(1)}|$.

Next consider a larger input voltage $v_i^{(2)}$. The bias line is B2 on Figure 7-29, and the gate–source voltage is $v_{GS}^{(2)}$. The dc value of v_o is still larger than the input, but the difference is less at the larger voltage.

The output dc voltage thus tracks the dc input voltage over a certain range, but the total output voltage is larger than the input by the magnitude of gate–source voltage required to establish the operating point.

To avoid possible confusion later, it will be mentioned here that the change in output voltage is less than the change in input voltage, and the small-signal or ac

Figure 7-29 Bias lines for two possible input voltages with source follower.

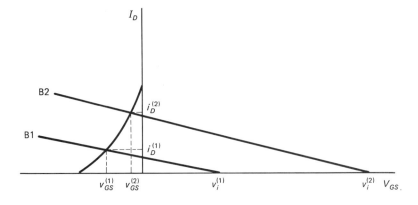

voltage gain, a concept considered later, is less than 1. However, for the source-follower circuit with an *n*-channel JFET, the total dc output voltage is greater than the input dc voltage, as we have shown.

It is not as easy to predict the dc output voltage with the source follower as it was for the emitter follower. The reason is that the BJT base–emitter voltage drop is more predictable than the FET gate–source bias voltage. However, the FET offers a higher input impedance than the BJT.

The two "follower" circuits that we have considered thus far (i.e., the emitter follower and the source follower) are used primarily for isolation between a source and a load. With the BJT, the dc output voltage tracks the input voltage with a difference of about 0.7 V. With the JFET, the dc output voltage tracks the input voltage with a difference of v_{GS} for the given operating point. Thus more accurate tracking is achieved when the magnitude of v_{GS} is very small.

DRILL PROBLEMS

7-1. The circuit of Figure P7-1(a) contains an *n*-channel JFET. The JFET characteristics are assumed to be idealized and are shown in Figure P7-1(b). Verify analytically that operation is in the beyond-pinchoff region, and determine I_D and V_{DS}.

(a)

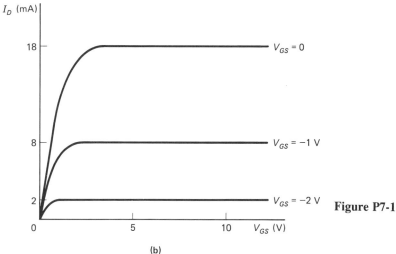

(b)

Figure P7-1

7-2. The circuit of Figure P7-2(a) contains an n-channel JFET. The JFET characteristics are assumed to be idealized and are shown in Figure P7-2(b). Verify analytically that operation is in the beyond-pinchoff region, and determine I_D and V_{DS}.

(a)

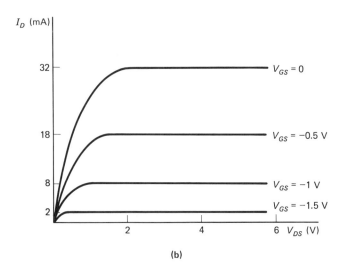

(b)

Figure P7-2

7-3. For the circuit of Figure P7-1(a), assume that the drain resistance is changed to $R_D = 2$ kΩ. Attempt to repeat the analysis of Problem 7-1.

7-4. For the circuit of Figure P7-2(a), assume that the drain resistance is changed to $R_D = 1.5$ kΩ. Attempt to repeat the analysis of Problem 7-2.

7-5. Construct a load line for the conditions of Problem 7-1.

7-6. Construct a load line for the conditions of Problem 7-2.

7-7. Construct a load line for the conditions of Problem 7-3.

FET-Resistor Circuit Applications   Chap. 7

7-8. Construct a load line for the conditions of Problem 7-4.

7-9. The self-bias circuit shown in Figure P7-9 contains the idealized n-channel JFET whose drain characteristics were given in Figure P7-1(b). Determine analytically the values of I_D, V_{GS}, V_{R_S}, V_{R_D} and V_{DS}.

Figure P7-9

7-10. The self-bias circuit shown in Figure P7-10 contains the idealized n-channel JFET whose drain characteristics were given in Figure P7-2(b). Determine analytically the values of I_D, V_{GS}, V_{R_S}, V_{R_D} and V_{DS}.

Figure P7-10

7-11. For the self-bias circuit of Problem 7-9, determine the operating point using the bias line.

7-12. For the self-bias circuit of Problem 7-10, determine the operating point using the bias line.

DESIGN PROBLEMS

7-13. It is desired to switch on and off the 10-kΩ load in the circuit of Figure P7-13. The FET characteristics are assumed to be those considered in Figure

+10 V

10 kΩ

+
v_i
−

Figure P7-13

P7-1(b). **(a)** Using a load line, show that the circuit is capable of operating as a switch. **(b)** Determine the required control levels for v_i. **(c)** Estimate the voltage across the load when the switch is on.

7-14. Design a voltage-controlled attenuator using the JFET whose characteristics were shown in Figure P7-2(b). The input voltage is a small positive voltage with a peak value of about 0.1 V. The desired transfer ratio $A = v_o/v_i$ is specified as 0.1 at $V_{GS} = -1$ V. After you have completed the design, determine A for the following values of V_{GS}: **(a)** 0; **(b)** −0.5 V; **(c)** −1.5 V; **(d)** −2 V.

7-15. A certain n-channel JFET has the idealized characteristics given in Figure P7-1(a). Design a self-bias circuit of the form of Figure 7-18 to achieve the following operating point conditions: $I_{DQ} = 9$ mA, $V_{R_D} = V_{DSQ}$. The power supply voltage is 24 V. List the exact values of resistances as computed.

7-16. A certain n-channel JFET has the idealized characteristics given in Figure P7-2(a). Design a self-bias circuit of the form of Figure 7-18 to achieve the following operating point conditions: $I_{DQ} = 16$ mA, $V_{R_D} = V_{DSQ}$. The power supply voltage is 30 V. List the exact values of resistances as computed.

7-17. Assume in the design of Problem 7-15 that I_{DSS} is subject to a variation from 12 to 24 mA and that $V_{GS(off)}$ is subject to a variation from −2 to −4 V. Construct a table showing the possible range of circuit variables.

7-18. Assume in the design of Problem 7-16 that I_{DSS} is subject to a variation from 24 to 40 mA and that $V_{GS(off)}$ is subject to a variation from −1.5 to −3 V. Construct a table showing the possible range of circuit variables.

TROUBLESHOOTING PROBLEMS

7-19. If the transistor in Figure 7-18 is initially biased in the beyond-pinchoff region, indicate whether the ground-referred drain voltage V_D would *increase*, *decrease*, or *remain nearly* the same for each of the following troubles. **(a)** R_G shorted; **(b)** R_D shorted; **(c)** R_D open; **(d)** R_S shorted; **(e)** R_S open.

7-20. If the transistor in Figure 7-22 is initially biased in the beyond-pinchoff region, indicate whether V_D would *increase*, *decrease*, or *remain nearly the same* for each of the following troubles. (a) R_1 shorted; (b) R_1 open; (c) R_2 shorted; (d) R_2 open; (e) R_D shorted; (f) R_D open; (g) R_S shorted; (h) R_S open.

7-21. You are troubleshooting an FET circuit of the form shown in Figure 7-22. A comparison of dc voltages to *ground* specified by the equipment manual and those measured is as follows:

	Specified	Measured
V_{DD} (V)	24	23.9
V_D (V)	11	3.5
V_S (V)	1	2.5

Which one of the following *could* be the cause of the difficulty? (a) R_1 open; (b) R_2 open; (c) R_2 shorted; (d) R_S open; (e) R_D open.

7-22. You are troubleshooting an FET circuit of the form shown in Figure 7-22. A comparison of dc voltages to *ground* specified by the equipment manual and those measured is as follows:

	Specified	Measured
V_{DD} (V)	15	15.1
V_D (V)	8	14
V_S (V)	1	0.3

Which one of the following *could* be the cause of the difficulty? (a) R_D open; (b) R_1 open; (c) R_1 shorted; (d) R_2 open; (e) R_S open.

COMPUTER PROBLEMS

7-23. Write a program to determine the operating point data in a self-bias JFET circuit (*n*-channel or *p*-channel) of the form shown in Figure 7-12 or 7-18.

Input data: $V_{GS(off)}$, I_{DSS}, R_S, R_D, V_{DD}
Output data: I_D, V_{GS}, V_{R_S}, V_{R_D}, V_{DS}

7-24. Write a program to design a self-bias circuit for a JFET (*n*-channel or *p*-channel) of the form shown in Figure 7-12 or 7-18.

Input data: $V_{GS(off)}$, I_{DSS}, I_{DQ}, V_{DD}, V_{DSQ}
Output data: R_S, R_D, V_{R_S}, V_{R_D}

Frequency Response

8-1 OVERVIEW OF SINUSOIDAL AC CIRCUIT ANALYSIS

In Chapter 1 the three basic passive circuit parameters were introduced. These are *resistance*, *capacitance*, and *inductance*. As we have seen, the circuit element that exhibits resistance is called a *resistor*. In a similar fashion, the circuit element that exhibits capacitance is called a *capacitor*, and the circuit element that exhibits inductance is called an *inductor*. Depending on the application, an inductor is sometimes denoted as a *coil* or as a *choke*. A review of the three basic circuit parameters and their symbols are shown in Figure 8-1.

Thus far in the book, the only passive element considered has been resistance. In all cases, the assumption has been that either the circuit has been operating at dc or, in the case of instantaneous voltages and currents, the effects of capacitance and

Figure 8-1 Three circuit parameters and their symbols.

Resistance   Capacitance   Inductance

inductance were negligible. However, we have now reached a point where the effects of capacitance and/or inductance must be considered in electronic circuits.

The manner in which capacitance and inductance is treated in basic electronic circuit analysis and design is through the concept of *sinusoidal steady-state ac circuit analysis*. This is an elegant body of circuit theory in which the source exciting the circuit is assumed to be sinusoidal in form, and the effects of all circuit elements to the input sinusoidal are determined. Although the ultimate application of the circuit may have nothing to do with a sinusoidal input, it is possible to infer many important properties of the circuit from the response to a sinusoid. The study of ac circuit theory utilizes certain mathematical forms of voltages and currents called *phasors* and the use of a so-called *j-operator*.

To accommodate readers who have not yet studied ac circuit theory, the use of the *j*-operator is deferred to Chapter 18. However, the concepts of ac analysis will be developed in this chapter in a manner that should allow a practical and intuitive approach to understanding the frequency-dependent properties of basic ac circuits. Readers who have already studied ac circuit theory using the *j*-operator may wish to fill in certain steps in this chapter.

Sinusoidal Function

The sinusoidal waveform is the most common function used in the analysis of electronic circuits. Consider the instantaneous voltage v shown in Figure 8-2. The equation of this waveform may be expressed as

$$v = V_p \sin \omega t \qquad \textbf{(8-1)}$$

The various quantities appearing in this equation are defined as follows:

V_p = *peak value* or *amplitude*, volts

ω = *angular frequency*, radians/second (rads/s)

   = $2\pi f$

f = *cyclic frequency*, hertz (Hz)

t = independent variable *time*, seconds (s)

Note that there are two definitions of frequency. The more common designation of frequency is the quantity f, which is the actual number of cycles of the wave-

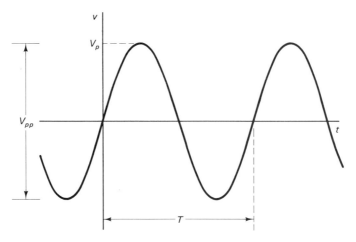

Figure 8-2   Sinusoidal voltage function.

form that occurs per second. The unit is hertz and 1 Hz = 1 cycle per second. However, it is the angular frequency ω that must appear as a multiplier of time in the equation to provide the correct dimensions for the angle of the sinusoidal function. In subsequent discussions, the term "frequency" without a modifier will be interpreted to be either the cyclic frequency f, or the reference could apply to either frequency form.

The *period T* is the duration of a full cycle in seconds. It is related to the cyclic frequency by

$$T = \frac{1}{f} \qquad \textbf{(8-2)}$$

In performing measurements on a sinusoid with an oscilloscope, it is often easier to measure the *peak-to-peak value* than the peak value. Let V_{pp} represent the peak-to-peak voltage. This quantity is related to the peak voltage simply as

$$V_{pp} = 2V_p \qquad \textbf{(8-3)}$$

Most of the various terms defined here are illustrated in Figure 8-2.

The function described by (8-1) and shown in Figure 8-2 is defined in terms of the basic sine function, and it passes through the origin with a positive slope. Many sinusoidal functions are shifted either to the left or to the right with respect to the assumed origin. For example, consider the voltage v shown in Figure 8-3. This function can be expressed as

$$v = V_p \sin(\omega t + \theta) \qquad \textbf{(8-4)}$$

The quantity θ in (8-4) represents an arbitrary phase shift, which for the example selected represents a shift to the left. Note that the time shift is θ/ω since the units of the horizontal scale must be units of time. All other quantities in (8-4) have the same meaning as in (8-2).

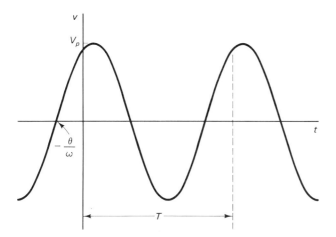

Figure 8-3 Sinusoidal voltage function with arbitrary phase angle.

The functions chosen thus far were voltages. However, currents appearing in steady-state ac circuits will have the same form. Thus a typical instantaneous current will have the form

$$i = I_p \sin(\omega t + \phi) \qquad (8\text{-}5)$$

where I_p is the peak value of the current and ϕ is some arbitrary phase angle.

The *sine function* has been chosen here as the basis for representation. However, the *cosine function* has the same basic shape, but it has a different time origin. Some books use the cosine function, but the practice in this book is to employ the sine function. Both the sine and cosine function, as well as shifts resulting from arbitrary phase angles, are collectively referred to as *sinusoidal functions*.

Example 8-1

A certain sinusoidal voltage v has a peak value of 20 V and a frequency of 1000 Hz. The time origin is chosen in the form of Figure 8-2 (i.e., the function crosses $t = 0$ with a positive slope). **(a)** Determine the period. **(b)** Write an equation for the voltage.

Solution

(a) The period T is

$$T = \frac{1}{f} = \frac{1}{1000} = 0.001 \text{ s} = 1 \text{ ms} \qquad (8\text{-}6)$$

(b) The radian frequency is $\omega = 2\pi f = 2\pi \times 1000 = 6283.2$ rad/s (to five significant figures). The function can be expressed as

$$v = 20 \sin 2\pi \cdot 1000t \qquad (8\text{-}7a)$$

$$= 20 \sin 6283.2t \qquad (8\text{-}7b)$$

Note the two alternative forms used in expressing the angles. The first form, with the 2π shown as a multiplier, is often used since it allows the cyclic frequency to be immediately evident.

Example 8-2

A certain current is described by the equation

$$i = 0.05 \sin 1600t \tag{8-8}$$

Determine **(a)** the peak value, **(b)** the radian frequency, **(c)** the cyclic frequency, and **(d)** the period.

Solution

(a) The peak value is determined by inspection as

$$I_p = 0.05 \text{ A} = 50 \text{ mA} \tag{8-9}$$

(b) The radian frequency is also determined by inspection as

$$\omega = 1600 \text{ rad/s} \tag{8-10}$$

(c) Since $\omega = 2\pi f$, the cyclic frequency is

$$f = \frac{\omega}{2\pi} = \frac{1600}{2\pi} = 254.65 \text{ Hz} \tag{8-11}$$

(d) The period is

$$T = \frac{1}{f} = \frac{1}{254.65} = 3.9270 \text{ ms} \tag{8-12}$$

RMS Value of Sinusoid

One very important quantity associated with a sinusoidal waveform is the *root-mean-square* (denoted as rms) value. For the moment, this quantity will be indicated by the subscript rms. The rms value of a voltage v will be denoted as V_{rms}, and the rms value of a current i will be denoted as I_{rms}.

The rms value of a sinusoidal voltage with peak value V_p is given by

$$V_{rms} = \frac{V_p}{\sqrt{2}} = 0.7071 V_p \tag{8-13}$$

Similarly, the rms value of a sinusoidal current with peak value I_p is

$$I_{rms} = \frac{I_p}{\sqrt{2}} = 0.7071 I_p \tag{8-14}$$

The significance of the rms value is that it is the quantity used when determining power produced by the ac waveform. If a sinusoidal voltage v exists across a resistance R, the power P is

$$P = \frac{V_{rms}^2}{R}$$

(8-15)

Similarly, if a sinusoidal current i is flowing in a resistance R, the power P is

$$P = I_{rms}^2 R$$

(8-16)

Thus the rms voltage and current are treated like dc voltage and current for computing power in a resistance.

The expressions for power in (8-15) and (8-16) represent *average* power, which is the quantity most often desired in ac circuits. The *peak power* produced by the ac waveform is much greater. We defer detailed discussion of peak power to the point in the book where it is needed. For present developments, power computations with an ac waveform are considered to produce average power.

Many ac voltmeters measure the rms value of a sinusoidal waveform. It is important to recognize this fact when correlating oscilloscopic measurements with voltmeter measurements. For example, suppose that the peak-to-peak value of a certain sinusoidal voltage has been measured as 4 V. This means that the peak value is 2 V. An rms-measuring ac voltmeter applied to the waveform would ideally read 1.414 V. However, some ac voltmeters respond to different properties of the waveform, so it is important to understand the type of ac voltmeter used.

The rms value is also denoted as the *effective value* of the waveform, and some books use that definition. We use the rms designation in this book.

The rms value has been introduced to complete the treatment of the basic sine function here. For the next few chapters, however, very little use of the rms value will be required. However, it will be very significant in the treatment of power amplifiers and power supplies in Chapters 19 and 20.

A final note is that the formulas given by (8-13) and (8-14) apply only to a *sinusoidal function*. In general, the relationship of the rms value to the peak value is different for different types of waveforms.

Example 8-3

(a) Determine the rms value of the sinusoidal voltage of Example 8-1.
(b) Determine the power dissipated in a 820-Ω resistor.

⌐ **Solution**

(a) The peak value is 20 V, and the rms value is

$$V_{rms} = \frac{20}{\sqrt{2}} = 14.142 \text{ V}$$

(8-17)

(b) The power dissipated in a 820-Ω resistor is

$$P = \frac{V_{\text{rms}}^2}{R} = \frac{(14.142)^2}{820} = 0.2439 \text{ W} \qquad (8\text{-}18)$$

Example 8-4

(a) Determine the rms value of the sinusoidal current of Example 8-2.
(b) Determine the power dissipated in a 150-Ω resistor.

Solution

(a) The peak value is 0.05 A, and the rms value in

$$I_{\text{rms}} = \frac{0.05}{\sqrt{2}} = 0.03536 \text{ A} \qquad (8\text{-}19)$$

(b) The power dissipated in a 150-Ω resistor is

$$P = I_{\text{rms}}^2 R = (0.03536)^2 \times 150 = 0.1875 \text{ W} \qquad (8\text{-}20)$$

Spectral Analysis

The reason that the response of electronic circuits to sinusoidal waveforms is so important is the following fact: *Arbitrary waveforms can be considered to be composed of a sum of sinusoidal waveforms at different frequencies.* The process of determining the frequency components that compose a given waveform is called *spectral analysis*. It is also called *Fourier analysis* as a tribute to the great French mathematician Fourier, who developed the techniques that were later applied to spectral analysis.

To relate this concept to an area familiar to many readers, consider the common audio amplifier used in stereo music systems. The human ear is capable of responding to sinusoidal frequencies from about 20 Hz or so up to nearly 20 kHz. Music waveforms are known to be composed of frequencies encompassing this range. Therefore audio amplifiers are tested with sinusoidal waveforms, and a good-quality amplifier should be capable of reproducing sinusoids equally well over that range. If the amplifier is defective and is not able to reproduce frequencies above 10 kHz, for example, there will be a loss of the higher-frequency notes, and the output signal will be degraded.

Steady-State Assumption

Consider a linear circuit having one or more sinusoidal sources of the same frequency. When the circuit is first excited by a sinusoid, there are certain *transient* effects occurring, which will eventually damp out, assuming that the circuit contains some resistance. The various voltages and currents in the circuit will then settle into what is called a *sinusoidal steady-state* condition.

A basic property of the *sinusoidal steady-state* for a *linear* constant parameter circuit is that *all voltages and currents are sinusoids having the same frequency as the source*. Thus all steady-state waveforms have the same basic sinusoidal shape. However, their peak values and phase angles will differ.

Since the various waveforms in the sinusoidal steady state are all sinusoids of the same frequency, only two quantities are required to specify each sinusoid: (1) peak value (or magnitude) and (2) phase angle. This is the basis of steady-state ac circuit analysis.

8-2 REACTANCE

A pure resistance presents the same opposition to current flow at all sinusoidal frequencies. Further, the voltage across the resistance is in phase with the current through the resistance at all frequencies.

The effects of capacitance and inductance on ac current flow are somewhat more involved. Both capacitance and inductance provide opposition to ac current flow, but the manner and extent of the opposition are dependent on the frequency.

The opposition to ac current flow exhibited by either a capacitance or an inductance is called *reactance*. A reactance can be further classified as either a *capacitive reactance* or an *inductive reactance,* depending on which parameter is responsible. The unit of reactance is the same as the unit of resistance (i.e., the *ohm*). The general symbol for reactance is X.

If they oppose current flow, why don't we call these effects "capacitive resistance" and "inductive resistive"? The reason is that they describe a totally different kind of phenomenon. In pure resistance, the voltage and current are in phase, and all the power is dissipated in the resistor. For purely reactive elements, however, voltage and current are 90° out of phase, and there is no real power dissipation. Instead, the energy is alternately stored in the reactance and released back to the circuit. Therefore, the term *reactance* describes a completely different type of physical behavior than resistance.

Capacitive Reactance

The capacitive reactance of a capacitance C is denoted by X_C and it is determined as

$$X_C = \frac{1}{\omega C} \tag{8-21a}$$

$$= \frac{1}{2\pi f C} \tag{8-21b}$$

where alternate forms using f and ω are given. Strictly speaking, a more "proper" definition would include a negative sign for capacitive reactance, but in accordance with widespread usage at this level, it will be defined for convenience as a positive quantity.

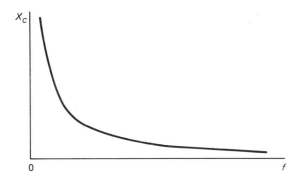

Figure 8-4 Capacitive reactance variation with frequency.

The capacitive reactance is inversely proportional to both frequency and the capacitance. Thus a larger capacitance exhibits a smaller reactance, and vice versa. The behavior of X_C as a function of frequency is illustrated in Figure 8-4. At very low frequencies, the capacitive reactance is very large (i.e., it presents a very large opposition to current flow). Conversely, at high frequencies, the capacitive reactance is very small.

Inductive Reactance

The inductive reactance of an inductance L is denoted by X_L, and it is determined as

$$X_L = \omega L \tag{8-22a}$$
$$= 2\pi f L \tag{8-22b}$$

where alternative forms using f and ω are given. *The inductive reactance is directly proportional to both frequency and the inductance.* Thus a larger inductance exhibits a larger reactance, and vice versa.

The behavior of X_L as a function of frequency is illustrated in Figure 8-5. At very low frequencies, the inductive reactance is very small (i.e., it presents a very small opposition to current flow). Conversely, at high frequencies, the inductive reactance is very large.

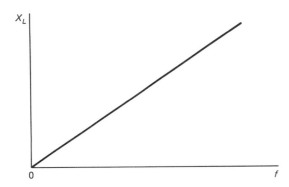

Figure 8-5 Inductive reactance variation with frequency.

Example 8-5

Determine the reactance of a 0.01-μF capacitance at each of the following frequencies: **(a)** 10 Hz; **(b)** 1 kHz; **(c)** 100 kHz; **(d)** 10 MHz.

Solution

The expression for X_C is

$$X_C = \frac{1}{2\pi f C} = \frac{1}{2\pi \times 0.01 \times 10^{-6} f} = \frac{1.5915 \times 10^7}{f} \qquad (8\text{-}23)$$

where $C = 0.01\ \mu$F has been substituted, and only f remains as a variable. This step was appropriate since several values of f are to be considered.

The remaining part of the problem amounts to substituting the four values of f and determining the four values of X_C. The different computations follow.

(a) For $f = 10$ Hz,

$$X_C = \frac{1.5915 \times 10^7}{10} = 1.5915\ \text{M}\Omega \qquad (8\text{-}24)$$

(b) For $f = 1$ kHz,

$$X_C = \frac{1.5915 \times 10^7}{10^3} = 15.915\ \text{k}\Omega \qquad (8\text{-}25)$$

(c) For $f = 100$ kHz,

$$X_C = \frac{1.5915 \times 10^7}{10^5} = 159.15\ \Omega \qquad (8\text{-}26)$$

(d) For $f = 10$ MHz,

$$X_C = \frac{1.5915 \times 10^7}{10^7} = 1.5915\ \Omega \qquad (8\text{-}27)$$

Observe the wide range of reactance as the frequency is varied. Observe how the reactance changes from a very large value at the lowest frequency to a very small value at the highest frequency.

Example 8-6

Determine the reactance of a 50-mH inductance at each of the following frequencies: **(a)** 10 Hz; **(b)** 1 kHz; **(c)** 100 kHz; **(d)** 10 MHz.

Solution

The expression for X_L is

$$X_L = 2\pi f L = 2\pi \times 50 \times 10^{-3} f = 0.31416f \qquad \text{(8-28)}$$

where $L = 50$ mH has been substituted, and only f remains as a variable. The different computations follow.

(a) For $f = 10$ Hz,

$$X_L = 0.31416 \times 10 = 3.1416\ \Omega \qquad \text{(8-29)}$$

(b) For $f = 1$ kHz,

$$X_L = 0.31416 \times 10^3 = 314.16\ \Omega \qquad \text{(8-30)}$$

(c) For $f = 100$ kHz,

$$X_L = 0.31416 \times 10^5 = 31.416\ \text{k}\Omega \qquad \text{(8-31)}$$

(d) For $f = 10$ MHz,

$$X_L = 0.31416 \times 10^7 = 3.1416\ \text{M}\Omega \qquad \text{(8-32)}$$

As in the case of the capacitive reactance of Example 8-5, we see a very wide range of inductive reactance as the frequency is varied. However, the inductive reactance changes from a very small value at the lowest frequency to a very large value at the highest frequency.

Example 8-7

In a certain application requiring a capacitor, the exact value of the reactance is not critical, but it is desired that the reactance be no greater than 100 Ω at a frequency of 50 Hz. Determine the minimum value of capacitance required.

Solution

This problem has been stated in a practical fashion typical of many electronic circuit requirements. Before explaining the "no greater" and "minimum" qualifiers, we will first determine the exact value of C required to produce a reactance of exactly 100 Ω at 50 Hz.

In general, we have

$$X_C = \frac{1}{2\pi f C} \qquad \text{(8-33)}$$

Solving for C, we have

$$C = \frac{1}{2\pi f X_C} = \frac{1}{2\pi \times 50 \times 100} = 31.83\ \mu\text{F} \qquad \text{(8-34)}$$

For the particular application, the exact value of X_C is not critical, so we need not be concerned about obtaining the exact value given by (8-34). This value of capacitance is fairly large, and the tolerances of standard capacitance values in that range are usually large. A larger standard value of capacitance would be acceptable for the application since its reactance would be even smaller than the 100-Ω maximum value permitted. In conclusion, the value 31.83 μF represents the *minimum* value for the particular application, but a larger stock value would be acceptable.

8-3 STEADY-STATE DC MODELS FOR REACTANCE

Whenever dc analysis is employed in an electronic circuit containing reactive elements, the limiting forms of the reactance functions at dc lead to some simplified circuit models. For certain purposes, *dc may be considered as a special case of ac with zero frequency (i.e., f = 0).*

From the expression for X_C as given by (8-21) or from Figure 8-4, it is observed that X_C "approaches ∞" as f approaches 0 (i.e., X_C increases without limit as f approaches 0). An *infinite reactance* corresponds to an *open circuit.*

From the expression for X_L as given by (8-22) or from Figure 8-5, it is observed that $X_L = 0$ for $f = 0$. A *zero reactance* corresponds to a *short circuit.*

The dc steady-state models of an ideal capacitor and an ideal inductor are shown in Figure 8-6. To summarize, for *steady-state dc analysis, an ideal capacitor is represented by an open circuit, and an ideal inductor is represented by a short circuit.*

Under steady-state dc conditions, the capacitor will charge to a certain voltage and store that voltage until circuit conditions change. To compute the voltage stored on the capacitor, solve the dc circuit for the voltage across the open circuit representing the capacitor. This will be the voltage stored by the capacitor.

In a similar fashion, the inductor will store magnetic flux, which is a result of

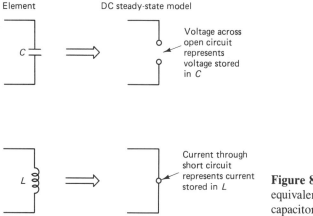

Figure 8-6 Steady-state dc equivalent-circuit models of ideal capacitor and inductor.

current flowing through it. To compute the steady-state dc current flowing in the inductor, solve for the current through the short circuit representing the inductor. This will be the steady-state dc current flow for the inductor.

It should be stressed that these models apply after all transients damp out and steady-state conditions remain. When a reactive circuit is first excited by a dc voltage, current flow in a capacitor will occur until the capacitor is fully charged, and voltage can exist across an inductor until a full level of flux is obtained. The general solutions of such transients is covered in advanced circuits and network analysis texts. However, certain important transient forms are considered in Chapter 16 as they relate to nonlinear electronic circuits.

In addition to dc, these limiting models may be applied, with very little loss of accuracy, to analyze ac conditions at very low frequencies. The key to applying the limiting models is that all capacitive reactances must be large compared with all resistances, and all inductive reactances must be small compared with all resistances.

Example 8-8

Determine the steady-state dc voltage across the capacitor in the simple *RC* circuit of Figure 8-7(a).

Figure 8-7 Circuit of Example 8-8.

Solution

The steady-state dc equivalent circuit is shown in Figure 8-7(b). The capacitor is replaced by an open circuit. Since no current is flowing in the loop, the voltage across the capacitor is simply

$$V_C = 12 \text{ V} \qquad \text{(8-35)}$$

If this simple "quick result" bothers the reader, one can form a loop and write

$$-12 + 0 + V_C = 0 \qquad \text{(8-36)}$$

in which the "0" term on the left represents the fact that the resistive voltage is zero since the current is zero. The same result is readily obtained.

We need to remind ourselves that this model is applicable only after steady-state dc conditions are reached. When an uncharged capacitor is con-

nected through a resistance to a dc source, current will flow in the circuit until
the capacitor charges to a voltage level sufficient to terminate current flow in
the capacitor (12 V in this case).

Example 8-9

Determine the steady-state dc voltages across the three capacitors in the circuit
of Figure 8-8(a).

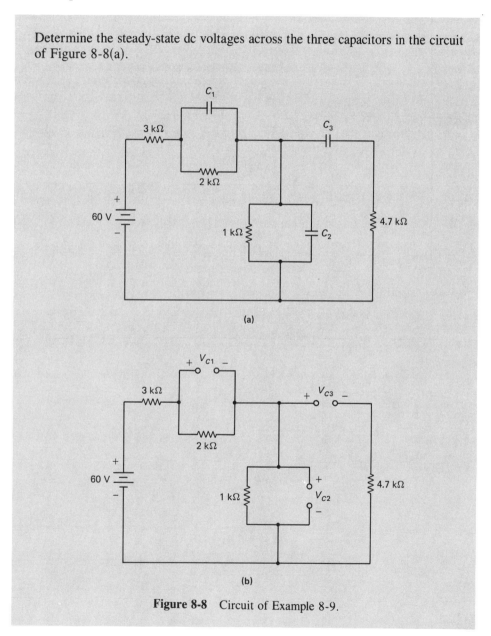

(a)

(b)

Figure 8-8 Circuit of Example 8-9.

Solution

The steady-state dc equivalent circuit is shown in Figure 8-8(b), in which all capacitors are replaced by open circuits. The right-hand loop is uncoupled from the left-hand loop by the open circuit representing C_3. This permits a simple application of the voltage-divider rule to the left-hand loop. We have

$$V_{C1} = \frac{2 \text{ k}\Omega}{3 \text{ k}\Omega + 2 \text{ k}\Omega + 1 \text{ k}\Omega} \times 60 = 20 \text{ V} \qquad \text{(8-37)}$$

and

$$V_{C2} = \frac{1 \text{ k}\Omega}{3 \text{ k}\Omega + 2 \text{ k}\Omega + 1 \text{ k}\Omega} \times 60 = 10 \text{ V} \qquad \text{(8-38)}$$

To determine V_{C3}, we note that the voltage across the 4.7-kΩ resistor is zero since the current in the right-hand loop is zero. This forces V_{C3} to be the same as V_{C2}, and we have

$$V_{C3} = V_{C2} = 10 \text{ V} \qquad \text{(8-39)}$$

A more "formal" way to deduce the first part of (8-39) is the loop equation

$$-V_{C2} + V_{C3} + 0 = 0 \qquad \text{(8-40)}$$

where the 0 term on the left represents the voltage drop of zero for the 4.7-kΩ resistor.

Example 8-10

The circuit of Figure 8-9(a) represents a complete single-stage RC-coupled BJT amplifier circuit, which is studied in detail in Chapter 10. Determine the

(a)

Figure 8-9   Circuit of Example 8-10.

steady-state dc voltages that will appear across the three capacitors. The function v_g represents an *ac* signal source, and by superposition, it can be modeled as a short-circuit for dc calculations.

Solution

A steady-state dc equivalent circuit, in which the three capacitors are represented by open circuits, is shown in Figure 8-9(b). We could, of course, represent the transistor by its dc current source model, but we have had plenty of practice already on that type of model. Consequently, the transistor variables will be determined directly by the simplified formulas.

First, V_{TH} as a step for computing I_C is determined.

$$V_{\text{TH}} = \frac{5.6 \text{ k}\Omega}{5.6 \text{ k}\Omega + 18 \text{ k}\Omega} \times 24 \text{ V} = 5.695 \text{ V} \tag{8-41}$$

I_C is then determined as

$$I_C = \frac{5.695 - 0.7}{3 \times 10^3} = 1.66 \text{ mA} \tag{8-42}$$

where some rounding has been performed.

The three transistor voltages are

$$V_{R_E} = 5.695 - 0.7 = 4.99 \text{ V} \tag{8-43}$$

$$V_{R_C} = 5.6 \times 10^3 \times 1.66 \times 10^{-3} = 9.30 \text{ V} \tag{8-44}$$

$$V_{CE} = 24 - 9.30 - 4.99 = 9.71 \text{ V} \tag{8-45}$$

Up to this point, the analysis has been the same as performed back in Chapter 5. However, observe that the complete circuit is more involved than just the bias circuits considered in Chapter 5. Yet the dc analysis on the transistor part of the circuit is the same.

The capacitors in the circuit serve to isolate the transistor dc circuit from the source on the left and the 10-kΩ load on the right. As we will see later, the ac signal is effectively coupled through the various capacitors. However, the capacitors "block" the dc from the source and load. If the capacitors were not used in this circuit, the bias point would be affected drastically and adversely by the source and load.

Returning to the analysis, the dc voltage V_{C_1} is the same as the voltage across the 5.6-kΩ resistor and is approximately

$$V_{C_1} = V_{\text{TH}} = 5.695 \text{ V} \tag{8-46}$$

The voltage V_{C_E} is the same as the voltage across the 3-kΩ resistor:

$$V_{C_E} = V_{R_E} = 4.99 \text{ V} \tag{8-47}$$

(Don't confuse V_{C_E}, the voltage across C_E, with V_{CE}, the collector–emitter voltage.) Finally, the voltage V_{C_2} is the same as the collector–ground voltage; that is,

$$V_{C_2} = V_C = V_{CE} + V_{R_E} = 9.71 + 4.99 = 14.70 \text{ V} \tag{8-48}$$

The analysis we have performed is helpful both in troubleshooting and in design. In troubleshooting, it leads to dc values to look for in checking the operation of a circuit. In design, it leads to the determination of the dc voltage ratings of capacitors in a circuit. In this context, the peak signal voltage should be added to the dc voltage to determine the maximum instantaneous voltage across a given capacitor. The voltage rating of the capacitor should then be selected to be somewhat larger than this value to provide additional leeway.

8-4 FLAT-BAND AC MODELS FOR REACTANCE

From the expression for X_C as given by (8-21) or from Figure 8-4, it is observed that X_C becomes very small as f increases without limit. Similarly, from the expression for X_L as given by (8-22) or from Figure 8-5, it is observed that X_L becomes very large as f increases without limit. Under certain conditions, X_C may be approximated as a short circuit and X_L may be approximated as an open circuit.

Many electrical and electronic circuits having reactive elements are characterized by the fact that the following two conditions are met over a certain frequency range: (1) all capacitive reactances are very small compared with resistances and inductive reactances present, and (2) all inductive reactances are very large compared with resistances and capacitive reactances present. When these assumptions are valid, capacitive reactances may be approximated as short circuits, and inductive reactances may be approximated as open circuits. These are just the opposite conditions assumed in the steady-state dc models.

When these approximations are valid, the given circuit reduces to a resistive circuit. Although voltages and currents are varying with time, simple algebraic methods may be used to determine voltages and currents in the circuit on an instantaneous basis.

An important point is that *over the given frequency range where the approximations are valid, the peak value of any given voltage or current will be independent of the frequency of any sources.* We will use the term *flat-band* to describe the range of frequencies over which the limiting models are valid.

A summary of the flat-band ac models for capacitance and inductance is shown in Figure 8-10. In all subsequent developments in the book, a reference to *flat-band ac model* will refer to conditions in which these models are valid.

The reader is not expected to understand at this level when such models are valid. As we progress through the text, the judgment for assessing the validity of the models will develop. Ultimately, ac circuit theory will be used as an analytical way of scrutinizing the models for questionable situations. For the moment, simply learn the models and how to work with them for analysis, troubleshooting, and design.

In this section and the preceding several sections, both capacitive and inductive reactances have been considered in order to provide a complete basic treatment. However, the particular order of development in the text leads first to amplifier circuits in which only capacitive effects are significant, and the absence of inductance will be conspicuous in the next several chapters. Inductance will appear again in the applications of Chapter 17.

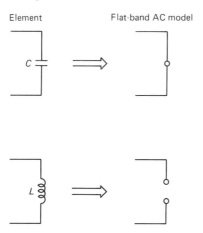

Element   Flat-band AC model

Figure 8-10 Steady-state flat-band ac models for ideal capacitor and inductor.

Example 8-11

Consider the simple *RC* circuit of Figure 8-11(a). Assume that the frequency range of the time-varying signal v_g is in the flat-band region of the circuit. Determine the voltage v_o in terms of v_g.

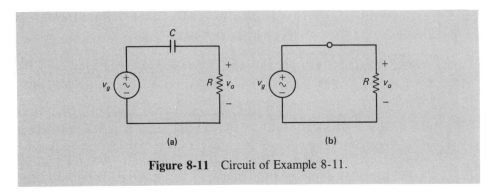

Figure 8-11 Circuit of Example 8-11.

Solution

The flat-band equivalent circuit is shown in Figure 8-11(b). The capacitor is replaced by a short circuit. Consequently, the voltage v_o in this circuit is simply the generator voltage v_g; that is,

$$v_o = v_g \qquad\qquad (8\text{-}49)$$

The circuit thus reduces to a trivial parallel connection of the source and load as far as the flat-band response is concerned. Understand that the frequency or frequencies of the source must be sufficiently high that the capacitive reactance is negligible.

Example 8-12

In the circuit of Figure 8-12(a), assume that the frequency range of the time-varying signal v_g is in the flat-band region of the circuit. Determine the voltage v_o in terms of v_g.

Figure 8-12 Circuit of Example 8-12.

Solution

The flat-band equivalent circuit is shown in Figure 8-12(b). The two capacitors are replaced by short circuits. The short around the 10-kΩ resistor effectively shunts all current around it, so that resistor has no effect in the flat-band region. Another way to deduce the same result is to recognize that the parallel combination of a short circuit and any element is a short circuit. Thus the 3-kΩ resistor and the 6-kΩ resistor are effectively in parallel.

The circuit is redrawn in Figure 8-12(c) in a simplified form. By considering the parallel combination on the right, the voltage-divider rule may readily be employed. We have

$$v_o = \frac{3 \text{ k}\Omega \| 6 \text{ k}\Omega}{3 \text{ k}\Omega \| 6 \text{ k}\Omega + 8 \text{ k}\Omega} \times v_g = \frac{2 \text{ k}\Omega}{2 \text{ k}\Omega + 8 \text{ k}\Omega} \times v_g = \frac{1}{5} v_g \qquad \textbf{(8-50)}$$

8-5 DC-PLUS-FLAT-BAND AC RESPONSE

In many electronic circuits, one or more dc voltages are used to establish bias and other pertinent reference levels, but the ultimate goal is to process a time-varying signal. For this purpose it is necessary to consider separately a dc model and a time-varying or ac model. Although we are not yet ready to study time-varying models completely because of the frequency dependency of the reactances, in many cases it is the flat-band region models that are of primary interest. By combining the concepts of the preceding several sections, we can study both the dc and flat-band responses of certain circuits.

The focus in this section will therefore be on obtaining two models for a given circuit: (1) dc model and (2) flat-band ac model. As we will see, the circuit form may change drastically as we convert from one form to the other.

The principle that allows separate models to be constructed for analysis is the principle of *superposition*, which was considered in Chapter 1. Strictly speaking, the principle of superposition applies only to linear circuits, and all electronic circuits have some degree of nonlinearity. However, as long as the behavior of the electronic devices can be represented by linear models over the range of operation, superposition may be applied with care. In this section, all examples considered will not contain electronic devices, so that the basic process may be developed. This will set the stage for working with linearized small-signal models for electronic devices in later chapters.

When the dc equivalent circuit is constructed, all ideal time-varying sources are replaced by short circuits (on paper). If the source contains an internal resistance, the resistance is retained in the model in the event that it could have some effect on various dc voltages and currents in the circuit. In many cases the circuit is designed so that internal resistances of time-varying sources do not affect dc operating points, but as a procedure at this point, we should consider any *possible* effects by keeping them in the models.

When the ac model is constructed, any dc sources are normally eliminated from the analysis. The possible drastic changes in configuration often confuse beginners, so the process will be illuminated.

AC Model for DC Source

Since a dc bias supply primarily establishes operating points, it is normally not considered as part of a signal. Therefore, it is simply modeled as a short circuit for ac signal analysis. This process is illustrated in Figure 8-13 two ways. In (a) the battery symbol is used, and it is replaced as a short circuit. In (b), a dc connection to a ground-referred voltage is shown. In this case, the point is grounded in the model as shown.

Two points need to be emphasized. First, this procedure with dc sources is simply a "paper model" and is not to be interpreted as a laboratory process. We obviously should not short any dc sources in the laboratory since the consequences may be serious! In addition, the circuit will not function properly without the dc bias.

The second point is that we are considering situations at the moment where the signal is time varying and where dc sources represent bias or reference levels rather than signals. Later in the text, some direct-coupled amplifiers will be considered in which the signal itself could be a dc voltage.

The voltages and currents computed from the dc model represent the various dc levels in the circuit, and they would be the quantities measured with ideal dc instruments. Similarly, the voltages and currents computed from the ac model represent the time-varying quantities in the circuit. Depending on the type of time-varying signal, appropriate ac instruments could be used to measure these quantities. The sum of a given dc variable and the corresponding ac variable represents the instantaneous form of the pertinent voltage or current.

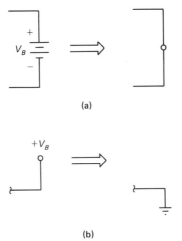

(a)

(b)

Figure 8-13 Steady-state ac equivalent circuits of ideal battery and power supply connection.

Example 8-13

The circuit of Figure 8-14(a) contains an ac signal v_g and a 15-V power supply. Assume that all frequency components of v_g are in the flat-band region. A number of possible variables of interest are identified. **(a)** Construct a dc model and compute the dc values of the variables. **(b)** Construct a flat-band ac model and compute the ac values of the variables. **(c)** Write expressions for the total instantaneous variables.

(a) Circuit

(b) DC model

(c) Flat-band AC model

Figure 8-14 Circuit of Example 8-13.

Solution

Along with the analysis of the circuits, this problem will provide an opportunity to review the notational conventions given earlier in the text. Note that all variables except the signal source in the basic circuit of Figure 8-14(a) are labeled for total instantaneous form (i.e., lowercase symbols and uppercase subscripts).

(a) The dc model is shown in Figure 8-14(b), in which the two capacitors are replaced by open circuits and the source v_g is replaced by a short circuit. All variables are now labeled for dc analysis (i.e., uppercase symbols and uppercase subscripts).

The capacitors uncouple the dc portions of the circuit on the left and right, so by a simple inspection we have

$$I_A = 0 \tag{8-51}$$

$$I_D = 0 \tag{8-52}$$

$$V_B = 0 \tag{8-53}$$

The 15-V source, the 12-kΩ resistor, and the 6-kΩ resistor constitute a simple closed loop. The current through this loop is

$$I_B = I_C = \frac{15 \text{ V}}{12 \text{ k}\Omega + 6 \text{ k}\Omega} = 0.8333 \text{ mA} \tag{8-54}$$

We could, of course, have "walked around the loop," but at this point in the book we assume that the reader has sufficient skill to analyze a simple loop such as this by a direct inspection process.

The voltage V_A is

$$V_A = 6 \times 10^3 \times 0.8333 \times 10^{-3} = 5 \text{ V} \tag{8-55}$$

Alternatively, the voltage-divider rule could have been used.

Although the left- and right-hand portions of the circuits are uncoupled at dc, voltages will appear across the open circuits represented by the capacitors. These voltages represent the dc values to which the capacitors will charge. It is observed that the two capacitor voltages are both equal to V_A, and we have

$$V_{C1} = V_{C2} = V_A = 5 \text{ V} \tag{8-56}$$

Note that the polarities were assumed in the directions resulting in positive values. With some practice, this can be done by intuition for many circuits.

(b) The flat-band ac model is shown in Figure 8-14(c). The two capacitors and the 15-V dc supply are replaced by short circuits. In the case of the dc supply, the circuit configuration is easier to visualize by turning the 12-kΩ resistor over and bringing it down to ground on the bottom. This process makes it obvious that three resistors are effectively in parallel as far as ac is concerned.

The variables are now labeled for ac analysis (i.e., lowercase symbols and lowercase subscripts). Note that the positive reference for i_b is opposite to that of the other currents on the right. The reason is that our original reference for i_B was from the power supply to the junction in the middle, and when this branch was inverted, it was necessary to keep the original reference direction.

Since the two capacitors are represented by short circuits, these two ac voltages may be written by a simple inspection as

$$v_{c1} = 0 \tag{8-57}$$

$$v_{c2} = 0 \tag{8-58}$$

The other variables may be determined by any standard resistive circuit analysis procedure. The approach used will be one of the "intuitive" methods of Chapter 1. The equivalent resistance R_{eq} seen by the source v_g is

$$R_{eq} = 1.2 \text{ k}\Omega + 12 \text{ k}\Omega\|6 \text{ k}\Omega\|1 \text{ k}\Omega$$

$$= 1200 \ \Omega + 800 \ \Omega = 2000 \ \Omega \tag{8-59}$$

The current i_a is then

$$i_a = \frac{v_g}{2000} = 5 \times 10^{-4} v_g \tag{8-60}$$

The voltages v_a and v_b are equal and may be determined as

$$v_a = v_b = 800 \times 5 \times 10^{-4} v_g = 0.4 v_g \tag{8-61}$$

Alternatively, the voltage-divider rule could have been employed. The three currents on the right are now determined as

$$i_b = \frac{-v_a}{12,000} = \frac{-0.4 v_g}{12,000} = -3.333 \times 10^{-5} v_g \tag{8-62}$$

$$i_c = \frac{v_a}{6000} = \frac{0.4 v_g}{6000} = 6.667 \times 10^{-5} v_g \tag{8-63}$$

$$i_d = \frac{v_a}{1000} = \frac{0.4 v_g}{1000} = 4 \times 10^{-4} v_g \tag{8-64}$$

Note that all ac voltages and currents are directly proportional to v_g and will have the same form. Thus if v_g is a sinusoid of a particular frequency, all ac voltages and currents will be sinusoids of the same frequency.

(c) The total instantaneous variables are determined by combining the dc quantities plus the ac quantities. The various steps follow.

$$i_A = I_A + i_a \tag{8-65a}$$

$$= 0 + 5 \times 10^{-4} v_g = 5 \times 10^{-4} v_g \tag{8-65b}$$

$$i_B = I_B + i_b \tag{8-66a}$$

$$= 0.8333 \times 10^{-3} - 3.333 \times 10^{-5} v_g \tag{8-66b}$$

$$i_C = I_C + i_c \tag{8-67a}$$

$$= 0.8333 \times 10^{-3} + 6.667 \times 10^{-5} v_g \tag{8-67b}$$

$$i_D = I_D + i_d \tag{8-68a}$$

$$= 0 + 4 \times 10^{-4} v_g = 4 \times 10^{-4} v_g \tag{8-68b}$$

$$v_{C1} = V_{C1} + v_{c1} \tag{8-69a}$$

$$= 5 + 0 = 5 \text{ V} \tag{8-69b}$$

$$v_{C2} = V_{C2} + v_{c2} \tag{8-70a}$$

$$= 5 + 0 = 5 \text{ V} \tag{8-70b}$$

$$v_A = V_A + v_a \tag{8-71a}$$

$$= 5 + 0.4 v_g \tag{8-71b}$$

$$v_B = V_B + v_b \tag{8-72a}$$

$$= 0 + 0.4 v_g = 0.4 v_g \tag{8-72b}$$

Reviewing some of these results, the following observations are pertinent: (1) The capacitors uncouple the source and the load from the dc levels in the middle. Thus there is no dc current flow through the source v_g, and the load voltage v_B has only an ac component. (2) The voltages across the capacitors have no time-varying components (i.e., they have only dc levels). This is exactly what coupling capacitors should do; that is, they should be effective shorts for ac and thus should have no ac voltages across them.

8-6 FREQUENCY RESPONSE

As we prepare to study the properties of electronic amplifier circuits, it is necessary to understand the concept of *frequency response*. The frequency response of a system is a measure of the extent to which sinusoids of different frequencies are reproduced by the system.

Consider the block diagram of Figure 8-15 with the input v_i and output v_o. The block diagram could represent any type of linear electronic circuit. Assume that the input is a sinusoid with peak value V_{pi} of the form

$$v_i = V_{pi} \sin \omega t \tag{8-73}$$

Figure 8-15 Block diagram illustrating frequency response concept.

From previous discussions, if the circuit is linear, the steady-state output voltage is also a sinusoid and can be expressed as

$$v_o = V_{po} \sin(\omega t + \theta) \qquad \text{(8-74)}$$

where V_{po} is the peak value of the output voltage, and θ is the phase shift introduced through the circuit. Thus the output is a sinusoid of the same frequency as the input, but the peak value and angle are different than the input.

The process being discussed could be monitored on an oscilloscope, or an ac voltmeter could be used. On an oscilloscope, either the peak or the peak-to-peak value of each signal could be measured directly. With a voltmeter, the quantity measured would depend on the type of meter (i.e., peak reading meter, rms meter, etc.).

Irrespective of the type of instrument used, the values of V_{pi} and V_{po} may be determined. However, instead of working with the actual quantities in volts, it is the *ratio of output voltage to input voltage* that is most significant. Thus if $V_{po}/V_{pi} = 0.5$, the output voltage has half the peak value of the input voltage at the given frequency. Let us denote the ratio simply as $|v_o|/|v_i|$ with the understanding that as long as the same property of each waveform is measured (peak, peak-to-peak, etc.), the ratio of the output quantity to the input quantity is the same.

The waveforms in Figure 8-16 illustrate how typical measurements of this type could be performed on a dual-trace oscilloscope. In (a), the voltage v_i has a peak-to-peak value of 6 V, and v_o has a peak-to-peak value of 4 V at a certain frequency. The ratio $|v_o|/|v_i| = 4/6 = 0.667$ at that frequency.

Assume now that the frequency is changed to a new value, as shown in Figure 8-16(b). [The sweep rate has been changed to show two cycles here as in (a).] The input voltage is maintained at a peak-to-peak value of 6 V. However, the output has now changed to a peak-to-peak value of 2 V. Thus $|v_o|/|v_i| = 2/6 = 0.333$ at the new frequency. This process could be continued at as many frequencies as desired.

It is convenient to define the ratio of output voltage to input voltage at various frequencies by the function $A(f)$ in which

$$\boxed{A(f) = \frac{|v_o|}{|v_i|}} \qquad \text{(8-75)}$$

The quantity f represents the frequency, and the form $A(f)$ emphasizes that the function may vary with frequency. To simplify the notation, $A(f)$ will be expressed in many places simply as A.

A plot of $A(f)$ as a function of frequency is called the *amplitude response function*. It is a measure of the relative weighting of the system to sinusoidal signals. Since complex waveforms can be considered to be composed of sinusoids, this information is very useful in characterizing an amplifier.

We will see in Chapter 9 that the same symbol A, along with the same definition of (8-75), will be used as the definition of amplifier gain. There is no conflict in these concepts since amplifier gain is, in general, a frequency-dependent function. In this chapter, however, only passive circuits are being considered.

(a)

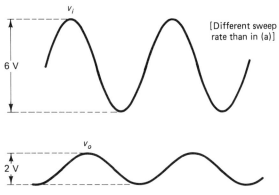

[Different sweep rate than in (a)]

(b)

Figure 8-16 Frequency response measurements on a dual-trace or dual-sweep oscilloscope.

From the oscilloscopic waveforms of Figure 8-16, observe that there is also a phase shift, which varies with frequency. A function $\theta(f)$ providing a measure of the phase shift at various frequencies can be constructed. A plot of $\theta(f)$ as a function of frequency is called the *phase response function*. To simplify the notation, $\theta(f)$ will be expressed in many places simply as θ.

In practice, the amplitude response function is more widely used than the phase response function, although the latter function is important in some applications. Strictly speaking, the term *frequency response* refers to the combination of both the *amplitude response and* the *phase response*. In casual usage, however, the term *frequency response* is often used to mean simply the *amplitude response* whenever phase considerations are not important.

Some typical amplitude response functions of importance are shown in Figure 8-17. The function of (a) is a *low-pass* function. It is characterized by the fact that dc and all frequencies lower than a certain reference frequency are passed with equal weighting. If the circuit involved is an amplifier, the level of the amplitude response in the low-frequency region will normally be greater than 1 as a result of the gain. If

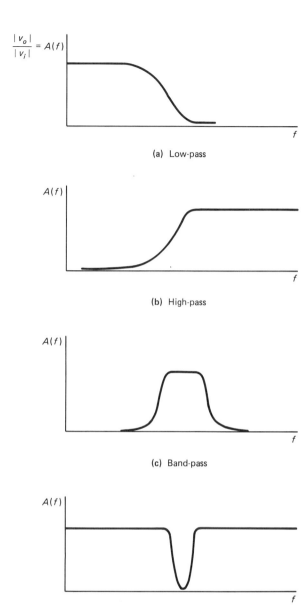

(a) Low-pass

(b) High-pass

(c) Band-pass

(d) Band-rejection

Figure 8-17 Four important types of frequency response functions.

the circuit is a filter or a coupling network, the level in the low-frequency region may be 1 or less than 1. However, the important point is that all frequencies in that range are weighted equally. The low-frequency region of transmission is called the *pass-band*.

The function of Figure 8-17(b) is called a *high-pass* function. It is character-ized by the fact that dc and all frequencies below a certain frequency are attenuated or reduced in amplitude, while higher frequencies are passed by the system. Thus the passband for a high-pass filter is above the reference frequency.

The function of Figure 8-17(c) is called a *bandpass* function. Frequencies within a certain passband range are passed, while frequencies lower and above this range are attenuated.

The function of Figure 8-17(d) is called a *band-rejection* function. Frequencies within a certain range are rejected, while frequencies lower and above this range are passed.

8-7 *SIMPLE* RC *CIRCUIT FORMS*

The amplitude response forms of certain simple *RC* circuits are investigated in this section on a somewhat descriptive basis. The purpose is to emphasize certain kinds of low-frequency and high-frequency behavior that occur in electronic amplifiers. While the actual models of complex electronic circuits are quite involved, they can often be reduced to combinations of the forms to be discussed here. Thus an understanding of the operation of these forms can lead to an understanding of similar effects in much more complex electronic circuits.

Certain useful formulas will be given with each of the circuits to be considered. These formulas may be readily derived using steady-state ac circuit analysis, and readers having that background are strongly encouraged to verify the results. Readers not having a background in ac circuits should still be able to understand the significance of the results and to apply the formulas to practical problems.

Certain "10% rules" will be given. These guidelines are very useful in analyzing and designing circuits, and they provide useful intuitive approaches to circuit understanding. It should be stressed, however, that the rules of thumb provided should not be interpreted as absolute facts or requirements. Generally, they provide useful information, but when requirements are very exact, one must resort to an actual mathematical analysis for precise guidelines.

RC *Low-Pass Model*

Consider the circuit shown in Figure 8-18(a). Assume that v_i is a sinusoidal voltage source whose frequency can be varied. The steady-state output voltage v_o is measured across the capacitor. This arrangement thus allows a frequency response measurement to be performed.

The dc steady-state model of the capacitor is an open circuit. Similarly, at very low frequencies, the capacitive reactance is very large, and the capacitor is acting nearly like an open circuit. Under these conditions, the output voltage is essentially the same as the input voltage, and $A \simeq 1$.

As the frequency continues to increase, the capacitive reactance decreases to a point where sufficient current is drawn from the source to cause a significant voltage drop across R. The output voltage then starts to drop.

At much higher frequencies, the capacitive reactance becomes very low. Almost all the input voltage is then dropped across R, and very little appears across C. This corresponds to the frequency range where, for some applications, the flat-band model (short circuit) for the capacitor might be used. However, for this application, we recognize that there will continue to be a small voltage across the capacitor. The

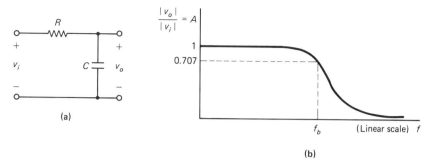

(a)

(b)

Figure 8-18 Simple *RC* low-pass circuit model and its amplitude response.

drop in response as the frequency continues to increase is called *high-frequency roll-off*.

The form of $A(f)$ for this circuit is shown in Figure 8-18(b). This response is a *low-pass* type of function. The frequency scale here is assumed to be *linear*. Later we investigate the shape of such a curve when the scale is *logarithmic*, but let's not confuse the situation now.

It can be shown using ac circuit theory that the amplitude response $A(f)$ is given by

$$A(f) = \frac{X_C}{\sqrt{R^2 + X_C^2}} \qquad (8\text{-}76)$$

A convenient frequency to use in characterizing this response is that frequency at which $R = X_C$. For reasons that will not be clear until Section 9-10, we will define such a frequency as the *break frequency*, denoted as f_b.

The frequency f_b is determined by setting $X_C = R$. Letting $\omega = 2\pi f$ and $f = f_b$ yields

$$\frac{1}{2\pi f_b C} = R \qquad (8\text{-}77)$$

This leads to

$$\boxed{f_b = \frac{1}{2\pi RC}} \qquad (8\text{-}78)$$

By some algebraic manipulations, (8-76) can be rewritten as

$$A(f) = \frac{1}{\sqrt{1 + (f/f_b)^2}} \qquad (8\text{-}79)$$

It is readily shown that when $f = f_b$, the amplitude response is

$$\boxed{A(f_b) = \frac{1}{\sqrt{2}} \simeq 0.707} \qquad (8\text{-}80)$$

This frequency is identified on the curve of Figure 8-18(b).

At frequencies well below f_b, $A(f) \simeq 1$ and there is very little reduction in magnitude of a signal in passing through the circuit. Conversely, at frequencies well above f_b, there is considerable attenuation, and the output signal will be reduced significantly. In the vicinity of f_b, there is some attenuation of a signal, so f_b provides some measure of the transition from nearly flat response to the attenuation region. Some references define f_b as the *cutoff frequency*, but it is not a sharp cutoff characteristic for this simple circuit.

Note how the break frequency is inversely dependent on the RC product. Thus an increase of either R or C will result in a reduction of the break frequency and a smaller passband region.

There are two separate points of view about this circuit that should be emphasized. First, the circuit has value as a simple low-pass filter, and it is often used for this purpose. As a low-pass filter, this circuit is about the most "elementary type" that can be implemented. Much sharper characteristics are available with more complex circuit structures, some of which are discussed later in the text. However, this simple circuit is widely used in minimal filtering applications where there is a wide band available between the desired signal and the undesired components to be attenuated.

The second point of view is that this circuit form arises frequently and naturally in electronic devices without intent on the part of the design. All electronic devices exhibit capacitance across the input, much of which arises from complex semiconductor properties. This capacitance, in conjunction with circuit resistance, results in an inherent frequency-limiting effect.

When flat transmission is desired through a circuit fitting the low-pass model, it is necessary that the signal frequency range of interest be well below f_b. If the capacitance is fixed (such as the input capacitance of an active device), this is achieved by ensuring that R be sufficiently small.

The actual amplitude response at any given frequency can be evaluated using (8-79). At a frequency $f = 0.1 f_b$, the value of $A(f)$ as determined from (8-79) is $A(0.1 f_b) = 0.995$. This represents a drop in amplitude response of 0.5%. This particular frequency is sometimes used as a measure of the upper-frequency limit for negligible effect on the amplitude response. At the frequency $0.1 f_b$, the resistance is one-tenth the value of the capacitive reactance.

In many active circuits, the input circuit to the electronic device will contain a shunt capacitance in parallel with a resistance, and the driving circuit will contain a series circuit. The model for such a case is shown in Figure 8-19(a). The resistance R_2 has been deliberately shown on the right to assist the reader in seeing all possible points of view. We first redraw the circuit so that the capacitance C is on the right, as shown in (b). A Thévenin equivalent circuit looking back from the capacitor is then determined as shown in (c). The circuit is now in the form of the simpler one given in Figure 8-18(a).

The amplitude response for this circuit is shown in Figure 8-19(d). The low-frequency amplitude response is $R_2/(R_1 + R_2)$ because of the resistive voltage-divider reduction. The break frequency f_b is

$$f_b = \frac{1}{2\pi R_{eq} C} \tag{8-81}$$

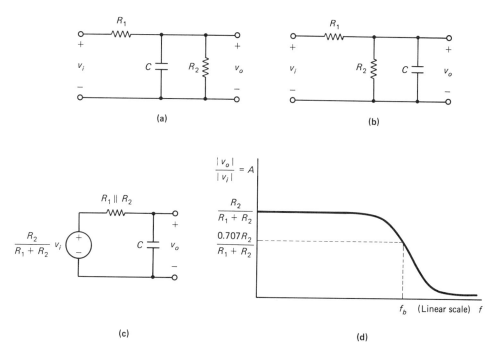

Figure 8-19 *RC* low-pass circuit containing both series and shunt resistors and simplification.

where R_{eq} is the parallel combination of R_1 and R_2; that is,

$$R_{eq} = R_1 \parallel R_2 = \frac{R_1 R_2}{R_1 + R_2} \tag{8-82}$$

RC *Coupling Circuit*

Consider the *RC* circuit shown in Figure 8-20(a). At dc, the capacitor is an open circuit, and there is no coupling between input and output. Similarly, at very low fre-

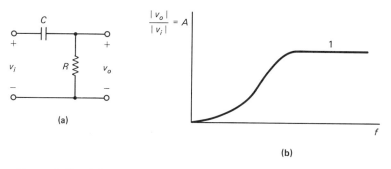

Figure 8-20 *RC* coupling circuit and form of the amplitude response.

Sec. 8-7 Simple RC Circuit Forms

quencies, the capacitive reactance is very large, and it blocks most of the signal. As the frequency increases, the capacitive reactance decreases, thus allowing more of the signal to reach the output. At sufficiently high frequencies, the flat-band model of the capacitor (short-circuit) is valid, and the input signal is effectively coupled to the output.

The form of $A(f)$ for this circuit is shown in Figure 8-20(b). This response is a *high-pass* type of function. The drop in response at low frequencies is called *low-frequency roll-off*.

It can be shown using ac circuit theory that the amplitude response $A(f)$ is given by

$$A(f) = \frac{R}{\sqrt{R^2 + X_C^2}} \qquad (8\text{-}83)$$

As in the case of the low-pass function, a *break frequency* f_b will be defined as the frequency at which $R = X_C$. The steps involved are exactly the same as for the low-pass case, so (8-77) and (8-78) are valid for the high-pass case. The result is exactly the same; that is,

$$\boxed{f_b = \frac{1}{2\pi RC}} \qquad (8\text{-}84)$$

By some algebraic manipulations, (8-83) can be expressed as

$$A(f) = \frac{1}{\sqrt{1 + (f_b/f)^2}} \qquad (8\text{-}85)$$

It is readily shown that when $f = f_b$, the amplitude response is

$$\boxed{A(f_b) = \frac{1}{\sqrt{2}} \simeq 0.707} \qquad (8\text{-}86)$$

Comparing low-pass and high-pass circuits, the expression for break frequency applies to both cases. For the *low-pass* case, the break frequency is at the *upper* end of the passband, while for the *high-pass* case, it is at the *lower* end. In both cases, the response has a level $1/\sqrt{2}$ at the break frequency. However, there is an important difference in the amplitude-response forms of (8-79) and (8-85). The difference is that f/f_b for the low-pass case is replaced by f_b/f for the high-pass case.

This circuit form is used extensively to pass a time-varying (or "ac") signal while blocking a dc level. For distortion-free transmission, it is necessary that all frequencies comprising the signal fall in the flat-band region of the signal. The break frequency is inversely proportional to the RC product, so the lower range frequency response can be improved by increasing the RC product.

The actual amplitude response at any given frequency can be evaluated using (8-85). At a frequency $f = 10f_b$, the value of $A(f)$ as determined from (8-85) is $A(10f_b) = 0.995$. This particular frequency is sometimes used as a measure of the lower-frequency limit for negligible effect on the amplitude. In many design problems, R will be fixed and C must be selected for flat coupling. If the preceding criterion based on $10f_b$ is used, a capacitance C is calculated to establish f_b at one-tenth of the lowest applicable frequency.

In many practical coupling circuits, there will be both a source resistance and a load resistance. The applicable circuit model is shown in Figure 8-21(a). The form of the amplitude response is shown in (b). Note that the flat-band amplitude response is now $R_2/(R_1 + R_2)$, resulting from the resistive voltage-divider action.

The break frequency f_b is now

$$f_b = \frac{1}{2\pi R_{eq} C} \tag{8-87}$$

where R_{eq} is the equivalent series resistance; that is,

$$R_{eq} = R_1 + R_2 \tag{8-88}$$

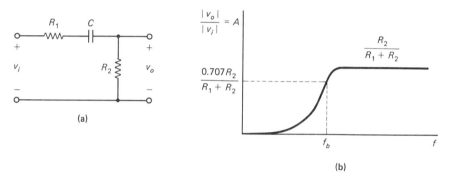

(a)

(b)

Figure 8-21 *RC* coupling network with attenuation in flat-band region.

RC *Bypass Circuit*

Consider the circuit shown in Figure 8-22 containing a resistance R in parallel with a capacitance C. This concept is best developed with a Norton model looking back from the capacitor. Although this form appears in a number of applications, one particular application arising in many electronic circuits is that of a *bypass circuit*. The concept is as follows: The resistor serves some dc function, such as establishing a transistor bias point. However, it is desired to pass the ac signal around the resistor so that the resistance does not affect the signal. If the capacitor can be considered as a short circuit in the pertinent frequency range, the objective will be accomplished.

A useful rule of thumb applicable to this circuit is the following: *For most prac-*

Figure 8-22 Basic form of an *RC* bypass circuit.

tical purposes, the capacitor can be considered to fully bypass the signal current at all ac frequencies in which the capacitive reactance is not greater than 10% of the net parallel resistance. Expressed mathematically,

$$X_C \leq 0.1R \qquad \text{for bypass} \qquad (8\text{-}89)$$

Note that it is the *net parallel resistance* that is used in this case. This rule is often misinterpreted in practical circuits whenever the *RC* combination shown is in parallel with some additional circuit whose equivalent resistance is not readily apparent.

Consider the circuit shown in Figure 8-23(a) in which a voltage source in series with R_1 appears on the left. Suppose that it is desired to bypass R_2 with the capacitor *C*. One might be tempted to base the rule on X_C and R_2 above. However, it is necessary to determine the net equivalent parallel resistance. This is achieved by looking back to the left of R_2 and using a source transformation. The modified circuit is shown in Figure 8-23(b). The net parallel resistance R_{eq} is then

$$R_{eq} = R_1 \| R_2 = \frac{R_1 R_2}{R_1 + R_2} \qquad (8\text{-}90)$$

Application of this rule yields

$$X_C \leq 0.1 R_{eq} \qquad (8\text{-}91)$$

at the lowest frequency of interest.

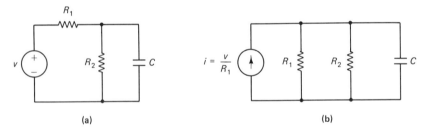

(a)   (b)

Figure 8-23 More complex bypass situation and equivalent circuit used to determine net parallel resistance.

Example 8-14

A simple low-pass filter of the form of Figure 8-18(a) is to be designed to eliminate some high-frequency noise. Design the circuit to achieve a break frequency of 20 kHz.

Solution

The break frequency is given by

$$f_b = \frac{1}{2\pi RC} \qquad (8\text{-}92)$$

The required RC product is then determined as

$$RC = \frac{1}{2\pi f_b} = \frac{1}{2\pi \times 20 \times 10^3} = 7.958 \times 10^{-6} \qquad (8\text{-}93)$$

This result indicates that any R and C whose product is equal to the value determined will achieve the objective. This situation illustrates a common occurrence in design. Since no other constraints are given, we are free to choose either R or C to be some common stock value and then determine the other value from the required product. It is usually better to select C as a stock value and determine R since resistance is usually easier to adjust.

As a particular solution, we will choose $C = 0.001~\mu F$, in which case R is determined as

$$R = \frac{7.958 \times 10^{-6}}{C} = \frac{7.958 \times 10^{-6}}{0.001 \times 10^{-6}} = 7.958~k\Omega \qquad (8\text{-}94)$$

However, many other combinations could be determined just as easily.

Example 8-15

A certain signal transmission system can be modeled by the circuit of Figure 8-24. The 100-pF capacitance represents the net circuit shunt capacitance, which cannot be eliminated. The voltage v_g and resistance R_g represent the Thévenin model of a source used to drive the load. **(a)** Determine the break frequency f_b if $R_g = 10~k\Omega$. **(b)** If the source is replaced with one having a much lower internal resistance, repeat the analysis based on $R_g = 50~\Omega$.

Figure 8-24 Circuit of Example 8-15.

Solution

(a) The break frequency based on $R_g = 10~k\Omega$ is

$$f_b = \frac{1}{2\pi R_g C} = \frac{1}{2\pi \times 10^4 \times 100 \times 10^{-12}} = 159.15~kHz \qquad (8\text{-}95)$$

(b) With $R_g = 50~\Omega$, the break frequency now changes to

$$f_b = \frac{1}{2\pi \times 50 \times 100 \times 10^{-12}} = 31.83~MHz \qquad (8\text{-}96)$$

Note the significant increase in the break frequency when the source resistance is reduced!

The actual usable bandwidth in each case would depend on how much deviation in the amplitude response would be acceptable. Recall that at the break frequency, the response is down to a level of 0.707. If, for example, the $0.1f_b$ rule based on 0.5% degradation in amplitude response were used, the respective bandwidths in (a) and (b) would be 15.915 kHz and 3.183 MHz.

Example 8-16

Determine the break frequency of the RC-coupling circuit of Figure 8-25.

Figure 8-25 Circuit of Example 8-16.

Solution

The break frequency is

$$f_b = \frac{1}{2\pi RC} = \frac{1}{2\pi \times 200 \times 10^3 \times 0.1 \times 10^{-6}} = 7.958 \text{ Hz} \quad \textbf{(8-97)}$$

Example 8-17

It is desired to couple two particular amplifiers together using a capacitor to block dc levels as shown in Figure 8-26(a). The ac Thévenin equivalent circuit at the output of the first stage is a voltage source in series with 2 kΩ, and the input resistance of the second stage is a 10-kΩ resistance. Determine the value of C required to achieve a break frequency of 10 Hz.

Figure 8-26 Circuit of Example 8-17.

Solution

An equivalent circuit at the point of coupling is shown in Figure 8-26(b). The net series resistance is

$$R_{eq} = 2 \text{ k}\Omega + 10 \text{ k}\Omega = 12 \text{ k}\Omega \qquad \text{(8-98)}$$

The break frequency is

$$f_b = \frac{1}{2\pi R_{eq} C} \qquad \text{(8-99)}$$

Solving for C, we have

$$C = \frac{1}{2\pi R_{eq} f_b} = \frac{1}{2\pi \times 12 \times 10^3 \times 10} = 1.326 \ \mu\text{F} \qquad \text{(8-100)}$$

8-8 BANDWIDTH AND RISE TIME

Many signals that must be transmitted or processed in modern electronic systems are digital or pulse-type signals. Such signals often have only two levels, representing a logical 1 and a logical 0, and perfect reproduction is not necessary as long as the two levels are distinguishable. In other pulse applications, it may be necessary to reproduce the pulse to the highest accuracy possible.

It turns out that the effect of a limited bandwidth is to impose a rise time on a pulse waveform. The basic concept will be developed with the RC low-pass circuit shown in Figure 8-27(a). Assume that the input is a dc voltage of magnitude V_i,

Figure 8-27 Low-pass RC circuit and its effect on a dc voltage switched on at $t = 0$.

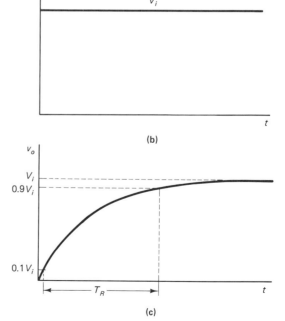

which is switched in to the *RC* circuit at $t = 0$. The voltage v_i to the right of the switch then changes instantaneously from 0 to V_i at $t = 0$, as shown in Figure 8-27(b).

A fundamental property of capacitance is that a voltage across it cannot change instantaneously. Instead, it takes some time for the capacitor voltage to charge or discharge. For the circuit given, the voltage buildup across the capacitance has the form shown in Figure 8-27(c), and this is the circuit output voltage.

The capacitor voltage can be described by an *exponential function*. As a function of time t, the voltage can be expressed mathematically as

$$v_o = V_i(1 - \epsilon^{-t/RC}) \tag{8-101}$$

In this equation, $\epsilon = 2.718$ is the base for the natural system of logarithms. More will be said about exponential functions in the context of nonlinear circuits in Chapter 16, but the form of (8-101) will suffice for the moment.

Rise Time

The quantity *rise time* is a measure of the duration of the charging time in a circuit such as this. However, it is often difficult to determine precisely where the output voltage starts and where it reaches a final value. For that reason, the standard rise-time definition is based on the time between the point where the output voltage is at 10% of its final value and the point where the output voltage is at 90% of its final value. This is illustrated in Figure 8-27(c).

For the simple *RC* circuit, it can be shown from the basic properties of the exponential function that the rise time T_R is given by

$$\boxed{T_R = 2.2RC} \tag{8-102}$$

The rise time is thus directly proportional to both R and C.

A relationship between rise time and bandwidth will now be obtained. The break frequency f_b is

$$\boxed{f_b = \frac{1}{2\pi RC}} \tag{8-103}$$

If the product RC, expressed in terms of f_b, is determined from (8-103) and substituted in (8-102), the following equation is obtained:

$$\boxed{T_R = \frac{0.35}{f_b}} \tag{8-104}$$

Although this equation is based on the simple *RC* low-pass circuit, the equation can be used to estimate the approximate rise time of more complex circuits. Let us denote the low-pass bandwidth of any circuit of interest as B, in which case (8-104) as an estimate becomes

$$T_R = \frac{0.35}{B} \qquad \textbf{(8-105)}$$

The results of this equation are very important and need to be emphasized. In general, *the rise time and bandwidth are inversely related*. Thus, to reproduce a perfect pulse without distortion, an infinite bandwidth is required.

Pulse Reproduction

The effects of finite bandwidth on a pulse train are illustrated in Figure 8-28. The waveform of (a) represents some input pulse train v_i. (It could be digital data, for example.) The width of a given pulse is assumed to be T_p. When the bandwidth is rela-

(a)

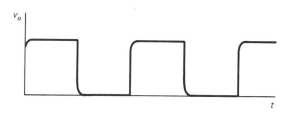

(b) Large bandwidth ($B \gg 1/T_p$)

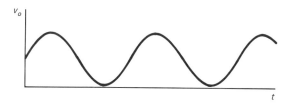

(c) Medium bandwidth ($B \approx 0.5/T_p$)

(d) Small bandwidth ($B \ll 1/T_p$)

Figure 8-28 Effects of finite bandwidth on pulse train.

tively large, the output pulse train is quite close in form to the input, as shown in Figure 8-28(b). Some rounding at the edges occurs, but this may not even be apparent if the bandwidth is sufficiently large.

As the bandwidth is reduced, the output pulse train eventually assumes the form of Figure 8-28(c). The two-level character of the output is quite evident, but considerable rounding has occurred. Depending on the application, this type of output may or may not be acceptable.

If the bandwidth is reduced further, the output pulses are broadened and smeared to the point where it is nearly impossible to distinguish between different levels. This situation is illustrated in Figure 8-28(d).

From a quantitative point of view, the bandwidth required to reproduce a pulse train *accurately* should satisfy the inequality $B \gg 1/T_p$. However, for many forms of pulse data, where the levels must be distinguishable but where accurate pulse reproduction is not necessary, the following "coarse reproduction" estimate is useful:

$$B \simeq \frac{0.5}{T_p} \qquad \text{(8-106)}$$

If the bandwidth is reduced much below the value $0.5/T_p$, the amplitude of the output pulses is degraded severely, and it will become nearly impossible to distinguish the levels involved.

Example 8-18

A certain *RC* low-pass circuit of the form shown in Figure 8-27(a) has $R = 10 \text{ k}\Omega$ and $C = 0.01 \ \mu\text{F}$. Calculate the rise time.

Solution

The rise time is

$$T_R = 2.2RC = 2.2 \times 10^4 \times 0.01 \times 10^{-6} = 220 \ \mu\text{s} \qquad \text{(8-107)}$$

Example 8-19

A certain *RC* low-pass circuit of the form shown in Figure 8-27(a) has $f_b = 2 \text{ kHz}$. Calculate the rise time.

Solution

The rise time is

$$T_R = \frac{0.35}{f_b} = \frac{0.35}{2 \times 10^3} = 175 \ \mu\text{s} \qquad \text{(8-108)}$$

Example 8-20

A certain digital data signal has bits with width 2 μs. Rounding the data according to the "coarse bandwidth" estimate is acceptable for the particular application involved. Estimate the approximate bandwidth required for signal transmission.

Solution

The bandwidth estimate is given by

$$B \simeq \frac{0.5}{T_p} = \frac{0.5}{2 \times 10^{-6}} = 250 \text{ kHz} \tag{8-109}$$

Understand that this is an estimate rather than an exact equation.

DRILL PROBLEMS

8-1. A certain sinusoidal voltage v has a peak value of 5 V and a frequency of 2 kHz. The time origin is chosen in the form of Figure 8-2. **(a)** Determine the period. **(b)** Write an equation for the voltage.

8-2. A certain sinusoidal current i has a peak value of 120 mA and a frequency of 500 Hz. The time origin is chosen in the form of Figure 8-2. **(a)** Determine the period. **(b)** Write an equation for the current.

8-3. A certain voltage is described by the equation

$$v = 25 \sin 2\pi \times 5000t$$

Determine **(a)** the peak value, **(b)** the radian frequency, **(c)** the cyclic frequency, and **(d)** the period.

8-4. A certain current is described by the equation

$$i = 2 \cos 1000t$$

Determine **(a)** the peak value, **(b)** the radian frequency, **(c)** the cyclic frequency, and **(d)** the period.

8-5. **(a)** Determine the rms value of the voltage of Problem 8-1. **(b)** Determine the power dissipated in a 180-Ω resistor.

8-6. **(a)** Determine the rms value of the current of Problem 8-2. **(b)** Determine the power dissipated in a 200-Ω resistor.

8-7. **(a)** Determine the rms value of the voltage of Problem 8-3. **(b)** Determine the power dissipated in a 2.2-kΩ resistor.

8-8. (a) Determine the rms value of the current of Problem 8-4. (b) Determine the power dissipated in a 10-Ω resistor.

8-9. Determine the reactance of a 200-pF capacitor at each of the following frequencies: (a) 100 Hz; (b) 50 kHz; (c) 4 MHz.

8-10. Determine the reactance of a 10-μF capacitor at each of the following frequencies: (a) 50 Hz; (b) 20 kHz; (c) 1 MHz.

8-11. Determine the reactance of a 40-μH inductance at each of the following frequencies: (a) 200 Hz; (b) 100 kHz; (c) 5 MHz.

8-12. Determine the reactance of a 2-H inductance at each of the following frequencies: (a) 25 Hz; (b) 10 kHz; (c) 200 kHz.

8-13. Determine the steady-state dc voltage V_C across the capacitor in the circuit of Figure P8-13.

Figure P8-13

8-14. Determine the steady-state dc voltage V_C across the capacitor in the circuit of Figure P8-14.

Figure P8-14

8-15. Determine the steady-state dc voltages V_{C1} and V_{C2} in the circuit of Figure P8-15.

Figure P8-15

8-16. Determine the steady-state dc voltage V_C in the circuit of Figure P8-16.

Figure P8-16

8-17. Determine the steady-state current I and voltage V_C for the circuit of Figure P8-17.

Figure P8-17

8-18. Determine the steady-state current I and voltages V_{C1} and V_{C2} for the circuit of Figure P8-18.

Figure P8-18

8-19. For the single-stage RC-coupled BJT amplifier of Figure P8-19, determine the steady-state dc voltages across the three capacitors. The voltage v_g is an ac signal source.

Figure P8-19

8-20. The circuit of Figure P8-20 represents an RC-coupled JFET amplifier. Determine the steady-state dc voltages across the three capacitors. The voltage v_g is an ac signal source.

+24 V

$I_D = 2$ mA

4.3 kΩ

470 kΩ

V_{C_2}

C_2

V_{C_1}

C_1

V_g

100 kΩ

8.2 kΩ

3 kΩ

C_S V_{CS}

Figure P8-20

8-21. For the *RC* circuit of Figure P8-21, assume that the frequency range of the signal v_g is in the flat-band region. Determine the voltage v_o in terms of v_g.

C 3 kΩ

V_g

2 kΩ v_o

+

−

Figure P8-21

8-22. For the *RC* circuit of Figure P8-22, assume that the frequency range of the signal v_g is in the flat-band region. Determine the voltage v_o in terms of v_g.

6 kΩ C_1 C_2

V_g

6 kΩ 12 kΩ v_o

+

−

Figure P8-22

8-23. The circuit of Figure P8-23 contains a time-varying signal v_g and a 24-V dc

+24 V

3 kΩ

8 kΩ C_1

v_{C_1} v_A v_{C_2}

V_g

6 kΩ v_B **Figure P8-23**

source. Assume that all frequency components of v_g are in the flat-band region. A number of possible variables of interest are identified. **(a)** Construct a dc model and compute the dc values of the variables. **(b)** Construct a flat-band model and compute the ac values of the variables. **(c)** Write expressions for the total instantaneous variables.

8-24. Repeat the analysis of Problem 8-23 for the circuit of Figure P8-24.

Figure P8-24

8-25. An *RC* low-pass circuit of the form of Figure 8-18(a) has $R = 10$ kΩ and $C = 0.001$ μF. Determine the break frequency.

8-26. An *RC* low-pass circuit of the form of Figure 8-18(a) has $R = 600$ Ω and $C = 200$ pF. Determine the break frequency.

8-27. An *RC* low-pass circuit of the form of Figure 8-19(a) has $R_1 = 12$ kΩ, $R_2 = 6$ kΩ, and $C = 0.01$ μF. Determine the break frequency.

8-28. An *RC* low-pass circuit of the form of Figure 8-19(a) has $R_1 = 200$ kΩ, $R_2 = 50$ kΩ, and $C = 2$ μF. Determine the break frequency.

8-29. An *RC* coupling network of the form of Figure 8-20(a) has $R = 20$ kΩ and $C = 0.01$ μF. Determine the break frequency.

8-30. An *RC* coupling network of the form of Figure 8-20(a) has $R = 470$ kΩ and $C = 50$ pF. Determine the break frequency.

8-31. An *RC* coupling network of the form of Figure 8-21(a) has $R_1 = 12$ kΩ, $R_2 = 24$ kΩ, and $C = 0.01$ μF. Determine the break frequency.

8-32. An *RC* coupling network of the form of Figure 8-21(a) has $R_1 = 150$ Ω, $R_2 = 220$ Ω, and $C = 100$ pF. Determine the break frequency.

8-33. A certain signal source has a Thévenin resistance of 600 Ω and it is driving a capacitive load of 800 pF. Calculate the rise time.

8-34. A certain signal source has a Thévenin resistance of 2 kΩ, and it is driving a capacitive load of 50 pF. Calculate the rise time.

8-35. Calculate the rise time for the circuit of Problem 8-25.

8-36. Calculate the rise time for the circuit of Problem 8-26.

8-37. Calculate the rise time for the circuit of Problem 8-27.

8-38. Calculate the rise time for the circuit of Problem 8-28.

8-39. The measured rise time of a certain low-pass circuit is 20 μs. Estimate the bandwidth B.

8-40. The measured rise time of a certain low-pass circuit is 50 ns. Estimate the bandwidth B.

DERIVATION PROBLEMS

8-41. Starting with (8-76), fill in the steps required to express it in the form of (8-79).

8-42. Starting with (8-83), fill in the steps required to express it in the form of (8-85).

8-43. From the expression for v_o in (8-101), determine the time t_1 at which $v_o = 0.1V_i$ and the time t_2 at which $v_o = 0.9V_i$. The value of rise time is then $T_R = t_2 - t_1$. Show that the result is (8-102).

8-44. Fill in the steps leading from (8-102) to (8-104).

DESIGN PROBLEMS

8-45. In a certain application requiring a capacitor, it is desired that the reactance be no greater than 250 Ω at a frequency of 20 Hz. Determine the minimum value of capacitance required.

8-46. In a certain application requiring a capacitor, it is desired that the reactance be no less than 500 Ω at a frequency of 2 kHz. Determine the maximum value of capacitance required.

8-47. In a certain application requiring an inductor, it is desired that the reactance be no less than 5 kΩ at a frequency of 200 kHz. Determine the minimum value of inductance required.

8-48. In a certain application requiring an inductor, it is desired that the reactance be no greater than 200 Ω at a frequency of 50 Hz. Determine the maximum value of inductance required.

8-49. A simple low-pass filter of the form of Figure 8-18(a) is to be designed to have a break frequency of 5 kHz. Determine the value of R required if the capacitance is selected as $C = 0.01$ μF.

8-50. For the design of Problem 8-49, determine the value of C required if the resistance is selected as $R = 10$ kΩ.

8-51. A coupling network of the form of Figure 8-20(a) is to be designed to have a break frequency of 50 Hz. Determine the value of R required if the capacitance is selected as $C = 12 \ \mu F$.

8-52. For the design of Problem 8-51, determine the value of C required if the resistance is selected as $R = 200 \ k\Omega$.

TROUBLESHOOTING PROBLEMS

8-53. You are troubleshooting the circuit of Figure P8-15 and you measure the following voltages with a dc voltmeter: source voltage = 30 V, $V_{C1} = 5$ V, $V_{C2} = 0$. Which *two* of the following conditions could represent possible difficulties? **(a)** 10-kΩ resistor shorted; **(b)** C_1 shorted; **(c)** 2-kΩ resistor shorted; **(d)** 2-kΩ resistor open; **(e)** C_2 shorted; **(f)** 3-kΩ resistor shorted.

8-54. You are troubleshooting the circuit of Figure P8-16 and you measure the following voltages with a dc voltmeter: source voltages = 60 V and 45 V (as shown), $V_C = 60$ V. Which *two* of the following conditions could represent possible difficulties? **(a)** 20-kΩ resistor open; **(b)** 10-kΩ resistor open; **(c)** 20-kΩ resistor shorted; **(d)** 10-kΩ resistor shorted; **(e)** C shorted.

8-55. You are troubleshooting a simple low-pass filter circuit of the form shown in Figure 8-18(a). The circuit is not suppressing high-frequency noise to the extent expected. Which *two* of the following conditions could represent possible difficulties? **(a)** R too small; **(b)** R too large; **(c)** C too small; **(d)** C too large.

8-56. You are troubleshooting a circuit of the form shown in Figure 8-20(a) and the frequency of the source is in the specified range of flat-band frequencies. With an ac rms voltmeter, you measure $v_g = 80$ mV and $v_o = 20$ mV. Which *two* of the following conditions could represent possible difficulties? **(a)** R too small; **(b)** R too large; **(c)** C too small; **(d)** C too large.

COMPUTER PROBLEMS

8-57. Write a program to compute the capacitive reactance of a capacitor.

Input data: C, f
Output data: X_C

8-58. Write a program to compute the inductive reactance of an inductor.

Input data: L, f
Output data: X_L

8-59. Write a program to compute the amplitude response of a simple RC low-pass filter of the type shown in Figure 8-18(a) over a range of frequencies from f_1 to f_2 in steps of Δf.

Input data: f_b, f_1, f_2, Δf
Output data: $A(f)$

8-60. Write a program to compute the amplitude response of a simple RC coupling circuit of the type shown in Figure 8-20(a) over a range of frequencies from f_1 to f_2 in steps of Δf.

Input data: f_b, f_1, f_2, Δf
Output data: $A(f)$

CHAPTER 9

Introduction to Amplifiers

OBJECTIVES

After completing this chapter, the reader should be able to:

1. Define the voltage gain of an amplifier and discuss its physical significance.
2. Explain various classifications of amplifiers such as noninverting or inverting, dc coupled or ac coupled, voltage or power, and so on.
3. Construct the flat-band model of an amplifier.
4. Define and explain the significance of the input and output resistances of an amplifier.
5. Determine the net gain of an amplifier when the generator has internal resistance, and/or an external load resistance is connected to the output.
6. Determine the net gain of a cascade of two or more amplifiers when there is interaction between the stages.
7. Define and calculate decibel gain and loss quantities from the actual gain and loss values.
8. Calculate actual gain and loss values from decibel gain and loss quantities.
9. Calculate the decibel gain for a cascade of amplifier stages.
10. Discuss the general concept of a Bode plot.
11. Construct the Bode plot amplitude and phase approximations for a simple low-pass circuit.
12. Same as objective 11 for a simple high-pass circuit.

9-1 OVERVIEW

One of the most important applications of electronic devices is in the implementation of *amplifier* circuits. An *amplifier* is a circuit used to increase the level of a signal to meet some desired objective. For example, when a phonograph needle (stylus) moves along a record, tiny signals are generated at the output of the cartridge. By use of electronic amplifiers, these signals are increased to levels that may be connected to speakers from which audible sound is generated. A second example is a radio receiver, in which tiny electromagnetic signals generated in the antenna are amplified and processed to produce useful output.

A major objective of the next several chapters is to develop the various basic individual circuits from which composite amplifiers are formed. In many books, the reader is immediately exposed to the details of various amplifier circuits without thorough grounding in the general properties that are sought in such circuits. Consequently, beginning students are often bewildered with amplifier details and do not understand the significance of the major amplifier properties.

The approach in this book is to provide this chapter as a basis for general amplifier properties without regard to specific configurations or devices. Consequently, the circuit models developed in this chapter are quite general and may be applied to virtually all amplifiers to be considered later. Therefore, study the basic properties presented here carefully and make sure that you understand the significance of each. You will see these properties used extensively in later chapters as particular amplifiers using BJTs, FETs, and operational amplifiers are discussed.

9-2 AMPLIFIERS

An ideal amplifier is a circuit that increases the level of a signal while preserving the exact form. Consider the block diagram shown in Figure 9-1. The voltage v_i represents an assumed input signal, and the voltage v_o represents an assumed output signal. The output voltage is related to the input voltage in an ideal amplifier by the equation

$$v_o = Av_i \qquad \text{(9-1)}$$

The quantity A represents the *voltage gain,* and it can be expressed as

$$A = \frac{v_o}{v_i} \qquad \text{(9-2)}$$

for the ideal case.

The key to the definition of the ideal amplifier is that the *exact shape of the signal is preserved.* If the signal shape is altered, it is not possible to express input and output relationships exactly in the simple algebraic forms of (9-1) and (9-2).

Any change in the signal shape resulting from the amplifier is called *distortion.* An ideal amplifier is said to be *distortion-free.* All electronic devices introduce some amount of distortion, but a good design results in a distortion level so low that it produces negligible degradation for the intended application. It is appropriate, there-

Figure 9-1 Block diagram of amplifier with assumed input and output voltages.

fore, to use the simple algebraic forms of (9-1) and (9-2) in amplifiers having very low distortion.

Although the output signal of an amplifier is normally larger than the input signal, it should be stressed that the amplifier is incapable of generating power. Rather, one or more dc power sources provide the power required for the unit to function properly. The small input signal acts as a control for the device and causes the power from the dc source to be converted to an output signal having the form of the input signal. In performing signal analysis of an amplifier, it is customary to omit the dc power supplies in many cases, but their presence is understood for proper operation.

In the explanations that follow, several methods for classifying amplifiers will be discussed. The terms and meanings should be noted as they will arise throughout the remainder of the book.

Noninverting and Inverting

Amplifiers can be classified as either noninverting or inverting. If $A > 0$ (positive), the device is said to be a *noninverting amplifier*, and if $A < 0$ (negative), the device is said to be an *inverting amplifier*.

The transfer characteristic of an ideal *noninverting* amplifier is shown in Figure 9-2. It is a straight line through the origin with a slope A, which is assumed to be positive in this case.

The transfer characteristic of an ideal *inverting* amplifier is shown in Figure 9-3. It is also a straight line through the origin, but the slope A is assumed to be negative in this case.

To illustrate the process of distortion-free transmission in a noninverting amplifier, consider the signal shown in Figure 9-4(a). The output of a noninverting amplifier with gain $A = 2$ is shown in Figure 9-4(b). All points on the curve in (b) have exactly twice the levels of the points in (a). This gain is relatively low for an amplifier but was chosen for ease of illustration. Because the amplifier is noninverting, the input and output signals are said to be *in phase*.

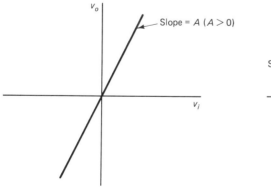

Figure 9-2 Transfer characteristic of ideal *noninverting* amplifier.

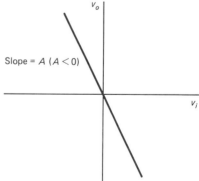

Figure 9-3 Transfer characteristic of ideal *inverting* amplifier.

(a)

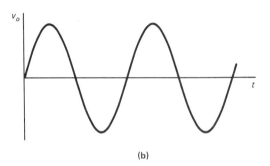

(b)

Figure 9-4 Illustration of input and output waveforms for ideal *noninverting* amplifier.

(a)

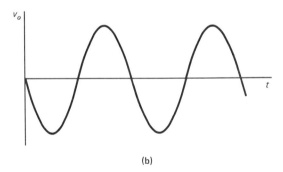

(b)

Figure 9-5 Input and output waveforms for ideal *inverting* amplifier.

A similar illustration for an inverting amplifier is shown in Figure 9-5. The gain is now $A = -2$, and all points on the output waveform of (b) have twice the magnitude but are inverted with respect to the input waveform of (a). Because of the inversion, the input and output waveforms are said to be *180° out of phase*. This simple inversion is usually not considered as a form of distortion since the basic waveform shape is preserved. Many amplifier circuits exhibit the inversion property. In some applications, it is desirable, while in other situations inversion is unacceptable. For certain applications, it is immaterial whether the amplifier is inverting or noninverting.

AC Coupled and DC Coupled

Amplifiers may be classified as either *ac coupled* or *dc coupled*. An ac-coupled amplifier has one or more coupling circuits between stages to block dc levels while passing the time-varying signal on to subsequent stages. The most common coupling circuit is the *RC* coupling circuit considered in Chapter 8, but transformers are also used in some amplifiers.

Strictly speaking, the "dc" in the designation *dc amplifier* represents *directly coupled,* but the term *dc coupled* is widely used due to one of its basic properties. The basic property is that it will amplify signals in frequency all the way down to dc because of the absence of coupling capacitors. The most common example of a dc amplifier is the operational amplifier, which will be studied in detail later in the book.

The major differences in the frequency responses between ac-coupled and dc-coupled amplifiers are shown in Figure 9-6. The plots given represent the amplitude response functions of the gains considered. The gain of the ac-coupled amplifier drops drastically at low frequencies, and there is no transmission at dc. However, the

(a) AC-coupled amplifier

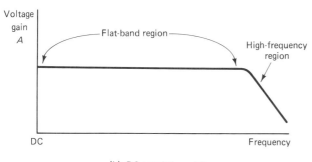

(b) DC-coupled amplifier

Figure 9-6 Frequency ranges for ac and dc coupled amplifiers.

gain of the dc amplifier is flat all the way to dc. The ac amplifier exhibits a flat mid-frequency range where the gain is essentially constant. Both types of amplifiers eventually show a decreasing gain characteristic as the frequency is increased. This high-frequency roll-off is caused primarily by shunt capacitive effects in the active elements providing the amplification.

Note again that the flat-band response of a dc-coupled amplifier extends all the way to dc, while the flat-band response of an ac-coupled amplifier starts at some frequency above dc and usually extends across a rather broad range of frequencies. When ac amplifiers are being discussed, the term *midband* is used to describe this broad range of frequencies. Although this midband varies considerably with the device and the circuit design, a typical range is from less than 100 Hz to more than 100 kHz.

The *high-frequency* region as shown in Figure 9-6 also varies considerably with the electronic device, the circuit design, and the type of configuration. It may range from 100 kHz or less to the range of tens of megahertz. Thus there are no absolute ranges of frequencies that can be associated with these definitions.

Untuned and Tuned Amplifiers

Amplifiers may be classified as untuned or tuned. An *untuned amplifier* is one in which no resonant circuits are employed. Most amplifiers designed for the frequency range below a few hundred kilohertz are untuned. Both frequency response curves of Figure 9-6 display untuned-type characteristics.

Tuned amplifiers are used primarily to create bandpass response curves that provide a uniform gain over a narrow band of frequencies. By employing resonant circuits consisting of inductance and capacitance combinations, the circuit capacitance actually becomes a part of the resonant function. In this fashion, operation at much higher frequencies is feasible. Tuned amplifiers are designed for operation well into the range of hundreds of megahertz. Most amplifiers operating above a few megahertz utilize tuned structures.

Common examples of tuned amplifiers appear in many stages of radio and television sets. The form of the amplitude response of a typical tuned amplifier is shown in Figure 9-7.

All amplifier circuits to be studied in Chapters 10 through 15 are untuned in the forms to be given, although some could be adapted to tuned forms. Examples of tuned amplifiers are considered in Chapter 18.

Figure 9-7 Typical form of the gain of a tuned amplifier.

Voltage and Power Amplifiers

Amplifiers are often classified as either *voltage amplifiers* or *power amplifiers*. The terms are somewhat ambiguous but have been established by long usage. The ambiguity arises because amplifiers classified as voltage amplifiers may also amplify power, and vice versa. It is the major function and intended application of the amplifier that determines the classification.

Voltage amplifiers are those amplifiers intended simply to raise the voltage level of a signal and for which there is no particular requirement for a specified amount of output power. The power level of the signal may be amplified in the process, but it is of secondary interest in the particular application. In many cases, the output of a voltage amplifier will be applied as the input to the next stage.

Power amplifiers are amplifiers designed to provide a specified output power to some given load. In some cases, the load may be an additional amplifier stage. However, most power amplifier circuits are used to provide power to an external load. A common example of a power amplifier is the output stage of an audio amplifier, which provides the substantial power required to drive the speakers. This electrical power is in turn converted to acoustical power. The output stage of a commercial broadcast transmitter is a power amplifier, which provides power to the antenna.

There is no universal standard on the power level which characterizes a power amplifier. However, some references classify amplifiers providing an output power of 0.5 W or more as power amplifiers. Power amplifiers providing levels of hundreds of thousands of watts have been constructed.

Most of the amplifiers to be considered in the next several chapters could be classified primarily as voltage amplifiers. Power amplifiers are treated in Chapter 19.

The classifications that have been provided are only a few of the possible types employed, but they will suffice for this point in the book. Other classifications will be considered as the need arises and as the conceptual basis is established.

Example 9-1

A certain inverting amplifier has a voltage gain $A = -20$. Determine the output signal voltage v_o when the input signal voltage has a value **(a)** $v_i = 0$, **(b)** $v_i = 0.5$ V, and **(c)** $v_i = -0.2$ V.

Solution

The input/output relationship is

$$v_o = -20v_i \qquad (9\text{-}3)$$

The three cases are analyzed as follows:

(a) For $v_i = 0$,

$$v_o = -20 \times 0 = 0 \qquad (9\text{-}4)$$

(b) For $v_i = 0.5$ V,

$$v_o = -20 \times 0.5 = -10 \text{ V} \qquad (9\text{-}5)$$

(c) For $v_i = -0.2$ V,

$$v_o = -20 \times (-0.2) = 4 \text{ V} \qquad (9\text{-}6)$$

Note that the output signal voltage is zero when the input signal voltage is zero. Note also that a positive input voltage results in an amplified negative output voltage, and vice versa.

Example 9-2

The gain of a certain amplifier is to be determined from an oscilloscope measurement using a sine-wave input. The input and output waveforms are shown in Figure 9-8. The vertical sensitivities of the two channels are indicated. The time base reference is established by channel 1. Determine the gain.

Figure 9-8   Waveforms of Example 9-2.

Solution

Gain determination on an oscilloscope is usually easier and more accurate if peak-to-peak values are used. Since the phase of channel 2 is established from the reference of channel 1, it is noted that the output is 180° out of phase with the input. Thus the amplifier is an inverting unit.

Let V_{ipp} and V_{opp} represent the peak-to-peak values of the input and output voltages, respectively. We have

$$V_{ipp} = 0.1 \text{ V/cm} \times 3 \text{ cm} = 0.3 \text{ V} \qquad (9\text{-}7)$$

and

$$V_{opp} = 5 \text{ V/cm} \times 4 \text{ cm} = 20 \text{ V} \qquad \textbf{(9-8)}$$

The voltage gain is

$$A = -\frac{20}{0.3} = -66.7 \qquad \textbf{(9-9)}$$

The peak-to-peak value of a quantity is always defined as a positive value, so the negative value in (9-9) was necessary to force the gain to be negative as required.

9-3 AMPLIFIER FLAT-BAND SIGNAL MODEL

In this section a signal model applicable to a wide variety of amplifier circuits is considered. The model is useful in predicting the behavior of a single amplifier stage, and when combined with similar models for several stages, the analysis of a multistage amplifier may be easily performed.

The model is basically a linear model; that is, the amplifier is assumed to be distortion-free. It is applicable only to signals, and any dc bias sources are not shown. The restriction to the flat-band region means that the only passive elements to be considered are resistances; that is, all capacitive and inductive reactances are either short or open circuits in the applicable frequency ranges.

Consider the block diagram shown in Figure 9-9(a). The quantities v_i and i_i represent the input signal voltage and current, respectively. The quantity v_o represents the output signal voltage, and i_o represents the output current into any external load (not shown here).

(a)

(b)

Figure 9-9 Block diagram of linear amplifier and a flat-band signal model.

Input Resistance

The signal model for this amplifier is shown in Figure 9-9(b). The resistance R_{in} appearing across the input is called the *input resistance,* and it is related to v_i and i_i by

$$R_{in} = \frac{v_i}{i_i} \qquad (9\text{-}10)$$

The input resistance is the resistance "seen" by any signal source connected to the input. Its value is important in determining any loading on the signal connected to the input.

Output Resistance

The resistance R_{out} represents the *output resistance* of the amplifier. Basically, R_{out} is the Thévenin equivalent resistance as viewed from the output terminals. It is important in determining the loading of any external load connected to the output. It is not possible to write a simple equation such as (9-10) at the output since there is a source in series with R_{out}.

Voltage Gain

The dependent voltage source Av_i in Figure 9-9(b) represents a signal voltage whose value is a constant times the input voltage. The quantity A represents the *open-circuit* or *unloaded voltage gain*. The term "open circuit" is necessary here since the effect of any external load will modify the gain, as we will see later. It is readily verified that if $i_o = 0$, $v_o = Av_i$, and the voltage gain under this open-circuit condition is A.

Example 9-3

A certain amplifier is characterized by the following parameters:

$$R_{in} = 100 \text{ k}\Omega$$
$$R_{out} = 75 \text{ }\Omega$$
$$A = 60$$

Draw the flat-band signal model.

Solution

The model will have the same form as the circuit of Figure 9-9(b). For the particular parameters given, the circuit is shown in Figure 9-10.

Figure 9-10   Amplifier model of Example 9-3.

Example 9-4

The circuit of Figure 9-11 represents the flat-band signal model of a certain amplifier. Determine the input resistance, the output resistance, and the open-circuit voltage gain.

Figure 9-11 Amplifier model of Example 9-4.

Solution

We immediately note that the input and output resistances are

$$R_{in} = 20 \text{ k}\Omega \tag{9-11}$$

and

$$R_{out} = 50 \text{ }\Omega \tag{9-12}$$

However, when the given circuit is compared with the general model of Figure 9-9(b), we see that the polarity of the voltage source in Figure 9-11 is opposite to that of the general form. Since the positive references of both v_i and v_o are at the top, this means that the amplified voltage is inverted with respect to the reference; that is, this amplifier is an *inverting* form. Thus the open-circuit voltage gain is

$$A = -25 \tag{9-13}$$

An alternative way to show the dependent source is to keep the positive terminal of the source at the top and to label it as $-25v_i$. However, the form given in Figure 9-11 is usually more convenient for inverting amplifiers.

9-4 AMPLIFIER, SOURCE, AND LOAD CONNECTIONS

We have seen that a linear amplifier can be represented in the flat-band frequency region by a model displaying input resistance, output resistance, and voltage gain. However, rarely does the amplifier stand alone in an application. A signal source must be connected to the input terminals, and in most cases some external load is connected to the output terminals. In general, the signal source contains an internal resistance which interacts with the input resistance of the amplifier to reduce the gain. Similarly, any external load connected to the output terminals will further act to reduce the gain.

In this section a number of possible situations involving source and/or load connections are considered. The best approach to emphasize to the reader is *not* to try to memorize the resulting formulas but rather to work them out as the need arises. It turns out that each case is a relatively simple exercise employing basic circuit laws. A pattern will emerge after a few cases are considered, and each subsequent situation can be dealt with from these principles.

The symbol A will be used to denote the voltage gain from the amplifier input terminals to the output terminals with no external load connected as in the preceding section. Under loaded conditions, subscripts will be added to represent various loaded gain parameters. The pattern should be clear as it is developed.

The term "signal source" is a popular term for describing the excitation to an amplifier, and we will use this term when appropriate. This would suggest the subscript s with this voltage indicated by v_s and its internal resistance as R_s. Indeed, many books use this terminology. There is, however, an ambiguity with this notation when FET amplifiers are considered. The reason is that one of the FET terminals is called the "source," as we already know. To avoid this ambiguity, the source will be viewed as a "generator" and the subscript g will be employed. In discussions, however, the terms "signal source" and "generator" will both be used, as appropriate.

Amplifier–Source Connection with No Loading

The most ideal situation to be considered is shown in Figure 9-12(a). The amplifier is excited by a signal source (or generator) v_g having zero internal resistance (i.e., the source is assumed to be an ideal voltage source). The output v_o is to be considered with no external load connected. In a practical sense, this means that whatever is used to monitor or respond to v_o presents no measurable loading.

(a)

(b)

Figure 9-12 Amplifier connection with no loading at either port and circuit model.

Introduction to Amplifiers   Chap. 9

The quantity A_{go} will be defined as the voltage gain between the generator v_g and the output voltage v_o; that is,

$$A_{go} = \frac{v_o}{v_g} \qquad (9\text{-}14)$$

The equivalent circuit model is shown in Figure 9-12(b). Since the input source is ideal, it is immediately obvious that the amplifier input voltage is the same as the generator voltage (i.e., $v_i = v_g$). The dependent source is thus $Av_i = Av_g$. The open-circuit output voltage is

$$v_o = Av_i = Av_g \qquad (9\text{-}15)$$

The gain between generator and output is

$$A_{go} = \frac{v_o}{v_g} = A \qquad (9\text{-}16)$$

This result should not come as any surprise, but it is worth emphasizing in the context of a complete amplifier-source connection. It says that the *composite amplifier circuit gain from signal generator to output is the same as the unloaded gain if the circuit is driven from an ideal voltage source and if there is no external load connected to the output.*

Amplifier with Input Loading

Consider the circuit of Figure 9-13(a), in which a generator v_g with internal resistance R_g excites an amplifier. However, the output has no external load connected. The equivalent-circuit model is shown in Figure 9-13(b).

(a)

(b)

Figure 9-13 Amplifier connection with input loading and circuit model.

In this case, the amplifier input voltage v_i differs from the generator voltage v_g due to the drop across R_g. By the voltage-divider rule, v_i is related to v_g by

$$v_i = \frac{R_{\text{in}}}{R_{\text{in}} + R_g} v_g \qquad (9\text{-}17)$$

Since there is no loading at the output, the output voltage v_o is

$$v_o = A v_i \qquad (9\text{-}18)$$

To determine the gain from generator to output, we need a relationship between v_o and v_g. This is achieved by substituting v_i in terms of v_g from (9-17) in (9-18). The result is

$$v_o = \frac{A R_{\text{in}}}{R_{\text{in}} + R_g} v_g \qquad (9\text{-}19)$$

The gain A_{go} is

$$A_{go} = \frac{v_o}{v_g} = \frac{R_{\text{in}}}{R_{\text{in}} + R_g} A \qquad (9\text{-}20)$$

where a slight rearrangement was made, for reasons that will be clear later.

Amplifier with Output Loading

Consider the circuit of Figure 9-14(a), in which a signal source with zero internal resistance excites an amplifier. However, the output is connected to an external load R_L. The equivalent circuit model is shown in Figure 9-14(b).

In this case the amplifier input voltage is the same as the generator voltage (i.e., $v_i = v_g$). The dependent source is $A v_i = A v_g$. However, the output voltage v_o is not equal to this dependent source since a portion of this voltage is dropped across R_{out}.

(a)

(b)

Figure 9-14 Amplifier connection with output loading and circuit model.

Introduction to Amplifiers   Chap. 9

By the voltage-divider rule, the output voltage v_o across the load is given by

$$v_o = \frac{R_L}{R_L + R_{\text{out}}} A v_i = \frac{R_L A v_g}{R_L + R_{\text{out}}} \qquad (9\text{-}21)$$

The gain between generator and output is

$$A_{go} = \frac{v_o}{v_g} = A \frac{R_L}{R_L + R_{\text{out}}} \qquad (9\text{-}22)$$

where a slight rearrangement was made, for reasons that will be clear later.

Amplifier with Loading at Input and Output

Consider the circuit of Figure 9-15(a), in which a generator v_g with internal resistance R_g excites an amplifier. Simultaneously, the output is connected to an external load R_L. The equivalent circuit is shown in Figure 9-15(b).

By the voltage-divider rule, the amplifier input voltage is related to the generator voltage by

$$v_i = \frac{R_{\text{in}}}{R_{\text{in}} + R_g} v_g \qquad (9\text{-}23)$$

Again using the voltage-divider rule, the load voltage is related to the dependent voltage source by

$$v_o = \frac{R_L}{R_L + R_{\text{out}}} A v_i \qquad (9\text{-}24)$$

(a)

(b)

Figure 9-15 Amplifier connection with both input and output loading and circuit model.

To obtain a relationship between v_o and v_g, the expression of (9-23) is substituted in (9-24), and we have

$$v_o = \frac{R_L}{R_L + R_{\text{out}}} A \frac{R_{\text{in}}}{R_{\text{in}} + R_g} v_g \qquad \textbf{(9-25)}$$

The gain between source and output is then determined as

$$A_{go} = \frac{v_o}{v_g} = \frac{R_{\text{in}}}{R_{\text{in}} + R_g} A \frac{R_L}{R_L + R_{\text{out}}} \qquad \textbf{(9-26)}$$

where a rearrangement was made, for reasons that will be clear shortly.

General Form

All the equations relating to the gain between generator voltage and output voltage of this section can be considered as variations of the following general equation:

$$A_{go} = (\text{input loading factor}) \times A \times (\text{output loading factor}) \qquad \textbf{(9-27)}$$

The "input loading factor" is simply the voltage-divider ratio representing the fraction of the open-circuit source or generator voltage appearing across the amplifier input; that is,

$$\text{input loading factor} = \frac{R_{\text{in}}}{R_{\text{in}} + R_g} \qquad \textbf{(9-28)}$$

When the amplifier is excited by an ideal voltage source, $R_g = 0$ and the "input loading factor" has a value of unity.

The "output loading factor" is simply the voltage-divider ratio representing the fraction of the amplifier open-circuit output voltage appearing across the load; that is,

$$\text{output loading factor} = \frac{R_L}{R_L + R_{\text{out}}} \qquad \textbf{(9-29)}$$

When the amplifier output is unloaded, R_L can be considered as approaching infinity ($R_L \to \infty$), and the output loading factor is unity. Thus, by appropriately setting the values of these factors, all expressions in this section can be considered as special cases of (9-27).

While the gain between the open-circuit generator voltage and the output is usually the primary parameter of interest, the gain between the amplifier input terminals and the output is also of interest. Let A_{io} represent this quantity, and it is defined as

$$A_{io} = \frac{v_o}{v_i} \qquad \textbf{(9-30)}$$

For this definition, the "input loading factor" does not have any effect since the reference is to the voltage across the amplifier input terminals rather than the source voltage. Thus, for any of the configurations, A_{io} is given by

$$\boxed{A_{io} = A \times \text{(output loading factor)}} \tag{9-31}$$

Comparing (9-27) and (9-31), A_{go} and A_{io} are related by

$$\boxed{A_{go} = \text{(input loading factor)} \times A_{io}} \tag{9-32}$$

From (9-32) it can readily be deduced that $A_{go} \leq A_{io}$. The two quantities are equal when the amplifier is driven by an ideal voltage source; but when there is a nonzero source resistance, $A_{go} < A_{io}$.

The significance of A_{io} is that it is usually easier to measure it than A_{go} in the laboratory under operating conditions. When the source is connected, the voltage v_g is "inside the source" and cannot be directly measured. However, v_i and v_o can usually be directly measured under operating conditions.

Ideal Voltage Amplifier

Reviewing the general expression from (9-26), but considering the realistic limitations of practical circuits, let us determine those conditions that will maximize the gain A_{go}. First, the input loading factor should be made as close to unity as possible. This mandates that $R_{in} \gg R_g$. Second, the output loading factor should also be made as close to unity as possible. This mandates that $R_{out} \ll R_L$. Thus *maximum voltage gain is achieved with an amplifier whose input resistance is very large compared with the source resistance and whose output resistance is very small compared with the load resistance.*

These conditions lead to the concept of the ideal voltage amplifier whose model is shown in Figure 9-16. This ideal amplifier model is characterized by $R_{in} = \infty$, which means that the input is essentially an open circuit. Conversely, $R_{out} = 0$, which means that the output dependent voltage source is an ideal voltage source with zero internal resistance. We will see later that many practical circuit designs work toward these goals. Although they can never be truly achieved, many practical circuits come sufficiently close that the ideal model may be assumed.

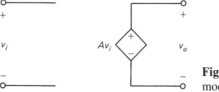

Figure 9-16 Ideal voltage amplifier model.

Example 9-5

The amplifier of Figure 9-17(a) has the following midband parameters: $A = 90$, $R_{in} = 6\ \text{k}\Omega$, and $R_{out} = 1.2\ \text{k}\Omega$. It is driven from a signal source with an internal resistance $R_g = 2\ \text{k}\Omega$, and a load $R_L = 2.4\ \text{k}\Omega$ is connected across the output. Determine **(a)** $A_{go} = v_o/v_g$ and **(b)** $A_{io} = v_o/v_i$.

(a)

(b)

Figure 9-17   Circuit of Example 9-5.

Solution

The equivalent circuit of the composite amplifier with source and load connected is shown in Figure 9-17(b).

(a) The presence of both a generator resistance and a load resistance results in loading factors at both the input and the output. The gain between source and output is determined as follows:

$$A_{go} = \frac{v_o}{v_g} = \frac{6\ \text{k}\Omega}{6\ \text{k}\Omega + 2\ \text{k}\Omega} \times 90 \times \frac{2.4\ \text{k}\Omega}{2.4\ \text{k}\Omega + 1.2\ \text{k}\Omega}$$

$$= \frac{3}{4} \times 90 \times \frac{2}{3} = 45 \tag{9-33}$$

The net effect of both factors is to reduce the voltage gain from the open-circuit value of 90 to a loaded gain of 45.

(b) To determine the gain between the amplifier input and the output, we consider only the output loading factor.

$$A_{io} = 90 \times \frac{2 \text{ k}\Omega}{2 \text{ k}\Omega + 1 \text{ k}\Omega} = 90 \times \frac{2}{3} = 60 \qquad (9\text{-}34)$$

Note that $A_{io} > A_{go}$, as expected.

Example 9-6

For the composite circuit of Example 9-5, assume that the signal voltage is $v_g = 0.2$ V. Determine **(a)** v_i and **(b)** v_o.

Solution

(a) The voltage v_i is simply the fraction of v_g appearing across the 6-kΩ input resistance. Thus

$$v_i = \frac{6 \text{ k}\Omega}{6 \text{ k}\Omega + 2 \text{ k}\Omega} \times 0.2 = 0.75 \times 0.2 = 0.15 \text{ V} \qquad (9\text{-}35)$$

(b) From (9-33),

$$v_o = 45v_g = 45 \times 0.2 = 9 \text{ V} \qquad (9\text{-}36)$$

As a check, since v_i is known, (9-34) can be used to determine v_o from v_i. We have

$$v_o = A_{io}v_i = 60v_i = 60 \times 0.15 = 9 \text{ V} \qquad (9\text{-}37)$$

9-5 CASCADE CONNECTIONS OF AMPLIFIERS

In this section the process of cascading two or more individual amplifier circuits is considered. The effects of loading at the source and/or the output are then combined with the cascade assumption. The result is a procedure for working with various combinations of cascaded stages as well as source and load connections.

To provide as simple an approach as possible in the development that follows, only two amplifier stages will be considered. However, the pattern that emerges should allow the reader to extend the result to an arbitrary member of cascaded stages without any difficulty.

Cascade Arrangement without Input or Output Loading

Consider first the cascade arrangement of two amplifiers shown in Figure 9-18(a). The parameters of the two amplifiers are labeled on the respective blocks. In this case, the input is an ideal source of value v_g, and the output has no assumed external load.

(a)

(b)

Figure 9-18 Cascade arrangement of two amplifiers without input or output loading and equivalent circuit.

An equivalent model for the arrangement is shown in Figure 9-18(b). Since there is no loading at the input, the voltage across the input of amplifier 1 is simply $v_{i1} = v_g$.

The voltage v_{i2} across the input to amplifier 2 is

$$v_{i2} = \frac{R_{in2}}{R_{in2} + R_{out1}} A_1 v_{i1} = \frac{R_{in2} A_1 v_g}{R_{in2} + R_{out1}} \tag{9-38}$$

Since there is no loading at the output, the output voltage v_o is simply the value of the dependent source on the right; that is,

$$v_o = A_2 v_{i2} \tag{9-39}$$

Substitution of (9-38) in (9-39) yields

$$v_o = \frac{A_2 R_{in2} A_1 v_g}{R_{in2} + R_{out1}} \tag{9-40}$$

Solving for the gain $A_{go} = v_o/v_g$ and rearranging in a particular order, we have

$$A_{go} = \frac{v_o}{v_g} = A_1 \frac{R_{in2}}{R_{in2} + R_{out1}} A_2 \tag{9-41}$$

This equation may be expressed as

$$\boxed{A_{go} = A_1 \times \text{(interaction factor)} \times A_2} \tag{9-42}$$

where

$$\boxed{\text{interaction factor} = \frac{R_{in2}}{R_{in2} + R_{out1}}} \tag{9-43}$$

The interaction factor between the stages is simply the voltage-divider rule involving the input resistance to the second stage and the output resistance of the first stage.

Conditions for No Interaction

Assume in (9-43) that $R_{in2} \gg R_{out1}$. In this case the denominator may be approximated by R_{in2}, and the interaction factor is approximately $R_{in2}/R_{in2} = 1$. The gain A_{go} is then

$$A_{go} = A_1A_2 \tag{9-44}$$

This result is important and needs to be emphasized. *If the input resistance of a given amplifier is very large compared to the output resistance of the amplifier driving it, and if there is no loading at input and output, the composite voltage gain is simply the product of the individual gains.* As a practical rule of thumb, if the input resistance is at least 100 times as large as the output resistance of the stage driving it, the error in using (9-44) is no greater than 1%.

Once again we see the desirability of high input resistance and low output resistance for a voltage amplifier. Without these desirable conditions, the voltage gain of a cascade amplifier will be less than it is capable of achieving.

Interaction and Loading Combined

Consider now the cascade arrangement in Figure 9-19. In this case, the generator contains an internal resistance R_g, and the output of the second amplifier is connected to a load R_L.

Figure 9-19 Cascade arrangement of two amplifiers with input and output loading.

The reader who has followed all the developments thus far should be able to write down the gain expression in this case. The composite gain is

$$A_{go} = \frac{v_o}{v_g} = \text{(input loading factor)} \times A_1 \times \text{(interaction factor)}$$
$$\times A_2 \times \text{(output loading factor)} \tag{9-45}$$

The various factors are defined as

$$\text{input loading factor} = \frac{R_{in1}}{R_{in1} + R_g} \qquad \textbf{(9-46)}$$

$$\text{interaction factor} = \frac{R_{in2}}{R_{in2} + R_{out1}} \qquad \textbf{(9-47)}$$

$$\text{output loading factor} = \frac{R_L}{R_L + R_{out2}} \qquad \textbf{(9-48)}$$

The reader should be able to generalize those results to more than two stages. In general, there will be a single input loading factor and a single output loading factor. However, each interface between stages will require an interaction factor. Thus, for three stages, there will be two interaction factors.

Example 9-7

Consider the cascade arrangement shown in Figure 9-20. Determine the net gain $A_{go} = v_o/v_g$.

Figure 9-20   System of Example 9-7.

Solution

There is no loading at either input or output, so only an interaction factor need be considered. Using the procedure developed, we can write by inspection

$$A_{go} = \frac{v_o}{v_g} = 40 \times \frac{800}{800 + 400} \times 30 = 40 \times \frac{2}{3} \times 30 = 800 \qquad \textbf{(9-49)}$$

If there were no interaction between stages, the net gain would be $A_1 A_2 = 40 \times 30 = 1200$.

Example 9-8

Consider the cascade arrangement shown in Figure 9-21. The amplifiers are the same as in Example 9-7, but the source now contains an internal resistance

Figure 9-21 System of Example 9-8.

and there is an external load connected. Determine the net gain $A_{go} = v_o/v_g$.

Solution

In this case we write

$$A_{go} = \frac{v_o}{v_g} = \frac{2000}{2000 + 500} \times 40 \times \frac{800}{800 + 400} \times 30 \times \frac{300}{300 + 100}$$

$$= \frac{4}{5} \times 40 \times \frac{2}{3} \times 30 \times \frac{3}{4} \tag{9-50}$$

$$= 480$$

9-6 DECIBEL POWER GAIN AND LOSS

The concepts of decibel power gain and loss are developed in this section. Decibel gain is a method of characterizing the gain of a circuit by means of a logarithmic ratio. This approach was developed very early in the history of the electronics industry and has widespread usage in many application areas. A beginner may tend to question the value of the concept, but it is impossible to understand and interpret many of the specifications and properties of electronic devices without utilizing decibel quantities. After some proficiency is established in using decibel quantities, their utility becomes apparent.

Decibel Power Gain

The original, and perhaps the "purest" form of a decibel measure, is that of the *decibel power gain*. To develop this concept, refer to the amplifier block diagram in Figure 9-22. Let P_{in} represent the power delivered to the input of the amplifier, and let P_{out} represent the power delivered to the load. The *power gain* of the amplifier will be denoted as G, and it is defined as

$$G = \frac{P_{out}}{P_{in}} \tag{9-51}$$

The *decibel power gain* will be denoted as G_{dB}, and it is defined as

$$G_{dB} = 10 \log_{10} G \tag{9-52}$$

Figure 9-22 Block diagram for development of decibel power gain.

where $\log_{10}$ means the "logarithm to the base 10." Even though G has no dimensions, it is customary to assign units to G_{dB}, and the units are *decibels* (abbreviated as *dB*). Thus, when G is specified, no units are given, but when G_{dB} is specified, the value given is stated in decibels, and this is a good way to recognize a decibel measure.

In effect, the decibel gain is a logarithmic conversion of the the actual gain. Many applications of electronic signals have responses that are logarithmic in nature. It turns out that this logarithmic conversion provides a better way of relating these responses to signal levels.

The choice of the constant 10 in the definition of (9-52) was made to arrive at a convenient size of the individual logarithmic units. In fact, a logarithmic definition has been used without this factor, and the resulting unit was known as the *bel*. However, the bel was judged to be rather awkward because of its size, and the decibel was found to be more convenient. Consequently, virtually all logarithmic measures today are based on the decibel (and its variations), and that will be the basis in this text.

The quantity G in (9-51) will be assumed to be a positive real number. From some basic properties of the logarithmic function, the following conditions may be deduced:

1. If $G > 1$, G_{dB} is positive.
2. If $G < 1$, G_{dB} is negative.
3. If $G = 1$, $G_{dB} = 0$.

Thus the gain in decibels turns out to be positive when there is a true power gain $(G > 1)$. However, when the output power is less than the input power $(G < 1)$, the gain in decibels turns out to be negative. A negative value in decibels implies, therefore, that the output power is less than the input power. When input and output power levels are equal, there is no decibel gain (i.e., $G_{dB} = 0$).

Although the definition has been established here with an assumed amplifier, it is widely used for all types of transmission networks. In many passive circuits, the output power will be less than the input power (i.e., the signal is *attenuated* by the circuit). The decibel gain will thus turn out to be negative. In working with circuits having *attenuation*, it is often more convenient to work with loss factors. Let L represent the *power loss factor* associated with a circuit. It is defined as

$$L = \frac{P_{in}}{P_{out}} \tag{9-53}$$

Comparing (9-53) with (9-51), we see that

$$L = \frac{1}{G} \qquad \text{(9-54)}$$

The *decibel power loss* will be denoted as L_{dB} and it is defined as

$$L_{dB} = 10 \log_{10} L \qquad \text{(9-55)}$$

As in the case of G_{dB}, L_{dB} is measured in decibels.

A basic property of logarithms is

$$\log_{10} \frac{1}{x} = -\log_{10} x \qquad \text{(9-56)}$$

By substituting (9-54) in (9-55) and utilizing (9-56), it is observed that

$$L_{dB} = -G_{dB} \qquad \text{(5-57)}$$

Thus, while L and G have a reciprocal relationship as given by (9-54), L_{dB} and G_{dB} are simply related by negation.

Do not let these relationships confuse you, because the pattern is quite simple. If there is a true power gain ($G > 1$), the decibel power gain has a positive value, and there is very little need even to utilize a loss factor. However, in situations where there is a true loss ($G < 1$), one can use a decibel power gain with negative values of decibels, or the decibel loss function can be used, in which case the decibel values are positive.

Example 9-9

A certain amplifier has a power gain $G = 1000$. Determine the decibel power gain G_{dB}.

Solution

The decibel power gain is

$$G_{dB} = 10 \log_{10} 1000 = 10 \times 3 = 30 \text{ dB} \qquad \text{(9-58)}$$

Example 9-10

In certain lossy transmission system, the input power is $P_{in} = 2$ W, and the output power is $P_{out} = 20$ mW. Determine (a) the power gain, (b) the decibel gain, (c) the power loss factor, and (d) the decibel loss.

Solution

(a) The power gain G is

$$G = \frac{P_{\text{out}}}{P_{\text{in}}} = \frac{0.02 \text{ W}}{2 \text{ W}} = 0.01 \tag{9-59}$$

(b) The decibel gain G_{dB} is

$$G_{\text{dB}} = 10 \log_{10} G = 10 \log_{10} 0.01 = 10 \times (-2) = -20 \text{ dB} \tag{9-60}$$

This is obviously not a true "gain," but it is perfectly proper to use the gain definitions. The conditions that $G < 1$ and $G_{\text{dB}} < 0$ imply an actual loss.

(c) The loss factor L is

$$L = \frac{P_{\text{in}}}{P_{\text{out}}} = \frac{2 \text{ W}}{0.02 \text{ W}} = 100 \tag{9-61}$$

(d) The decibel loss factor L_{dB} is

$$L_{\text{dB}} = 10 \log_{10} L = 10 \log_{10} 100 = 10 \times 2 = 20 \text{ dB} \tag{9-62}$$

It is more convenient to refer to a *loss of* 20 *dB* than a *gain of* -20 *dB*, but both references are correct.

Converting Decibel Power Gain to Power Gain

Suppose that the decibel gain (or loss) is given and we desire to determine the corresponding value of G (or L). We start with the definition of the decibel quantity, but rearrange it so that the unknown is on the left. For gain, we have

$$10 \log_{10} G = G_{\text{dB}} \tag{9-63}$$

First, both sides of (9-63) are divided by 10. This gives

$$\log_{10} G = \frac{G_{\text{dB}}}{10} \tag{9-64}$$

To extract G from the logarithmic function, both sides of (9-64) are raised as powers of 10. This step yields

$$10^{\log_{10} G} = 10^{G_{\text{dB}}/10} \tag{9-65}$$

By definition, however, the left side of (9-65) is simply G. Thus we have

$$\boxed{G = 10^{G_{\text{dB}}/10}} \tag{9-66}$$

A similar process can be used to determine L from L_{dB}, and the result is

$$\boxed{L = 10^{L_{\text{dB}}/10}} \tag{9-67}$$

The reader is encouraged to perform the various steps required rather than memorizing the formulas.

Example 9-11

A certain amplifier has a decibel power gain $G_{dB} = 26$ dB. Determine the value of G.

Solution

Since

$$10 \log_{10} G = 26 \tag{9-68}$$

$$\log_{10} G = 2.6 \tag{9-69}$$

and

$$G = 10^{2.6} = 398.1 \tag{9-70}$$

Most modern scientific calculators have functions of the form y^x and 10^x, each of which could be used to perform the computation of (9-70).

Example 9-12

A certain transmission system has a decibel power loss $L_{dB} = 30$ dB. **(a)** Determine L directly from L_{dB}. **(b)** Convert to a power gain basis and determine G.

Solution

(a) Since

$$10 \log_{10} L = 30, \tag{9-71}$$

$$\log_{10} L = 3 \tag{9-72}$$

and

$$L = 10^3 = 1000 \tag{9-73}$$

(b) The corresponding power gain is

$$G_{dB} = -L_{dB} = -30 \text{ dB} \tag{9-74}$$

We now have

$$10 \log_{10} G = -30 \tag{9-75}$$

$$\log_{10} G = -3 \tag{9-76}$$

and

$$G = 10^{-3} = 0.001 \tag{9-77}$$

We readily note that $G = 1/L$.

Although the "gain approach" and the "loss approach" are equally valid, it is more convenient to work with positive logarithms and ratios greater than 1. Thus the "gain approach" is more convenient when there is a true gain, and the "loss approach" is more convenient when there is a true loss.

9-7 DECIBEL VOLTAGE AND CURRENT GAIN

As indicated in the preceding section, the original concept for decibel gain was based on power ratios. However, in modern usage, the definition has been extended to include both voltage and current gain. The development here will focus first on voltage gain, and the extension to current gain will then be evident.

To bring together the concepts of power and voltage gain, consider the block diagram of Figure 9-23. The input and output instantaneous voltages are denoted as v_i and v_o, respectively.

The input instantaneous power p_i is the power that is delivered to R_{in}, and it is

$$p_i = \frac{v_i^2}{R_{in}} \tag{9-78}$$

Similarly, the output instantaneous power p_o is the power delivered to R_L, and it is

$$p_o = \frac{v_o^2}{R_L} \tag{9-79}$$

The power gain is

$$G = \frac{p_o}{p_i} = \frac{v_o^2/R_L}{v_i^2/R_{in}} = \frac{R_{in}}{R_L}\left(\frac{v_o}{v_i}\right)^2 \tag{9-80}$$

Substituting $A = v_o/v_i$, we have

$$\boxed{G = \frac{R_{in}}{R_L}A^2} \tag{9-81}$$

Assume momentarily that $R_{in} = R_L$; that is, the load and input resistances are equal. The power gain then reduces to

$$\boxed{G = A^2 \qquad \text{for } R_{in} = R_L} \tag{9-82}$$

Thus *the power gain is simply the square of the voltage gain when the load and input resistances are equal.* This situation will be referred to as a *matched condition* in the development that follows.

Figure 9-23 Block diagram for development of decibel voltage gain.

Next, the decibel power gain is determined. This quantity is

$$G_{dB} = 10 \log_{10}G = 10 \log_{10}A^2 \qquad (9\text{-}83)$$

A fundamental property of logarithms is

$$\log_{10}x^2 = 2 \log_{10}x \qquad (9\text{-}84)$$

for any variable x. Utilization of (9-84) in (9-83) results in

$$\boxed{G_{dB} = 20 \log_{10}A} \qquad (9\text{-}85)$$

Comparing this result with (9-52), we see that the decibel power gain can be determined directly from the voltage gain under matched conditions. However, the constant multiplier is now 20 instead of 10. Equation (9-85) thus provides an alternative way to compute the decibel power gain when the voltage gain is given and when matched conditions exist.

Suppose, however, that the system is not matched (i.e., $R_L \neq R_{in}$). Although decibel power gain could still be determined in some cases by calculating input and load power values or by using (9-81), the definition may not even make any sense in many cases. For example, it was indicated earlier that an ideal voltage amplifier was characterized by infinite input resistance (i.e., an open circuit). In that case, the input power is zero, and the calculation of power gain would lead to an infinite value. Similarly, if the output voltage is measured with no external load, there is no output power.

To resolve this dilemma, a quantity we will define as the *decibel voltage gain* will be used. This quantity will be denoted as A_{dB} and it will be defined as

$$\boxed{A_{dB} = 20 \log_{10}A} \qquad (9\text{-}86)$$

where A is the voltage gain under consideration. The quantity A can be replaced by any of the voltage-gain definitions encountered previously (i.e., A_{go} and A_{io}). The definition is applied as given by (9-86) irrespective of whether input and load resistances are equal.

What is the significance of A_{dB} as compared to G_{dB}? The quantity G_{dB} represents a true power gain expressed as a decibel ratio. The quantity A_{dB}, in general, simply represents a voltage gain expressed in a "power-like" logarithmic ratio.

Note that in the special matched case, the two quantities are equal, that is,

$$\boxed{A_{dB} = G_{dB} \qquad \text{for } R_L = R_{in}} \qquad (9\text{-}87)$$

In general, however, A_{dB} is simply a logarithmic expression of a voltage ratio rather than a true power ratio.

Converting Decibel Voltage Gain to Voltage Gain

The process of determining A from a given A_{dB} is nearly identical to that of the corresponding power quantities except for one constant. We start with

$$20 \log_{10}A = A_{dB} \qquad (9\text{-}88)$$

First, both sides of (9-88) are divided by 20. This yields

$$\log_{10} A = \frac{A_{dB}}{20} \tag{9-89}$$

Both sides are raised as powers of 10, and we have

$$\boxed{A = 10^{A_{dB}/20}} \tag{9-90}$$

Example 9-13

A certain amplifier has $A = 200$. Determine the decibel voltage gain.

Solution

The decibel voltage gain is

$$A_{dB} = 20 \log_{10} 200 = 20 \times 2.301 = 46.02 \text{ dB} \tag{9-91}$$

Example 9-14

A certain amplifier has $A_{dB} = 40$ dB. Determine A.

Solution

Start with

$$20 \log_{10} A = 40 \tag{9-92}$$

Division of both sides by 20 yields

$$\log_{10} A = 2 \tag{9-93}$$

and

$$A = 10^2 = 100 \tag{9-94}$$

9-8 DECIBEL GAIN FOR CASCADED SYSTEMS

One of the primary advantages of decibel forms is the ease of characterizing a system containing a cascade of individual stages. The concept will be developed here on the basis of three stages, but the result can be readily extended to any arbitrary number.

Consider the system of Figure 9-24. For the initial phase of the development, we will assume that the three amplifiers are ideal (i.e., the input resistances are infinite, and the output resistances are zero). Thus there are no loading or interaction factors to consider.

Figure 9-24 Cascade connection of three amplifiers.

A series of equations relating the outputs of the various stages to the inputs may be written. We have

$$v_1 = A_1 v_i \tag{9-95}$$

$$v_2 = A_2 v_1 \tag{9-96}$$

$$v_o = A_3 v_2 \tag{9-97}$$

To determine a relationship between v_o and v_i, (9-95) is substituted in (9-96) and this result is substituted in (9-97). The net result is

$$v_o = A_1 A_2 A_3 v_i \tag{9-98}$$

Finally, the net gain A is

$$A = \frac{v_o}{v_i} = A_1 A_2 A_3 \tag{9-99}$$

This result comes as no surprise, since a similar form was developed in Section 9-5 for two sections.

Next, the decibel voltage gain A_{dB} is determined. This quantity is

$$A_{dB} = 20 \log_{10} A = 20 \log_{10} A_1 A_2 A_3 \tag{9-100}$$

A fundamental property of logarithms is

$$\log_{10} xyz = \log_{10} x + \log_{10} y + \log_{10} z \tag{9-101}$$

for any variables x, y, and z. Application of (9-101) to (9-100) yields

$$A_{dB} = 20 \log_{10} A_1 + 20 \log_{10} A_2 + 20 \log_{10} A_3 \tag{9-102}$$

We now define

$$A_{1dB} = 20 \log_{10} A_1 \tag{9-103}$$

$$A_{2dB} = 20 \log_{10} A_2 \tag{9-104}$$

$$A_{3dB} = 20 \log_{10} A_3 \tag{9-105}$$

The final form is then

$$\boxed{A_{dB} = A_{1dB} + A_{2dB} + A_{3dB}} \tag{9-106}$$

Stated in words and generalizing, the *net decibel voltage gain of a cascade system of ideal amplifiers is the sum of the individual decibel gains*.

Suppose that the amplifiers are not ideal and that generator and output loading conditions are present. Decibel forms can still be applied provided that decibel measures for the loading and/or interaction effects are used. Since the basis of the analysis is decibel gain, the decibel measures of these factors must be treated as *negative* dB gains. Thus their effects will be to reduce the overall gain.

Example 9-15

Consider the cascade of three amplifiers in Figure 9-25. All amplifiers are assumed to be ideal. **(a)** Determine the net voltage gain $A = v_o/v_i$, and determine A_{dB} from A. **(b)** From the individual decibel gains, determine A_{dB}.

Figure 9-25 Cascade system of Example 9-15.

Solution

(a) Observe that the voltage gains of the first two stages are given, and the decibel gain of the third stage is given. The latter quantity must first be converted to a voltage gain. We have

$$20 \log_{10} A_{3dB} = 18 \qquad (9\text{-}107)$$

which leads to

$$A_{3db} = 10^{0.9} = 7.943 \qquad (9\text{-}108)$$

The net voltage gain A is

$$A = A_1 A_2 A_3 = 50 \times 40 \times 7.943 = 15{,}886 \qquad (9\text{-}109)$$

The net decibel voltage gain A_{dB} is

$$A_{dB} = 20 \log_{10} A = 20 \log_{10} 15{,}886 = 84.02 \text{ dB} \qquad (9\text{-}110)$$

(b) To work with all stages in decibels, we must first convert A_1 and A_2 to dB form. We have

$$A_{1dB} = 20 \log_{10} 50 = 33.98 \text{ dB} \qquad (9\text{-}111)$$

$$A_{2dB} = 20 \log_{10} 40 = 32.04 \text{ dB} \qquad (9\text{-}112)$$

The net decibel gain can then be determined as

$$A_{dB} = A_{1dB} + A_{2dB} + A_{3dB} = 33.98 + 32.04 + 18 = 84.02 \text{ dB} \qquad (9\text{-}113)$$

which agrees with (9-110).

The basic application of decibel values is to express power or voltage gains as logarithmic ratios, as we have seen in the last several sections. In this context a decibel value defines a *ratio* of an output quantity to an input quantity.

In many segments of the electronics industry, it has become widespread practice to extend decibel measures to represent actual signal levels. However, it is necessary that a reference level be used in the definition since a decibel measure is always a ratio. Such levels are denoted by adding an additional letter to the unit abbreviation dB. Several of the most common decibel-level definitions are summarized in Table 9-1.

For the three cases shown, the reference levels are 1 W, 1 mW, and 1 fW (10^{-15}W), respectively. The use of the 1-W reference is probably the most natural and is used in many higher-power communications-related areas. However, it is awkward for small-signal applications. The use of 1 mW is very common in many electronic small-signal applications and in telephone transmission systems. The use of 1 fW as a reference is relatively new and is being used in specifying antenna signal levels in communications receivers.

For each of the forms, the formula for computing the decibel signal level is the same as developed earlier in the book, with some additional stipulations. The power level must be expressed in the units given in parentheses for each case, and the corresponding reference is one unit for the particular case. The corresponding units are dBW, dBm, and dBf for the three cases, respectively.

Whenever there is confusion about whether a decibel value represents a signal level or a gain value, note the units. If dB is used without an additional letter, it is referring to a gain (or a loss). However, if it is accompanied by an additional letter, it represents a signal. Unfortunately for beginners, sometimes experienced personnel in a given area will use the basic term "dB" when they really mean one of the signal level units.

We will now investigate the advantage obtained from using decibel signal levels. For the development that follows, we use decibels referred to 1 mW as the basis because of its widespread usage.

TABLE 9-1 SOME COMMON dB-LEVEL DEFINITIONS

Reference power level	Defining formula	Unit abbreviation
1 W	$10 \log_{10} \dfrac{P(\text{W})}{1 \text{ W}}$	dBW
1 mW	$10 \log_{10} \dfrac{P(\text{mW})}{1 \text{ mW}}$	dBm
1 fW	$10 \log_{10} \dfrac{P(\text{fW})}{1 \text{ fW}}$	dBf

Assume that the input power P_i and output power P_o are both expressed in milliwatts. For a power gain G, the system relationship is

$$P_o(\text{mW}) = GP_i(\text{mW}) \qquad \textbf{(9-114)}$$

Taking the logarithms of both sides of (9-114), multiplying by 10, and employing the logarithmic property of (9-101), we have

$$10 \log_{10} P_o(\text{mW}) = 10 \log_{10} GP_i(\text{mW}) = 10 \log_{10} G + 10 \log_{10} P_i(\text{mW}) \qquad \textbf{(9-115)}$$

Let $P_i(\text{dBm})$ and $P_o(\text{dBm})$ represent the respective decibel levels of input and output expressed in dBm. Comparing (9-115) with the formulas of Table 9-1 and utilizing the definition of G_{dB} as developed earlier in the chapter, we have

$$\boxed{P_o(\text{dBm}) = P_i(\text{dBm}) + G_{\text{dB}}} \qquad \textbf{(9-116)}$$

Stated in words, the result of (9-116) says that *the output level in dBm is the sum of the input level in dBm plus the decibel gain of the system*. This formula may be adapted to any of the decibel signal definitions of Table 9-1 simply by replacing dBm for the two power levels by the desired dB-level form. However, the input and output levels must be expressed in the same units. Note that the value of G_{dB} is not affected in any way by the input and output signal definitions since it is based on a dimensionless ratio. Thus G_{dB} is expressed in decibels irrespective of the other units in the equation. This result often bothers students who have been taught that all quantities to be added must be expressed in the same units. In the present situation, both additive terms are really dimensionless, but the units added provide clarification of the process involved.

Example 9-16

Assume that an amplifier with a power gain of 30 dB is excited by a signal source which supplies an input power level of 250 μW to the amplifier. Express the input in dBm, and determine the corresponding output level in dBm.

Solution

Let P_i (dBm) and P_o(dBm) represent the respective input and output power levels in dBm. First, P_i is expressed in mW as $P_i(\text{mW}) = 0.25$ mW. The value of $P_i(\text{dBm})$ is

$$P_i(\text{dBm}) = 10 \log_{10} \frac{P_i(\text{mW})}{1 \text{ mW}} = 10 \log_{10} \frac{0.25}{1} = -6.02 \text{ dBm} \qquad \textbf{(9-117)}$$

Note that this particular level is smaller than 1 mW, so the corresponding decibel level is negative. The dBm value of the output is

$$P_o(\text{dBm}) = P_i(\text{dBm}) + G_{\text{dB}} = -6.02 + 30 = 23.98 \text{ dBm} \qquad \textbf{(9-118)}$$

9-10 BODE PLOTS

In general, amplifier gain will vary with frequency. The concept of frequency response was introduced in Chapter 8, and several representative frequency-dependent circuits containing reactances were provided. Amplifier circuits contain complex combination of these reactance components, so the frequency response is an important consideration.

A widely employed method for displaying the amplitude response as a function of frequency is through the use of a *Bode plot,* in which A_{dB} versus frequency is shown. The method employs a *semilog* coordinate system. An example of semilog graph paper is shown in Figure 9-26. One scale is linear, and the other has a logarithmic form. Frequency is represented with the logarithmic scale, and the decibel amplitude response is represented with the linear scale. The particular sample shown has what is called by paper manufacturers "two cycles" on the logarithmic

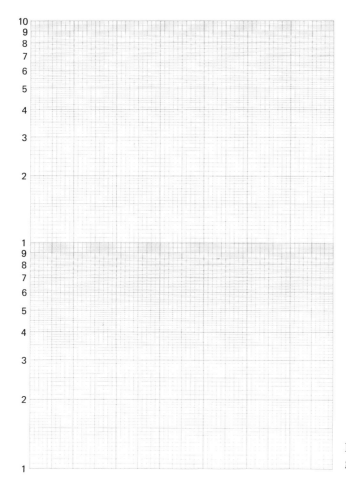

Figure 9-26 "Two-cycle" semilog graph scales.

scale. Strictly speaking, it should be called "two decades" since each segment provides a 10:1 frequency range.

The use of Bode plots for involved frequency response functions is a major topic in more advanced circuit analysis books.* The treatment here will provide sufficient information to deal with certain amplifier functions to be encountered later in the text.

RC Low-Pass Model

Consider the simple *RC* low-pass circuit shown in Figure 9-27. This circuit was considered in Chapter 8, and some analysis was performed at that time. Recall that a so-called *break frequency* f_b was defined as

$$f_b = \frac{1}{2\pi RC} \tag{9-119}$$

The form of the Bode plot for the amplitude response A_{dB} is shown in Figure 9-28. The frequency scale is labeled as f/f_b, so the result can thus be extended to any frequency range. The exact curve is labeled as E. It starts at a level of 0 dB at very low frequencies and gradually starts to drop off. Eventually, it rolls off at a rate approximating a straight line on a logarithmic frequency scale. The actual response is down 3 dB at $f/f_b = 1$ (i.e., at $f = f_b$).

The curve labeled BP is the *Bode breakpoint amplitude* approximation. It is formed as a horizontal line with a level 0 dB until $f = f_b$, at which point it breaks downward as a straight line. Note the appropriateness of the term "break frequency."

The slope of the straight-line approximation beyond the break frequency is -6 dB/octave or -20 dB/decade. An octave is a doubling of the frequency, and a decade is the multiplication of the frequency by 10. Thus, for each range in which the frequency if doubled, the response drops 6 dB, and in each range in which the frequency is increased by a factor of 10, the response drops 20 dB.

The reader may now compare the Bode plot amplitude functions of Figure 9-28 with the linear sketch of *A* for the same circuit shown back in Figure 8-18(b). From this point on, the Bode plot forms will be used. In some cases the exact curve will be displayed, but in many cases the breakpoint approximation will suffice.

The phase shift θ for this circuit is shown in Figure 9-29 using a semilog scale again. The exact phase shift is the curve labeled E. There is also a Bode breakpoint approximation for the phase, and it is shown as the curve BP in Figure 9-29. This

*See, for example, W. D. Stanley, *Network Analysis with Applications* (Englewood Cliffs, N.J.: Reston, 1985), Chap. 9.

Figure 9-27 Simple *RC* low-pass circuit.

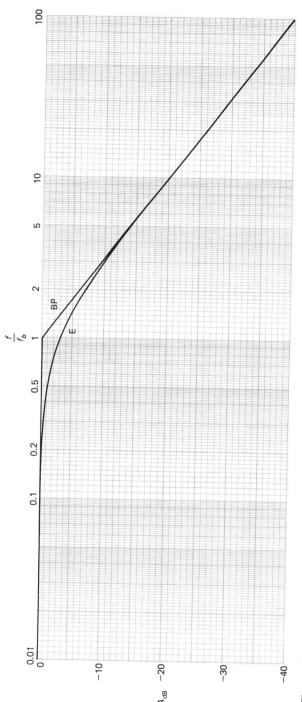

Figure 9-28 Bode plot of amplitude response for *RC* low-pass circuit.

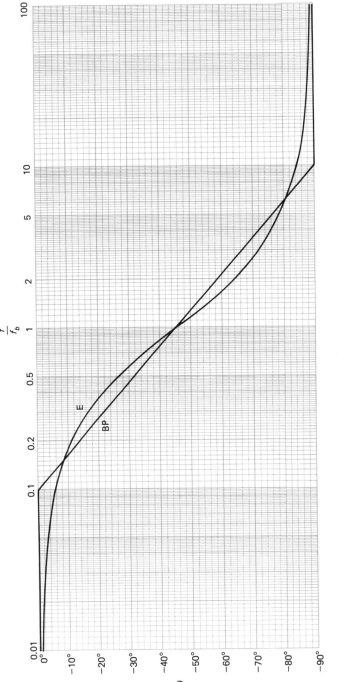

Figure 9-29 Bode plot of phase response for *RC* low-pass circuit.

Introduction to Amplifiers   Chap. 9

approximation assumes $0°$ up to $f/f_b = 0.1$, at which point the curve breaks to a slope of $-45°$ per decade. It passes through $-45°$ at $f/f_b = 1$ and reaches a level of $-90°$ at $f/f_b = 10$. The actual curve differs from the approximation by $5.73°$ at the points $f/f_b = 0.1$ and $f/f_b = 10$.

The curves shown were developed on the basis of a simple RC low-pass circuit. However, many amplifiers are dominated by a frequency response of this form. The major difference is that the low-frequency flat level will normally have a positive decibel level as opposed to the 0-dB level of the RC circuit. The curves of Figure 9-28 may be used in such a case simply be interpreting the dB level as the level relative to the maximum dB gain.

In its general form, any low-pass function displaying a -6-dB/octave high-frequency roll-off rate is referred to as a *one-pole low-pass* model. The term "pole" relates to mathematical definitions used in advanced network theory.

Example 9-17

Consider the simple RC low-pass circuit of Figure 9-30(a). Sketch the Bode breakpoint approximations to the amplitude and phase response functions.

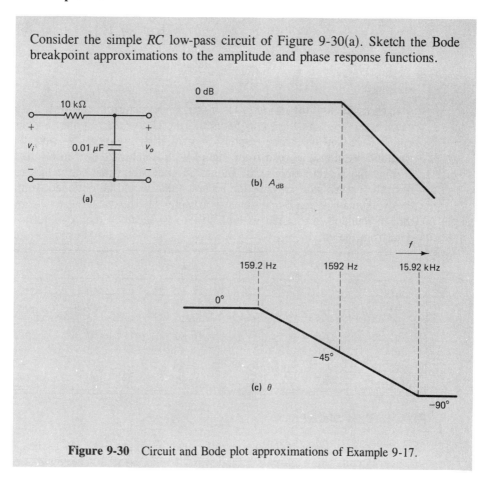

Figure 9-30 Circuit and Bode plot approximations of Example 9-17.

The break frequency is computed as

$$f_b = \frac{1}{2\pi RC} = \frac{1}{2\pi \times 10^4 \times 0.01 \times 10^{-6}} = 1592 \text{ Hz} \qquad \textbf{(9-120)}$$

The breakpoint amplitude approximation is shown in Figure 9-30(b). The actual response differs from the approximation by 3 dB at 1592 Hz. The breakpoint phase approximation is shown in Figure 9-30(c).

Example 9-18

A certain amplifier has a gain function of the form of the one-pole low-pass model. The low-frequency gain is 60 dB, and the break frequency is 10 kHz. Using the amplitude and phase curves of Figures 9-28 and 9-29, determine the amplitude response in dB and the phase response at each of the following frequencies: **(a)** 1 kHz; **(b)** 2 kHz; **(c)** 5 kHz; **(d)** 10 kHz; **(e)** 20 kHz; **(f)** 50 kHz; **(g)** 100 kHz.

Solution

At each frequency, the ratio $f/f_b = f/10^4$ is first determined so that the proper abscissa on the two curves may be used. The relative dB response is read from Figure 9-28 and the phase response is read from Figure 9-29. Note that the amplitude response as read from Figure 9-28 is subtracted from 60 dB to determine the actual response. It should be understood that the values as read from the curves will not have the accuracy that would be obtained from using the mathematical functions given in Chapter 8. The values listed below, although rounded, are stated with slightly more accuracy than can normally be read from the curves.

	f (kHz)	f/f_b	A_{dB} (dB)	A_{dB} (dB)	θ
(a)	1	0.1	≈ 0	≈ 60	$-5.7°$
(b)	2	0.2	-0.2	59.8	$-11.3°$
(c)	5	0.5	-1	59	$-26.6°$
(d)	10	1	-3	57	$-45°$
(e)	20	2	-7	53	$-63.4°$
(f)	50	5	-14	46	$-78.7°$
(g)	100	10	-20	40	$-84.3°$

RC *High-Pass Model*

Consider the simple *RC* high-pass circuit shown in Figure 9-31. As in the case of the low-pass circuit, this high-pass circuit was also discussed in Chapter 8. Let us now

Figure 9-31 Simple *RC* high-pass circuit.

consider the form of the Bode plot for this circuit. Recall that the break frequency is

$$f_b = \frac{1}{2\pi RC} \qquad (9\text{-}121)$$

The form of the Bode plot for the amplitude response A_{dB} is shown in Figure 9-32. The frequency scale is labeled as f/f_b. The exact curve is labeled as E. One

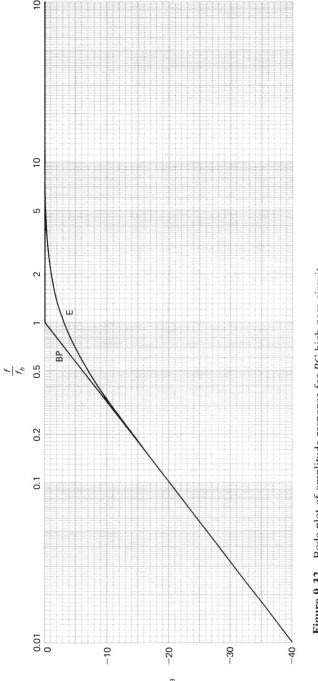

Figure 9-32 Bode plot of amplitude response for RC high-pass circuit.

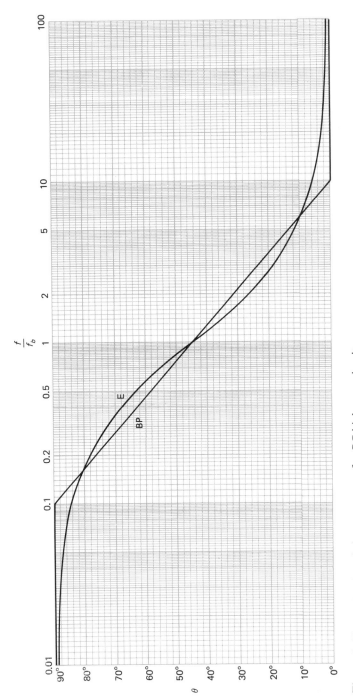

Figure 9-33 Bode plot of phase response for *RC* high-pass circuit.

can never get to the origin on a logarithmic scale, but this response at zero frequency would be a negative infinite dB level. The slope is positive at low frequencies, but it gradually tapers off and flattens out. The actual response is -3 dB at $f/f_b = 1$ or $f = f_b$.

The curve labeled BP is the Bode breakpoint approximation to the high-pass function. It is formed as a straight line in the frequency range below f_b with a slope of $+6$ dB/octave or $+20$ dB/decade. At the frequency $f = f_b$, it reaches a 0-dB level and then breaks to a horizontal line.

The phase shift for this circuit is shown in Figure 9-33 using a semilog scale again. The exact phase shift is the curve labeled E. The Bode breakpoint approximation for the phase is labeled as BP. The approximation assumes $+90°$ up to $f/f_b = 0.1$, at which point the curve breaks to a slope of $-45°$ per decade. It passes through $+45°$ at $f/f_b = 1$. The curve reaches $0°$ at $f/f_b = 10$, at which point it breaks to a zero slope again.

Example 9-19

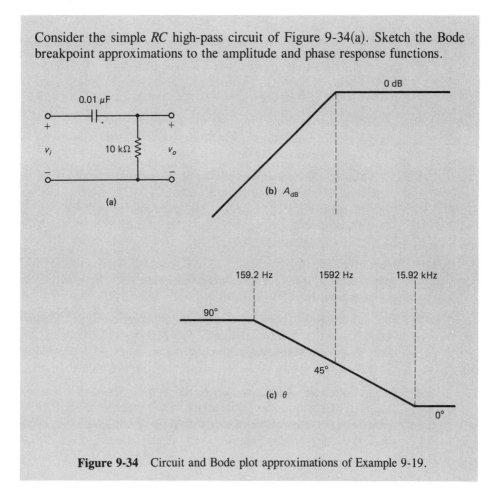

Consider the simple *RC* high-pass circuit of Figure 9-34(a). Sketch the Bode breakpoint approximations to the amplitude and phase response functions.

Figure 9-34 Circuit and Bode plot approximations of Example 9-19.

Solution

The break frequency is calculated as

$$f_b = \frac{1}{2\pi RC} = \frac{1}{2\pi \times 10^4 \times 0.01 \times 10^{-6}} = 1592 \text{ Hz} \qquad \textbf{(9-122)}$$

The breakpoint amplitude approximation is shown in Figure 9-34(b). The actual response differs from the approximation by 3 dB at 1592 Hz. The breakpoint phase approximation is shown in Figure 9-34(c).

The reader may find it instructive to compare the circuit and Bode plot of this example with those of Example 9-17. The element values and break frequencies of the two circuits are identical. However, the circuit of Example 9-17 is arranged in a low-pass form, while the circuit of this example is arranged in a high-pass form.

DRILL PROBLEMS

9-1. A certain inverting amplifier has a voltage gain $A = -50$. Determine the output signal voltage v_o when the input signal voltage is (**a**) $v_i = 0$, (**b**) $v_i = 20$ mV, and (**c**) $v_i = -0.1$ V.

9-2. A certain noninverting amplifier has a voltage gain $A = 25$. Determine the output signal voltage v_o when the input signal voltage is (**a**) $v_i = 0$, (**b**) $v_i = 80$ mV, and (**c**) $v_i = -0.4$ V.

9-3. The gain of a certain amplifier is to be determined from an oscilloscope measurement. The peak-to-peak input voltage is 0.2 V, and the peak-to-peak output voltage is 8 V. The output is 180° out of phase with the input. Determine the voltage gain A.

9-4. The gain of a certain amplifier is to be determined from an oscilloscope measurement. The peak-to-peak input voltage is 0.5 V, and the peak-to-peak output voltage is 12 V. The output is in phase with the input. Determine the voltage gain A.

9-5. An oscilloscope is connected across the input terminals of a certain inverting amplifier with a sinusoidal signal. The peak-to-peak voltage is measured as 0.2 V. An ac voltmeter across the output reads 4 V rms. Determine the gain A.

9-6. An ac voltmeter is connected across the input terminals of a certain noninverting amplifier with a sinusoidal signal. The voltmeter reads 50 mV rms. The peak-to-peak value of the output voltage is determined by an oscilloscope to be 5 V. Determine the gain A.

9-7. A certain amplifier is characterized by the following parameters:

$$R_{in} = 200 \text{ k}\Omega$$

$$R_{out} = 50 \text{ }\Omega$$

$$A = 80$$

Draw the flat-band signal model.

9-8. A certain amplifier is characterized by the following parameters:

$$R_{in} = 500 \text{ k}\Omega$$

$$R_{out} = 200 \text{ }\Omega$$

$$A = -60$$

Draw the flat-band signal model.

9-9. The circuit of Figure P9-9 represents the flat-band signal model of a certain amplifier. Determine the input resistance, the output resistance, and the open-circuit voltage gain.

Figure P9-9

9-10. The circuit of Figure P9-10 represents the flat-band signal model of a certain amplifier. Determine the input resistance, the output resistance, and the open-circuit voltage gain.

Figure P9-10

9-11. The circuit of Figure P9-11 represents the flat-band signal model of a certain amplifier with the output arranged in Norton equivalent circuit form. Determine the input resistance, the output resistance, and the open-circuit voltage gain.

Figure P9-11

9-12. The circuit of Figure P9-12 represents the flat-band signal model of a certain amplifier with output arranged in a Norton equivalent circuit form and with input current expressed as the controlling variable. Determine the input resistance, the output resistance, and the open-circuit voltage gain.

Figure P9-12

9-13. An ideal voltage source signal is applied to the input of the amplifier of Problem 9-7, and there is negligible loading across the output. Determine the gain A_{go}.

9-14. An ideal voltage source signal is applied to the input of the amplifier of Problem 9-8, and there is negligible loading across the output. Determine the gain A_{go}.

9-15. A signal source with an internal resistance of 50 kΩ is applied to the input of the amplifier of Problem 9-7, and there is negligible loading across the output. Determine the gain A_{go}.

9-16. A signal source with an internal resistance of 100 kΩ is applied to the input of the amplifier of Problem 9-8, and there is negligible loading across the output. Determine the gain A_{go}.

9-17. An ideal voltage source signal is applied to the input of the amplifier of Problem 9-7, and a resistance of 150 Ω is connected across the output. Determine the gain A_{go}.

9-18. An ideal voltage source signal is applied to the input of the amplifier of Problem 9-8, and a resistance of 400 Ω is connected across the output. Determine the gain A_{go}.

9-19. A signal source with an internal resistance of 50 kΩ is applied to the input of the amplifier of Problem 9-7, and a resistance of 150 Ω is connected across the output. Determine the gain A_{go}.

9-20. A signal source with an internal resistance of 100 kΩ is applied to the input of the amplifier of Problem 9-8, and a resistance of 400 Ω is connected across the output. Determine the gain A_{go}.

9-21. Determine the net gain A_{go} of the cascade amplifier configuration of Figure P9-21.

Figure P9-21

9-22. Determine the net gain A_{go} of the cascade amplifier configuration of Figure P9-22.

Figure P9-22

9-23. For the system of Problem 9-21, assume that the generator has an internal resistance of 200 Ω and that a 2-kΩ load is connected across the output. Determine the gain A_{go}.

9-24. For the system of Problem 9-22, assume that the generator has an internal resistance of 2 kΩ and that a 1-kΩ load is connected across the output. Determine the gain A_{go}.

9-25. A certain amplifier has a power gain $G = 8000$. Determine the decibel power gain G_{dB}.

9-26. A certain amplifier has a power gain $G = 500$. Determine the decibel power gain G_{dB}.

9-27. A certain amplifier has a decibel power gain $G_{dB} = 48$ dB. Determine G.

9-28. A certain amplifier has a decibel power gain $G_{dB} = 33$ dB. Determine G.

9-29. In a certain amplifier, the input power is 5 mW, and the output power is 2 W. Determine G and G_{dB}.

9-30. In a certain amplifier, the input power is 2 mW, and the output power is 1.2 W. Determine G and G_{dB}.

9-31. In a certain transmission system, the input power is 50 mW, and the output power is 2 mW. Determine G, G_{dB}, L, and L_{dB}.

9-32. In a certain transmission system, the input power is 200 mW, and the output power is 40 mW. Determine G, G_{dB}, L, and L_{dB}.

9-33. A certain amplifier has $A = 120$. Determine the decibel voltage gain A_{dB}.

9-34. A certain amplifier has $A = 800$. Determine the decibel voltage gain A_{dB}.

9-35. A certain amplifier has $A_{dB} = 30$ dB. Determine A.

9-36. A certain amplifier has $A_{dB} = 68$ dB. Determine A.

9-37. Consider the system of Problem 9-21 (Figure P9-21). Determine the decibel voltage gains of the two stages and the interaction factor, and use these results to determine the value of A_{dB} for the complete system.

9-38. Consider the system of Problem 9-22 (Figure P9-22). Determine the decibel voltage gains of the two stages and the interaction factor, and use these results to determine the value of A_{dB} for the complete system.

9-39. Express the following power levels in dBm: **(a)** 1 W; **(b)** 5 W; **(c)** 28 mW; **(d)** 60 μW; **(e)** 20 fW.

9-40. Determine the power in mW corresponding to the following dBm values: **(a)** 20 dBm; **(b)** 50 dBm; **(c)** 1 dBm; **(d)** −16 dBm; **(e)** −50 dBm.

9-41. Assume that an amplifier with a power gain of 50 dB is excited by a signal source which supplies an input power level of 60 μW to the amplifier. Express the input in dBm, and determine the corresponding output level in dBm.

9-42. The output power of a certain line amplifier is 2 W. It is connected to a transmission system that has a loss of 40 dB. Express the line input in dBm, and determine the corresponding line output level in dBm.

9-43. Consider the *RC* low-pass circuit of Figure P9-43. Sketch the Bode breakpoint approximations to the amplitude and phase response functions.

Figure P9-43

9-44. Consider the *RC* low-pass circuit of Figure P9-44. Sketch the Bode breakpoint approximations to the amplitude and phase response functions.

Figure P9-44

9-45. Consider the simple *RC* high-pass circuit of Figure P9-45. Sketch the Bode breakpoint approximations to the amplitude and phase response functions.

Figure P9-45

9-46. Consider the simple *RC* high-pass circuit of Figure P9-46. Sketch the Bode breakpoint approximations to the amplitude and phase response functions.

Figure P9-46

400

9-47. Consider an amplifier with open-circuit voltage gain A. Assume that it is connected to a generator whose resistance is matched to the amplifier input resistance (i.e., $R_g = R_{in}$). Show that

$$A_{go} = \tfrac{1}{2} A$$

9-48. Consider an amplifier with open-circuit voltage gain A. Assume that it is connected to a generator whose resistance is matched to the amplifier input resistance and that the output is connected to a matched load (i.e., $R_g = R_{in}$ and $R_{out} = R_L$). Show that

$$A_{go} = \tfrac{1}{4} A$$

9-49. Show that the power level in dBm is related to the power level in dBW by

$$P(\text{dBm}) = P(\text{dBW}) + 30$$

9-50. Show that the power level in dBf is related to the power level in dBW by

$$P(\text{dBf}) = P(\text{dBW}) + 150$$

DESIGN PROBLEMS

9-51. An amplifier is to be selected to connect between a source with internal resistance of 50 Ω and a 2-kΩ load. The following criteria are specified:

> Gain reduction due to input loading factor ≤ 0.1 dB
> Gain reduction due to output loading factor ≤ 0.1 dB
> Loaded voltage gain between source and load ≥ 40 dB under worst-case conditions of A and B

Specify the minimum value of A, the minimum value of R_{in}, and the maximum value of R_{out} for the amplifier to be selected.

9-52. Determine new design specifications for the amplifier of Problem 9-51 if the three criteria are changed as follows:

> Input loading factor ≥ 0.9
> Output loading factor ≥ 0.9
> Loaded voltage gain between source and load ≥ 100 under worst-case conditions of A and B

9-53. Design an RC circuit whose Bode breakpoint amplitude approximation is as follows:

> 0 dB level from dc to 2 kHz
> -6 dB/octave slope above 2 kHz

Select $C = 0.01~\mu\text{F}$.

9-54. Design an *RC* circuit whose Bode breakpoint amplitude approximation is as follows:

> 6 dB/octave slope below 500 Hz
>
> 0 dB above 500 Hz

Select $C = 0.1\ \mu\text{F}$.

9-55. Design an *RC* circuit whose Bode breakpoint amplitude approximation is as follows:

> -6 dB level from dc to 5 kHz
>
> -6 dB/octave slope above 5 kHz

Select $C = 0.002\ \mu\text{F}$.

9-56. Design an *RC* circuit whose Bode breakpoint amplitude approximation is as follows:

> 6 dB/octave slope from dc to 1 kHz
>
> -6 dB level above 1 kHz

TROUBLESHOOTING PROBLEMS

9-57. You are testing an *RC* low-pass circuit of the form shown in Figure 9-27, and you find that the break frequency is lower than specified. Which *two* of the following could be possible troubles? **(a)** *R* too small; **(b)** *R* too large; **(c)** *C* too small; **(d)** *C* too large.

9-58. You are testing an *RC* high-pass circuit of the form shown in Figure 9-31, and you find that the break frequency is higher than specified. Which *two* of the following could be possible troubles? **(a)** *R* too small; **(b)** *R* too large; **(c)** *C* too small; **(d)** *C* too large.

COMPUTER PROBLEMS

9-59. Write a program to determine the loaded gain A_{go} of the general single-amplifier configuration of Figure 9-15(a).

> Input data: R_{in}, R_{out}, R_g, R_L, A
>
> Output data: A_{go}

[*Note*: An "infinite" value for a resistance may be approximated by a value much larger than all other resistances (e.g., 1000 times as large).]

9-60. Write a program to determine the loaded gain A_{go} of a cascaded configuration of two amplifiers as shown in Figure 9-18(a).

> Input data: R_{in1}, R_{out1}, A_1, R_{in2}, R_{out2}, A_2, R_g, R_L
>
> Output data: A_{go}

CHAPTER 10

BJT Small-Signal Amplifier Circuits

OBJECTIVES

After completing this chapter, the reader should be able to:

1. Explain the process by which amplification occurs in a BJT and illustrate the process on a load line.
2. Discuss the concept of a small-signal or ac model of a transistor.
3. Define and be able to determine for a BJT the input dynamic resistance, the dynamic current gain, and the transconductance.
4. Construct for a BJT the small-signal flat-band circuit models.
5. Draw the schematic diagram for and explain the operation of a complete common-emitter amplifier using a stable bias design and with emitter resistance bypassed.
6. Analyze the circuit of objective 5 for gain, input resistance, and output resistance.
7. Draw the schematic diagram for and explain the operation of a common-emitter amplifier with emitter feedback.
8. Analyze the circuit of objective 7 for gain, input resistance, and output resistance.
9. Draw the schematic diagram for and explain the operation of a common-base amplifier.
10. Analyze the circuit of objective 9 for gain, input resistance, and output resistance.
11. Draw the schematic diagram for and explain the operation of a common-collector amplifier.
12. Analyze the circuit of objective 11 for gain, input resistance, and output resistance.

10-1 OVERVIEW

General properties that apply to all amplifiers have been considered in the preceding two chapters in terms of gain, input and output resistances, and frequency response considerations. We are now in a position to begin considering the implementation of specific types of amplifiers utilizing various electronic devices.

In this chapter we return to the bipolar junction transistor (BJT) to study how amplification can be achieved with this device. The dc properties of the BJT were considered in Chapters 4 and 5, but all applications considered there were based on either the dc model or on total instantaneous signals. In this chapter we focus on the properties of small-signal amplifier circuits implemented with BJTs.

A detailed treatment of BJT small-signal or ac parameters will be considered. These parameters are widely used to predict the performance of small-signal amplifiers. The small-signal models are developed, and typical specifications are studied.

The three basic amplifier configurations utilizing a BJT are denoted respectively as the *common-emitter amplifier*, the *common-base amplifier,* and the *common-collector amplifier*. The latter circuit is also known as the *emitter-follower circuit,* and a dc version of this circuit was considered in Chapter 5. All of these basic amplifier circuits will be considered in detail in this chapter. In addition, an improved form of the common-emitter amplifier utilizing emitter feedback will be considered.

The basic amplifier configurations to be discussed represent some of the major circuit forms used in the formulation of linear integrated-circuit modules. To understand and fully utilize linear integrated circuits, it is important to comprehend the operation of the basic amplifier configurations. In addition, many high-frequency amplifiers still utilize these basic forms for implementation.

Most of the sections in this chapter contain a number of derivations. Among the formulas to be derived are voltage gain, input resistance, and output resistance for various amplifier configurations. Some of the derivations are straightforward and others are a bit tricky. Depending on the objectives sought from the text, it may or may not be necessary to be able to reproduce the details of all the derivations. If you are taking a formal course using this text, the instructor will assist in identifying the expectations on derivations.

Even if you are not expected to be able to reproduce the details of the derivations, it is recommended that you at least study the general approaches taken. However, you can learn to use the results of the derivations intelligently if you understand the assumptions and limitations.

All the small-signal results derived in this chapter, as well as similar results derived in Chapter 11, are summarized and "tied together" in detail in Chapter 12.

10-2 SIMPLE SMALL-SIGNAL COMMON-EMITTER AMPLIFIER

The process of amplification with a BJT will be illustrated with the simple common-emitter ac-coupled circuit of Figure 10-1(a). This circuit has some serious practical limitations, as will be noted later, but its simplicity serves to illustrate the amplification process more clearly than the more elaborate circuits to be developed later.

The voltage v_i represents an input signal voltage, and v_o represents the output signal voltage. The input voltage v_i could represent the voltage across an ideal voltage source, or it could represent the terminal voltage when there is an internal

(a) Amplifier circuit   (b) DC equivalent circuit

Figure 10-1 Simple BJT common-emitter amplifier and dc equivalent circuit.

generator drop. From the work of Chapter 9, a loading factor could be used to account for loading effects between the signal generator and the amplifier, so only the amplifier input terminals will be considered now.

The output voltage v_o could represent a voltage across an open circuit (infinite resistance load), or there may be a load added for some application. As in the case of the input, a loading factor could be used when there is an external load connected. At this point only, the open-circuit ac voltage across the output will be considered.

The dc equivalent circuit is shown in Figure 10-1(b) based on the work of Chapter 8. The two capacitors are represented as open circuits, and this uncouples both the input signal and an output load from the dc path for the transistor. If v_i had been identified as an ac voltage source, the point to the left of the input capacitor would have been grounded. However, since it may or may not be a source, the point will be left uncommitted in this analysis since it does not affect the dc circuit of the transistor.

The dc circuit is recognized as one that was considered early in Chapter 5 to study basic transistor models. The base current is given by

$$I_B = \frac{V_{CC} - V_{BE}}{R_B} = \frac{V_{CC} - 0.7}{R_B} \tag{10-1}$$

where $V_{BE} = 0.7$ V has been assumed. The collector current I_C is

$$I_C = \beta_{dc} I_B \tag{10-2}$$

The collector–emitter voltage V_{CE} is

$$V_{CE} = V_{CC} - R_C I_C \tag{10-3}$$

Before considering amplification, a property concerning V_{BE} needs to be emphasized. From a dc point of view, it is usually reasonable (as we have done many times) to assume that this voltage is a constant (i.e., $V_{BE} = 0.7$ V for many purposes). However, for amplification to occur, it is necessary that V_{BE} change when an input signal is applied. The amount by which it changes is usually quite small compared with the assumed constant value. In a typical application, V_{BE} might change from 0.69 V to 0.71 V or so, for example. When amplification is analyzed mathematically, these changes will have to be considered. However, the dc circuit may continue to assume a constant value in the process. Do not be disturbed by this apparent contradiction since it is strictly a matter of focusing on the important effect for the purpose desired.

Load-Line Analysis

The load line introduced in Chapter 5 serves as a very useful concept for illustrating the basic amplification process qualitatively. Refer to Figure 10-2, in which some idealized collector characteristics for an *npn* transistor are shown. The particular values of V_{CE}, I_C, and I_B at the dc operating point (Q-point) are designated as V_{CEQ}, I_{CQ}, and I_{BQ}.

In accordance with the load-line analysis of Chapter 5, an expression for I_C from (10-3) is determined as

$$I_C = \frac{V_{CC}}{R_C} - \frac{1}{R_C} V_{CE} \tag{10-4}$$

The vertical intercept is V_{CC}/R_C, the horizontal intercept is V_{CC}, and the slope is $-1/R_C$.

The load line is shown as the straight line on the collector characteristics in Figure 10-2. The intersection of the load line with the curve $I_B = I_{BQ}$ defines the Q-point, and the values of V_{CE} and I_C at that point are V_{CEQ} and I_{CQ}, respectively, as shown.

Amplification from Load Line

Next, the process for amplification will be studied. The basic idea is to couple a time-varying signal to the base circuit in a manner that will cause the base–emitter

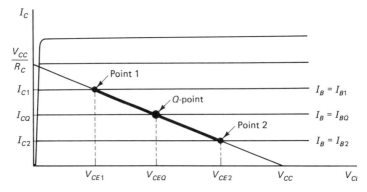

Figure 10-2   Load line for simple BJT common-emitter transistor amplifier.

voltage and base current to vary with the input signal. One might naively be tempted simply to connect the signal directly to the base. However, if the signal has a dc path to ground, the dc operating point would shift immediately (even with zero signal level), and this would adversely affect transistor operation. The basic rule for ac-coupled circuits is that *the input and output connection paths of the signal should not affect the dc operating point*. The input and output coupling capacitors achieve the required isolation.

Assume that the frequency range of the signal is sufficiently high that the flat-band models of the two capacitors (short circuits) are valid. However, assume that the frequency is sufficiently low that all the transistor high-frequency effects are negligible. We are thus assuming *midfrequency* operation of the circuit, which might be from about 100 Hz to 100 kHz or so in a typical case.

In the discussion that follows, refer both to the load line of Figure 10-2 and to the waveforms shown in Figures 10-3 and 10-4. It is customary in analyzing am-

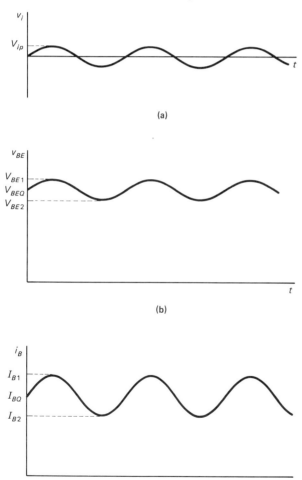

Figure 10-3 Idealized base waveforms of simple amplifier circuit. (Scale is different from that of Figure 10-4.)

(a)

(b)

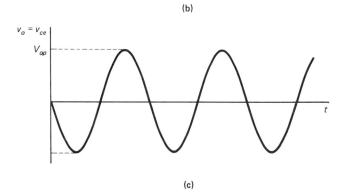

(c)

Figure 10-4 Idealized collector waveforms of simple amplifier circuit. (Scale is different from that of Figure 10-3.)

plifiers to assume a sinusoidal input signal in many cases. The assumed sinusoidal input signal v_i is shown in Figure 10-3(a). The various other waveforms to be studied have been idealized in this development and are also shown as sinusoids. In actual practice *significant distortion* could appear in this simple amplifier, but it will not be considered at this point in the book.

The input signal v_i is coupled through the capacitor C_1 and adds algebraically to the Q-point base–emitter voltage. The resulting idealized instantaneous voltage v_{BE} is shown in Figure 10-3(b). This voltage varies up to some maximum value V_{BE1} on the positive half-cycle and down to a minimum value V_{BE2} on the negative half-cycle.

As v_{BE} varies about V_{BEQ}, i_B varies correspondingly about I_{BQ}. The resulting instantaneous current i_B is shown in Figure 10-3(c). This current varies up to some maximum value I_{B1} on the positive half-cycle and down to a minimum value I_{B2} on the negative half-cycle. Note that i_B is *in phase* with v_{BE}; that is, the positive half-cycle of v_{BE} corresponds to the positive half-cycle of i_B, and vice versa.

Next, refer back to the load line in Figure 10-2. On the positive input half-cycle, operation shifts up to point 1, which corresponds to the intersection of the load line with the curve $I_B = I_{B1}$. On the negative input half-cycle, operation shifts down to point 2, which corresponds to the intersection of the load line with the curve $I_B = I_{B2}$. Note that compared to I_{CQ}, point 1 corresponds to a larger collector current I_{C1}, and point 2 corresponds to a smaller collector current I_{C2}. This certainly seems reasonable since collector current increases with base current.

In contrast to collector current, note that point 1 corresponds to a smaller collector voltage V_{CE1}, and point 2 corresponds to a larger collector voltage V_{CE2}. The reason for this is that a larger drop across R_C occurs for a larger collector current, and the remaining voltage across the transistor is decreased. The opposite situation is true for a smaller collector current.

The various waveforms in the collector circuit are shown in Figure 10-4. Understand that the vertical scale is different here than in Figure 10-3. It is necessary to use a different scale since typical collector current and signal voltage levels could be many times larger than the corresponding base levels. The origin in Figure 10-4 is assumed to be aligned exactly with that of Figure 10-3; that is, the first half-cycle of each waveform corresponds to the first half-cycle of the input voltage of Figure 10-3.

The instantaneous collector current waveform i_C is shown in Figure 10-4(a), and the limits I_{C1} and I_{C2} correspond to those shown on the load line. The instantaneous collector–emitter voltage v_{CE} is shown in Figure 10-4(b), and the limits V_{CE1} and V_{CE2} also correspond to those shown on the load line. Observe here how the variation in collector voltage is opposite to that of the collector current.

The output voltage waveform is shown in Figure 10-4(c). Comparing the output signal v_o with the input signal v_i of Figure 10-3(a), and assuming the much larger scale, two things may be concluded: (1) the input signal voltage has been amplified, and (2) the output signal is inverted with respect to the input signal (i.e., the *common-emitter amplifier is an inverting amplifier*).

From the load line and the figures, it appears that the Q-point in Figure 10-2 is at the "halfway" point; that is, the dc voltages are divided equally between the transistor and the load resistance R_C. This allows the output to swing symmetrically a maximum amount in both directions. If an input signal were sufficiently large to move operation over to the intersection with the I_C axis, saturation would occur. Conversely, if an input signal were sufficiently large to move operation over to the intersection with the V_{CE} axis, cutoff would occur.

Example 10-1

Consider a simple BJT amplifier of the form shown in Figure 10-1 with waveforms as shown in Figures 10-3 and 10-4. Assume that for a specific set of operating conditions, the peak values of input and output voltages are measured as $V_{ip} = 12$ mV, and $V_{op} = 2.16$ V. Determine the voltage gain A between input and output terminals.

Solution

In general, the gain A may be expressed as

$$A = \frac{v_o}{v_i} \qquad (10\text{-}5)$$

In measuring this gain for an ac amplifier, only the signal components at input and output are of interest for this purpose. For sinusoidal waveforms, either the peak values, the peak-to-peak values, or the rms values could be used as long as *both* input and output are expressed in the same forms. Oscilloscopic measurements are often best performed using peak-to-peak values.

In this example we are given the peak values of input and output voltages. We also know that the voltage gain of a common-emitter amplifier is inverting, so a negative sign will be added. Thus

$$A = -\frac{V_{op}}{V_{ip}} = -\frac{2.16 \text{ V}}{12 \times 10^{-3} \text{ V}} = -180 \qquad (10\text{-}6)$$

10-3 SMALL-SIGNAL BJT MODELS

The load line approach in the preceding section has provided some useful insight into the process of amplification for a BJT. If the collector characteristics were given for a particular transistor, it would be possible to determine various amplifier properties, such as gain, directly from the load line. Unfortunately, the collector characteristics vary so widely between individual transistors that such an approach is quite limited in practice.

The most meaningful approach for predicting amplifier performance with transistors is through the use of so-called *small-signal* models. These equivalent-circuit models are used to predict the levels of various *signal* voltages and currents within the transistor. These signals are the *changes* that occur in the dc levels when an input signal is applied. From the concept of superposition, the dc levels are not shown in the small-signal models. However, their presence is always understood in order to make the active devices function properly.

The adjective "small" in the term *small signal* refers to the assumption that the equivalent circuits are generally valid only when the shift of a given variable is small compared with the dc operating point. Most electronic devices are basically nonlinear, but the linear models are reasonably correct in a narrow region about a given operating point. However, when exacting requirements must be met in advanced applications, more complete nonlinear models may be required.

The term *ac model* will be used here to mean the same thing as *small-signal model*. The adjective *dynamic* is often used to describe an ac or small-signal quantity, while the term *static* could refer to a corresponding dc quantity.

Consider the *npn* BJT shown in Figure 10-5. No external connections are shown here since the interest at this point is to focus on the transistor above. In accordance with notation established in Chapter 5, the total instantaneous base current, base–emitter voltage, collector current, and collector–emitter voltage are identified. To review that notation in reference to this situation, the total base current i_B can be expressed as

$$i_B = I_B + i_b \qquad (10\text{-}7)$$

where I_B is the dc component and i_b is the small-signal or ac component. The total base–emitter voltage v_{BE} is

$$v_{BE} = V_{BE} + v_{be} \qquad (10\text{-}8)$$

where V_{BE} is the dc component and v_{be} is the ac component. On the collector side, i_C is the total current and it is

$$i_C = I_C + i_c \qquad (10\text{-}9)$$

where I_C is the dc component and i_c is the ac component. Finally, v_{CE} is the total collector–emitter voltage, and it is

$$v_{CE} = V_{CE} + v_{ce} \qquad (10\text{-}10)$$

where V_{CE} is the dc component and v_{ce} is the ac component.

Dynamic Input Resistance

Refer to Figure 10-6(a), in which a typical BJT input characteristic is shown. As we have done on other occasions in graphical analysis, the subscript Q has been added to distinguish the dc operating point values from the abscissa and ordinate variables. Thus the transistor is assumed to be biased at operating dc voltage and current V_{BEQ} and I_{BQ}, respectively.

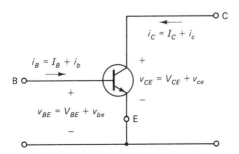

Figure 10-5 Bipolar junction transistor and variables used in establishing circuit models.

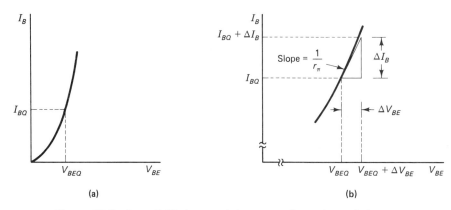

Figure 10-6 Input BJT characteristic at operating point and changes.

An enlarged portion of the input characteristic is shown in Figure 10-6(b). Assume that the base–emitter voltage is changed by a positive amount ΔV_{BE} as shown. The base current changes by a positive amount ΔI_B. Although the curve is basically nonlinear, if the changes are small, the slope of the straight line between the two points will be nearly the same as the slope of the characteristic at the operating point. This slope can be expressed as

$$\text{slope} = \frac{\Delta I_B}{\Delta V_{BE}} \qquad \textbf{(10-11)}$$

The slope has the dimensions of conductance (amperes/volt = siemens). It is more convenient to work with units of resistance. A resistance quantity will be defined as

$$\boxed{r_\pi = \frac{1}{\text{slope}} = \frac{\Delta V_{BE}}{\Delta I_B}} \qquad \textbf{(10-12)}$$

The quantity r_π is defined as the *dynamic input resistance* of the BJT. It is clear that r_π will vary considerably as the operating point is changed, but at a given operating point, it can be approximated as a constant value.

For a *pnp* transistor ΔV_{BE} in (10-11) is replaced by ΔV_{EB} to maintain a positive sense of the change. The important point to remember is that r_π is a positive quantity for both types of BJTs.

Dynamic Current Gain

The effect of a change in voltage and current at the base of a BJT has been represented in terms of a dynamic resistance. Let us now consider the corresponding effect at the collector.

Refer to Figure 10-7. Assume that the curve $I_B = I_{BQ}$ corresponds to the dc operating-point base current, and the corresponding collector current is $I_C = I_{CQ}$. Assume that the base current increases by a positive value ΔI_B, which shifts opera-

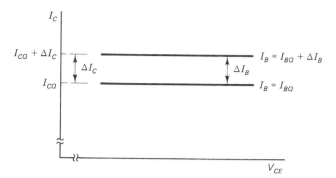

Figure 10-7 Output BJT characteristics at operating point and changes.

tion to the curve corresponding to $I_B = I_{BQ} + \Delta I_B$. The corresponding change in the collector current is a positive shift ΔI_C, so the new collector current is $I_C + \Delta I_C$.

A dynamic flat-band current gain will now be defined. We use the term *flat band* to refer to the fact that none of the frequency-limiting aspects of the transistor are considered at this point. The flat-band dynamic current gain in denoted as β_o, and is defined as

$$\beta_o = \frac{\Delta I_C}{\Delta I_B} \qquad (10\text{-}13)$$

Strictly speaking, this definition applies at a constant value of V_{CE}, but since the ideal curves are assumed to show no variation with V_{CE}, this point need not concern us at the moment.

How does β_o compare with β_{dc}? If the transistor characteristics were ideal, and if β_{dc} were constant for all active-region conditions, the two quantities would be equal. However, since β_{dc} is defined in terms of *total* collector and base currents, while β_o is defined for *small changes* about an operating point, the two quantities usually differ somewhat for realistic transistors. Manufacturers usually provide separate data for the two quantities. Usually, however, the values of β_{dc} and β_o for a given transistor tend to be reasonably close to each other.

Small-Signal Circuit Models

The input dynamic resistance and current gain have been defined in terms of changes. In a practical amplifier, these changes occur as a result of *signals*. Thus the Δ-quantities used to define the parameters in this section become the ac or small-signal variables in an amplifier configuration.

Consider the BJT shown in Figure 10-8(a). Only signal or ac quantities are shown in this particular case, and all dc levels are understood. The first form of a flat-band ac model is shown in Figure 10-8(b). This model will be referred to as the β-*model*, for reasons that will be clear shortly.

The resistance r_π is connected between base and emitter, and it relates the ac base–emitter voltage to the ac base current by Ohm's law; that is,

$$v_{be} = r_\pi i_b \qquad (10\text{-}14)$$

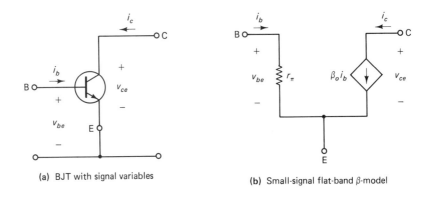

(a) BJT with signal variables

(b) Small-signal flat-band β-model

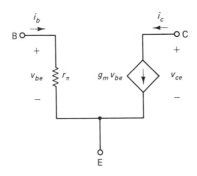

(c) Small-signal flat-band g_m-model

Figure 10-8 Bipolar junction transistor and two flat-band small-signal models.

Typical values of r_π for small-signal transistors range from a few hundred ohms to several thousand ohms.

The process of ac current gain is accounted for by placing a dependent current source with the value $\beta_o i_b$ between collector and base. Note that $i_c = \beta_o i_b$ in the model.

If this ac equivalent circuit is compared with the corresponding dc equivalent circuit considered extensively earlier in the text, the main difference in form is on the base side. The dc circuit employs a fixed voltage drop (usually 0.7 V), which is a reasonable assumption for dc analysis. However, for ac analysis, the dc drop is unimportant, but the dynamic resistance is most significant.

Although the ac collector–emitter signal voltage v_{ce} is labeled on the model, it is not possible to determine this quantity until external circuit parameters are connected. The collector acts nearly like a current source, and the voltage depends on how this current reacts to external circuit parameters.

In most applications of the BJT, it is more desirable to express the dependent current source in terms of the base–emitter voltage than the base current. The base current is readily expressed in terms of the base–emitter voltage as

$$i_b = \frac{v_{be}}{r_\pi} \tag{10-15}$$

The dependent current source is

$$\beta_o i_b = \frac{\beta_o v_{be}}{r_\pi} \tag{10-16}$$

The dependent source is now a *voltage-controlled current source*. A transconductance parameter g_m can then be defined as

$$g_m = \frac{\beta_o}{r_\pi} \tag{10-17}$$

or

$$\boxed{\beta_o = g_m r_\pi} \tag{10-18}$$

The model modified for v_{be} as the controlling variable is shown in Figure 10-8(c). This circuit will be referred to as the g_m-*model* in subsequent work.

Traditionally, the β-model has been used more than the g_m-model in basic texts, probably because of earlier conventions used in transistor modeling and because it correlates more closely with the dc-model. However, the g_m-model has recently begun to emerge as a better model for a number of reasons that will be clearer later. Consequently, the latter model will serve as the basis for much of our work through the book.

It turns out that the g_m-model is a simplified form of a more elaborate circuit, called the *hybrid-pi model,* which is considered in Chapter 18. The hybrid-pi model can be readily employed to predict the high-frequency performance of transistors when frequency dependency is significant. Indeed, the subscript π in r_π refers to the hybrid-pi model notation. We will also see later that the g_m-model results in a closer comparison between BJT and FET small-signal models.

The parameters β_o and r_π at a given operating point vary considerably from one transistor to another for a given type. A typical range may be $4:1$ or greater. However, the parameter g_m tends to remain reasonably close for a given operating point. It can be shown from advanced semiconductor physics that for an assumed junction temperature of $25°C$, the theoretical ideal value of g_m is $g_m = 38.9I_C$. Because of the nonideal nature of the junction and the temperature variation, this constant has been rounded to several values in the literature. We choose the fairly common and more easily remembered value of 40. Consequently, the following reasonable estimate for g_m will be used here:

$$\boxed{g_m = 40I_C} \tag{10-19}$$

If I_C is given in amperes, g_m will be expressed in siemens. However, if I_C is expressed in milliamperes, g_m will be expressed in millisiemens.

From (10-17) and (10-18), we see that β_o, r_π, and g_m are all related. If r_π and β_o vary by as much as 4:1 or more, how does g_m tend to be much closer to a constant value at a given operating point? The answer is determined by inspecting (10-17). The ratio of β_o to r_π tends to remain nearly the same at a given operating point. This means that a transistor having a higher value of β_o will normally have a higher value of r_π, and vice versa. However, the ratio tends to remain nearly the same.

One more point about the direction of the dependent current source needs to be made. Back in Chapter 4 we saw that the dependent current source in the dc model of a *pnp* transistor was opposite to that of an *npn* transistor because of the opposite direction of current flow. Our development of the amplifier circuit in this chapter has assumed an *npn* transistor, and the current source small-signal equivalent circuit has been observed to have the same direction as in the dc circuit. One might then be led to believe that the current source in the ac model for a *pnp* transistor should have the opposite direction of the current source in the *npn* case. This is *not* the case, however.

The current source in the ac model represents a *change* in total collector current rather than total current. In an *npn* transistor, a positive input signal causes an increase in positive base current and this increases positive collector current. In a *pnp* transistor, however, a positive input signal has the effect of decreasing the base current (whose positive direction is out of the base), and this decreases the collector current, whose positive direction is out of the collector. A decrease in positive current out of the collector is equivalent to an increase in signal current from collector to emitter. The final result is that the positive collector signal current is directed downward.

To summarize, *either of the small-signal equivalent circuits of Figure 10-8 applies to both npn and pnp transistors as given.*

Example 10-2

Some fine dc measurements performed in the base circuit of a particular *npn* transistor with fixed collector–emitter voltage yield the following data:

V_{BE} (V)	I_B (μA)
0.65	100
0.66	150

Determine the approximate value of r_π at the operating point.

Solution

The first point to note is that V_{BE} is no longer assumed to be our "standard" value of 0.7 V. When small-signal models are being determined, it is necessary to measure carefully the actual voltage and changes about that value.

The concept of (10-12) is used for this purpose. The change ΔV_{BE} is given by

$$\Delta V_{BE} = 0.66 \text{ V} - 0.65 \text{ V} = 0.01 \text{ V} \qquad \textbf{(10-20)}$$

The corresponding change ΔI_B is

$$\Delta I_B = 150 \; \mu A - 100 \; \mu A = 50 \; \mu A \qquad (10\text{-}21)$$

The value of r_π is

$$r_\pi = \frac{0.01 \; V}{50 \times 10^{-6} \; A} = 200 \; \Omega \qquad (10\text{-}22)$$

This example was formulated more as a learning process than as a practical procedure. The measurement of a small change in a dc voltage about a much larger dc voltage is difficult to achieve with any degree of accuracy. In practice, the parameter r_π is easier to determine using ac small-signal techniques.

Example 10-3

Some dc measurements performed on a particular BJT with fixed collector–emitter voltage yield the following data:

$I_B \; (\mu A)$	$I_C \; (mA)$
100	9
120	11.2

Determine the approximate value of β_o at the operating point.

Solution

The concept of (10-13) is used for this purpose. The change ΔI_B is

$$\Delta I_B = 120 \; \mu A - 100 \; \mu A = 20 \; \mu A \qquad (10\text{-}23)$$

The corresponding change ΔI_C is

$$\Delta I_C = 11.2 \; mA - 9 \; mA = 2.2 \; mA \qquad (10\text{-}24)$$

The value of β_o is

$$\beta_o = \frac{2.2 \times 10^{-3} \; A}{20 \times 10^{-6} \; A} = 110 \qquad (10\text{-}25)$$

Incidentally, the value of β_{dc} could be computed at either of the two points. The reader is invited to show that neither value of β_{dc} is equal to β_o since the phenomena involved are different.

Example 10-4

A certain transistor is operating at $I_C = 2$ mA and it is known that $\beta_o = 120$. Determine **(a)** g_m and **(b)** r_π.

Solution

(a) The value of g_m is estimated from (10-19) as

$$g_m = 40I_C = 40 \times 2 \times 10^{-3} = 80 \times 10^{-3} \text{ S} = 80 \text{ mS} \qquad \textbf{(10-26)}$$

(b) From (10-18), we have

$$r_\pi = \frac{\beta_o}{g_m} = \frac{120}{80 \times 10^{-3}} = 1500 \ \Omega \qquad \textbf{(10-27)}$$

Example 10-5

For the transistor of Example 10-4 at the given operating point, construct the β-model and the g_m-model for small signals.

Solution

The models have the forms given in Figure 10-8 and they are shown in Figure 10-9 for the particular parameters of Example 10-4.

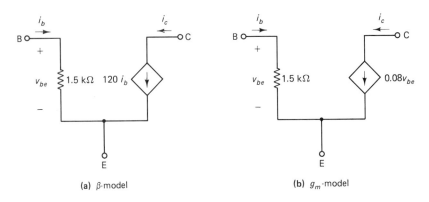

(a) β-model (b) g_m-model

Figure 10-9 Small-signal models of Example 10-5.

Example 10-6

A certain transistor is characterized by the fact that β_o remains nearly constant with the value $\beta_o \simeq 100$ as I_C varies from 1 to 8 mA. For this assumption, determine g_m and r_π at the following values of I_C: **(a)** 1 mA; **(b)** 2 mA; **(c)** 4 mA; **(d)** 8 mA.

Solution

At any value of I_C, we have

$$g_m = 40I_C \qquad \textbf{(10-28)}$$

The corresponding value of r_π is then determined as

$$r_\pi = \frac{\beta_o}{g_m} = \frac{100}{g_m} \qquad \text{(10-29)}$$

For the given values of I_C, the results can be readily calculated and tabulated as follows:

	I_C (mA)	g_m (mS)	β_o	r_π
(a)	1	40	100	2.5 kΩ
(b)	2	80	100	1.25 kΩ
(c)	4	160	100	625 Ω
(d)	8	320	100	312.5 Ω

In practice, β_o does not remain constant as I_C varies, but it often varies little for a particular transistor over a reasonably wide range of collector current. Therefore, this example illustrates a very important point: *The value of r_π decreases very rapidly as the collector current increases. If β_o were exactly constant (as assumed in this example), the value of r_π would vary inversely with collector current.*

10-4 COMMON-EMITTER CIRCUIT WITH STABLE BIAS

The simple common-emitter circuit previously considered has provided insight into the amplification process. That basic circuit, however, is subject to a wide variation in the operating point as a result of transistor parameter variations and has only limited usefulness in practice.

We now consider a more practical common-emitter circuit employing stabilized bias. This circuit is analyzed with the small-signal model developed in the preceding section.

The basic form of the stabilized common-emitter ac-coupled amplifier using an *npn* transistor is shown in Figure 10-10(a). Effects of possible source resistance and external load resistance will not be considered at this time.

The dc circuit is shown in Figure 10-10(b). This circuit was developed in some detail in Chapter 5 and is the standard stabilized bias circuit for a BJT. Recall that if the bias circuit is designed properly, the collector current can be closely approximated as

$$I_C = \frac{V_{\text{TH}} - V_{BE}}{R_E} = \frac{V_{\text{TH}} - 0.7}{R_E} \qquad \text{(10-30)}$$

where

$$V_{\text{TH}} = \frac{R_2 V_{CC}}{R_1 + R_2} \qquad \text{(10-31)}$$

+V_{CC}

R_1 R_C C_2

C_1

+

v_o

v_i R_2

−

R_E C_E

(a) Amplifier

+V_{CC}

$\downarrow I_C$

R_1 R_C

R_2 R_E

(b) DC circuit

Figure 10-10 Common-emitter amplifier with stabilized bias and dc equivalent circuit.

The corresponding *midfrequency* ac equivalent circuit for small signals is shown in Figure 10-11. This circuit combines some of the concepts of Chapter 8 with the ac signal model of Section 10-3. The actual transistor is replaced by the small-signal model between the three transistor terminals as noted. The value of g_m is determined from the dc operating-point current previously calculated and is approximated as

$$g_m = 40I_C \tag{10-32}$$

The functions of C_1 and C_2 have already been discussed for the simpler amplifier, and they serve to couple the signal in and out while blocking dc from flowing into the signal source or any external load. The function of C_E, however, is to bypass the signal around the emitter resistance R_E. Without C_E, the effect of R_E would be to reduce the gain considerably, as will be discussed in a later section. In the flat-band region of operation, C_E acts as a short, and all signal current flows directly from emitter to ground. Thus R_E does not even appear in the midfrequency equivalent circuit.

i_i B i_b C

+   +   +

v_i R_B v_{be} r_π $g_m v_{be}$ R_C v_o

−   −   −

$R_B = R_1 \parallel R_2$ E

Figure 10-11 Small-signal midfrequency model of common-emitter amplifier.

In accordance with the principles discussed in Chapter 8, the dc bias supply is replaced by a short circuit to ground as far as signals are concerned. In effect, this brings both R_1 and the collector resistance R_C to ground. Since R_1 and R_2 appear as a simple parallel combination of two resistors, for convenience a resistance R_B is defined as

$$R_B = R_1 \| R_2 = \frac{R_1 R_2}{R_1 + R_2} \tag{10-33}$$

Note that two separate signal voltage labels v_i and v_{be} are shown on the diagram even though they are equal in this case. The voltage v_i is the input signal voltage to the amplifier, and v_{be} is the signal base–emitter voltage, which controls the collector current. In many configurations, these two voltages are not the same, but for the common-emitter configuration in the flat-band region with emitter resistance completely bypassed, $v_{be} = v_i$.

The voltage gain A will now be determined. The output ac voltage v_o must be determined in terms of the input ac voltage v_i. Since all of the available ac current flows into R_C, with the resulting positive flow into the lower terminal, we have

$$v_o = -g_m v_{be} R_C = -g_m v_i R_C \tag{10-34}$$

The voltage gain A is

$$\boxed{A = \frac{v_o}{v_i} = -g_m R_C} \tag{10-35}$$

Some other quantities of immediate interest for the amplifier are the input and output resistances. However, it is convenient to define two separate values for each of these quantities. The first value for each will represent the quantities associated with the transistor itself without regard to external resistances. The second value for each will represent the corresponding first value coupled with the effects of external resistances.

Let R'_{in} and R'_{out} represent the ac input and output resistances of the transistor *without the effects of external resistances*. From the equivalent circuit of Figure 10-11, we see that

$$\boxed{R'_{in} = \frac{v_{be}}{i_b} = r_\pi} \tag{10-36}$$

and

$$\boxed{R'_{out} = \infty} \tag{10-37}$$

The result of (10-37) is based on the assumption of an ideal current source for the $g_m v_{be}$ generator. In practice, there is a large dynamic resistance in parallel with this current source, but it will be approximated as an infinite resistance.

Let R_{in} and R_{out} represent the input and output resistances of the *composite circuit*. We have

$$\boxed{R_{\text{in}} = \frac{v_i}{i_i} = R_B \| r_\pi = R_1 \| R_2 \| r_\pi} \qquad \text{(10-38)}$$

and

$$\boxed{R_{\text{out}} = R_C} \qquad \text{(10-39)}$$

The result of (10-38) indicates that the net input resistance of the complete amplifier circuit is simply the parallel combination of the three resistances R_1, R_2, and r_π. Although the transistor output is assumed to be an ideal current source, the result of (10-39) indicates that the output resistance of the composite circuit is the collector resistance R_C. Because of the typical moderate size of R_C, the assumption of (10-37) is usually valid since the effect of the transistor dynamic resistance on R_{out} will often be insignificant.

The quantities R_{in} and R_{out} are generally the most meaningful of the input and output resistance quantities since they predict the net effects of the complete circuit. However, when special attention is focused only on the transistor itself, R_{in}' and R_{out}' are useful.

For this amplifier and other configurations to be discussed in this chapter and in Chapter 11, most of the effects of source and load interactions will be momentarily disregarded until the basic properties of the amplifiers themselves are established. The effects of source and load interactions, as well as combinations of the basic amplifier configurations, are considered in Chapter 12.

Recall the unstabilized common-emitter amplifier of Section 10-2, which was used to qualitatively analyze amplifier operation with a load line. If a small-signal ac circuit were constructed for that circuit, it would have exactly the same form as the stabilized circuit except that R_B in the simpler circuit represents one resistance instead of a parallel combination. This equivalence is due to the fact that the emitter stabilizing resistance R_E is bypassed for signals, and thus the emitter appears at signal ground.

Example 10-7

The schematic diagram of a certain *RC*-coupled common-emitter amplifier is shown in Figure 10-12(a). **(a)** Determine the dc operating point current I_C and the three voltages V_{R_E}, V_{R_C} and V_{CE}. **(b)** Construct the midfrequency small-signal equivalent circuit. It is known that $\beta_o = 240$. **(c)** Determine the gain A and the input and output resistances R_{in}', R_{out}', R_{in}, and R_{out}.

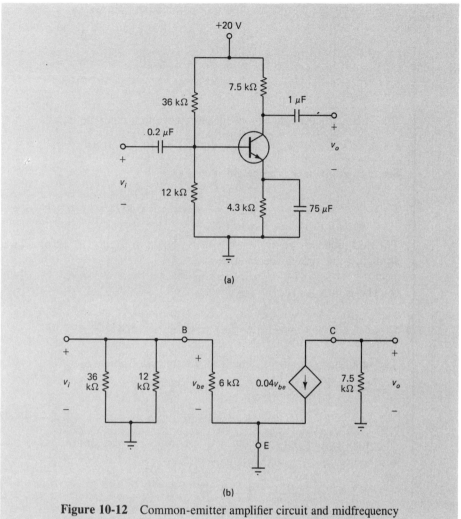

Figure 10-12 Common-emitter amplifier circuit and midfrequency model of Example 10-7.

Solution

(a) The first part is simply a rehash of the work of Chapter 5 but is a necessary part of a complete amplifier analysis. The steps involved in determining I_C follow.

$$V_{\text{TH}} = \frac{12 \text{ k}\Omega}{12 \text{ k}\Omega + 36 \text{ k}\Omega} \times 20 \text{ V} = 5 \text{ V} \qquad \textbf{(10-40)}$$

$$I_C = \frac{5 - 0.7}{4.3 \times 10^3} = 1 \text{ mA} \qquad \textbf{(10-41)}$$

Various voltages in the circuit are determined as follows:

$$V_{R_E} = 4.3 \times 10^3 \times 1 \times 10^{-3} = 4.3 \text{ V} \tag{10-42}$$

$$V_{R_C} = 7.5 \times 10^3 \times 1 \times 10^{-3} = 7.5 \text{ V} \tag{10-43}$$

$$V_{CE} = 20 - 4.3 - 7.5 = 8.2 \text{ V} \tag{10-44}$$

(b) To construct the ac equivalent circuit we first compute g_m as

$$g_m = 40I_C = 40 \times 1 \times 10^{-3} = 0.04 \text{ S} \tag{10-45}$$

Since β_o is given, r_π can be calculated as

$$r_\pi = \frac{\beta_o}{g_m} = \frac{240}{0.04} = 6 \text{ k}\Omega \tag{10-46}$$

The midfrequency equivalent circuit is shown in Figure 10-12(b). For clarity, R_1 and R_2 are shown separately.

(c) The gain A is

$$A = \frac{v_o}{v_i} = -g_m R_C = -0.04 \times 7.5 \times 10^3 = -300 \tag{10-47}$$

The input and output resistances of the transistor above are

$$R'_{\text{in}} = r_\pi = 6 \text{ k}\Omega \tag{10-48}$$

$$R'_{\text{out}} = \infty \tag{10-49}$$

The input and output resistances of the composite circuit are

$$R_{\text{in}} = R_1 \| R_2 \| r_\pi = 36 \text{ k}\Omega \| 12 \text{ k}\Omega \| 9 \text{ k}\Omega = 4.5 \text{ k}\Omega \tag{10-50}$$

$$R_{\text{out}} = R_C = 7.5 \text{ k}\Omega \tag{10-51}$$

It is not always necessary to construct the small-signal model to determine these various parameters since the formulas can often be applied directly. However, the small-signal model provides some meaningful insight into the process involved, and it permits the desired properties to be determined directly from the circuit parameters.

Because of the uncertainties associated with transistor parameters and the approximations involved, these results should be interpreted as "good estimates" rather than exact values. In particular, the magnitude of the gain of an actual transistor circuit based on these conditions would tend to be somewhat lower than the calculated value of (10-47) because of the transistor dynamic output resistance in parallel with R_C and because g_m was rounded upward in earlier work.

The common-emitter amplifier of the preceding section is capable of assuming a very stable dc operating point, but its ac or signal characteristics have some serious limitations. The most serious limitation of the circuit is the degree of distortion present in the output when the input signal is nearly an ideal voltage source and when the desired output signal has an appreciable level.

The reason for the serious distortion will be explained with Figure 10-13. Certain of the curves here will be exaggerated to demonstrate the situation fully. In (a) an ideal sinusoidal signal is assumed across the amplifier input (i.e., the

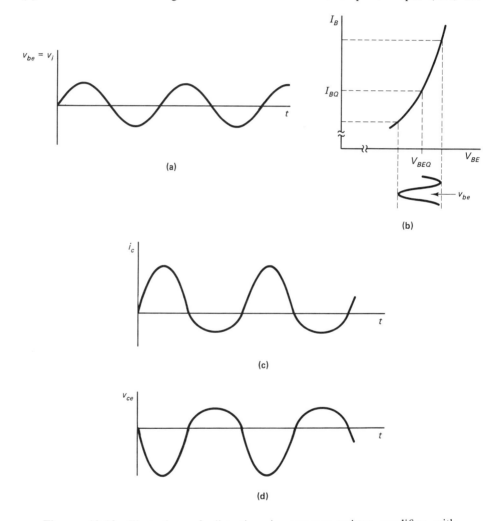

Figure 10-13 Illustration of distortion in common-emitter amplifier with voltage-driven input.

waveform is $v_{be} = v_i$). The nonlinear form of the input characteristic is shown in Figure 10-13(b), and the bias point corresponds to V_{BEQ} and I_{BQ}.

The sinusoidal voltage causes an equal swing on either side of V_{BEQ}. However, observe that the change in I_B is much greater on the positive half-cycle of v_{be} than on the negative half-cycle. If β_o is assumed to be constant over the operating range, this means that the positive peak collector signal current will be greater in magnitude than the corresponding negative peak as shown in Figure 10-13(c). The resulting collector voltage variation will be of the form shown in Figure 10-13(d). The signal is no longer a pure sinusoid, and it exhibits appreciable distortion.

An additional limitation of this circuit is the relatively wide variation in amplification that results from differences in transistor operating characteristics. Thus, while the circuit has been stabilized for dc, it is not particularly stable for ac.

What is needed is a way of stabilizing the ac characteristics in a manner similar to that of the dc circuit. A major factor in stabilizing the circuit for dc is the presence of the emitter resistance. However, in the amplifier configuration of Section 10-4, the emitter resistance was bypassed with a capacitor so that its effect on ac was eliminated.

Suppose that the emitter bypass capacitor were completely eliminated so that R_E functioned for both dc and ac. In that extreme case, the circuit would be very stable for dc and ac. Unfortunately, however, the gain would be reduced so low that the amplifier would be virtually useless. In effect, the stability imposed on the ac characteristics would be so extreme that almost no signal variation could occur.

A realistic and practically feasible compromise is to implement R_E with two separate resistances R_{E1} and R_{E2} and bypass one of the two resistances. This process is referred to as *emitter feedback*, and the resulting circuit is shown in Figure 10-14.

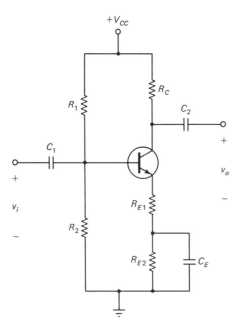

Figure 10-14 Common-emitter amplifier with emitter feedback.

For dc purposes, the circuit is still the stabilized bias circuit, and the required emitter resistance R_E for bias is simply

$$R_E = R_{E1} + R_{E2} \tag{10-52}$$

The resistance R_{E2} is bypassed with the capacitor C_E, but R_{E1} is left unbypassed. In a typical situation, R_{E1} is much smaller than R_{E2}, so that only a small fraction of the total emitter resistance is left unbypassed. As we will see shortly, however, the relatively small value of R_{E1} will have a pronounced effect on the circuit operating characteristics.

Formulation of the small-signal midfrequency model follows the same procedure previously used with R_{E2} assumed to be bypassed completely by C_E. However, a major difference is that R_{E1} appears between the emitter terminal and ground as shown in Figure 10-15. The control voltage v_{be} is no longer equal to the input voltage v_i. Indeed, the input and output circuits are coupled not only through g_m but also through R_{E1}. A KVL equation from ground to the base, through r_π, and back to ground through R_{E1} reads

$$-v_i + v_{be} + R_{E1}i_e = 0 \tag{10-53}$$

or

$$v_{be} = v_i - R_{E1}i_e \tag{10-54}$$

A brief explanation of the process will now be made with reference to the model and (10-54). As v_i increases, v_{be} starts to increase, and the collector signal current $g_m v_{be}$ also starts to increase. This causes the emitter signal current i_e to increase, but this change tends to decrease v_{be} as noted by (10-54). The result is that the net increase in v_{be} is considerably less than when the signal voltage is applied directly between base and emitter.

Gain with Feedback

An expression for the gain A of the amplifier with emitter feedback will now be

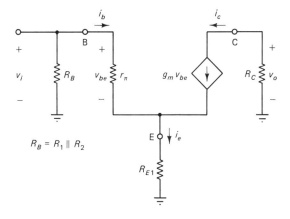

$R_B = R_1 \parallel R_2$

Figure 10-15 Small-signal midfrequency model of common-emitter amplifier with emitter feedback.

derived. Refer to the small-signal equivalent circuit of Figure 10-15. The collector signal current i_c is

$$i_c = g_m v_{be} \qquad \text{(10-55)}$$

Since $v_{be} \neq v_i$, we must next substitute an expression for v_{be} in terms of v_i. Such an expression has already been determined and was given in (10-54). In practice, the signal emitter current i_e is very nearly equal to the signal collector current i_c (i.e., $i_e \simeq i_c$). Thus (10-54) can be approximated as

$$v_{be} \simeq v_i - R_{E1} i_c \qquad \text{(10-56)}$$

Substitution of (10-56) in (10-55) results in

$$i_c = g_m(v_i - R_{E1} i_c) = g_m v_i - g_m R_{E1} i_c \qquad \text{(10-57)}$$

Note in (10-57) that i_c appears both on the left and the right sides of the equation. Regrouping and solving for i_c, we obtain

$$i_c = \frac{g_m v_i}{1 + g_m R_{E1}} \qquad \text{(10-58)}$$

The output voltage v_o is then

$$v_o = -R_C i_C = \frac{-g_m R_C v_i}{1 + g_m R_{E1}} \qquad \text{(10-59)}$$

Finally, the gain A is

$$\boxed{A = \frac{v_o}{v_i} = \frac{-g_m R_C}{1 + g_m R_{E1}}} \qquad \text{(10-60)}$$

Comments on this result will be made after other parameters are derived.

Input and Output Resistances

The input resistance will be derived in a two-step process. First, the resistance R'_{in} will be determined, in which case the effect of R_B will be ignored. The net resistance R_{in} will than be determined by combining R'_{in} with R_B.

For the circuit as given, the input current i_b can be expressed as

$$i_b = \frac{v_i - R_{E1} i_e}{r_\pi} \qquad \text{(10-61)}$$

We next assume that $i_e \approx i_c$ and substitute the expression for i_c from (10-58) in (10-61). After a slight rearrangement, we have

$$i_b = \frac{v_i}{r_\pi}\left(1 - \frac{R_{E1} g_m}{1 + g_m R_{E1}}\right) = \frac{v_i}{r_\pi} \frac{1}{(1 + g_m R_{E1})} \qquad \text{(10-62)}$$

The input resistance R'_{in} is than determined as

$$R'_{in} = \frac{v_i}{i_b} = r_\pi(1 + g_m R_{E1}) \qquad \textbf{(10-63)}$$

The net input resistance R_{in} is simply the parallel combination of R'_{in} and R_B; that is,

$$R_{in} = R_B \parallel R'_{in} = R_B \parallel r_\pi(1 + g_m R_{E1}) \qquad \textbf{(10-64)}$$

It can be shown that for the idealized model assumed, the output resistances R'_{out} and R_{out} are the same as for the amplifier circuit with emitter completely bypassed; that is,

$$R'_{out} = \infty \qquad \textbf{(10-65)}$$

and

$$R_{out} = R_C \qquad \textbf{(10-66)}$$

Comparison

A comparison between the common-emitter amplifier with the emitter resistance completely bypassed and with emitter feedback will now be made. To aid in this comparison, a summary of the key formulas for the two circuits is provided in Table 10-1.

From the first line of the table, it is observed that the magnitude of the gain is reduced by the factor $1/(1 + g_m R_{E1})$. This fact alone might be interpreted as a detrimental effect. However, as will be seen shortly, this is a small price to pay for the benefits that will result. A significant gain can still be achieved by a careful choice of R_{E1}. Note that the expression for gain with feedback reduces to the expression for gain without feedback when $R_{E1} = 0$.

TABLE 10-1 COMPARISON OF COMMON-EMITTER PROPERTIES WITH EMITTER RESISTANCE BYPASSED AND WITH EMITTER FEEDBACK

	Emitter resistance bypassed	Emitter feedback
A	$-g_m R_C$	$\dfrac{-g_m R_C}{1 + g_m R_{E1}}$
R'_{in}	r_π	$r_\pi(1 + g_m R_{E1})$
R_{in}	$R_B \parallel r_\pi$	$R_B \parallel (1 + g_m R_{E1})r_\pi$
R'_{out}	∞	∞
R_{out}	R_C	R_C

From the second line of the table, note that the input resistance R'_{in} increases by a factor $(1 + g_m R_{E1})$. This is a very desirable effect and results in less loading at the input. The net resistance R_{in} for both cases is the parallel combination of R_B and R'_{in} so the increase in the net input resistance depends on the relationship between R_B and R'_{in}.

The output resistance based on the idealized model assumed is not affected by the feedback. It should be noted that if there is any shunt resistance associated with the dependent current source in the output, there will be a change in the output resistance resulting from feedback. However, that effect is not significant in most applications of this circuit and will be ignored.

Some of the major improvements resulting from feedback are not apparent from these equations. The most significant improvement is a drastic reduction in distortion. The increased input resistance at the base results in a more linear relationship between the input signal voltage and the input base current. Assuming a constant value of β_o, the resulting collector signal current is more nearly a replica of the input signal voltage.

A different way of explaining the same mechanism can be deduced by comparing the gain equations without and with feedback. We have

Without feedback:

$$A = -g_m R_C \qquad (10\text{-}67)$$

With feedback:

$$A = \frac{-g_m R_C}{1 + g_m R_{E1}} \qquad (10\text{-}68)$$

The transconductance g_m changes linearly with current (i.e., $g_m \simeq 40 I_C$). Without feedback the gain from (10-67) then varies around the operating point as the current changes, thus distorting the signal.

To observe the effect with feedback, let us momentarily assume that $g_m R_{E1} \gg 1$. With that assumption, (10-68) can be approximated as

$$A \approx \frac{-g_m R_C}{g_m R_{E1}} = \frac{-R_C}{R_{E1}} \qquad (10\text{-}69)$$

The effect of g_m has been canceled out! Said differently, the denominator of (10-68) changes nearly in the same fashion as the numerator, so the net gain is stabilized. The approximation that $g_m R_{E1} \gg 1$ is not always valid, but even when it is not, the variation in the denominator of (10-68) definitely reduces the distortion.

Incidentally, the result of (10-69) provides a quick check of the approximate gain of the circuit without knowing the exact operating point. Although it may or may not be a very close approximation, depending on the product $g_m R_{E1}$, it does at least provide a "ballpark" estimate for rough checks.

To summarize, the common-emitter circuit with emitter feedback has less distortion and higher input resistance than the corresponding circuit with all the emitter resistance bypassed. It is also more stable from an ac point of view, and the gain is generally more predictable. The "price" that is paid for these improvements is a reduction in gain. As a general trade-off, the greater the sacrifice in gain, the greater are the improvements obtained.

Example 10-8

The common-emitter amplifier of Example 10-7, which was shown in Figure 10-12(a), is modified to provide emitter feedback. The resulting circuit is shown in Figure 10-16. **(a)** Verify that the dc operating point is essentially the same as in Example 10-7. **(b)** Determine the midfrequency gain A and the input and output resistances R'_in, R'_out, R_in, and R_out.

Figure 10-16 Amplifier circuit of Example 10-8.

Solution

(a) The only change in the dc circuit is that the original emitter resistance of 4.3 kΩ has now been split into two resistances of 3.9 kΩ and 390 Ω. Actually, $3900 + 390 = 4290$ Ω, which appears to be 10 Ω short. However, both implementations have been based on assumed stock 5% resistance values, and when the uncertainty of the resistances and the approximations involved in the operating-point formulas are considered, the dc circuits are essentially equivalent. Thus we assume that $I_C = 1$ mA and $g_m = 40$ mS as in Example 10-7.

(b) In the steps that follow, the results of Table 10-1 will be freely used. The unbypassed emitter resistance is $R_{E1} = 390 \ \Omega$. In addition, we assume that $r_\pi = 6 \ \text{k}\Omega$ as in Example 10-7. The gain A is

$$A = \frac{-g_m R_C}{1 + g_m R_{E1}} = \frac{-0.04 \times 7.5 \times 10^3}{1 + 0.04 \times 390} = \frac{-300}{1 + 15.6} = -18.1 \quad \textbf{(10-70)}$$

The input resistance R'_{in} is

$$R'_{in} = r_\pi(1 + 0.04 \times 390) = 6 \times 10^3 (16.6) \simeq 99.6 \ \text{k}\Omega \quad \textbf{(10-71)}$$

The net circuit input resistance is

$$R_{in} = R_B \| R'_{in} = 9 \ \text{k}\Omega \| 99.6 \ \text{k}\Omega = 8.25 \ \text{k}\Omega \quad \textbf{(10-72)}$$

The output resistances R'_{out} and R_{out} are the same as before, that is,

$$R'_{out} = \infty \quad \textbf{(10-73)}$$

$$R_{out} = 7.5 \ \text{k}\Omega \quad \textbf{(10-74)}$$

A comparison of the amplifier properties without and with feedback are listed below.

	Without feedback	With feedback
A	-300	-18.1
R'_{in} (kΩ)	6	99.6
R_{in} (kΩ)	4.5	8.25
R'_{out} (kΩ)	∞	∞
R_{out} (kΩ)	7.5	7.5

Several comments concerning these results are in order. The transistor input resistance R'_{in} is raised by the same factor as the gain is reduced $(1 + g_m R_{E1} = 16.6)$. When this resistance is combined with R_B, the net input resistance is raised only from 4.5 kΩ to 8.25 kΩ, which seems like less of an improvement. However, this value of 8.25 kΩ is much more predictable and stable since the effect of r_π is far less significant on the total input resistance when feedback is present. As discussed in the text, the distortion with the feedback circuit will be reduced dramatically, and this effect is not apparent from the data.

As one final check, let's see how closely (10-69) comes to predicting the actual gain. The ratio $-R_C/R_{E1}$ is

$$\frac{-R_C}{R_{E1}} = \frac{-7.5 \times 10^3}{390} = -19.23 \quad \textbf{(10-75)}$$

This value differs from the actual gain by about 6%, so the utility of this result as a quick check is evident. Remember that this estimate always predicts a gain somewhat larger in magnitude than the actual gain.

10-6 COMMON-BASE AMPLIFIER

All BJT small-signal amplifier circuits considered up to this point have utilized the common-emitter configuration. However, it is also possible to implement amplifiers with either the base or the collector as the common terminal between the input and the output. Such configurations have properties that can be useful in certain applications. The *common-base* configuration is considered in this section, and the *common-collector* configuration is considered in the next section.

An ac-coupled common-base amplifier circuit is shown in Figure 10-17(a). This layout represents the usual manner in which the circuit is drawn. However, it will be useful at this point to show the circuit momentarily in a form which most easily relates to the common-emitter circuit, and this is provided in Figure 10-17(b). The dc circuit of the common-base configuration is equivalent to the

(a)

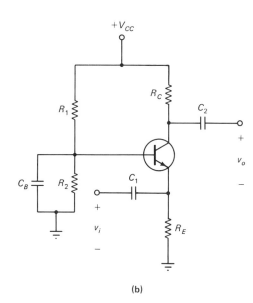

(b)

Figure 10-17 Common-base circuit drawn two ways.

common-emitter configuration. Thus the operating point is established with R_1, R_2, R_E, and R_C exactly as for a common-emitter circuit.

The common-base circuit is different from the common-emitter circuit in two ways: (1) the signal is coupled to the emitter terminal by the capacitor C_1, and (2) the base terminal is established at signal ground by means of the capacitor C_B. However, the output signal is taken at the collector terminal in the same manner as in the common-emitter circuit.

Study the two forms in Figure 10-17 so that you can recognize the common-base circuit in either form. From this point on, the form of Figure 10-17(a) will be used.

The midfrequency small-signal equivalent circuit of the common-base amplifier is shown in Figure 10-18. Note how the base terminal of the BJT is connected to ground and the control voltage v_{be} has its positive reference at ground. The input signal appears at the emitter terminal and the dependent current source appears along the top.

The voltage gain A will now be derived. From Figure 10-18, the output voltage v_o is

$$v_o = -g_m v_{be} R_C \tag{10-76}$$

However, the control voltage v_{be} is simply

$$v_{be} = -v_i \tag{10-77}$$

Substitution of (10-77) in (10-76) yields

$$v_o = -g_m(-v_i)R_C = g_m R_C v_i \tag{10-78}$$

The gain A is

$$\boxed{A = \frac{v_o}{v_i} = g_m R_C} \tag{10-79}$$

This value is exactly the same as the magnitude of A for the common-emitter circuit. However, the gain of the common base is *noninverting*.

Next, the input resistance of the common-base amplifier will be derived. To that end, i_e is first calculated in terms of v_i. Application of KCL at the node to the right of the emitter terminal yields

$$-i_e + \frac{v_i}{r_\pi} - g_m v_{be} = 0 \tag{10-80}$$

Figure 10-18 Small-signal midfrequency model of common-base circuit.

Substitution of $v_{be} = -v_i$ in (10-80) and rearrangement yields

$$v_i\left(\frac{1}{r_\pi} + g_m\right) = i_e \tag{10-81}$$

The input resistance R'_{in} can be expressed as

$$R'_{in} = \frac{v_i}{i_e} = \frac{1}{(1/r_\pi) + g_m} \tag{10-82}$$

The result of (10-82) is in the form of the reciprocal of the sum of two conductances, which is the formula for obtaining the parallel combination of two resistances. One resistance is r_π, and the other resistance is an equivalent resistance $1/g_m$. This quantity is defined in many texts as the dynamic emitter resistance r'_e, which is given by

$$\boxed{r'_e = \frac{1}{g_m}} \tag{10-83}$$

However, we will simply refer to it as $1/g_m$.

The result of the preceding discussion is that R'_{in} can be expressed as the parallel combination of two resistances r_π and $1/g_m$, that is,

$$\boxed{R'_{in} = r_\pi \parallel \frac{1}{g_m}} \tag{10-84}$$

The net circuit input resistance can be expressed as

$$\boxed{R_{in} = R_E \parallel R'_{in} = R_E \parallel r_\pi \parallel \frac{1}{g_m}} \tag{10-85}$$

In many cases, the resistances R_E and r_π are much greater than $1/g_m$. In this common situation, the net input resistance is very close to

$$R_{in} \approx \frac{1}{g_m} \tag{10-86}$$

This value is usually quite small compared to the input resistance of the common-emitter circuit.

The output resistances R'_{out} and R_{out} are

$$\boxed{R'_{out} = \infty} \tag{10-87}$$

and

$$\boxed{R_{out} = R_C} \tag{10-88}$$

The gain magnitude and the output resistance of the common-base circuit are the same as for the common-emitter circuit, but the input resistance is much lower.

Because of this last property, the net voltage gain from source to output may be considerably lower than for the common-emitter circuit when the source resistance is appreciable, due to the input loading factor. Thus the low input resistance may be a serious limitation in many situations. For this reason, the common-base circuit finds only limited usage in low- and midfrequency applications.

The common-base circuit does offer some attributes at high frequencies that are not apparent from the low-frequency small-signal model. It turns out that there is an internal capacitance between the base and the collector of a transistor that tends to limit the upper frequency range in the common-emitter configuration. When the base is grounded for signals, this capacitance is isolated from the input, and the effect is reduced. By proper circuit design, it is possible to achieve stable amplifier gains at high frequencies in the common-base configuration.

Example 10-9

Consider the RC-coupled common-base circuit shown in Figure 10-19. **(a)** Verify that the dc bias circuit is exactly the same as that of the common-emitter circuit of Example 10-7, which was shown in Figure 10-12(a). **(b)** Determine the midfrequency gain A and the input and output resistances R'_{in}, R'_{out}, R_{in}, and R_{out}. Assume that the transistor has the same characteristics as in Example 10-7.

Figure 10-19   Common-base circuit of Example 10-9.

Solution

(a) By either rearranging the dc circuit form or by comparing the two circuit diagrams of Figure 10-17, it is readily verified that the dc circuit is the same as in Figure 10-12(a). The dc collector current and the various dc voltages then are the same as in Example 10-7. The reader may wish to review the calculations from (10-40) through (10-46).

(b) Assuming that $g_m = 40$ mS, the gain A is

$$A = \frac{v_o}{v_i} = g_m R_C = 0.04 \times 7.5 \times 10^3 = 300 \qquad \textbf{(10-89)}$$

As in the preceding two examples, we will assume that $r_\pi = 6\text{ k}\Omega$. The dynamic emitter resistance is $r'_e = 1/g_m = 1/0.04 = 25\ \Omega$. The transistor input resistance R'_{in} is

$$R'_{in} = r_\pi \parallel \frac{1}{g_m} = 6000 \parallel 25 = 24.9\ \Omega \qquad \textbf{(10-90)}$$

The circuit input resistance R_{in} is

$$R_{in} = R_E \parallel R'_{in} = 4300 \parallel 24.9 \simeq 24.8\ \Omega \qquad \textbf{(10-91)}$$

Note how close this result is to $1/g_m = 25\ \Omega$, as given by the approximation of (10-86).

The output resistances are

$$R'_{out} = \infty \qquad \textbf{(10-92)}$$

and

$$R_{out} = 7.5\text{ k}\Omega \qquad \textbf{(10-93)}$$

10-7 COMMON-COLLECTOR AMPLIFIER

The third basic BJT configuration is the *common-collector amplifier*, which is also referred to as the *emitter-follower* circuit. The reader will recognize that we have already considered a dc-coupled version of this circuit in Chapter 5 as an isolation circuit. The circuit will now be considered in ac-coupled form and from the small-signal point of view.

The basic form of an ac-coupled common-collector circuit is shown in Figure 10-20. The dc biasing circuit is the same as for the common-emitter and common-base circuits except for one difference. There is no collector resistance in the common-collector circuit, and this establishes the collector at ground potential as far as ac signals are concerned (but obviously not for dc). The input signal is

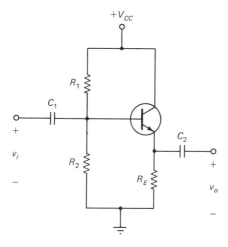

Figure 10-20 Common-collector (emitter-follower) ac-coupled amplifier circuit.

applied at the base terminal through a coupling capacitor, and the output signal is taken through a capacitor at the emitter terminal.

Determination of the operating point proceeds as follows: The voltage V_{TH} is

$$V_{TH} = \frac{R_2 V_{CC}}{R_1 + R_2} \qquad (10\text{-}94)$$

For a stable bias design, the collector current is approximated as

$$I_C = \frac{V_{TH} - V_{BE}}{R_E} = \frac{V_{TH} - 0.7}{R_E} \qquad (10\text{-}95)$$

The dc voltage V_{R_E} across the emitter resistance is approximately

$$V_{R_E} = R_E I_C \qquad (10\text{-}96)$$

The dc collector–emitter voltage V_{CE} is

$$V_{CE} = V_{CC} - R_E I_C \qquad (10\text{-}97)$$

The major difference between the dc circuits of the common-collector and the common-emitter circuits is that the supply voltage V_{CC} is only divided into two parts for the common-collector circuit. Furthermore, since the output voltage is taken across the emitter, a logical division for many applications is about half of the available supply voltage across each of the two parts.

The midfrequency small-signal equivalent circuit of the common-collector circuit is first drawn as in Figure 10-21(a). Note how the collector terminal is

(a)

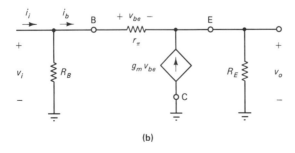

(b)

Figure 10-21 Small-signal midfrequency model of common-collector circuit as used to derive gain and input resistance.

grounded. It is best to redraw the circuit as in Figure 10-21(b) for the analysis that follows.

The gain A will now be derived. A KVL loop around the complete circuit yields

$$-v_i + v_{be} + v_o = 0 \tag{10-98}$$

or

$$v_{be} = v_i - v_o \tag{10-99}$$

A KCL equation at the right-hand upper node reads

$$\frac{v_o}{R_E} + \frac{v_o - v_i}{r_\pi} - g_m v_{be} = 0 \tag{10-100}$$

When the expression of v_{be} from (10-99) is substituted in (10-100), the resulting equation now contains only the variables v_i and v_o. After several steps of manipulation, the following result is obtained:

$$A = \frac{v_o}{v_i} = \frac{R_E/r_\pi + g_m R_E}{1 + R_E/r_\pi + g_m R_E} = \frac{R_E(1/r_\pi + g_m)}{1 + R_E(1/r_\pi + g_m)} \tag{10-101}$$

For virtually any transistor, it turns out that $g_m \gg 1/r_\pi$. A very reasonable approximation for (10-101) is then

$$\boxed{A = \frac{v_o}{v_i} = \frac{g_m R_E}{1 + g_m R_E}} \tag{10-102}$$

This expression will be retained as the basic gain equation for the common-collector configuration. However, it should be noted that in many applications, $g_m R_E \gg 1$, in which case A can be further approximated as

$$\boxed{A \simeq 1} \tag{10-103}$$

When this approximation is not used, the calculated gain turns out to be slightly less than unity.

Inasmuch as the voltage gain is slightly less than unity, one may logically question whether the common-collector circuit is really an "amplifier" or not. However, the virtue of the common-collector circuit is its isolation property. This will be evident from the results of input and output resistances.

The input resistance R'_{in} will now be derived. Referring to Figure 10-21(b), the base current i_b is

$$i_b = \frac{v_i - v_o}{r_\pi} \tag{10-104}$$

From the gain equation of (10-102), v_o may be expressed in terms of v_i, and when this expression is substituted in (10-104), some simplification and rearrangement yield

$$i_b = \frac{1}{r_\pi} \frac{1}{(1 + g_m R_E)} v_i \tag{10-105}$$

The transistor input resistance is

$$R'_{in} = \frac{v_i}{i_b} = r_\pi(1 + g_m R_E) \qquad \textbf{(10-106)}$$

The net circuit input resistance is

$$R_{in} = R_B \parallel R'_{in} = R_B \parallel r_\pi(1 + g_m R_E) \qquad \textbf{(10-107)}$$

Normally, the output resistance is determined with the input signal source replaced by a short circuit. Since we are assuming an ideal voltage source at the input at this time, the short circuit in the model would be in parallel with R_B and would, in effect, eliminate any effects of R_B. The result could be misleading since many common-collector circuits are used in conjunction with high-resistance sources. To circumvent this difficulty and to permit the later inclusion of a source resistance, *the output resistance of the emitter follower will be determined with the source replaced by an open circuit*.

The circuit used to derive the output resistance is shown in Figure 10-22. As indicated, the input source is replaced by an open circuit, and this leaves R_B across the input. Although the circuit will not function without R_E, the small-signal model permits us to determine the resistance to the left of R_E, and this result can then be combined in parallel with R_E.

A fictitious source v_x is applied across the output, and the current i_x is to be determined. First, the current i_a is determined as

$$i_a = \frac{v_x}{r_\pi + R_B} \qquad \textbf{(10-108)}$$

The control voltage v_{be} is

$$v_{be} = -r_\pi i_a \qquad \textbf{(10-109)}$$

The current i_x is

$$i_x = i_a - g_m v_{be} \qquad \textbf{(10-110)}$$

Substitution of (10-109) in (10-110) yields

$$i_x = i_a + g_m r_\pi i_a = i_a(1 + g_m r_\pi) \qquad \textbf{(10-111)}$$

Next, (10-111) can be used to express i_a in terms of i_x, and this can be substituted in (10-108) to yield

$$\frac{i_x}{1 + g_m r_\pi} = \frac{v_x}{r_\pi + R_B} \qquad \textbf{(10-112)}$$

Figure 10-22   Midfrequency model used to derive R'_{out} for common-collector circuit.

The output resistance R'_{out} is then determined as

$$R'_{out} = \frac{v_x}{i_x} = \frac{r_\pi + R_B}{1 + g_m r_\pi} \qquad (10\text{-}113)$$

The quantity $g_m r_\pi = \beta_o$ normally satisfies the inequality $\beta_o \gg 1$. With a slight approximation R'_{out} can be expressed as

$$\boxed{R'_{out} = \frac{1}{g_m} + \frac{R_B}{\beta_o}} \qquad (10\text{-}114)$$

If this signal source had been replaced by a short circuit, R'_{out} in (10-114) would simply be $1/g_m$, and this is the approximate value when the circuit is driven by an ideal voltage source.

The composite output resistance R_{out} is given by

$$\boxed{R_{out} = R_E \,\|\, R'_{out}} \qquad (10\text{-}115)$$

Example 10-10

Consider the common-collector circuit shown in Figure 10-23. **(a)** Determine the operating-point dc collector current and circuit voltages. **(b)** Determine the gain A and input and output resistances. It is known that $\beta_o = 240$.

Figure 10-23   Common-collector circuit of Example 10-10.

Solution

(a) The Thévenin voltage V_{TH} is

$$V_{TH} = \frac{47 \text{ k}\Omega}{47 \text{ k}\Omega + 39 \text{ k}\Omega} \times 15 \text{ V} = 8.20 \text{ V} \qquad (10\text{-}116)$$

The collector current is

$$I_C = \frac{V_{TH} - V_{BE}}{R_E} = \frac{8.20 - 0.7}{7.5 \times 10^3} = 1 \text{ mA} \qquad \textbf{(10-117)}$$

The emitter dc voltage is

$$V_E = V_{R_E} = V_{TH} - 0.7 = 7.5 \text{ V} \qquad \textbf{(10-118)}$$

The collector–emitter voltage is

$$V_{CE} = 15 - 7.5 = 7.5 \text{ V} \qquad \textbf{(10-119)}$$

To the degree of accuracy of the approximations involved, the dc supply voltage is evenly distributed between the transistor and the emitter resistance.

(b) Since $I_C = 1$ mA, $g_m = 40 \times 10^{-3}$ S.
The value of the gain A is

$$A = \frac{g_m R_E}{1 + g_m R_E} = \frac{40 \times 10^{-3} \times 7.5 \times 10^3}{1 + 40 \times 10^{-3} \times 7.5 \times 10^3} = 0.997 \qquad \textbf{(10-120)}$$

This result is very close to unity.
The value of r_π is

$$r_\pi = \frac{\beta_o}{g_m} = \frac{240}{0.04} = 6 \text{ k}\Omega \qquad \textbf{(10-121)}$$

The input resistance R'_{in} is

$$R'_{in} = r_\pi(1 + g_m R_E) = 6 \times 10^3(301) = 1.81 \times 10^6 \ \Omega \qquad \textbf{(10-122)}$$

This large input resistance is one of the desirable properties of the common-collector circuit.
The output resistance R'_{out} with the *input open* is

$$R'_{out} = \frac{1}{g_m} + \frac{R_B}{\beta_o} = \frac{1}{0.04} + \frac{39 \times 10^3 \| 47 \times 10^3}{240} = 25 + \frac{21,314}{240} \qquad \textbf{(10-123)}$$

$$= 25 + 88.81 = 113.8 \ \Omega$$

This output resistance is quite small compared to the typical values obtained with the other configurations. Note again that the output resistance has been determined with the input terminals *open* rather than *shorted* as was the case for the other configurations. As we will see in Chapter 12, it will be much easier to incorporate the effect of a source resistance into the expression for output resistance from our assumed open-circuit condition.
The circuit input resistance is

$$R_{in} = R_B \| R'_{in} = R_1 \| R_2 \| R'_{in} = 39 \text{ k}\Omega \| 47 \text{ k}\Omega \| 1.81 \text{ M}\Omega = 21 \text{ k}\Omega$$
$$\textbf{(10-124)}$$

The circuit output resistance is

$$R_{out} = R_E \| R'_{out} = 7.5 \text{ k}\Omega \| 113.8 \ \Omega = 112.1 \ \Omega \qquad \textbf{(10-125)}$$

10-8 BJT AC SPECIFICATIONS

Now that we have analyzed a number of basic amplifier configurations, it is appropriate to inspect some of the specifications pertinent to the analysis and design of such amplifiers. Refer again to Appendix D for the 2N3904 specifications sheets. We now observe certain of the ac or small-signal data values.

Some of the notation used in specifying small-signal data utilize the so-called *hybrid parameters*. The hybrid parameters were used extensively in the early developments of transistor theory and modeling. As the utilization of complete integrated circuits has increased, other methods have become more useful to the average technician and technologist. Consequently, hybrid parameters are included only as a supplement (Appendix C). However, since the data provided on specifications sheets still utilize the hybrid terminology, we must learn to recognize the symbols and convert the appropriate data to the symbols used in the text.

Refer to the section entitled "Dynamic Characteristics" on the second page of the 2N3904 specifications. Some of the data contained here will be considered in Chapter 18, when the complete hybrid-pi equivalent circuit is developed.

At this time, refer to the portion of the section entitled "Hybrid Parameters." The data provided here are based on $I_E = 1$ mA, $V_{CE} = 10$ V, and $f = 1$ kHz. This would correspond to midband operation in most cases.

The value of β_o is listed in the table as h_{fe}. Recall that β_{dc} was listed as h_{FE}, so there is some consistency between the corresponding symbols.

The range of h_{fe} is shown to be from 100 to 400 for the 2N3904. Thus, out of a large stock of these transistors, the value of $h_{fe}(\beta_o)$ is expected to be within this wide range. This compares with a range of 100 to 300 for $h_{FE}(\beta_{dc})$ as noted on the first page of data. Thus the range of the two quantities is similar, but the maximum value of h_{fe} can be greater.

The value of r_π is listed in the table as h_{ie}. The range of h_{ie} is shown to be from 1 to 10 kΩ for a 2N3904.

The quantities h_{re} and h_{oe} are neglected in our small-signal model. Their significance is discussed in Appendix C dealing with hybrid parameters. Considering the range of approximations involved with transistors, their effects are relatively small in most cases and are considered only in the most exacting analysis.

DRILL PROBLEMS

10-1. A simple BJT amplifier of the form shown in Figure 10-1 has waveforms as shown in Figures 10-3 and 10-4. The peak-to-peak values of input and output voltages are measured as $V_{ipp} = 40$ mV and $V_{opp} = 2$ V. Determine the voltage gain A.

10-2. A simple BJT amplifier of the form shown in Figure 10-1 has waveforms as shown in Figures 10-3 and 10-4. The rms values of input and output voltages are measured as $V_{i(rms)} = 10$ mV and $V_{o(rms)} = 1.5$ V. Determine the voltage gain A.

10-3. Input voltage–current data for a certain *npn* transistor are as follows:

V_{BE} (V)	I_B (μA)
0.68	120
0.70	200

Determine the approximate value of r_π.

10-4. Input voltage–current data for a certain *pnp* transistor are as follows:

V_{EB} (V)	I_B (μA)
0.64	90
0.65	110

Determine the approximate value of r_π.

10-5. Some dc measurements on a BJT with a fixed collector–emitter voltage yield the following data:

I_B (μA)	I_C (mA)
70	6.1
100	7.9

Determine the approximate value of β_o.

10-6. Some dc measurements on a BJT with fixed collector–emitter voltage yield the following data:

I_B (μA)	I_C (mA)
40	4.6
50	8.2

Determine the approximate value of β_o.

10-7. A certain *npn* BJT is biased at $I_C = 8$ mA, and it is known that $\beta_o = 150$. Determine **(a)** g_m and **(b)** r_π.

10-8. A certain *pnp* BJT is biased at $I_C = 5$ mA, and it is known that $\beta_o = 200$. Determine **(a)** g_m and **(b)** r_π.

10-9. For the transistor of Problem 10-7 at the given operating point, construct the β-model and the g_m-model for small signals.

10-10. For the transistor of Problem 10-8 at the given operating point, construct the β-model and the g_m-model for small signals.

10-11. The schematic diagram of a certain RC-coupled common-emitter amplifier is shown in Figure P10-11. **(a)** Determine the dc operating point current I_C and the three voltages V_{R_E}, V_{R_C}, and V_{CE}. **(b)** Construct the midfrequency small-signal equivalent circuit. It is known that $\beta_o = 200$. **(c)** Determine the gain A and the input and output resistances R'_{in}, R'_{out}, R_{in}, and R_{out}.

+30 V

5.6 kΩ

24 kΩ

C_2

C_1

+

v_o

−

v_i

6.8 kΩ

3 kΩ

C_E

Figure P10-11

10-12. The schematic diagram of a certain RC-coupled common-emitter amplifier employing a *pnp* transistor and a positive bias supply is shown in Figure P10-12. **(a)** Determine the dc operating-point current I_C, and the three voltages V_{R_E}, V_{R_C}, and V_{EC}. (Assume reference directions for first two voltages that will yield positive values.) **(b)** Construct the midfrequency small-signal equivalent circuit. It is known that $r_\pi = 4$ kΩ. **(c)** Determine the gain A and the input and output resistances R'_{in}, R'_{out}, R_{in}, and R_{out}.

+15 V

3.9 kΩ

C_E = 33 μF

5.6 kΩ

C_1

+

v_i

10 kΩ

C_2

+

4.7 kΩ

v_o

−

Figure P10-12

10-13. The common-emitter amplifier of Problem 10-11 is modified to provide emitter feedback, and the resulting circuit is shown in Figure P10-13. **(a)** Verify that the dc operating point is the same as in Problem 10-11. **(b)** Determine the midfrequency gain A and the input and output resistances R'_{in}, R'_{out}, R_{in}, and R_{out}.

+30 V

5.6 kΩ

24 kΩ

C_2

C_1

+

v_o

+

v_i

300 Ω

−

6.8 kΩ

2.7 kΩ

C_E

−

Figure P10-13

10-14. The common-emitter amplifier of Problem 10-12 is modified to provide emitter feedback, and the resulting circuit is shown in Figure P10-14. **(a)** Verify that the dc operating point is the same as in Problem 10-12. **(b)** Determine the midfrequency gain A and the input and output resistances R'_{in}, R'_{out}, R_{in}, and R_{out}.

+15 V

3.6 kΩ

C_E

5.6 kΩ

300 Ω

C_1

+

C_2

v_i

10 kΩ

+

v_o

−

4.7 kΩ

Figure P10-14

−

10-15. Consider the *RC*-coupled common-base circuit shown in Figure P10-15. **(a)** Verify that the dc bias circuit is exactly the same as that of the common-emitter circuit of Problem 10-11. **(b)** Determine the midfrequency gain A and the input and output resistances R'_{in}, R'_{out}, R_{in}, and R_{out}.

Figure P10-15

10-16. Consider the *RC*-coupled common-base circuit shown in Figure P10-16. **(a)** Verify that the dc bias circuit is the same as that of the common-emitter circuit of Problem 10-12. **(b)** Determine the midfrequency gain A and the input and output resistances R'_{in}, R'_{out}, R_{in}, and R_{out}.

Figure P10-16

10-17. Consider the common-collector circuit shown in Figure P10-17. **(a)** Determine the dc operating-point collector current I_C and the voltages V_{R_E} and V_{CE}. **(b)** Determine the gain A and the input and output resistances R'_{in}, R'_{out}, R_{in}, and R_{out}.

Figure P10-17

447

10-18. Consider the common-collector circuit shown in Figure P10-18. **(a)** Determine the dc operating-point collector current I_C and the voltages V_{R_E} and V_{EC}. **(b)** Determine the gain A and the input and output resistances R'_{in}, R'_{out}, R_{in}, and R_{out}.

Figure P10-18

DERIVATION PROBLEMS

10-19. Consider the common-emitter circuit of Figure 10-10(a). Assume that the dc operating-point voltages are established as follows: $V_{CE} = V_{R_C} = V_{R_E}$. Show that the gain A is given by

$$A = -13.33V_{CC}$$

10-20. Consider the common-emitter circuit of Figure 10-10(a). Assume that the dc operating-point voltages are established as follows: $V_{CEQ} = V_{R_C} = 2V_{R_E}$. Show that the gain A is given by

$$A = -26.67V_{CC}$$

DESIGN PROBLEMS

10-21. One possible approach to designing a common-emitter amplifier with feedback to achieve a specified gain is as follows: First, design a stable dc biasing circuit based on a desired I_C using the techniques of Chapter 5. One of the parameters determined from this process is R_E. Next, use (10-60) to determine the value of R_{E1} required to achieve the gain A. R_E can then be divided into two parts: R_{E1} and $R_{E2} = R_E - R_{E1}$. (In some cases, R_{E1} may be so small that the approximation $R_{E2} \approx R_E$ may be valid.) Assume that the following design constraints are given:

$$V_{CC} = 30 \text{ V}$$

$$V_{CE} = V_{R_C} = V_{R_E} = 10 \text{ V}$$

$$I_C = 1 \text{ mA}$$

$$A = -12$$

Assume $\beta_{dc}(\text{min}) = 50$ and $K = 20$. Perform the design by specifying all resistance values.

10-22. Repeat the design of Problem 10-21 if the dc voltage distribution and current are changed to the following:

$$V_{CE} = V_{R_C} = 12 \text{ V}$$

$$V_{R_E} = 6 \text{ V}$$

$$I_C = 2 \text{ mA}$$

TROUBLESHOOTING PROBLEMS

10-23. You are troubleshooting a common-emitter amplifier of the type shown in Figure P10-11 with an applied signal source. All dc levels seem satisfactory. The ac level at the base seems satisfactory, but the ac level at the collector is much too small. Which one of the following would likely be the trouble? **(a)** Collector resistance too small; **(b)** emitter resistance too large; **(c)** shorted emitter bypass capacitor; **(d)** open emitter bypass capacitor; **(e)** defective transistor.

10-24. You are troubleshooting a common-emitter amplifier of the type shown in Figure P10-11 with a applied signal source. The source has a dc path to ground through its output. The signal appears at the base but there is no output signal. The dc voltages measured at the base and collector terminals are 0 and 30 V, respectively. Which one of the following would likely be the trouble? **(a)** Shorted collector resistor; **(b)** open 6.8-kΩ resistor; **(c)** shorted input coupling capacitor; **(d)** open emitter bypass capacitor; **(e)** shorted emitter resistor.

COMPUTER PROBLEMS

10-25. Write a program to determine the midband small-signal gain parameters of a common-emitter amplifier as shown in Figure 10-10(a).

Input data: R_1, R_2, R_C, R_E, V_{CC}, β_o
Output data: A, R'_{in}, R'_{out}, R_{in}, R_{out}

10-26. Write a program to determine the midband small-signal gain parameters of a common-emitter amplifier with emitter feedback as shown in Figure 10-14(a).

Input data: R_1, R_2, R_C, R_E, V_{CC}, β_o
Output data: A, R'_{in}, R'_{out}, R_{in}, R_{out}

10-27. Write a program to determine the midband small-signal gain parameters of a common-base amplifier as shown in Figure 10-17(a).

Input data: R_1, R_2, R_C, R_E, V_{CC}, β_o
Output data: A, R'_{in}, R'_{out}, R_{in}, R_{out}

10-28. Write a program to determine the midband small-signal gain parameters of a common-collector amplifier as shown in Figure 10-20.

Input data: R_1, R_2, R_E, V_{CC}, β_o
Output data: A, R'_{in}, R'_{out}, R_{in}, R_{out}

CHAPTER 11

FET Small-Signal Amplifier Circuits

OBJECTIVES

After completing this chapter, the reader should be able to:

1. Explain the process by which amplification occurs in an FET and compare it with the process for a BJT.
2. Define and be able to determine transconductance for both depletion-mode and enhancement-mode FETs.
3. Construct for an FET the small-signal flat-band circuit model.
4. Draw the schematic diagram for and explain the operation of a complete common-source amplifier with source resistance bypassed.
5. Analyze the circuit of objective 4 for gain, input resistance, and output resistance.
6. Draw the schematic diagram for and explain the operation of a common-source amplifier with source feedback.
7. Analyze the circuit of objective 6 for gain, input resistance, and output resistance.
8. Draw the schematic diagram for and explain the operation of a common-gate amplifier.
9. Analyze the circuit of objective 8 for gain, input resistance, and output resistance.
10. Draw the schematic diagram for and explain the operation of a common-drain amplifier.
11. Analyze the circuit of objective 10 for gain, input resistance, and output resistance.

11-1 OVERVIEW

This chapter has essentially the same objectives for FET amplifiers that Chapter 10 had for BJT amplifiers. A detailed treatment of FET small-signal or ac parameters is considered, and a small-signal model is developed.

The three basic amplifier configurations utilizing an FET are denoted as the

451

common-source amplifier, the *common-gate amplifier*, and the *common-drain amplifier*, respectively. The latter circuit is also known as the *source-follower circuit*, and a dc version of this circuit is considered in Chapter 7. All of these basic amplifier circuits will be considered in some detail. In addition, an improved version of the common-source amplifier utilizing source feedback will be considered.

Each of the basic FET configurations has very close similarities to one of the basic BJT configurations. Because of these close similarities, the results of the various detailed derivations developed in Chapter 10 for BJT circuits can be adapted to the corresponding FET circuits. Except for a few of the more basic and simpler cases, which will be developed fully in this chapter, a common approach of this chapter is to adapt the results of Chapter 10 to FET circuits.

11-2 SMALL-SIGNAL FET MODELS

In developing the basic form of a BJT amplifier in Chapter 10, a load-line approach was used to convey a graphical description of the process. A similar approach could be used for an FET amplifier, but the basic concept is virtually the same. The major difference in such an approach for an FET would be that the variations in input signal would be directly related to changes in gate–source voltage rather than to changes in base current, which was the case for the BJT. We sidestep that approach for the FET and go directly to the small-signal models.

Consider the *n*-channel JFET shown in Figure 11-1. No external connections are shown here since the intent at this point is to focus on the transistor alone. In accordance with earlier notation, the instantaneous gate–source voltage, drain current, and gate–source voltage are identified.

The gate–source voltage can be expressed as

$$v_{GS} = V_{GS} + v_{gs} \tag{11-1}$$

where V_{GS} is the dc component and v_{gs} is the ac component. The drain current is expressed as

$$i_D = I_D + i_d \tag{11-2}$$

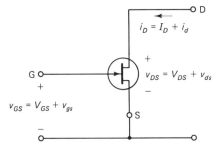

Figure 11-1 Junction field-effect transistor and total instantaneous variables used in establishing circuit models.

where I_D is the dc component and i_d is the ac component. Finally, v_{DS} is the total drain–source voltage, and it is

$$v_{DS} = V_{DS} + v_{ds} \qquad \textbf{(11-3)}$$

where V_{DS} is the dc component and v_{ds} is the ac component.

One noticeable difference between the FET and the BJT is that no input current is assumed for the FET. While the base current in a BJT is both sizable and significant to the operation of the device, the gate current is, for most applications, insignificant in an FET, and it serves no major function in the operation. At this point it will be assumed to be zero.

Transconductance

Refer to Figure 11-2, in which a portion of two curves of the drain characteristics of an FET are shown. Assume that the curve $V_{GS} = V_{GSQ}$ corresponds to the dc operating-point gate–source voltage, and the corresponding drain current is $I_D = I_{DQ}$. Assume that the gate–source voltage increases by a positive value ΔV_{GS}, which shifts operation to the curve corresponding to $V_{GS} = V_{GSQ} + \Delta V_{GS}$. The corresponding change in the drain current is a positive shift ΔI_D, so the new drain current is $I_{DQ} + \Delta I_D$.

A dynamic flat-band transconductance g_m is defined as

$$g_m = \frac{\Delta I_D}{\Delta V_{GS}} \qquad \textbf{(11-4)}$$

A constant value for V_{DS} is assumed in this definition, but since the ideal curves are assumed to show no variation with V_{DS} in the beyond-pinchoff region, this point need not concern us now.

Recalling the procedure used for a BJT, we first determined β_o and then changed the reference to g_m. In the case of the FET, however, there is no β_o parame-

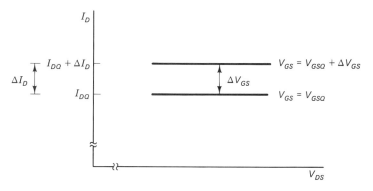

Figure 11-2 FET drain characteristics at operating point and changes.

ter since the input current is assumed to be zero. Thus the only meaningful approach is that of the g_m parameter. This leads to another argument for the use of the g_m-model for a BJT since, as we will see shortly, it results in a very close similarity between BJT and FET circuit models.

Small-Signal Circuit Model

A small-signal model for the FET will now be developed. Consider the n-channel JFET shown in Figure 11-3(a). Only signal or ac quantities are shown in this particular case, and all dc levels are understood.

The flat-band small-signal or ac model is shown in Figure 11-3(b). This form is a g_m-model employing a voltage-controlled current source.

Let us now compare the g_m-model of the BJT from Figure 10-8(c) with the corresponding model of the FET from Figure 11-3(b). Basically, the forms have the same structure. In each case, a voltage across the input terminals controls a current in the output. This similarity results in a great deal of analysis flexibility in studying complex circuit configurations.

One noticeable difference is that the BJT contains a resistance r_π across the input while the ideal FET has an open circuit across the input. The dynamic input resistance of a BJT at the base is sufficiently low that the input signal current plays a significant role in the overall analysis of the circuit. However, an FET has a very high input resistance which, for many applications, may be considered as an open circuit or infinite resistance.

Depletion-Mode FET Transconductance

The drain current I_D as a function of V_{GS} for a *depletion-mode FET* was considered in Chapter 6, and it was shown to be

$$I_D = I_{DSS}\left[1 - \frac{V_{GS}}{V_{GS(\text{off})}}\right]^2 \qquad \textbf{(11-5)}$$

Recall that the classification *depletion-mode FET* includes both *JFETs* and *depletion-mode IGFETS*. An expression for g_m may be derived from (11-5) using differential

Figure 11-3 Field-effect transistor and the flat-band small-signal g_m-model.

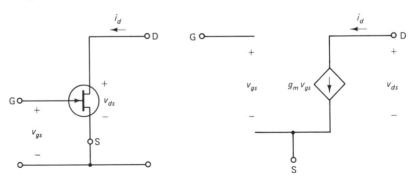

454   (a) FET with signal variables   (b) Small-signal flat-band g_m-model

calculus. The result will be stated here, and it is

$$g_m = g_{mo}\left[1 - \frac{V_{GS}}{V_{GS(off)}}\right] \tag{11-6}$$

where g_{mo} is the zero-bias value of g_m (i.e., the value of g_m at $V_{GS} = 0$). This quantity is

$$g_{mo} = \frac{-2I_{DSS}}{V_{GS(off)}} \tag{11-7}$$

The negative sign in (11-7) may be misleading since it arises in the mathematical formulation of the model. For an *n*-channel unit, $I_{DSS} > 0$ and $V_{GS(off)} < 0$. The latter negative sign cancels the negative sign in front, resulting in a positive value of g_{mo}. For a *p*-channel unit, $I_{DSS} > 0$ and $V_{GS} > 0$, which would imply a negative value of g_m. However, recall that the positive direction of I_D is *into* the drain for an *n*-channel unit and *out* of the drain for a *p*-channel. It is more convenient to establish the same current directions in the small-signal models for both *n*-channel and *p*-channel units. To accomplish this, the direction of the small-signal model for the *p*-channel unit will also be considered into the drain, in which case the sign will be changed to a positive value.

If the preceding discussion seems confusing, it will be summarized very simply as follows: The direction of the small-signal current source for *both* n-channel and *p*-channel units may be assumed from drain to source as shown by Figure 11-3(b). The value of g_m for *n*-channel units with negative gate–source voltage and *p*-channel units with positive gate–source voltage may be considered to be positive according to the following equation:

$$g_m = g_{mo}\left[1 - \frac{|V_{GS}|}{|V_{GS(off)}|}\right] \tag{11-8}$$

where

$$g_{mo} = \frac{2|I_{DSS}|}{|V_{GS(off)}|} \tag{11-9}$$

The variation of g_m with V_{GS} for a depletion-mode FET has the form shown in Figure 11-4. Observe that g_m varies from g_{mo} to zero as the gate–source bias voltage varies from zero to cutoff.

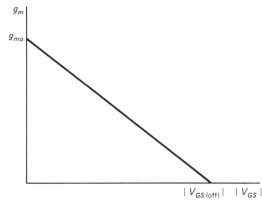

Figure 11-4 Variation of g_m with $|V_{GS}|$ for a depletion-mode FET.

Enhancement-Mode FET Transconductance

The drain current I_D as a function of V_{GS} for an n-channel enhancement-mode IGFET was considered in Chapter 6, and it was shown to be

$$I_D = K\left[\frac{V_{GS}}{V_{GS(th)}} - 1\right]^2 \qquad (11\text{-}10)$$

An expression for g_m may be derived from (11-10) using differential calculus. The result will be stated here, and it is

$$g_m = \frac{2K}{V_{GS(th)}}\left[\frac{V_{GS}}{V_{GS(th)}} - 1\right] \qquad (11\text{-}11)$$

For an n-channel unit, all parameters and variables in (11-11) are positive in the beyond-pinchoff region, so g_m is necessarily positive. For a p-channel unit, the value of g_m as given literally by (11-11) in the beyond-pinchoff region would be negative. However, by the same type of argument previously used for depletion-mode devices, the direction of the small-signal current will be made to conform to that of n-channel units, and in the process, the value of g_m will be made to be positive.

In summary, the equivalent-circuit model for enhancement-mode IGFETs is the same form as that for depletion-mode FETs, and the value of g_m for both n-channel and p-channel enhancement-mode devices can be expressed as

$$g_m = \frac{2K}{|V_{GS(th)}|}\left[\frac{|V_{GS}|}{|V_{GS(th)}|} - 1\right] \qquad (11\text{-}12)$$

The quantity K has been defined all along as being positive, so no magnitude bars are required for it. The variation of g_m with V_{GS} for an enhancement-mode IGFET has the form shown in Figure 11-5.

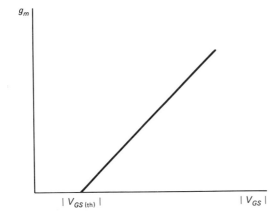

Figure 11-5 Variation of g_m with $|V_{GS}|$ for an enhancement-mode IGFET.

Example 11-1

Some dc measurements performed on an n-channel JFET with fixed drain–source voltage yield the following data:

V_{GS} (V)	I_D (mA)
-2	8
-1.9	8.6

Determine the approximate value of g_m in the region of operation.

Solution

The concept of (11-4) is used for this purpose. The change ΔV_{GS} is

$$\Delta V_{GS} = -1.9 \text{ V} - (-2 \text{ V}) = 0.1 \text{ V} \qquad \textbf{(11-13)}$$

The corresponding change ΔI_D is

$$\Delta I_D = 8.6 \text{ mA} - 8 \text{ mA} = 0.6 \text{ mA} \qquad \textbf{(11-14)}$$

The value of g_m is

$$g_m = \frac{0.6 \times 10^{-3} \text{A}}{0.1 \text{ V}} = 6 \times 10^{-3} \text{ S} = 6 \text{ mS} \qquad \textbf{(11-15)}$$

Example 11-2

A certain idealized n-channel JFET with $I_{DSS} = 16$ mA and $V_{GS(\text{off})} = -4$ V was used as the basis for several examples in Chapters 6 and 7. **(a)** Determine g_{mo}. **(b)** Determine an expression for g_m as a function of V_{GS}. **(c)** Compute g_m at the following values of V_{GS}: 0, -1 V, -2 V, -3 V, -4 V.

Solution

(a) The value of g_{mo} is determined from (11-9) as

$$g_{mo} = \frac{2|I_{DSS}|}{|V_{GS(\text{off})}|} = \frac{2 \times 16 \times 10^{-3} \text{ A}}{4 \text{ V}} = 8 \times 10^{-3} \text{ S} = 8 \text{ mS} \qquad \textbf{(11-16)}$$

(b) The value of g_m as a function of V_{GS} is determined from (11-8) as

$$g_m = 8 \times 10^{-3}\left(1 - \frac{|V_{GS}|}{4}\right) \qquad \textbf{(11-17)}$$

(c) The values of V_{GS} are substituted in (11-17), and the results are summarized as follows:

V_{GS} (V)	g_m (mS)
0	8
−1	6
−2	4
−3	2
−4	0

As expected, $g_m = g_{mo} = 8$ mS for $V_{GS} = 0$. The value of g_m then decreases linearly with increasing negative voltage until it reaches a value of zero at cutoff.

Example 11-3

For the n-channel JFET of Example 11-2, construct the small-signal g_m-model at an operating point corresponding to $V_{GS} = -2$ V.

Solution

From the results of Example 11-2, the value of g_m at $V_{GS} = -2$ V is $g_m = 4$ mS. The small-signal model has the form of Figure 11-3(b), and it is shown in Figure 11-6.

Figure 11-6 Small-signal model of Example 11-3.

11-3 COMMON-SOURCE AMPLIFIER CIRCUIT

The form of a practical common-source amplifier circuit utilizing an n-channel JFET is shown in Figure 11-7(a). This circuit utilizes a combination of fixed plus self-bias. However, the analysis to be employed here can be adapted to the simpler case of

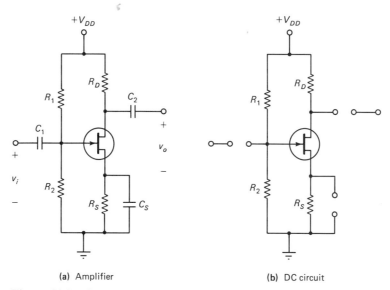

(a) Amplifier   **(b)** DC circuit

Figure 11-7 Common-source amplifier with stabilized bias and dc equivalent circuit.

self-bias just by eliminating R_1 in the circuit model. Effects of possible source resistance and external load resistance will be considered later.

The dc circuit is shown in Figure 11-7(b). As a result of the fixed- and self-bias combination, a dc operating point for the transistor is established.

The midfrequency ac equivalent circuit for small signals is shown in Figure 11-8. The flat-band small-signal model replaces the transistor between the three terminals. The capacitors C_1 and C_2 are used for input and output coupling, and they act as short circuits in the midfrequency range. The function of C_S is to bypass signals around R_S. Without C_S, the gain would be reduced considerably. In the midfrequency region C_S appears as a short circuit.

The resistances R_1 and R_D are both brought to ground for signal analysis purposes. Since R_1 and R_2 appear as a parallel combination, a resistance R_G is defined as

$$R_G = R_1 \parallel R_2 = \frac{R_1 R_2}{R_1 + R_2} \qquad \textbf{(11-18)}$$

$$R_G = R_1 \parallel R_2$$

Figure 11-8 Small-signal midfrequency model of common-source amplifier.

The voltage gain A will now be determined. The output voltage v_o is expressed in terms of the input voltage as

$$v_o = -g_m v_{gs} R_D = -g_m v_i R_D \qquad \textbf{(11-19)}$$

where $v_{gs} = v_i$ was noted. The voltage gain A is

$$\boxed{A = \frac{v_o}{v_i} = -g_m R_D} \qquad \textbf{(11-20)}$$

As in the case of the BJT, input and output resistances of the transistor and of the transistor plus external circuitry will be determined. Let R'_{in} and R'_{out} represent the input and output resistances of the transistor itself. We have

$$\boxed{R'_{in} = \infty} \qquad \textbf{(11-21)}$$

$$\boxed{R'_{out} = \infty} \qquad \textbf{(11-22)}$$

The infinite input resistance of the transistor is a result of the assumed open circuit across the input, and the infinite output resistance is a result of the assumed ideal current source model.

Let R_{in} and R_{out} represent the input and output resistances of the composite circuit. We have

$$\boxed{R_{in} = \frac{v_i}{i_i} = R_G = R_1 \parallel R_2} \qquad \textbf{(11-23)}$$

and

$$\boxed{R_{out} = R_D} \qquad \textbf{(11-24)}$$

A comparison of the common-source amplifier of this section with the common-emitter amplifier of Section 10-4 will lead to a number of close similarities. With the exception of component subscripts, the gain expressions and output resistances have exactly the same forms. The input resistances differ only with respect to the presence of r_π for the BJT.

Example 11-4

A certain ac-coupled common-source amplifier is shown in Figure 11-9(a). At the particular operating point established, $g_m = 20$ mS. **(a)** Construct the mid-frequency small-signal equivalent circuit. **(b)** Determine the voltage gain A and the input and output resistances R'_{in}, R'_{out}, R_{in}, and R_{out}.

Figure 11-9 Common-source circuit and midfrequency model of Example 11-4.

Solution

(a) Since we are given the value of g_m, it is not necessary in this case to determine the dc operating point, nor is it necessary to know R_S. The small-signal midfrequency model is shown in Figure 11-9(b). For clarity, R_1 and R_2 are shown separately.

(b) The gain A is determined as

$$A = \frac{v_o}{v_i} = -g_m R_D = -0.02 \times 3 \times 10^3 = -60 \qquad \textbf{(11-25)}$$

The ideal input and output resistances of the transistor are

$$R'_{in} = \infty \tag{11-26}$$

$$R'_{out} = \infty \tag{11-27}$$

The input and output resistances of the composite circuit are

$$R_{in} = 220 \text{ k}\Omega \parallel 470 \text{ k}\Omega = 150 \text{ k}\Omega \tag{11-28}$$

$$R_{out} = 3 \text{ k}\Omega \tag{11-29}$$

11-4 COMMON-SOURCE AMPLIFIER WITH SOURCE FEEDBACK

The common-source amplifier with source resistance completely bypassed is similar in many respects to the common-emitter amplifier with emitter resistance bypassed. Like the common-emitter amplifier, the common-source amplifier is subject to serious distortion except at the very smallest signal levels. Recall that the relationship between gate–source voltage and drain current is a second-degree function and that the transconductance varies with operating point. Thus, as the signal level varies, the gain varies, which results in a distorted output.

It turns out that the distortion may be reduced by a process very similar to that of the common-emitter amplifier. In the case of the FET, a portion of the source resistance is left unbypassed. This process is referred to as *source feedback,* and the circuit is shown in Figure 11-10.

For dc bias circuit analysis, the net source resistance R_S is

$$R_S = R_{S1} + R_{S2} \tag{11-30}$$

For ac, the resistance R_{S2} is bypassed with the capacitor C_S, but R_{S1} is much smaller

Figure 11-10 Common-source amplifier with source feedback.

FET Small-Signal Amplifier Circuits   Chap. 11

than R_{S2}, so that only a small fraction of the total source resistance is left unby-passed.

The small-signal midfrequency model of this amplifier is shown in Figure 11-11. The control voltage v_{gs} is no longer equal to the input voltage v_i. A KVL equation from ground to the gate, down to the source, and back to ground reads

$$-v_i + v_{gs} + R_{S1}i_d = 0 \qquad (11\text{-}31)$$

or

$$v_{gs} = v_i - R_{S1}i_d \qquad (11\text{-}32)$$

The process that occurs here is virtually the same as we encountered in the common-emitter with feedback. As v_i increases, v_{gs} starts to increase, and the drain signal current $g_m v_{gs}$ also starts to increase. This causes the voltage across R_{S1} to in-crease, and this change tends to decrease v_{gs} as noted by (11-32). The result is that the net increase in v_{gs} is considerably less than when the signal voltage is applied di-rectly between gate and source.

Because of the very close similarity between the g_m-models of the BJT and the FET, many of the BJT results of Chapter 10 can readily be adapted to the corre-sponding FET circuits. Along with notational changes, it will be necessary to re-place r_π in any circuit by an open circuit. This is achieved in various equations by allowing r_π to increase without limit (i.e., $r_\pi = \infty$).

An expression for the voltage gain of a BJT amplifier with emitter feedback was given by (10-60). By replacing R_C with R_D and R_{E1} with R_{S1}, the corresponding gain for the FET with source feedback can be determined. This result is

$$\boxed{A_{io} = \frac{v_o}{v_i} = \frac{-g_m R_D}{1 + g_m R_{S1}}} \qquad (11\text{-}33)$$

The input and output resistance of the transistor alone are assumed to be

$$\boxed{R'_{in} = \infty} \qquad (11\text{-}34)$$

and

$$\boxed{R'_{out} = \infty} \qquad (11\text{-}35)$$

$$R_G = R_1 \parallel R_2$$

Figure 11-11 Small-signal midfrequency model of common-source amplifier with source feedback.

The corresponding input and output resistances of the composite circuit are

$$R_{\text{in}} = \frac{v_i}{i_i} = R_G = R_1 \| R_2 \qquad (11\text{-}36)$$

and

$$R_{\text{out}} = R_D \qquad (11\text{-}37)$$

Comparison

A comparison between the common-source amplifier with the source resistance completely bypassed and the corresponding amplifier with source feedback can now be made. A summary of the key formula for the two circuits is provided in Table 11-1. It may also be helpful for the reader to refer back to Table 10-1, in which a similar comparison was made for the common-emitter amplifier.

The magnitude of the gain with source feedback is reduced by the factor $1/(1 + g_m R_{S1})$. Note that this expression reduces to the expression for gain without feedback when $R_{S1} = 0$.

The input resistance for the ideal assumptions employed is infinite in both cases. Strictly speaking, if the very large resistance between gate and source were shown, it would be multiplied by the factor $(1 + g_m R_{S1})$ in a manner analogous to the BJT. However, with the assumption of an open circuit from the outset, this factor does not appear. In practice, the external resistance R_G dominates the net circuit input resistance.

In some situations, $g_m R_{S1} \gg 1$. In this case, the gain A may be approximated as

$$A \simeq -\frac{g_m R_D}{g_m R_{E1}} = -\frac{R_D}{R_{E1}} \qquad (11\text{-}38)$$

The gain for this assumption then becomes virtually independent of transistor parameters.

TABLE 11-1 COMPARISON OF COMMON-SOURCE PROPERTIES WITH EMITTER RESISTANCE BYPASSED AND WITH EMITTER FEEDBACK

	Source resistance bypassed	Source feedback
A_{io}	$-g_m R_D$	$\dfrac{-g_m R_D}{1 + g_m R_{S1}}$
R'_{in}	∞	∞
R_{in}	R_G	R_G
R'_{out}	∞	∞
R_{out}	R_D	R_D

Example 11-5

A certain common-source amplifier with source feedback is shown in Figure 11-12. At the particular operating point established, $g_m = 20$ mS. Determine the midfrequency voltage gain A and the input and output resistances R'_{in}, R'_{out}, R_{in}, and R_{out}.

Figure 11-12 Amplifier circuit of Example 11-5.

$g_m = 20$ mS

Solution

The various results established in this section will be used in the analysis. The gain A is

$$A = \frac{v_o}{v_i} = \frac{-g_m R_D}{1 + g_m R_{S1}} = \frac{-0.02 \times 3 \times 10^3}{1 + 0.02 \times 220} = \frac{-60}{5.4} = -11.11 \qquad \textbf{(11-39)}$$

The transistor input and output resistances are

$$R'_{in} = \infty \qquad \textbf{(11-40)}$$

and

$$R'_{out} = \infty \qquad \textbf{(11-41)}$$

The net circuit input and output resistances are

$$R_{in} = 220 \text{ k}\Omega \parallel 470 \text{ k}\Omega = 150 \text{ k}\Omega \qquad \textbf{(11-42)}$$

and

$$R_{out} = 3 \text{ k}\Omega \qquad \textbf{(11-43)}$$

A check on the closeness of (11-38) will now be made. This ratio yields $-3 \times 10^3/220 = -13.64$. This result differs from the actual gain by about 23%.

11-5 COMMON-GATE AMPLIFIER

The *common-gate* amplifier will be considered in this section. The common-gate amplifier for an FET corresponds to the common-base amplifier for a BJT and has many similar properties.

An ac-coupled common-gate amplifier is shown in Figure 11-13(a). This layout represents the most common form. However, it is useful at this point to show the circuit in a form easily related to the common-source circuit, and this is shown in Figure 11-13(b). The dc circuit of the common-gate amplifier is equivalent to that of

(a)

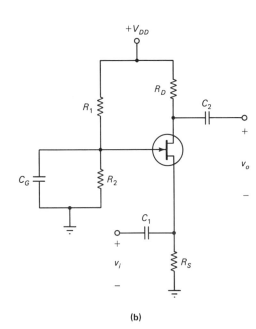

(b)

Figure 11-13 Common-gate circuit drawn two ways.

the common-source circuit. Thus the operating point is established with R_1, R_2, R_S, and R_D as for a common-source circuit.

The signal is coupled to the source terminal by the capacitor C_1, and the gate terminal is established at signal ground by means of the capacitor C_G. However, the output signal is taken at the drain terminal in the same manner as in the common-source circuit.

A small-signal midfrequency model of the common-gate circuit is shown in Figure 11-14. Note how the gate terminal of the FET is connected to ground, and the control voltage v_{gs} has its positive reference at ground. The input signal appears at the source terminal, and the dependent current source appears along the top.

By comparing Figure 11-14 for the common-gate circuit with Figure 10-18 for the common-base circuit, the similarities in the circuit models are evident. The various circuit properties for the common-gate circuit may be determined from those of the common-base circuit by appropriate notational changes.

The voltage gain is determined from (10-79) by replacing R_C with R_D. We have

$$A = \frac{v_o}{v_i} = g_m R_D \tag{11-44}$$

The transistor input resistance R'_{in} is determined from (10-82) by assuming $r_\pi = \infty$. This means that $1/r_\pi = 0$. The result is

$$R'_{in} = \frac{v_i}{i_s} = \frac{1}{g_m} \tag{11-45}$$

The corresponding output resistance R'_{out} is

$$R'_{out} = \infty \tag{11-46}$$

The composite circuit input and output resistances are now determined as

$$R_{in} = R_S \parallel \frac{1}{g_m} \tag{11-47}$$

and

$$R_{out} = R_D \tag{11-48}$$

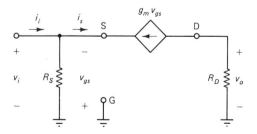

Figure 11-14 Small-signal midfrequency model of common-gate circuit.

Example 11-6

A certain common-gate amplifier is shown in Figure 11-15. At the particular operating point established, $g_m = 0.008$ S. Determine the midfrequency values of A, R'_{in}, R'_{out}, R_{in}, and R_{out}.

Figure 11-15 Common-gate circuit of Example 11-6.

Solution

The results established in this section will be applied directly in the analysis. The gain A is

$$A = \frac{v_o}{v_i} = g_m R_D = 0.008 \times 2 \times 10^3 = 16 \qquad \textbf{(11-49)}$$

The transistor input and output resistances are

$$R'_{in} = \frac{1}{g_m} = \frac{1}{0.008} = 125 \ \Omega \qquad \textbf{(11-50)}$$

and

$$R'_{out} = \infty \qquad \textbf{(11-51)}$$

The net circuit input and output resistances are

$$R_{in} = R_E \parallel \frac{1}{g_m} = 470 \parallel 125 = 98.7 \ \Omega \qquad \textbf{(11-52)}$$

and

$$R_{out} = R_D = 2 \ k\Omega \qquad \textbf{(11-53)}$$

The third basic FET configuration is the *common-drain amplifier*, which is also referred to as the *source-follower* circuit. A dc-coupled version of this circuit was considered in Chapter 7. The circuit will now be considered in ac-coupled form and with the small-signal approach.

The basic form of an ac-coupled common-drain amplifier circuit using an *n*-channel JFET is shown in Figure 11-16. There is no drain resistance in the common-drain circuit, so the drain terminal is established at ground potential for ac signals.

The midfrequency small-signal equivalent circuit is first drawn as in Figure 11-17(a). Note how the drain terminal is grounded. It is best to redraw the circuit as in Figure 11-17(b) for the discussion that follows.

A comparison of Figure 11-17(b) with Figure 10-21(b) reveals a very close similarity between the common-drain FET circuit and the common-collector circuit. Aside from terminology, the only real difference is that r_π for the BJT is replaced by an open circuit for the FET. As done earlier for several other cases, the results of the BJT analysis will be adapted to the FET case.

An expression for the voltage gain of the common-collector circuit was given in (10-102). By replacing R_E with R_S, the corresponding gain for the common-drain circuit is

$$A = \frac{v_o}{v_i} = \frac{g_m R_S}{1 + g_m R_S}$$ (11-54)

This expression will be retained as the basic gain equation for the common-drain circuit. However, in many applications, $g_m R_S \gg 1$, in which case A may be further approximated as

$$A \simeq 1$$ (11-55)

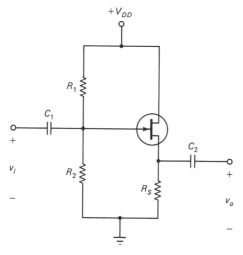

Figure 11-16 Common-drain (source-follower) ac-coupled amplifier circuit employing an *n*-channel JFET.

Figure 11-17 Small-signal midfrequency model of common-drain circuit.

The transistor input resistance is assumed to be

$$R'_{\text{in}} = \infty \tag{11-56}$$

The net circuit input resistance is

$$R_{\text{in}} = R_G = R_1 \parallel R_2 \tag{11-57}$$

The output resistance of the common-collector circuit was given by (10-113) based on replacing the source by an open circuit. To adapt this expression to the common-drain circuit, R_B is replaced by R_G and r_π is allowed to increase without limit. When the latter process is performed, $r_\pi \gg R_G$ and $g_m r_\pi \gg 1$. The numerator can then be approximated as r_π, and the denominator can be approximated as $g_m r_\pi$. The ratio can then be closely approximated as $r_\pi / g_m r_\pi = 1/g_m$. The resulting value of R'_{out} is

$$R'_{\text{out}} = \frac{1}{g_m} \tag{11-58}$$

The assumed infinite gate-source resistance thus uncouples the effect of R_G on the output resistance in the ideal case.

The net circuit output resistance is given by

$$R_{\text{out}} = R_S \parallel R'_{\text{out}} = R_S \parallel \frac{1}{g_m}$$

(11-59)

Example 11-7

A certain common-drain amplifier is shown in Figure 11-18. At the particular operating point established, $g_m = 12$ mS. Determine the midfrequency values of A, R'_{in}, R'_{out}, R_{in}, and R_{out}.

Figure 11-18 Common-drain circuit of Example 11-7.

$g_m = 12$ mS

Solution

The results established in this section will be used in the analysis. The gain A is

$$A = \frac{v_o}{v_i} = \frac{g_m R_S}{1 + g_m R_S} = \frac{0.012 \times 2 \times 10^3}{1 + 0.012 \times 2 \times 10^3} = \frac{24}{25} = 0.96$$

(11-60)

The transistor input and output resistances are

$$R'_{\text{in}} = \infty$$

(11-61)

and

$$R'_{\text{out}} = \frac{1}{g_m} = \frac{1}{0.02} = 50 \ \Omega$$

(11-62)

The net circuit input and output resistances are

$$R_{in} = 680 \text{ k}\Omega \parallel 560 \text{ k}\Omega = 307 \text{ k}\Omega \qquad \textbf{(11-63)}$$

$$R_{out} = \frac{1}{g_m} \parallel R_S = 50 \parallel 2000 = 48.8 \ \Omega \qquad \textbf{(11-64)}$$

11-7 FET AC SPECIFICATIONS

Some FET small-signal specifications will now be studied. Refer to the first page of the 2N4220 through 2N4222 data sheets in Appendix D. The first block in the "small-signal characteristic" section is entitled "forward transfer admittance common source." The symbol used there is $|y_{fs}|$, and this is equivalent to g_m. The symbol y is based on a set of parameters called the "short-circuit admittance" parameters which have been employed in some references for FET specifications. The subscript fs refers to *forward* (gain) and common *source*. The value of $|y_{fs}|$ depends on the particular transistor type, but for each one, there is more than a $2:1$ range in the possible value. The older unit of admittance "μmho" is used here.

We have assumed that the effective resistance in parallel with the output current source is infinite. When it is necessary to use an actual resistance in the model, the "output admittance common source" data may be useful. This admittance value is denoted as $|y_{os}|$ and only the maximum value is given for each transistor. The reciprocal of this quantity is the minimum value of the shunt resistance.

The three additional blocks provide information on three FET capacitances. Their significance is discussed in Chapter 18.

DRILL PROBLEMS

11-1. Some dc measurements performed on an n-channel JFET with fixed drain–source voltage yield the following data:

V_{GS} (V)	I_D (mA)
-2.5	4
-2.3	5.8

Determine the approximate value of g_m in the region of operation.

11-2. Some dc measurements performed on a p-channel JFET with fixed drain–source voltage yield the following data:

V_{GS} (V)	I_D (mA)
3	2
2.4	4.4

Determine the approximate value of g_m in the region of operation.

11-3. Some dc measurements performed on an n-channel enhancement-mode MOSFET yield the following data:

V_{GS} (V)	I_D (mA)
2.7	8
2.8	9.2

Determine the approximate value of g_m in the region of operation.

11-4. Some dc measurements on a p-channel enhancement-mode MOSFET yield the following data:

V_{GS} (V)	I_D (mA)
−3.6	11.5
−3.8	14.3

Determine the approximate value of g_m in the region of operation.

11-5. A certain idealized n-channel JFET has $I_{DSS} = 18$ mA and $V_{GS(off)} = -3$ V. **(a)** Determine g_{mo}. **(b)** Compute g_m at the following values of V_{GS}: 0, −1 V, −2 V, −3 V.

11-6. A certain idealized p-channel JFET has $I_{DSS} = 32$ mA and $V_{GS(off)} = 2$ V. **(a)** Determine g_{mo}. **(b)** Compute g_m at the following values of V_{GS}: 0, 1 V, 2 V.

11-7. For the n-channel JFET of Problem 11-5, construct the small-signal g_m-model at an operating point of $V_{GS} = -1$ V.

11-8. For the p-channel JFET of Problem 11-6, construct the small-signal g_m-model at an operating point of $V_{GS} = 1$ V.

11-9. Refer to the characteristics of the idealized n-channel enhancement-mode IGFET of Problem 6-19 (Figure P6-19). **(a)** Determine an expression for g_m. **(b)** Compute g_m at the following values of V_{GS}: 3 V, 4 V, 5 V, 6 V.

11-10. Refer to the characteristics of the idealized n-channel enhancement-mode IGFET of Problem 6-20 (Figure P6-20). **(a)** Determine an expression for g_m. **(b)** Compute g_m at the following values of V_{GS}: 2 V, 2.5 V, 3 V, 3.5 V, 4 V.

11-11. A certain ac-coupled common-source amplifier is shown in Figure P11-11. At the particular operating point established, $g_m = 12$ mS. **(a)** Construct the midfrequency small-signal equivalent circuit. **(b)** Determine the voltage gain A and the input and output resistances R'_{in}, R'_{out}, R_{in}, and R_{out}.

+V_{DD}

2 kΩ

390 kΩ

C_2

C_1

+

v_o

+

v_i

180 kΩ

R_s

C_s

−

Figure P11-11

11-12. A certain ac-coupled common-source amplifier is shown in Figure P11-12. At the particular operating point established, $g_m = 18$ mS. **(a)** Construct the midfrequency small-signal equivalent circuit. **(b)** Determine the voltage gain A and the input and output resistances R'_{in}, R'_{out}, R_{in}, and R_{out}.

−V_{DD}

1.5 kΩ

560 kΩ

C_2

C_1

+

v_o

+

v_i

300 kΩ

R_E

C_s

−

Figure P11-12

11-13. A common-source amplifier with source feedback is shown in Figure P11-13. At the operating point, $g_m = 12$ mS. Determine the midfrequency voltage gain A and the input and output resistances R'_{in}, R'_{out}, R_{in}, and R_{out}.

Figure P11-13

11-14. A common-source amplifier with source feedback is shown in Figure P11-14. At the operating point, $g_m = 18$ mS. Determine the midfrequency voltage gain A and the input and output resistances R'_{in}, R'_{out}, R_{in}, and R_{out}.

Figure P11-14

11-15. A certain common-gate amplifier is shown in Figure P11-15. At the operating point, $g_m = 15$ mS. Determine the midfrequency values of A, R'_{in}, R'_{out}, R_{in}, and R_{out}.

Figure P11-15

11-16. A certain common-gate amplifier is shown in Figure P11-16. At the operating point, $g_m = 10$ mS. Determine the midfrequency values of A, R'_{in}, R'_{out}, R_{in}, and R_{out}.

Figure P11-16

11-17. A certain common-drain amplifier is shown in Figure P11-17. At the operating point, $g_m = 12$ mS. Determine the midfrequency values of A, R'_{in}, R'_{out}, R_{in}, and R_{out}.

Figure P11-17

11-18. A certain common-drain amplifier is shown in Figure P11-18. At the operating point, $g_m = 18$ mS. Determine the midfrequency values of A, R'_{in}, R'_{out}, R_{in}, and R_{out}.

FET Small-Signal Amplifier Circuits   Chap. 11

$-V_{DD}$

560 kΩ

C_1

$+$

v_i

470 kΩ

1.5 kΩ

C_2

$+$

v_o

$-$

Figure P11-18

DERIVATION PROBLEMS

11-19. (Requires calculus.) Starting with (11-5), determine the derivative of I_D with respect to V_{GS}, and show that (11-6) is obtained using the definition of (11-7).

11-20. (Requires calculus.) Starting with (11-10), determine the derivative of I_D with respect to V_{GS}, and show that (11-11) is obtained.

DESIGN PROBLEMS

11-21. The FET in a certain common-source amplifier circuit has $g_m = 20$ mS. Determine the value of drain resistance required to achieve a gain $A = -40$.

11-22. The FET in a certain common-gate amplifier circuit has $g_m = 30$ mS. Determine the value of drain resistance required to achieve a gain $A = 15$.

11-23. The FET in a certain common-source amplifier with source feedback has $g_m = 25$ mS. For proper drain–source bias voltage, the value of drain resistance is selected as 2 kΩ. Determine the value of unbypassed source resistance required to establish a gain $A = -12$.

11-24. The output resistance of a certain common-drain amplifier is required to be no greater than 200 Ω. Determine the approximate value of g_m required in the FET.

TROUBLESHOOTING PROBLEMS

11-25. You are troubleshooting a common-source amplifier of the type shown in Figure P11-11 using an oscilloscope and with an applied signal. With the input signal within the specified voltage level, the output signal is very distorted on the negative peak. Which one of the following *could* be the cause of the difficulty? **(a)** 180-kΩ resistor shorted; **(b)** 180-kΩ resistor open; **(c)** 2-kΩ resistor shorted; **(d)** 390-kΩ resistor open; **(e)** C_S open.

11-26. Suppose in Problem 11-25 that the output signal appears satisfactory on the negative peak but that it clips on the positive peak. Which one of the following could be the cause of the difficulty? **(a)** 180-kΩ resistor open; **(b)** 390-kΩ resistor shorted; **(c)** 2-kΩ resistor shorted; **(d)** R_S too small; **(e)** R_S too large.

COMPUTER PROBLEMS

11-27. Write a program to determine g_m for a depletion-mode FET.

> Input data: I_{DSS}, $V_{GS(off)}$, V_{GS}
> Output data: g_{mo}, g_m

11-28. Write a program to determine g_m for an enhancement-mode IGFET.

> Input data: K, $V_{GS(th)}$, V_{GS}
> Output data: g_m

11-29. Write a program to determine the midband small-signal gain parameters of a common-source amplifier as shown in Figure 11-7(a).

> Input data: R_1, R_2, R_D, g_m
> Output data: A, R_{in}, R_{out}

11-30. Write a program to determine the midband small-signal gain parameters of a common-source amplifier with source feedback as shown in Figure 11-10.

> Input data: R_1, R_2, R_D, R_{S1}, g_m
> Output data: A, R_{in}, R_{out}

11-31. Write a program to determine the midband small-signal gain parameters of a common-gate amplifier as shown in Figure 11-13(a).

> Input data: R_S, R_D, g_m
> Output data: A, R'_{in}, R_{in}, R_{out}

11-32. Write a program to determine the midband small-signal gain parameters of a common-drain amplifier as shown in Figure 11-16.

> Input data: R_1, R_2, R_S, g_m.
> Output data: A, R_{in}, R'_{out}, R_{out}

CHAPTER 12

Multistage Amplifiers

OBJECTIVES

After completing this chapter, the reader should be able to:

1. Compare the general properties of the common-emitter, common-base, and common-collector BJT amplifier circuits.
2. Compare the general properties of the common-source, common-gate, and common-drain FET amplifier circuits.
3. Indicate which of the FET configurations has similar characteristics to a given BJT configuration.
4. Use tabulated results to determine voltage gain, input resistance, and output resistance for any of the basic configurations.
5. Construct a standard amplifier model using the results of objective 4.
6. Analyze any of the basic configurations with loading at the input and/or the output.
7. Develop models for and analyze multistage amplifier configurations.

12-1 COMPARISON OF AMPLIFIER CIRCUITS

In the preceding two chapters, various BJT and FET single-stage amplifier circuits were analyzed, and various properties of these amplifiers were derived. We are now in a position to review and compare the various properties and relationships.

As an initial qualitative comparison, Table 12-1 is provided for a general overview. This table is intended to provide some general guidelines as opposed to exact quantitative bounds. Consequently, the various ranges shown are intended as *typical* results rather than *exact* boundaries. Exceptions to these bounds are fairly common and can probably be found in this book. Furthermore, the transistors are assumed to be small-signal types rather than large power units.

Two properties that have not been considered yet in the book, but which are included in the comparison, are *current gain* and *power gain*. The major gain

TABLE 12-1 RELATIVE COMPARISON OF DIFFERENT AMPLIFIER CIRCUITS

BJT Circuits

	R_{in}	R_{out}	Voltage gain magnitude	Current gain magnitude	Power gain	Inverting (I) or noninverting (N)
Common emitter	Medium	Medium	>1	>1	>1	I
Common base	Low	Medium	>1	<1	>1	N
Common collector	Medium to high	Low	<1	>1	>1	N

FET Circuits

	R_{in}	R_{out}	Voltage gain magnitude	Current gain magnitude	Power gain	Inverting (I) or noninverting (N)
Common source	Very high	Medium	>1	>1	>1	I
Common gate	Low to medium	Medium	>1	<1	>1	N
Common drain	Very high	Low to medium	<1	>1	>1	N

Typical ranges are:
Low $<100\ \Omega$
Medium $100\ \Omega$ to $10\ k\Omega$
High $10\ k\Omega$ to $100\ k\Omega$
Very high $>100\ k\Omega$

parameter considered thus far has been *voltage gain,* and this will continue to be the dominant parameter. However, by relating output signal current to input signal current or output signal power to input signal power, current gain and power gain measures can be defined. The quantitative relationships for current and power gains are developed in the next section.

First, let us note the properties of the common-emitter amplifier. Based on the arbitrary ranges indicated, the common emitter has typical "medium" levels of input and output resistances. The voltage gain is inverting and its magnitude is normally greater than unity. Likewise, the current gain and power gain are normally greater than unity.

The common-base amplifier has a "low" input resistance and a "medium" output resistance. The voltage gain is noninverting and normally greater than unity. The power gain is also greater than unity but the current gain is less than unity. The latter result is due to the fact that collector current is less than emitter current.

The common-collector amplifier typically has a "medium" to "high" input resistance and a "low" output resistance. Its voltage gain is noninverting but is less than unity. (We often approximate the voltage gain as unity.) Its current gain and power gain are both greater than unity.

Reviewing the preceding results, only the common-emitter amplifier has all three gains (voltage, current, and power) greater than unity. However, all three configurations have power gains exceeding unity.

Turn now to the FET entries. In general, there is a strong association in gains and other results between the following corresponding amplifiers:

BJT		FET
Common emitter	$\longleftrightarrow$	Common source
Common base	$\longleftrightarrow$	Common gate
Common collector	$\longleftrightarrow$	Common drain

This pattern is immediately evident in the inverting and noninverting comparisons as well as the voltage gain, current gain, and power gain. Thus a common-source amplifier has properties very similar to a common-emitter amplifier, and so on. The major difference is in the relative level of certain resistances. For example, a common-emitter input resistance is classified as "medium," while a common-source input resistance is classified as "very high." This is a result of the fact that an FET has a much higher input resistance than a BJT.

12-2 AMPLIFIER SUMMARY

In this section the major results concerning both BJT and FET single-stage amplifier circuits in the midfrequency range are summarized. Tables of pertinent formulas will be provided.

For BJT amplifiers, we have considered the three basic amplifier structures: *common emitter*, *common base*, and *common collector* (or *emitter follower*). For the common-emitter circuit, we have considered both the case with the emitter resistance bypassed and with emitter feedback.

For FET amplifiers, we have also considered the three basic amplifier configurations: *common source*, *common gate*, and *common drain* (or *source follower*). For the common-source circuit, we have considered both the case with the source resistance bypassed and with source feedback.

A summary of the basic schematic diagrams for the BJT amplifiers is shown in Figure 12-1, and a similar summary for the FET amplifiers is shown in Figure 12-2. None of these configurations yet show the effects of generator resistance or of external load resistance. These effects will be considered shortly.

The standard basic amplifier form applicable to all cases is shown in Figure 12-3. It contains the following three parameters: input resistance R_{in}, output resistance R_{out}, and voltage gain $A = v_o/v_i$.

A general tabulation of the midfrequency small-signal gain and resistance relationships for all of the ac-coupled amplifier configurations is given in Table 12-2.

(a) Common emitter

(b) Common emitter with emitter feedback

(c) Common base

(d) Common collector

Figure 12-1 Summary of BJT amplifier configurations.

(a) Common source

(b) Common source with source feedback

(c) Common gate

(d) Common drain

Figure 12-2 Summary of FET amplifier configurations.

Figure 12-3 Standard model that can represent any of the basic amplifier configurations.

TABLE 12-2 SUMMARY OF MIDFREQUENCY SMALL-SIGNAL AMPLIFIER RELATIONSHIPS

BJT Circuits

	Common Emitter		Common Base	Common Collector
A	$-g_m R_C$	$\dfrac{-g_m R_C}{1 + g_m R_{E1}}$	$g_m R_C$	$\dfrac{g_m R_E}{1 + g_m R_E}$
R_{in}	$R_B \parallel r_\pi$	$R_B \parallel r_\pi (1 + g_m R_{E1})$	$R_E \parallel r_\pi \parallel \dfrac{1}{g_m}$	$R_B \parallel r_\pi (1 + g_m R_E')$
R_{out}	R_C	R_C	R_C	$R_E \parallel \left(\dfrac{1}{g_m} + \dfrac{R_B \parallel R_g}{\beta_o} \right)$

R_E bypassed   Emitter feedback

$R_B = R_1 \parallel R_2$

$R_E' = R_E \parallel R_L$

R_g = generator resistance

FET Circuits

	Common Source		Common Gate	Common Drain
A	$-g_m R_D$	$\dfrac{-g_m R_D}{1 + g_m R_{S1}}$	$g_m R_D$	$\dfrac{g_m R_S}{1 + g_m R_S}$
R_{in}	R_G	R_G	$R_S \parallel \dfrac{1}{g_m}$	R_G
R_{out}	R_D	R_D	R_D	$R_S \parallel \dfrac{1}{g_m}$

R_S bypassed   Source feedback

$R_G = R_1 \parallel R_2$

For the complete amplifier circuits, R'_{in} and R'_{out} are no longer of prime interest, so the resistance parameters tabulated are R_{in} and R_{out}. For a given amplifier configuration, formulas for A, R_{in}, and R_{out} are provided. When these three parameters are computed, the basic form of Figure 12-3 may then represent the equivalent signal model for any of the basic circuits. The "generic" form of Figure 12-3 is particularly convenient when a cascade of several stages is analyzed.

The formulas given in Table 12-2 are the same as derived earlier except for two modifications. In the common-collector circuit, the input resistance depends on the actual load resistance, and the output resistance depends on the generator resistance. The formulas have been modified to allow these effects.

In the expression for R_{in} for the common-collector circuit, note the term R'_E. Assume that the output is connected to a load resistance R_L through a coupling capacitor. For midfrequency analysis, the value of R'_E is defined as

$$R'_E = R_E \| R_L = \frac{R_E R_L}{R_E + R_L} \qquad (12\text{-}1)$$

Note that R_E appears without modification in the expression for A. The effect of R_L on A will be accounted for by a loading factor in the normal fashion.*

In the expression for R_{out} for the common-collector circuit, note that a parallel combination of R_B and the generator resistance R_g appears (i.e., $R_B \| R_g$). When the formula was derived in Chapter 10, the input terminals were left open since no generator resistance was considered. With an external generator resistance, the modification shown accounts for the effect of generator resistance on output resistance.

12-3 SINGLE STAGES WITH LOADING

In the actual application of most single-stage amplifier circuits, there will usually be loading effects at the input and the output that must be considered. A model depicting the interconnection is shown in Figure 12-4. The voltage v_g represents a signal

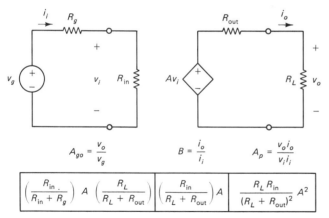

$$A_{go} = \frac{v_o}{v_g} \qquad B = \frac{i_o}{i_i} \qquad A_p = \frac{v_o i_o}{v_i i_i}$$

$$\left(\frac{R_{in}}{R_{in} + R_g}\right) A \quad \left(\frac{R_L}{R_L + R_{out}}\right)\left(\frac{R_{in}}{R_L + R_{out}}\right) A \quad \frac{R_L R_{in}}{(R_L + R_{out})^2} A^2$$

Figure 12-4 Model for connection of any single-stage amplifier with generator and load and additional gain relationships.

*Because of these additional coupling effects between generator and load resistances, a small error is introduced in common collector loaded gain calculations. However, the results should be sufficiently close for most practical purposes.

generator (or source) and R_g represents its internal resistance. The resistance R_L represents some additional load resistance connected across the output. The basic amplifier form is shown again in the middle. The parameters for any one of the basic BJT or FET circuits apply to this standard form. Several formulas for additional gain measures are given below the circuit diagram of Figure 12-4. These formulas will now be discussed.

Loaded Voltage Gain

The loaded voltage gain will be defined as the ratio of the output voltage across an external load resistance R_L to the open-circuit generator voltage v_g. This parameter will be denoted as A_{go}. Referring to Figure 12-4, the work of Chapter 9 enables us to write down this expression by inspection. We have

$$A_{go} = \frac{R_{in}}{R_{in} + R_g} A \frac{R_L}{R_L + R_{out}}$$ (12-2)

The values of A, R_{in}, and R_{out} for the various configurations are tabulated in Table 12-2.

Current Gain

The current gain for a given amplifier has not been considered up to this point in the text. However, it is a straightforward procedure to determine this gain in terms of the voltage gain and other pertinent amplifier parameters.

In this text the amplifier current gain is denoted as B, and it is defined as the ratio of the signal current i_o flowing out of the amplifier into a load to the input current i_i flowing into the amplifier. Referring to Figure 12-4, we have

$$B = \frac{i_o}{i_i}$$ (12-3)

The current i_o may be expressed in terms of v_o and R_L.

$$i_o = \frac{v_o}{R_L}$$ (12-4)

Similarly, i_i may be expressed in terms of v_i and R_{in}:

$$i_i = \frac{v_i}{R_{in}}$$ (12-5)

Substitution of (12-4) and (12-5) in (12-3) and simplification result in

$$B = \frac{v_o/R_L}{v_i/R_{in}} = \frac{v_o}{v_i} \frac{R_{in}}{R_L}$$ (12-6)

The ratio v_o/v_i under loaded output conditions can be expressed as

$$\frac{v_o}{v_i} = \frac{R_L}{R_L + R_{out}} A$$ (12-7)

Substitution of (12-7) in (12-6) results in

$$B = \frac{R_{\text{in}}}{R_L + R_{\text{out}}} A \qquad (12\text{-}8)$$

We see that the current gain is proportional to the voltage gain, but it includes an additional resistance ratio.

Power Gain

There are a number of different definitions of power gain used in various electronic applications. For our purposes, we consider only the most intuitive form. It will be denoted as A_p and it is defined as

$$A_p = \frac{P_L}{P_{\text{in}}} \qquad (12\text{-}9)$$

where P_L is defined as the output signal power accepted by the load P_L, and P_{in} is the signal power accepted by the amplifier input.

The power P_L is

$$P_L = v_o i_o \qquad (12\text{-}10)$$

where v_o is defined under loaded conditions. The input power P_{in} is

$$P_{\text{in}} = v_i i_i \qquad (12\text{-}11)$$

The power gain is then

$$A_p = \frac{P_L}{P_{\text{in}}} = \frac{v_o i_o}{v_i i_i} = \frac{v_o}{v_i} \frac{i_o}{i_i} \qquad (12\text{-}12)$$

Applying the relationship of (12-7) and the expression for $i_o/i_i = B$ from (12-8), we obtain

$$A_p = \frac{R_L R_{\text{in}}}{(R_L + R_{\text{out}})^2} A^2 \qquad (12\text{-}13)$$

The power gain is thus proportional to the square of the voltage gain.

Example 12-1

A certain common-emitter circuit with generator and external load connected is shown in Figure 12-5(a). An analysis is to be performed in the midfrequency range. **(a)** Determine the following properties of the amplifier: A, R_{in}, and R_{out}. **(b)** Draw a general standard equivalent circuit for the amplifier showing source and load connections. **(c)** From the model of (b), determine A_{go}, B, and A_p.

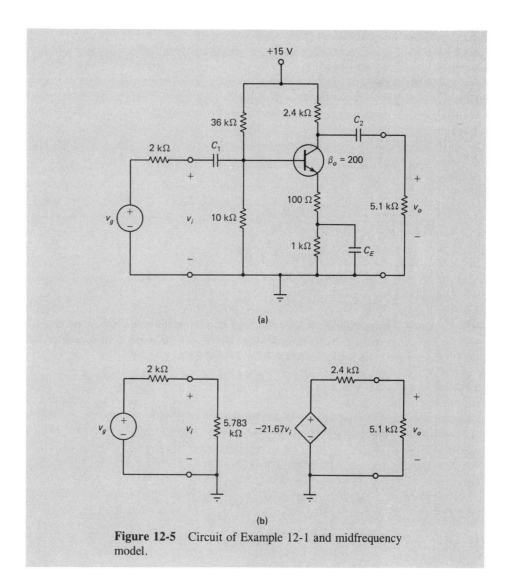

Figure 12-5 Circuit of Example 12-1 and midfrequency model.

Solution

(a) This part of the solution is a review of earlier work. First, the dc collector current must be determined. The voltage V_{TH} is

$$V_{TH} = \frac{10 \text{ k}\Omega}{10 \text{ k}\Omega + 36 \text{ k}\Omega} \times 15 = 3.261 \text{ V} \qquad \text{(12-14)}$$

The net dc emitter resistance is $1 \text{ k}\Omega + 100 \text{ }\Omega = 1100 \text{ }\Omega$, and the dc current I_C is

$$I_C = \frac{3.261 - 0.7}{1100} = 2.33 \text{ mA} \qquad \text{(12-15)}$$

The value of g_m is

$$g_m = 40I_C = 40 \times 2.33 \times 10^{-3} = 0.093 \text{ S} \qquad \text{(12-16)}$$

The value of r_π is

$$r_\pi = \frac{\beta_o}{g_m} = \frac{200}{0.093} = 2.15 \text{ k}\Omega \qquad \text{(12-17)}$$

Referring to Table 12-2, the three parameters of the basic amplifier circuit *without loading* are determined as follows:

$$A = \frac{-g_m R_C}{1 + g_m R_{E1}} = \frac{-0.093 \times 2.4 \times 10^3}{1 + 0.093 \times 100} = -21.67 \qquad \text{(12-18)}$$

$$R_{in} = R_B \| r_\pi (1 + g_m R_{E1})$$

$$= 36{,}000 \| 10{,}000 \| 2150(1 + 9.3) \qquad \text{(12-19)}$$

$$= 36{,}000 \| 10{,}000 \| 22{,}150 \text{ k}\Omega = 5.783 \text{ k}\Omega$$

$$R_{out} = R_C = 2.4 \text{ k}\Omega \qquad \text{(12-20)}$$

(b) The standard equivalent circuit depicting the amplifier properties is shown in Figure 12-5(b). If desired, the polarity of the dependent source could be reversed, and its value would then be $+21.67v_i$.

(c) The properties of the complete amplifier circuit are now determined as follows:

$$A_{go} = \frac{R_{in}}{R_{in} + R_g} A \frac{R_L}{R_L + R_{out}}$$

$$= \frac{5.783 \text{ k}\Omega}{5.783 \text{ k}\Omega + 2 \text{ k}\Omega} \times (-21.67) \times \frac{5.1 \text{ k}\Omega}{5.1 \text{ k}\Omega + 2.4 \text{ k}\Omega} \qquad \text{(12-21)}$$

$$= 0.743 \times (-21.67) \times 0.68 = -10.95$$

$$B = \frac{R_{in}}{R_L + R_{out}} A = \frac{5783}{5100 + 2400} \times (-21.67) = -16.71 \qquad \text{(12-22)}$$

$$A_P = \frac{R_L R_{in}}{(R_L + R_{out})^2} A^2 = \frac{5100 \times 5783}{(5100 + 2400)^2} \times (-21.67)^2 = 246.2 \qquad \text{(12-23)}$$

Note that the magnitude of the loaded voltage gain A_{go} is considerably less than that of the unloaded gain A. Note also that the magnitude of the current gain and the power gain of this amplifier are both greater than unity.

Example 12-2

A certain common-collector circuit with source and external load connected is shown in Figure 12-6(a). An analysis is to be performed in the midfrequency range. **(a)** Determine the following parameters of the amplifier: A, R_{in}, and

R_{out}. **(b)** Draw a general standard equivalent circuit for the amplifier showing source and load connections. **(c)** From the model of (b), determine A_{go}, B, and A_p.

(a)

(b)

Figure 12-6 Circuit of Example 12-2 and midfrequency model.

Solution

(a) The dc collector current is first determined by the standard procedure. The voltage V_{TH} is

$$V_{TH} = \frac{110 \text{ k}\Omega}{110 \text{ k}\Omega + 100 \text{ k}\Omega} \times 24 = 12.57 \text{ V} \qquad \textbf{(12-24)}$$

The collector current is

$$I_C = \frac{12.57 - 0.7}{2 \text{ k}\Omega} = 5.94 \text{ mA} \qquad \textbf{(12-25)}$$

The value of g_m is

$$g_m = 40I_C = 40 \times 0.00594 \times 10^{-3} = 0.238 \text{ S} \qquad \textbf{(12-26)}$$

The value of r_π is

$$r_\pi = \frac{\beta_o}{g_m} = \frac{300}{0.238} = 1261 \ \Omega \qquad \textbf{(12-27)}$$

Referring to Table 12-2, the amplifier parameters are now determined. The gain A is

$$A = \frac{g_m R_E}{1 + g_m R_E} = \frac{0.238 \times 2 \times 10^3}{1 + 0.238 \times 2 \times 10^3} = 0.998 \qquad \textbf{(12-28)}$$

To determine R_{in}, it is first necessary to determine R_E', which is

$$R_E' = R_E \| R_L = 2 \text{ k}\Omega \| 1 \text{ k}\Omega = 666.7 \ \Omega \qquad \textbf{(12-29)}$$

The input resistance is then

$$\begin{aligned}
R_{\text{in}} &= R_B \| r_\pi (1 + g_m R_E') \\
&= 10^5 \| 110,000 \| 1261(1 + 0.238 \times 666.7) \qquad \textbf{(12-30)} \\
&= 10^5 \| 110,000 \| 201,300 = 41,570 \ \Omega
\end{aligned}$$

The output resistance is determined as

$$\begin{aligned}
R_{\text{out}} &= R_E \| \left(\frac{1}{g_m} + \frac{R_B \| R_g}{\beta_o} \right) \\
&= 2000 \| \left(\frac{1}{0.238} + \frac{10^5 \| 110,000 \| 20,000}{300} \right) \\
&= 2000 \| (4.202 + 48.246) \qquad \textbf{(12-31)} \\
&= 51.11 \ \Omega
\end{aligned}$$

(b) The standard equivalent circuit displaying the amplifier properties is shown in Figure 12-6(b).

(c) The properties of the complete amplifier circuit will now be determined. The voltage gain A_{go} is

$$\begin{aligned}
A_{go} &= \frac{R_{\text{in}}}{R_{\text{in}} + R_g} A \frac{R_L}{R_L + R_{\text{out}}} \\
&= \frac{41,570}{41,570 + 20,000} \times 0.998 \times \frac{1000}{1000 + 51.11} \qquad \textbf{(12-32)} \\
&= 0.675 \times 0.998 \times 0.951 = 0.641
\end{aligned}$$

The current gain is

$$B = \frac{R_{\text{in}}}{R_L + R_{\text{out}}} A = \frac{41,570}{1000 + 51.11} \times 0.998 = 39.4 \qquad \textbf{(12-33)}$$

$$A_p = \frac{R_L R_{in}}{(R_L + R_{out})^2} A^2 = \frac{1000 \times 41,570}{(1000 + 51.11)^2} \times (0.998)^2 = 37.4 \qquad \textbf{(12-34)}$$

Note that while A is very close to unity, A_{go} is considerably less than unity. The reduction in net voltage gain is largely a result of the interaction at the input. Although the voltage gain is less than unity, both the current gain and the power gain are greater than unity.

Example 12-3

A certain common-source circuit with source and external load connected is shown in Figure 12-7(a). An analysis is to be performed in the midfrequency range. **(a)** Determine the following properties of the amplifier: A, R_{in}, and R_{out}. **(b)** Draw a general standard equivalent circuit for the amplifier showing source and load connections. **(c)** From the model of (b), determine A_{go}, B, and A_p.

Figure 12-7 Circuit of Example 12-3 and midfrequency model.

(a)

(b)

Solution

(a) Referring to Table 12-3, the amplifier properties without loading are determined as follows:

$$A = -g_m R_D = -12 \times 10^{-3} \times 2 \times 10^3 = -24 \qquad \text{(12-35)}$$

$$R_{\text{in}} = R_G = 200 \text{ k}\Omega \parallel 680 \text{ k}\Omega = 154.5 \text{ k}\Omega \qquad \text{(12-36)}$$

$$R_{\text{out}} = 2 \text{ k}\Omega \qquad \text{(12-37)}$$

(b) The standard equivalent circuit showing the amplifier properties is given in Figure 12-7(b).

(c) The properties of the complete amplifier will now be determined.

$$A_{go} = \frac{R_{\text{in}}}{R_{\text{in}} + R_g} A \frac{R_L}{R_L + R_{\text{out}}} = \frac{154.5 \text{ k}\Omega}{154.5 \text{ k}\Omega + 50 \text{ k}\Omega} \times (-24) \times \frac{5 \text{ k}\Omega}{5 \text{ k}\Omega + 2 \text{ k}\Omega}$$

$$= 0.756 \times (-24) \times 0.714 = -13.0 \qquad \text{(12-38)}$$

$$B = \frac{R_{\text{in}}}{R_L + R_{\text{out}}} A = \frac{154.5 \text{ k}\Omega}{5 \text{ k}\Omega + 2 \text{ k}\Omega} \times (-24) = -530 \qquad \text{(12-39)}$$

$$A_p = \frac{R_L R_{\text{in}}}{(R_L + R_{\text{out}})^2} A^2 = \frac{5000 \times 154{,}500}{(5000 + 2000)^2} \times (-24)^2 = 9081 \qquad \text{(12-40)}$$

The very large values of current gain and power gain may be misleading. They are a result of the fact that since the input current is so small, the ratio of output current to input current is a very large value.

Example 12-4

A certain common-gate circuit with source and external load connected is shown in Figure 12-8(a). An analysis is to be performed in the midfrequency range. **(a)** Determine the following properties of the amplifier: A, R_{in}, and R_{out}. **(b)** Draw a general standard equivalent circuit for the amplifier showing source and load connections. **(c)** From the model of (b), determine A_{go}, B, and A_p.

(a)

(b)

Figure 12-8 Circuit of Example 12-4 and midfrequency model.

Solution

(a) Referring to Table 12-2, the amplifier properties are determined as follows:

$$A = g_m R_D = 8 \times 10^{-3} \times 3 \times 10^3 = 24 \qquad \textbf{(12-41)}$$

$$R_{in} = R_S \left\| \frac{1}{g_m} = 1 \text{ k}\Omega \right\| \frac{1}{0.008} = 1 \text{ k}\Omega \| 125 \text{ } \Omega = 111.1 \text{ } \Omega \qquad \textbf{(12-42)}$$

$$R_{out} = R_D = 3 \text{ k}\Omega \qquad \textbf{(12-43)}$$

(b) The standard equivalent circuit showing the amplifier properties is given in Figure 12-8(b).

(c) The properties of the complete amplifier will now be determined.

$$\begin{aligned} A_{go} &= \frac{R_{in}}{R_{in} + R_g} A \frac{R_L}{R_L + R_{out}} \\ &= \frac{111.1}{111.1 + 50} \times 24 \times \frac{8 \text{ k}\Omega}{8 \text{ k}\Omega + 3 \text{ k}\Omega} \qquad \textbf{(12-44)} \\ &= 0.690 \times 24 \times 0.727 = 12.04 \end{aligned}$$

$$B = \frac{R_{in}}{R_L + R_{out}} A = \frac{111.1}{8000 + 3000} \times 24 = 0.242 \qquad \textbf{(12-45)}$$

$$A_p = \frac{R_L R_{in}}{(R_L + R_{out})^2} A^2 = \frac{8000 \times 111.1}{(8000 + 3000)^2} \times (24)^2 = 4.23 \qquad \textbf{(12-46)}$$

A few comments are in order. First, even though the generator has a relatively low output resistance (50 Ω), there is still a significant loading effect at the input due to the low input resistance of the amplifier (111.1 Ω). If the generator output resistance had been much larger, all of the potential voltage gain could have been lost due to attenuation at the input. Thus the common-gate amplifier and its BJT counterpart, the common-base amplifier, require low-resistance driving sources to achieve any real gain. In high-frequency circuits, this can be accomplished with transformer coupling at the input.

A second comment is to note that the current gain is less than unity, a characteristic of both the common-gate and common-base circuits. However, the power gain is greater than unity as expected.

12-4 CASCADE AMPLIFIERS

In the preceding section, source and load connections were made to various single-stage amplifiers. In this section the process is extended to include a cascade of two or more amplifier stages.

Depending on the intended application, almost any conceivable cascade combination of amplifier stages is possible. In fact, BJT and FET active devices may be used together when the properties of each are to be exploited.

All the basic material required for this purpose has been covered in preceding chapters, and the individual stage properties have been summarized earlier in this chapter. The process of analysis involves first determining the properties of each individual stage. By determining interaction factors between stages, along with source and output loading factors, the complete properties of the composite amplifier can be determined.

The process is developed through several representative examples. Not all possible cases can be covered, but the general approaches for analyzing other circuits should evolve from these examples.

All feedback considered in this section is limited to feedback within a stage, such as the common emitter with emitter feedback, for example. We are not yet ready to deal with feedback around several stages.

Example 12-5

The cascade amplifier of Figure 12-9(a) consists of two *npn* BJT common-emitter stages, each having emitter feedback. Determine for the composite circuit the net midfrequency gain between source voltage v_g and loaded output voltage v_o (i.e., $A_{go} = v_o/v_g$).

(a)

(b)

Figure 12-9 Circuit of Example 12-5 and midfrequency model.

Solution

It can be readily observed that the two common-emitter stages are identical. Thus it will be necessary to analyze only one of the stages in detail. First, the dc operating point must be determined. The voltage V_{TH} is determined as

$$V_{\text{TH}} = \frac{7.5 \text{ k}\Omega}{7.5 \text{ k}\Omega + 27 \text{ k}\Omega} \times 30 = 6.52 \text{ V} \qquad \textbf{(12-47)}$$

The net dc emitter resistance is $1200 + 100 = 1300\ \Omega$. The dc collector current is

$$I_C = \frac{6.52 - 0.7}{1300} = 4.48\ \text{mA} \tag{12-48}$$

The value of g_m is

$$g_m = 40I_C = 40 \times 4.48 \times 10^{-3} = 0.179\ \text{S} \tag{12-49}$$

The value of r_π is

$$r_\pi = \frac{\beta_o}{g_m} = \frac{200}{0.179} = 1117\ \Omega \tag{12-50}$$

The gain of each stage is

$$A = -\frac{g_m R_C}{1 + g_m R_{E1}} = -\frac{0.179 \times 2.7 \times 10^3}{1 + 0.179 \times 100} = -25.57 \tag{12-51}$$

The input resistance of each stage is

$$R_{\text{in}} = R_B \parallel r_\pi(1 + g_m R_{E1}) = 27{,}000 \parallel 7500 \parallel 1117(18.9)$$

$$= 4593\ \Omega \tag{12-52}$$

The output resistance of each stage is

$$R_{\text{out}} = 2.7\ \text{k}\Omega \tag{12-53}$$

It should be noted that the input resistance of the composite amplifier as seen by the source is the input resistance of the first stage (i.e., 4593 Ω). Further, the output resistance looking back from the load is the output resistance of the second stage (i.e., 2.7 kΩ).

A block diagram depicting the interaction between stages, source, and load is shown in Figure 12-9(b). The total gain A_{go} may be expressed as follows:

$$A_{go} = \frac{4593}{4593 + 1000} \times (-25.57) \times \frac{4593}{4593 + 2700} \times (-25.57)$$

$$\times \frac{5000}{5000 + 2700} \tag{12-54}$$

$$= 0.8212 \times (-25.57) \times 0.6298 \times (-25.57) \times 0.6494 = 219.6$$

Note that the net gain is noninverting since there is an even number of inversions.

Example 12-6

The three-stage amplifier of Figure 12-10 consists of a common-source input stage, a common emitter with emitter feedback stage, and a common-collector output stage. Determine the net loaded gain $A_{go} = v_o/v_g$ in the midfrequency range.

Figure 12-10   Circuit of Example 12-6.

Solution

Each of the three stages will be analyzed individually. The calculations are grouped separately as follows:

Common source:

$$A = -g_m R_D = -15 \times 10^{-3} \times 2.2 \times 10^3 = -33 \qquad \text{(12-55)}$$

$$R_{\text{in}} = R_G = 220 \text{ k}\Omega \,\|\, 680 \text{ k}\Omega = 166 \text{ k}\Omega \qquad \text{(12-56)}$$

$$R_{\text{out}} = 2.2 \text{ k}\Omega \qquad \text{(12-57)}$$

Common emitter:

$$V_{\text{TH}} = \frac{12 \text{ k}\Omega}{12 \text{ k}\Omega + 47 \text{ k}\Omega} \times 30 = 6.102 \text{ V} \qquad \text{(12-58)}$$

$$I_C = \frac{6.102 - 0.7}{3000 + 120} = 1.73 \text{ mA} \qquad \text{(12-59)}$$

$$g_m = 40 I_C = 40 \times 1.73 \times 10^{-3} = 0.0692 \text{ S} \qquad \text{(12-60)}$$

$$A = \frac{-g_m R_C}{1 + g_m R_{E1}} = \frac{-0.0692 \times 5.6 \times 10^3}{1 + 0.0692 \times 120} = -41.65 \qquad \text{(12-61)}$$

$$R_{\text{in}} = R_B \,\|\, r_\pi (1 + g_m R_{E1})$$

$$= 12{,}000 \parallel 47{,}000 \parallel 2000(1 + 0.0692 \times 120) = 6315 \ \Omega \qquad \textbf{(12-62)}$$

$$R_{\text{out}} = R_C = 5.6 \ \text{k}\Omega \qquad \textbf{(12-63)}$$

Common collector:

$$V_{\text{TH}} = \frac{100 \ \text{k}\Omega}{100 \ \text{k}\Omega + 100 \ \text{k}\Omega} \times 30 = 15 \ \text{V} \qquad \textbf{(12-64)}$$

$$I_C = \frac{15 - 0.7}{3.3 \times 10^3} = 4.33 \ \text{mA} \qquad \textbf{(12-65)}$$

$$g_m = 40 I_C = 40 \times 4.33 \times 10^{-3} = 0.173 \ \text{S} \qquad \textbf{(12-66)}$$

$$A = \frac{g_m R_E}{1 + g_m R_E} = \frac{0.173 \times 3.3 \times 10^3}{1 + 0.173 \times 3.3 \times 10^3} = 0.998 \qquad \textbf{(12-67)}$$

$$R'_E = R_E \parallel R_L = 3.3 \ \text{k}\Omega \parallel 1.5 \ \text{k}\Omega = 1.031 \ \text{k}\Omega \qquad \textbf{(12-68)}$$

$$R_{\text{in}} = R_B \parallel r_\pi (1 + g_m R'_E)$$
$$= 10^5 \parallel 10^5 \parallel 1000(1 + 0.173 \times 1031) = 39.1 \ \text{k}\Omega \qquad \textbf{(12-69)}$$

Before computing R_{out}, it is necessary to determine β_o. We have

$$\beta_o = g_m r_\pi = 0.173 \times 10^3 = 173 \qquad \textbf{(12-70)}$$

In computing R_{out}, the "generator" resistance R_g is the output resistance of the common emitter stage (i.e., 5.6 kΩ).

$$R_{\text{out}} = R_E \parallel \left(\frac{1}{g_m} + \frac{R_B \parallel R_g}{\beta_o} \right) \qquad \textbf{(12-71)}$$

$$= 3300 \parallel \left(\frac{1}{0.173} + \frac{50{,}000 \parallel 5600}{173} \right)$$

$$= 3300 \parallel (5.78 + 29.11) = 34.52 \ \Omega$$

Although an equivalent circuit showing the parameters of the three stages could be readily constructed, that process will be sidestepped in this problem. Instead, an expression for the net gain will be written as follows:

$$A_{go} = \frac{v_o}{v_g}$$

$$= \frac{166 \ \text{k}\Omega}{166 \ \text{k}\Omega + 5 \ \text{k}\Omega} \times (-33) \times \frac{6315}{6315 + 2200} \times (-41.65)$$
$$\qquad \textbf{(12-72)}$$

$$\times \frac{39.1 \ \text{k}\Omega}{39.1 \ \text{k}\Omega + 5.6 \ \text{k}\Omega} \times 0.998 \times \frac{1500}{1500 \ \Omega + 34.52}$$

$$= 0.9708 \times (-33) \times 0.7416 \times (-41.65) \times 0.8747 \times 0.998 \times 0.9775$$

$$= 844$$

12-5 CASCODE AMPLIFIER

A circuit that offers some significant advantages at high frequencies is the *cascode amplifer*. The basic form of the circuit utilizing JFETs is shown in Figure 12-11(a). The circuit may be considered as a common-source amplifier stage driving a common-gate amplifier stage. The input resistance of the common-gate stage acts as the load for the common-source stage.

A small-signal midfrequency equivalent circuit, after some slight approximations have been made, is shown in Figure 12-11(b). Neglecting gate current, the input resistance of the common-emitter stage is R_3. Since there is no source resistor for the common-gate stage, its input resistance is simply $1/g_{m2}$, as shown. The dependent current source $g_{m1}v_i$ for the first stage is effectively connected directly to this input resistance as shown.

(a)

(b)

Figure 12-11   Cascode amplifier using JFETs and small-signal model.

The gate–source controlling voltage v_{gs} for the common-gate stage is $v_{gs} = -v_a$, where v_a is the signal at the junction point between stages with respect to ground. Thus the dependent source in the second stage directed toward ground would be $g_{m2} v_{gs} = -g_{m2} v_a$. By reversing the current-source direction, the value of the current source is $g_{m2} v_a$ as shown.

An analysis of the circuit will now be performed. The voltage v_a is

$$v_a = -g_{m1} v_i \times \frac{1}{g_{m2}} = \frac{-g_{m1}}{g_{m2}} v_i \qquad (12\text{-}73)$$

The voltage v_o is given by

$$v_o = g_{m2} v_a R_D \qquad (12\text{-}74)$$

Substituting (12-73) in (12-74) results in

$$v_o = g_{m2} \left(\frac{-g_{m1}}{g_{m2}} v_i \right) R_D = -g_{m1} R_D v_i \qquad (12\text{-}75)$$

The gain A is

$$\boxed{A = \frac{v_o}{v_i} = -g_{m1} R_D} \qquad (12\text{-}76)$$

The expression for gain has the same form as that of a single-stage common-source circuit, but the FET transconductance is that of the common-source stage, and the drain resistance is that of the common-gate stage.

The prime virtue of this circuit is not evident from the midfrequency analysis. A concept known as the *Miller effect input capacitance* is developed in Chapter 18. This phenomenon is most pronounced in common-source (and common-emitter) stages and results in serious gain degradation at high frequencies. It turns out that the low input resistance of the common-gate circuit loads the common-source circuit so heavily that the Miller effect capacitance at the input of the cascode circuit is reduced dramatically. The cascode circuit thus offers the high input resistance and gain associated with the common-source circuit, but with improved high-frequency performance. Cascode circuits are often used in the radio-frequency stages of television sets.

DRILL PROBLEMS

Note: In Problems 12-1 through 12-8, all analysis is to be performed in the midfrequency range. For each circuit, determine, A_{go}, B, and A_p.

12-1. See Figure P12-1.

Figure P12-1

12-2. See Figure P12-2.

Figure P12-2

12-3. See Figure P12-3.

Figure P12-3

12-4. See Figure P12-4.

Figure P12-4

12-5. See Figure P12-5.

Figure P12-5

12-6. See Figure P12-6.

Figure P12-6

12-7. See Figure P12-7.

Figure P12-7

12-8. See Figure P12-8.

Figure P12-8

12-9. Determine the net loaded gain $A_{go} = v_o/v_g$ in the midfrequency range for the two-stage amplifier of Figure P12-9.

Figure P12-9

12-10. Determine the net loaded gain $A_{go} = v_o/v_g$ in the midfrequency range for the two-stage amplifier of Figure P12-10.

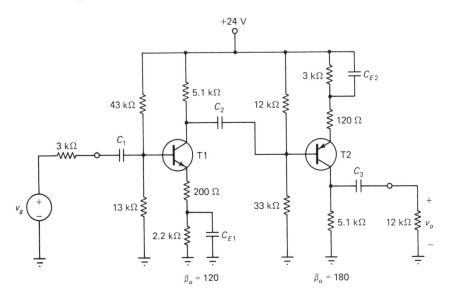

Figure P12-10

12-11. Determine the net loaded gain $A_{go} = v_o/v_g$ in the midfrequency range for the two-stage amplifier of Figure P12-11.

Figure P12-11

12-12. Determine the net loaded gain $A_{go} = v_o/v_g$ in the midfrequency range for the two-stage amplifier of Figure P12-12.

Figure P12-12

12-13. A cascode amplifier of the form shown in Figure 12-11(a) has $g_{m1} = 12$ mS, $g_{m2} = 10$ mS, and $R_D = 2$ kΩ. **(a)** Determine A. **(b)** If $v_i = 20$ mV, determine v_a and v_o.

12-14. A cascode amplifier of the form shown in Figure 12-11(a) has $g_{m1} = 20$ mS, $g_{m2} = 25$ mS, and $R_D = 1.5$ kΩ. **(a)** Determine A. **(b)** If $v_i = 50$ mV, determine v_a and v_o.

TROUBLESHOOTING PROBLEMS

12-15. You are troubleshooting the circuit of Figure P12-9 with an oscilloscope set for dc coupling. The dc levels at all points are close to specified values. A signal appears at the collector of T1, but there is no signal at the collector of T2. Which one of the following is the probable trouble? **(a)** C_{E2} open; **(b)** C_3 shorted; **(c)** C_2 shorted; **(d)** C_2 open; **(e)** defective transistor.

12-16. You are troubleshooting the circuit of Figure P12-9 with an oscilloscope set for dc coupling. The dc levels at all points are close to specified values. A signal of normal level appears at the collector of T1, but the signal level at the collector of T2 is much smaller than expected. Which one of the following is the probable trouble? **(a)** C_{E2} open; **(b)** C_3 shorted; **(c)** C_3 open; **(d)** C_2 open; **(e)** defective transistor.

COMPUTER PROBLEMS

12-17. Write a computer program to determine the net gain parameters of a cascade of two amplifier stages (denoted by subscripts 1 and 2) driven by a source with internal resistance R_g and connected to a load R_L.

Input data: R_{in1}, A_1, R_{out1}, R_{in2}, A_2, R_{out2}, R_g, R_L.
Output data: A_{go}, B, A_p

12-18. Write a computer program to determine the net gain parameters of a cascade of three amplifier stages (denoted by subscripts 1, 2, and 3) driven by a source with internal resistance R_g and connected to a load R_L.

Input data: R_{in1}, A_1, R_{out1}, R_{in2}, A_2, R_{out2}, R_{in3}, A_3, R_{out3}, R_g, R_L.
Output data: A_{go}, B, A_p

CHAPTER 13

Differential and Operational Amplifiers

OBJECTIVES

After completing this chapter, the reader should be able to:

1. Explain the general operation of a discrete-component differential amplifier.
2. Define differential input voltage.
3. Define common-mode input voltage.
4. Define differential gain.
5. Define common-mode gain.
6. Define common-mode rejection ratio.
7. Explain the operation of and analyze a current mirror.
8. Explain the operation of an active load.
9. Explain the general operation of an operational amplifier.
10. Show the manner in which dual power supplies are used to provide dc bias for an op-amp.
11. Draw the schematic symbol for an op-amp and explain the functions of the three signal terminals.
12. Explain saturation in an op-amp, and estimate the approximate levels.
13. Define input offset voltage for an op-amp.
14. Define bias and offset currents in an op-amp.
15. Discuss how the output dc offset for an op-amp can be minimized.

13-1 OVERVIEW

The differential amplifier is one of the most important circuits encountered in modern electronic systems. It represents a basic "building block" for many important linear and nonlinear integrated-circuit modules. It is an important key in the transition between discrete amplifier circuits and integrated circuits.

A major portion of this chapter is devoted to analyzing a two-transistor discrete-component differential amplifier. This detailed analysis is intended to assist the reader in comprehending the basic differential amplifier properties. Once this operation is understood, the transition to the integrated-circuit version may be achieved with minimum difficulty.

The last portion of the chapter is devoted to the consideration of the integrated-circuit operational amplifier. An operational amplifier is a high-gain dc-coupled differential amplifier. Emphasis at that point will be on the overall circuit modeling and operation rather than the detailed circuit elements. Typical operational amplifier specifications are investigated in detail.

13-2 DIFFERENTIAL AMPLIFIER

The simplest form of a discrete version of the basic differential amplifier circuit employing two *npn* bipolar junction transistors and two power supplies is shown in Figure 13-1. The differential amplifier normally has no coupling capacitors (i.e., it is *directly coupled*). It is therefore capable of processing frequencies all the way down to dc.

The circuit may have two input voltages (or *dual* inputs), and these voltages as referred to ground are designated as v_{i1} and v_{i2}, respectively. (When a single-input voltage is desired, one input may be connected to ground.) This version shown has a *double-ended* output; that is, both collector voltages v_{C1} and v_{C2} as referred to ground are possible outputs. Either one of the collector resistances may be omitted, and the output would be at the single collector with the resistance R_C. This latter form is denoted as a *single-ended* output. Because it is more general, the double-ended output form will be retained until the integrated-circuit form is introduced later.

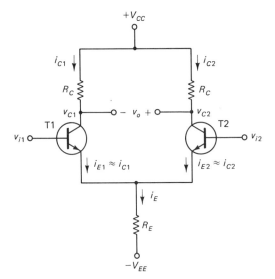

Figure 13-1 Basic differential amplifier circuit.

Another possible output voltage of interest is the difference between the collector voltages, denoted as v_o, and it is defined as

$$v_o = v_{C2} - v_{C1} \tag{13-1}$$

Some difficulty arises in utilizing v_o in this simplified circuit form since it is "floating" (i.e., it is *not* referred to ground). This problem is circumvented in the later integrated-circuit versions, so we will proceed to consider v_o without undue concern at this point.

The two emitters are connected together, and this point is connected through a common resistance R_E to a negative supply voltage $-V_{EE}$. The two transistor collector currents are represented as i_{C1} and i_{C2}, respectively. Although the bases are shown with no connection in this illustration, it is necessary that there be dc connections to ground for bias currents to flow in order for the circuit to function properly. Bias current flows from the positive terminal of the emitter power supply to ground and up to the bases through the external dc base connections.

It is assumed that the two transistors are matched; that is, corresponding dc and dynamic characteristics are essentially the same. Without such a match, the circuit may not function as intended. It is also assumed that β_{dc} for each transistor is very large. In the discussion that follows, it will be assumed in many steps that the two emitter currents i_{E1} and i_{E2} are nearly the same as the two collector currents (i.e., $i_{E1} \approx i_{C1}$ and $i_{E2} \approx i_{C2}$).

The operation of the differential amplifier is somewhat tricky and requires some detailed study for full comprehension. Before developing general equations, we move through a series of circuit diagrams with representative voltages and currents chosen to illustrate some of the key features of circuit operation. These results represent some typical (rather than general) results, but they should provide the reader with some insight about circuit operation.

The first step in this process is to consider only one of the transistors in the differential amplifier with the other completely removed. As we will see, the function is completely different with one transistor removed, and this will provide a way of comparing operation when both transistors are present.

Consider the circuit of Figure 13-2(a) containing only T1. The two power supply voltages are $+15$ V and -15 V. The base of T1 is momentarily grounded, so $v_{i1} = 0$. The emitter resistance is 14.3 kΩ, and the collector resistance is 8 kΩ.

A KVL loop will be formed by starting at the grounded base, moving over to the emitter, down through the emitter resistance to the negative power supply terminal, and back to ground again. This equation reads

$$v_{BE1} + 14.3 \times 10^3 i_{E1} - 15 = 0 \tag{13-2}$$

We then solve for i_{E1} and assume that $v_{BE1} = 0.7$ V. The result is

$$i_{E1} \approx i_{C1} = \frac{15 - v_{BE1}}{14.3 \times 10^3} = \frac{15 - 0.7}{14.3 \times 10^3} = 1 \text{ mA} \tag{13-3}$$

The voltage drop across R_C is

$$v_{R_C} = 1 \times 10^{-3} \times 8 \times 10^3 = 8 \text{ V} \tag{13-4}$$

Figure 13-2 Illustration of representative values for one transistor in differential amplifier with other transistor removed.

The collector voltage v_{C1} referred to ground is

$$v_{C1} = 15 - 8 = 7 \text{ V} \tag{13-5}$$

The emitter voltage is designated as v_E, and since $v_{BE1} = v_{B1} - v_E$, the value of v_E is

$$v_E = v_{B1} - v_{BE1} = 0 - v_{BE1} = -0.7 \text{ V} \tag{13-6}$$

where $v_{B1} = 0$ has been used. Finally, the collector–emitter voltage v_{CE1} is

$$v_{CE1} = v_{C1} - v_{E1} = 7 - (-0.7) = 7.7 \text{ V} \tag{13-7}$$

The various voltages and currents are labeled on Figure 13-2(a).

The single-transistor circuit as shown can be viewed in two ways. First, it has the form of a common-emitter circuit with feedback (i.e., with unbypassed emitter resistance). However, the value of this emitter resistance is very large relative to the values normally left unbypassed. Consequently, the gain of the circuit as a common-emitter circuit is very low as we shall see shortly. In fact, the gain magnitude is less than unity.

A second interpretation of the single-transistor circuit is that it can be viewed as an emitter follower between the base and emitter. Although the emitter-follower circuit does not normally have a collector resistance, the presence of the collector resistance does not change the function of the circuit as an emitter follower at low frequencies except to limit the range of active-region operation (i.e., saturation occurs at a lower input voltage due to the drop across R_C).

With the preceding points of view in mind, assume that the base is connected to a very small dc voltage as shown in Figure 13-2(b). The value $v_{i1} = 20$ mV has been selected for illustration. Because of the emitter-follower action of the circuit, the emitter voltage "follows" the input, and it increases by about 20 mV. The new value of v_E is thus $v_E = -0.7 + 0.02 = -0.68$ V.

The new value of emitter current is computed most easily by first determining the voltage across the emitter resistance and employing Ohm's law. We have

$$i_{E1} \approx i_{C1} = \frac{-0.68 - (-15)}{14.3 \times 10^3} \approx 1.001 \text{ mA} \qquad \text{(13-8)}$$

The change in emitter current is very small and could hardly be sensed with normal instrumentation. The slight increase in collector current causes a slight increase in the drop across the 8-kΩ resistor, so the collector voltage also drops slightly. The new approximate collector voltage is $v_{C1} \approx 6.99$ V.

From the point of view of a common-emitter circuit, the slight changes in current and collector voltage are a result of the large unbypassed emitter resistance, which has reduced the gain to a very small value. However, from the emitter-follower point of view, the emitter voltage has changed approximately the same amount as the input, and the base–emitter voltage has remained essentially constant.

Now, let us add the second transistor T2 as shown in Figure 13-3 and return to the initial point of discussion. In this case, both bases are grounded again (i.e., $v_{i1} = v_{i2} = 0$). To restore the original operating point, we do have to make one change in a circuit parameter. For symmetrical conditions, $i_{C1} = i_{C2} = 1$ mA. The current through the common resistor is $i_{C1} + i_{C2} = 2$ mA. To preserve the same voltage drop, this resistor is changed from 14.3 kΩ to 7.15 kΩ. Thus 2 mA flowing through 7.15 kΩ produces the same drop as 1 mA through 14.3 kΩ. All other points in the circuit have the same operating conditions as the single-transistor circuit of Figure 13-2.

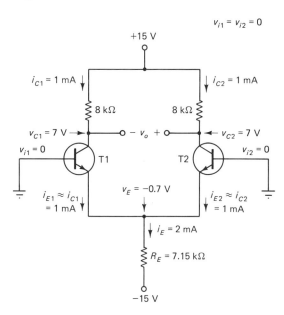

Figure 13-3 Representative differential amplifier voltages and currents when $v_{i1} = v_{i2} = 0$.

To generalize these dc results for this basic configuration, the dc emitter current I_E corresponding to an emitter resistance R_E with both base terminals grounded and a power supply voltage $-V_{EE}$ is

$$I_E = \frac{V_{EE} - 0.7}{R_E} \qquad \textbf{(13-9)}$$

where $V_{BE} = 0.7$ V has been assumed. As we will see shortly, this current remains essentially constant for different values of v_{i1} and v_{i2}.

For $v_{i1} = 0$ and $v_{i2} = 0$ the emitter current divides equally between the transistors (i.e., $i_{E1} = i_{E2} \approx i_{C1} = i_{C2} = I_E/2$). However, for various conditions of v_{i1} and v_{i2}, i_{E1} and i_{E2} will be shown to change.

The voltage v_o between the two collectors is

$$v_o = v_{C2} - v_{C1} = 7 - 7 = 0 \qquad \textbf{(13-10)}$$

The reader should be reminded that the assumed base–emitter voltage of 0.7 V is approximate and is a convenient assumption. However, once we assume this value as a reference, it will be necessary to follow variations up and down from that value, as we will shortly see.

Let us now assume that the small dc voltage of 20 mV is applied to the base of T1 (i.e., $v_{i1} = 20$ mV), as shown in Figure 13-4. However, we retain the ground connection for the base of T2 (i.e., $v_{i2} = 0$). In general, the two base–emitter voltages v_{BE1} and v_{BE2} can be expressed as

$$v_{BE1} = v_{i1} - v_E \qquad \textbf{(13-11)}$$

and

$$v_{BE2} = v_{i2} - v_E \qquad \textbf{(13-12)}$$

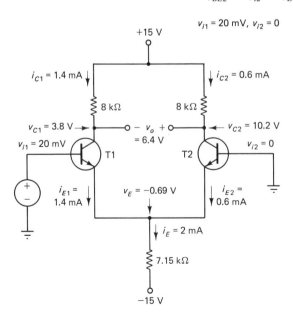

Figure 13-4 Representative differential amplifier voltages and currents when $v_{i1} = 20$ mV and $v_{i2} = 0$.

Differential and Operational Amplifiers   Chap. 13

For the values of v_{i1} and v_{i2} of Figure 13-4, we have

$$v_{BE1} = 0.02 - v_E \qquad \text{(13-13)}$$

$$v_{BE2} = 0 - v_E = -v_E \qquad \text{(13-14)}$$

The effect of a positive voltage for v_{i1} is to cause v_E to become more positive. When there was only T1 in the circuit, v_E essentially "followed" v_{BE1}. However, when T2 is present, the effect of an increase in v_{BE1} is to cause a decrease in v_{BE2}. This can be inferred from (13-14) by noting that v_E is a negative quantity, and when it becomes more positive, its magnitude decreases. Thus an increase in v_{BE1} is simultaneously accompanied by a decrease in v_{BE2}. An increase in v_{BE1} corresponds to an increase in i_{E1}, and a decrease in v_{BE2} corresponds to a decrease in i_{E2}.

The net current i_E through the emitter resistance is

$$\boxed{i_E = i_{E1} + i_{E2}} \qquad \text{(13-15)}$$

The combination of the large emitter resistance and the negative power supply acts nearly like a constant current source as far as the emitters are concerned. Thus i_E assumes nearly a constant value. This last property was also true when only T1 was present but the single current in that case was the only variable. In this case, however, i_{E1} and i_{E2} both vary but the sum remains nearly constant. Therefore, *an increase in current in one transistor is accompanied by a corresponding decrease in the other transistor, but the sum remains essentially constant.*

The combined effect of the two transistors working in opposition coupled with the constant current constraint is to prevent the emitter from simply following the base as it did for the single-transistor case. Instead, the emitter voltage changes enough to cause an increase in i_{E1} and a decrease in i_{E2} such that the sum of the currents remains the same. For $v_{i1} = 20$ mV, v_E changes by 10 mV. This has the effect of increasing v_{BE1} by 10 mV and decreasing v_{BE2} by 10 mV. Thus v_{BE1} increases to 0.71 V and v_{BE2} decreases to 0.69 V for the original assumed values.

The formulas for predicting various circuit variables will be developed later. For the conditions in Figure 13-4, it turns out that $i_{E1} \approx i_{C1} = 1.4$ mA and $i_{E2} \approx i_{C2} = 0.6$ mA. The two collector voltages may be readily verified to be $v_{C1} = 3.8$ V and $v_{C2} = 10.2$ V. The voltage v_o is

$$v_o = 10.2 - 3.8 = 6.4 \text{ V} \qquad \text{(13-16)}$$

The changes in this case are rather pronounced and represent a totally different situation than when the single transistor was used. Although the emitter resistance is unbypassed, it does not suppress changes in signal voltages and currents as it did for the single-transistor case. In effect, the presence of the second transistor "pulling" in opposition to the first transistor permits a large gain to be achieved even with the presence of a large emitter resistance. The reduction in gain that would be caused by the large emitter resistance for a single transistor is effectively canceled out when two transistors are used.

Next, let us consider the effect when equal voltages are applied to both input terminals. Refer to Figure 13-5, in which $v_{i1} = v_{i2} = 20$ mV is assumed. In this case the emitter voltage v_E may easily follow the inputs since the two inputs are

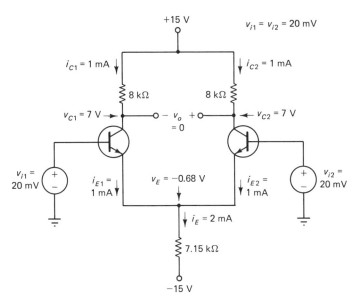

Figure 13-5 Representative differential amplifier voltages and currents when $v_{i1} = v_{i2} = 20$ mV.

equal. In a sense, the circuit here is acting almost like that of the single-transistor case of Figure 13-2. Indeed, there is even a very slight increase in collector current as was shown in Figure 13-2, but it will be ignored here. Because neither of the collector currents changes appreciably, the collector voltages are assumed to be the same as when the two inputs were grounded. The voltage v_o is

$$v_o = 7 - 7 = 0 \qquad (13\text{-}17)$$

The final situation to be considered for this circuit is shown in Figure 13-6. In this case, two different voltages are applied to the separate inputs. The specific values chosen are $v_{i1} = 60$ mV and $v_{i2} = 40$ mV. In this case, the "pulling" action between the transistors occurs as it did in the circuit of Figure 13-4. The effect is to cause the emitter voltage to shift upward by an amount equal to the midpoint between the two input voltages. Thus v_E increases by 50 mV from its original value of -0.7 V to a value of -0.65 V. The reader can readily verify that $v_{BE1} = 0.71$ V and $v_{BE2} = 0.69$ V. These are the same as for the circuit of Figure 13-4.

The collector and emitter currents and the collector voltages are the same as for the circuit of Figure 13-4. The voltage v_o is obviously also the same (i.e., $v_o = 6.4$ V).

The conditions of Figures 13-4 and 13-6 produce identical results for the collector voltages and the voltage v_o. Yet in one case, $v_{i1} = 20$ mV and $v_{i2} = 0$, and in the other case, $v_{i1} = 60$ mV and $v_{i2} = 40$ mV. On the other hand, when we set $v_{i1} = v_{i2} = 20$ mV, the results were the same as when $v_{i1} = v_{i2} = 0$. *The implication is that the circuit is responding to the difference between the input voltages.* Thus, when $v_{i1} = 60$ mV and $v_{i2} = 40$ mV, the difference is 20 mV, which is the

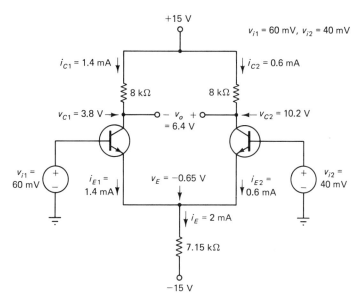

Figure 13-6 Representative differential amplifier voltages and currents when $v_{i1} = 60$ mV and $v_{i2} = 40$ mV.

same as when $v_{i1} = 20$ mV and $v_{i2} = 0$. When the two input voltages are equal, the difference is zero, and there is virtually no change in the circuit variables.

In analyzing differential amplifiers, the term *differential input voltage* is a convenient definition. Representing this quantity as v_d, it is defined as

$$v_d = v_{i1} - v_{i2} \qquad (13\text{-}18)$$

It should be clear from the symmetry of the circuit that if any input conditions are reversed between the bases, the resulting collector and emitter conditions are similarly reversed. For example, if $v_{i1} = 0$ and $v_{i2} = 20$ mV, it is readily deduced that $i_{C1} = 0.6$ mA, $i_{C2} = 1.4$ mA, $v_{C1} = 10.2$ V, and $v_{C2} = 3.8$ V. The voltage v_o would then be $v_o = 3.8 - 10.2 = -6.4$ V.

When the two inputs are equal, there is very little change in circuit conditions, and the output v_o is essentially zero. When both inputs are equal, the amplifier is said to be excited with *common-mode* input signals. Thus a differential amplifier responds primarily to differential input signals and offers very little gain to common-mode signals.

Why is this property important? It turns out that many spurious noise components and undesired pickup signals tend to be of the common-mode type. However, an actual signal may be applied as a differential input. Consequently, the differential amplifier tends to provide a high degree of suppression for common-mode pickup, and this is one of its many virtues.

Thus far, the reader has been guided through a series of voltage–current conditions for a differential amplifier for the purpose of establishing a "feel" for the process. Because no general equations for the circuit have been established, we are not yet in a position to work out example problems.

The next two sections are used to establish the general gain and transfer characteristics for the basic differential amplifier circuit. Examples considered then return to the particular amplifier configuration considered here.

13-3 CASCADE ANALYSIS OF DIFFERENTIAL AMPLIFIER

One particular interpretation of the differential amplifier using small-signal analysis is made in this section. This particular interpretation is limited in predicting the complete analysis of the circuit, but it does provide an interesting insight, and it ties in nicely with the work of Chapter 12.

Consider the two-transistor differential amplifier configuration of Figure 13-7(a) with a single input v_{i1} and with the second base grounded (i.e., $v_{i2} = 0$). The only output of interest for this purpose is v_{c2}, the signal portion of v_{C2}. Assume that the transistors are *identical* and have small-signal parameters g_m and r_π.

One way to visualize this circuit is to consider it as an emitter follower in cascade with a common-base circuit, as shown in Figure 13-7(b). The emitter resistance could be considered as a part of either circuit, but it has been arbitrarily grouped with the emitter-follower circuit.

From the work of Chapter 12, the small-signal analysis of the complete circuit can be represented by a cascade arrangement in which the individual small-signal parameters of each circuit are identified. The composite form of such an arrangement is shown in Figure 13-7(c). The parameters here represent *signals*, so the instantaneous voltages v_E and v_{C2} have been replaced by small-signal voltages v_e and v_{c2}, respectively.

Because of the fact that the input resistance of the emitter follower is affected by the load of the common-base circuit, the parameters of the latter circuit are determined first. From the work of either Chapter 10 or the summary of Chapter 12, the input resistance of the common-base circuit, identified here as R_{in2}, is

$$R_{in2} = r_\pi \| \frac{1}{g_m} \simeq \frac{1}{g_m} \tag{13-19}$$

where the assumption $r_\pi \gg 1/g_m$ has been made.

The gain A_2 of the common-base circuit is

$$A_2 = g_m R_C \tag{13-20}$$

The output resistance of the common-base circuit is simply R_C and is labeled as such on Figure 13-7(c).

The input resistance of the emitter-follower circuit is

$$R_{in1} = r_\pi (1 + g_m R_E') \tag{13-21}$$

where

$$R_E' = R_E \| R_{in2} \tag{13-22}$$

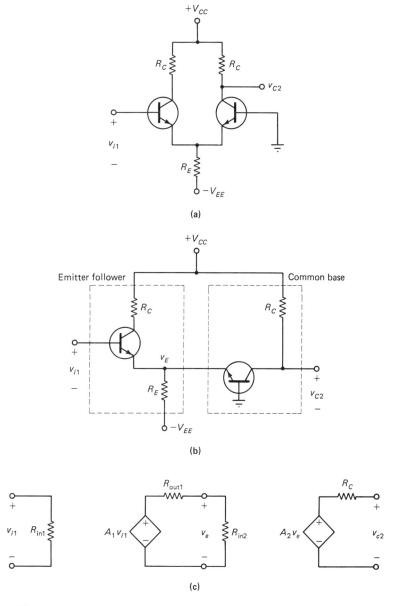

Figure 13-7 Interpretation of differential amplifier as emitter follower in cascade with common-base circuit.

Normally, $R_{in2} \ll R_E$, so (13-22) can be approximated as

$$R_E' \simeq R_{in2} = \frac{1}{g_m} \tag{13-23}$$

where (13-19) was employed. Substitution of (13-23) in (13-21) yields

$$R_{in1} = r_\pi \left(1 + g_m \frac{1}{g_m} \right) = 2r_\pi \tag{13-24}$$

The unloaded gain A_1 of the emitter-follower circuit is approximately

$$A_1 \simeq 1 \tag{13-25}$$

The output resistance R_{out1} of the common-collector circuit based on $v_{i1} = 0$ is

$$R_{out1} = R_E \| \frac{1}{g_m} \simeq \frac{1}{g_m} \tag{13-26}$$

where the assumption $R_E \gg 1/g_m$ has been made.

From Figure 13-7 we see that there is neither a generator nor an output loading factor to consider. There is, however, an interaction factor between the stages. The net gain between v_{i1} and v_{c2} will be denoted as A_{12} and it is

$$A_{12} = \frac{v_{c2}}{v_{i1}} = A_1 \frac{R_{in2}}{R_{in2} + R_{out1}} A_2 \tag{13-27}$$

Substitution of the various parameters in (13-27) results in

$$A_{12} = 1 \times \frac{1/g_m}{1/g_m + 1/g_m} g_m R_C = \frac{g_m R_C}{2} \tag{13-28}$$

This result is quite interesting in that it is one-half the gain of a common-base circuit (or one-half the magnitude of a common-emitter circuit). The factor of one-half can be thought of as arising from the interaction of the output resistance of the emitter-follower circuit and the input resistance of the common-base circuit. For equal transistor characteristics, these resistances are equal, and the signal voltage is halved at this point.

13-4 SMALL-SIGNAL ANALYSIS OF DIFFERENTIAL AMPLIFIER

In this section a rather detailed small-signal analysis of the two transistor differential amplifier is performed. The derivations involved are about the most unwieldy of any in the text, so do not feel intimidated if you do not understand all the steps at first. Like many electronic circuit derivations, there are some "tricks" and some approximations involved in the various steps. Try to focus on the basic concepts involved and the significance of the results obtained. If you are taking a course using this text, the instructor may choose to hold you responsible only for applying the results rather than deriving them.

Differential and Common-Mode Signals

Consider a differential amplifier of the form given back in Figure 13-1 with input signals v_{i1} and v_{i2}. The *differential input voltage* v_d was defined earlier, but for completeness, the definition is repeated here as

$$v_d = v_{i1} - v_{i2}$$

(13-29)

We now define the *common-mode input voltage* v_{cm} as

$$v_{cm} = \frac{v_{i1} + v_{i2}}{2}$$

(13-30)

The differential input voltage is the *difference* between the two input signals, and the common-mode input voltage is the *mean* or *average* of the two input signals. For example, if $v_{i1} = 8$ mV and $v_{i2} = 6$ mV, $v_d = 2$ mV and $v_{cm} = 7$ mV.

Equations (13-29) and (13-30) provide expressions for v_d and v_{cm} in terms of v_{i1} and v_{i2}. However, the latter two quantities can be treated as dependent variables, and by solving the two equations simultaneously, v_{i1} and v_{i2} can be expressed in terms of v_d and v_{cm}. The reader is invited to verify that the results of this process yield

$$v_{i1} = v_{cm} + \frac{v_d}{2}$$

(13-31)

$$v_{i2} = v_{cm} - \frac{v_d}{2}$$

(13-32)

For the numerical example considered earlier, substitution of $v_{cm} = 7$ mV and $v_d = 2$ mV, yields $v_{i1} = 8$ mV and $v_{i2} = 6$ mV as expected.

These forms may seem rather strange at this point. However, as we will see shortly, the effects of the amplifier on the quantities v_d and v_{cm} will be quite different, and some of the major properties of the differential amplifier can be most easily deduced from these forms.

Small-Signal Analysis

The small-signal equivalent circuit of the differential amplifier is shown in Figure 13-8. Note that the two input signals are represented in terms of differential and common-mode components.

The analysis will focus first on the determination of the emitter signal v_e. Once v_e is known, all other voltages and the currents may then be readily determined.

A KCL equation will now be written at the mode representing v_e. The currents can be expressed in terms of the various voltages. Assuming positive currents leaving the node, we have

$$\frac{v_e}{R_E} + \frac{v_e - v_{i1}}{r_\pi} + \frac{v_e - v_{i2}}{r_\pi} - g_m v_{be1} - g_m v_{be2} = 0$$

(13-33)

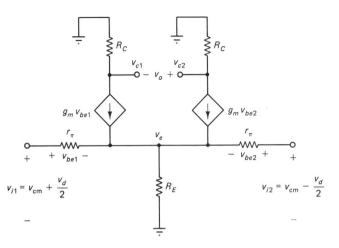

Figure 13-8 Small-signal equivalent circuit of differential amplifier.

The final form of the equation should contain no variables other than v_e, v_d, and v_{cm}. The voltages v_{be1} and v_{be2} may be first expressed in terms of v_{i1} and v_{i2} by applying KVL to the two loops at the bottom of the circuit. After simplification, we have

$$v_{be1} = v_{i1} - v_e \tag{13-34}$$

$$v_{be2} = v_{i2} - v_e \tag{13-35}$$

These relationships for v_{be1} and v_{be2} may first be substituted in (13-33). Next, the expressions for v_{i1} and v_{i2} in terms of v_d and v_{cm} may be substituted. The expression of (13-33) then becomes

$$\frac{v_e}{R_E} + \frac{v_e}{r_\pi} - \frac{v_{cm}}{r_\pi} - \frac{v_d}{2r_\pi} + \frac{v_e}{r_\pi} - \frac{v_{cm}}{r_\pi} + \frac{v_d}{2r_\pi} - g_m v_{cm} - \frac{g_m v_d}{2} + g_m v_e$$

$$- g_m v_{cm} + \frac{g_m v_d}{2} + g_m v_e = 0 \tag{13-36}$$

Note that the terms involving v_d cancel. The terms involving v_e are grouped on the left and those involving v_{cm} are grouped on the right. This results in

$$v_e\left(\frac{1}{R_E} + \frac{2}{r_\pi} + 2g_m\right) = \left(\frac{2}{r_\pi} + 2g_m\right)v_{cm} \tag{13-37}$$

Solving for v_e and simplifying the expression, we obtain

$$v_e = \frac{2(1 + g_m r_\pi)v_{cm}}{2(1 + g_m r_\pi) + r_\pi/R_E} \tag{13-38}$$

An approximation will be made later to simplify this expression. However, a significant point to make at this time is that the emitter signal voltage is affected only by the common-mode input voltage; that is, the effect of the differential voltage cancels as far as the emitter signal voltage is concerned.

Next we determine v_{c1} and v_{c2}. Since the signal currents flow from ground through the two collector resistances, we can write

$$v_{c2} = -g_m R_C v_{be2} \tag{13-39}$$

$$v_{c1} = -g_m R_C v_{be1} \tag{13-40}$$

It is necessary to express v_{be1} and v_{be2} in terms of v_d and v_{cm}. By employing the relationships of (13-34) and (13-35) and substituting v_e from (13-38), the following results can be determined:

$$v_{be1} = v_{i1} - v_e \tag{13-41a}$$

$$= v_{cm} + \frac{v_d}{2} - \frac{2(1 + g_m r_\pi)v_{cm}}{2(1 + g_m r_\pi) + r_\pi/R_E} \tag{13-41b}$$

$$v_{be2} = v_{i2} - v_e \tag{13-42a}$$

$$= v_{cm} - \frac{v_d}{2} - \frac{2(1 + g_m r_\pi)v_{cm}}{2(1 + g_m r_\pi) + r_\pi/R_E} \tag{13-42b}$$

When the preceding expressions are simplified over a common denominator, they become

$$v_{be1} = \frac{(1 + g_m r_\pi + r_\pi/2R_E)v_d + (r_\pi/R_E)v_{cm}}{2(1 + g_m r_\pi + r_\pi/R_E)} \tag{13-43}$$

$$v_{be2} = \frac{-(1 + g_m r_\pi + r_\pi/2R_E)v_d + (r_\pi/R_E)v_{cm}}{2(1 + g_m r_\pi + r_\pi/R_E)} \tag{13-44}$$

The preceding two expressions may be substituted in (13-39) and (13-40) to permit the determination of v_{c1} and v_{c2} in terms of v_d and v_{cm}. The results are

$$v_{c2} = \frac{g_m R_C(1 + g_m r_\pi + r_\pi/2R_E)v_d - g_m R_C(r_\pi/R_E)v_{cm}}{2(1 + g_m r_\pi + r_\pi/R_E)} \tag{13-45}$$

$$v_{c1} = \frac{-g_m R_C(1 + g_m r_\pi + r_\pi/2R_E)v_d - g_m R_C(r_\pi/R_E)v_{cm}}{2(1 + g_m r_\pi + r_\pi/R_E)} \tag{13-46}$$

At this point, we have carried these bulky expressions far enough, and some reasonable approximations are in order. In practice, $g_m r_\pi = \beta_o \gg 1$ and $= \beta_o \gg r_\pi/2R_E$. With these expressions, (13-45), (13-46), and the expression for v_e can be simplified to

$$v_e \simeq v_{cm} \tag{13-47}$$

$$v_{c2} \simeq \frac{g_m R_C}{2} v_d - \frac{R_C}{2R_E} v_{cm} \tag{13-48}$$

$$v_{c1} \simeq -\frac{g_m R_C}{2} v_d - \frac{R_C}{2R_E} v_{cm} \tag{13-49}$$

The two collector voltages are each a linear combination of the differential input voltage and the common-mode input voltage. The differential output voltage v_o measured between the collectors is

$$v_o = v_{c2} - v_{c1} \tag{13-50}$$

When (13-48) and (13-49) are substituted in (13-50), the common-mode components cancel, and we have

$$v_o = g_m R_C v_d \tag{13-51}$$

Let

$$v_o = A_d' v_d \tag{13-52}$$

where A_d' is the gain factor relating the differential output voltage to the differential input voltage. This quantity is

$$A_d' = \frac{v_o}{v_d} = g_m R_C \tag{13-53}$$

This gain has the same value as for a single-stage common-base circuit and the same magnitude as for a common-emitter circuit.

Although the differential output voltage is theoretically free of any common-mode components, the differential amplifier is most often used in a single-ended form with a single output. For v_{c2} as the output, we can write

$$v_{c2} = v_{cd} + v_{ccm} \tag{13-54}$$

where v_{cd} is the differential component of the voltage v_{c2} and v_{ccm} is the common-mode component of the voltage v_{c2}. Comparing (13-54) with (13-48), we can write

$$v_{cd} = A_d v_d \tag{13-55}$$

and

$$v_{ccm} = A_{cm} v_{cm} \tag{13-56}$$

The quantity A_d is the differential gain relating the differential portion of the single-ended output to the differential input, and it is

$$A_d = \frac{v_{cd}}{v_d} = \frac{g_m R_C}{2} \tag{13-57}$$

The quantity A_{cm} is the common-mode gain relating the common-mode portion of the single-ended output to the common-mode input, and it is

$$A_{cm} = \frac{v_{ccm}}{v_{cm}} = -\frac{R_C}{2R_E} \tag{13-58}$$

The differential gain for a single-ended output is one-half the value for a full differential output. This result is seen to be in agreement with the cascade interpretation of Section 13-3.

Common-Mode Rejection Ratio

A quantity called the *common-mode rejection ratio* (CMRR) will now be defined. In general, it is the ratio of the differential gain to the common-mode gain; that is,

$$\text{CMRR} = \frac{|A_d|}{|A_{\text{cm}}|} \qquad \text{(13-59)}$$

where the magnitude signs indicate that inverting gains are converted to positive values in the definition. Substitution of (13-57) and (13-58) for the circuit just analyzed yields

$$\text{CMRR} = g_m R_E \qquad \text{(13-60)}$$

The criterion for a differential amplifier is that the value of CMRR should be as large as possible. Note that the CMRR varies linearly with the value of R_E employed.

Example 13-1

Consider the differential amplifer of Section 13-2. Apply the small-signal relationships of this section to the conditions of Figure 13-4 to determine **(a)** the differential and common-mode input signals, **(b)** the differential gain (both for differential and single-ended output v_{c2}), common-mode gain, and common-mode rejection ratio, **(c)** the differential and common-mode output signals, and **(d)** the emitter signal voltage.

Solution

(a) The two input signals in Figure 13-4 are $v_{i1} = 20$ mV and $v_{i2} = 0$. The differential and common-mode input signals are

$$v_d = v_{i1} - v_{i2} = 20 - 0 = 20 \text{ mV} \qquad \text{(13-61)}$$

and

$$v_{\text{cm}} = \frac{v_{i1} + v_{i2}}{2} = \frac{20 + 0}{2} = 10 \text{ mV} \qquad \text{(13-62)}$$

(b) Before determining the various gains, we need to determine g_m. For $I_C = 1$ mA, the value of g_m is

$$g_m = 40 I_C = 40 \times 1 \times 10^{-3} = 40 \text{ mS} \qquad \text{(13-63)}$$

The differential gain A_d' relating the full differential output voltage to the differential input voltage is

$$A_d' = g_m R_C = 0.04 \times 8 \times 10^3 = 320 \qquad \text{(13-64)}$$

The differential gain A_d relating the differential component of v_{c2} to the differential input voltage is

$$A_d = \frac{g_m R_C}{2} = \frac{320}{2} = 160 \qquad (13\text{-}65)$$

The common-mode gain A_{cm} is

$$A_{cm} = \frac{-R_C}{2R_E} = \frac{-8 \times 10^3}{2 \times 7.15 \times 10^3} = -0.559 \qquad (13\text{-}66)$$

The common-mode rejection ratio is

$$\text{CMRR} = \frac{|A_d|}{|A_{cm}|} = \frac{160}{0.559} = 286 \qquad (13\text{-}67)$$

(c) For the differential and common-mode components of (a), the various output quantities are determined as follows:

$$v_o = A_d' v_d = 320 \times 0.02 = 6.4 \text{ V} \qquad (13\text{-}68)$$

$$v_{cd} = A_d v_d = 160 \times 0.02 = 3.2 \text{ V} \qquad (13\text{-}69)$$

$$v_{ccm} = A_{cm} v_{cm} = -0.559 \times 0.01 \approx -5.6 \text{ mV} \qquad (13\text{-}70)$$

$$v_{c2} = v_{cd} + v_{ccm} = 3.2 \text{ V} - 5.6 \text{ mV} \approx 3.2 \text{ V} \qquad (13\text{-}71)$$

(d) The emitter signal voltage v_e is

$$v_e = v_{cm} = 10 \text{ mV} \qquad (13\text{-}72)$$

Several comments concerning these results are in order. The value of v_o from Figure 13-4 is seen to be in agreement with (13-68). For the particular levels of v_{i1} and v_{i2} in this example, the common-mode component of v_{c2} is, for most practical purposes, negligible and $v_{c2} \approx 3.2$ V as noted from (13-71). (Example 13-2 will illustrate a different situation, however.) The value of $v_{c2} = 3.2$ V is the *signal* voltage at the collector of T2. Consequently, this is recognized on Figure 13-4 as a change from the quiescent value of 7 V to a value of 10.2 V when the signal is applied. Similarly, the emitter signal voltage of 10 mV is recognized as a change in the emitter voltage from -0.7 V to -0.69 V in Figure 13-4.

With this discrete-component circuit, the single-ended output thus contains a dc level, which would limit its use to ac coupling for most applications. However, we will later consider the use of dc-level translation circuitry to change the dc level so that dc coupling can be employed.

Example 13-2

Consider again the differential amplifier of Section 13-2. However, assume now that the two input voltages are $v_{i1} = 2.01$ V and $v_{i2} = 1.99$ V. Determine (a) the differential and common-mode input signals, (b) the differential and common-mode output signals, and (c) the emitter signal voltage.

Solution

(a) The differential and common-mode input signals are

$$v_d = v_{i1} - v_{i2} = 2.01 - 1.99 = 0.02 \text{ V} = 20 \text{ mV} \qquad \textbf{(13-73)}$$

and

$$v_{cm} = \frac{2.01 + 1.99}{2} = 2 \text{ V} \qquad \textbf{(13-74)}$$

The differential input voltage is the same as in Example 13-1, but the common-mode input voltage is considerably larger.

(b) The gain values computed in Example 13-1 will be used freely here. The various output quantities are

$$v_o = A_d' v_d = 320 \times 0.02 = 6.4 \text{ V} \qquad \textbf{(13-75)}$$

$$v_{cd} = A_d v_d = 160 \times 0.02 = 3.2 \text{ V}$$

$$v_{ccm} = A_{cm} v_{cm} = -0.559 \times 2 = -1.12 \text{ V} \qquad \textbf{(13-76)}$$

$$v_{c2} = v_{cd} + v_{ccm} = 3.2 - 1.12 = 2.08 \text{ V} \qquad \textbf{(13-77)}$$

(c) The emitter signal voltage is

$$v_e = v_{cm} = 2 \text{ V} \qquad \textbf{(13-78)}$$

Comparing these results with those of Example 13-1, the common-mode component is seen to have a significant effect on v_{c2} in the present case.

13-5 SOME IMPORTANT INTEGRATED-CIRCUIT FUNCTIONS

Now that we have analyzed the operation of a discrete component differential amplifier, the next step is to discuss the evolution of that basic circuit to its integrated circuit form. However, it is desirable to first observe a few basic functions that are achieved with integrated-circuit components. It turns out that some of the bias circuit and resistance loads for transistors are implemented with additional transistors. From a fabrication point of view, transistors are easier to realize in many cases than resistors and diodes. The resulting circuits appear to be overwhelmed with transistors, but many of these transistors are serving as active loads and bias sources.

Current Mirror

The first circuit that will be considered is the *current mirror*. This circuit has been used both in discrete component circuits and in integrated circuits.

Refer to Figure 13-9(a). The objective of the circuit is to establish a specified current I_C in the collector of the transistor. No particular connection is shown for the collector terminal as it may be connected in a number of ways depending on the application. However, it is assumed that it is connected through some appropriate load to a proper bias circuit.

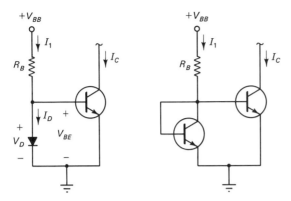

(a) Using a diode   (b) Using a transistor connected as a diode   **Figure 13-9** Current mirror.

The base of the transistor is connected to the diode circuit on the left, which is returned through a resistance R_B to a positive voltage V_{BB}. This voltage may or may not be the voltage used to bias the collector of the transistor.

The current I_1 through the resistance is readily expressed as

$$I_1 = \frac{V_{BB} - V_D}{R_B} = \frac{V_{BB} - 0.7}{R_B} \tag{13-79}$$

where $V_D = 0.7$ V has been assumed. Now assume that the transistor base–emitter diode has the same characteristics as that of the diode on the left; i.e., the emitter current (and in turn the collector current) corresponding to a given base–emitter voltage is the same as the diode current for a given diode voltage. The two voltages are clearly the same since they are connected in parallel (i.e., $V_D = V_{BE}$). For identical characteristics, therefore, the collector current is forced to be essentially the same as the diode current, that is,

$$I_C \approx I_1 = \frac{V_{BB} - 0.7}{R_B} \tag{13-80}$$

where the base current of the transistor has been neglected. Thus R_B is selected to produce a specified current on the left, and the *mirror image* of that current flows in the collector on the right.

In many integrated-circuit chips, it is easier to create a diode from a transistor. This is achieved by trying the base and collector terminals together, and this configuration is shown in Figure 13-9(b). The transistor on the left is thus functioning as a diode. On an integrated circuit, it is easier to approach identical characteristics than with discrete components since the two transistors are formed from the same semiconductor material. Thus the expression for I_C in Figure 13-9(b) is the same as in Figure 13-9(a).

Normally, $V_{BB} \gg V_{BE}$, so variations of V_{BE} with temperature have very little effect on I_B and I_C. The current established is primarily a function of V_{BB} and R_B, and variations of V_{BE} with temperature tend to be compensated for by the mirroring process.

Active Load

The current mirror circuit may be used to establish a resistive load to function in the same manner as a resistance. The resulting circuit is referred to as an *active load*.

One form of an active load that can be used as the collector resistance in a differential amplifier is shown in Figure 13-10(a). Note that the BJT is a *pnp* type, but as shown, it will serve as the collector load for an *npn* transistor. The corresponding form employing a transistor acting as a diode is shown in Figure 13-10(b). The latter form is used extensively in integrated circuits. Both forms are also referred to as *current sources*.

The reason that these circuit forms behave like resistors is the fact that the dynamic collector resistance is very large; that is, the slope of the collector voltage versus current characteristic in the operating region is very large. The active load is particularly useful where a large dynamic resistance is required but where only a moderate dc voltage drop across the component is feasible.

Integrated-Circuit Differential Amplifier Stage

A representative integrated-circuit differential amplifier stage is shown in Figure 13-11. The circuit shown here does not represent any specific amplifier, but it is representative of a number of different types.

The *npn* transistors T1 and T2 represent the two input stages of the differential amplifier. The *pnp* transistors T3 and T4 form an active load pair. Transistor T3 is functioning as the diode for T4, so the drop across T3 is constant. However, T4 acts as a dynamic load for T2. Since only T2 has an active load, the output v_{C2} is a single-ended type.

The transistor pair T5 and T6 act as a current source for the differential pair (i.e., they replace the emitter resistance R_E). Because of the large dynamic resistance compared with an actual practical resistor, the common-mode rejection ratio can be made to be very large.

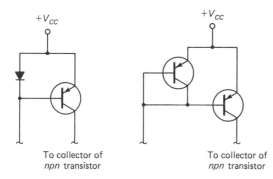

(a) Using a diode   (b) Using a transistor connected as a diode

To collector of *npn* transistor

To collector of *npn* transistor

Figure 13-10 Active collector load obtained with a current mirror.

+V_CC

T3   T4

v_{C2}

T1 v_{i1} v_{i2} T2

T6   T5

−V_EE

Figure 13-11 Typical integrated-circuit differential amplifier stage.

Example 13-3

Determine the value of I_C established by the current mirror of Figure 13-12.

+15 V

I_C

5.1 kΩ   2 kΩ

Figure 13-12 Circuit of Example 13-3.

Solution

The transistor on the left is functioning as an active diode, and its collector current establishes the collector current for the transistor on the right. Neglecting base currents and assuming that $V_{BE} = 0.7$ V, we have

$$I_C = \frac{15 - 0.7}{5.1 \times 10^3} = 2.80 \text{ mA} \qquad \text{(13-81)}$$

Strictly speaking, this result is valid only if the transistor on the right is not saturated. The voltage drop across the 2-kΩ resistor is readily determined as $2.80 \times 10^{-3} \times 2 \times 10^3 = 5.60$ V. The collector–emitter voltage is $15 - 5.60 = 9.40$ V. Active-region operation is thus assured.

13-6 OPERATIONAL AMPLIFIER

We now turn our attention to the modern integrated-circuit *operational amplifier*, hereafter referred to frequently by the abbreviated term *op-amp*. From an external point of view, an op-amp may be considered as a high-gain dc-coupled differential amplifier. Depending on the op-amp, it will generally contain the following circuits: (1) one or more cascade stages of basic differential amplifier circuits, (2) dc-level shifting or translation circuits to establish a reference-zero output voltage without coupling capacitors, (3) a compensation circuit to prevent instability and potential oscillations, and (4) a low-resistance output stage (often an emitter follower or source follower).

At this point in the development, it is not meaningful to continue working with the detailed circuit diagrams as we have done with the simpler two-transistor differential amplifier. Instead, the basic integrated-circuit operational amplifier will be accepted as a "building block" for many circuit functions, and a simplified approach to the modeling and layout will be assumed.

Before we develop this simplified approach, the actual schematic diagram of a typical operational amplifier will be observed, and this is shown in Figure 13-13. Do

Figure 13-13 Schematic diagram of typical integrated-circuit operational amplifier. (Courtesy of Fairchild Semiconductor Corporation.)

not be intimidated by this circuit diagram since we will not analyze it in detail. Actually, most of the circuitry shown here either directly or indirectly establishes the operation of various differential amplifier stages, whose functions are the same as those of the much simpler discrete-component type considered earlier. If you inspect the various transistors closely, you will see a number of transistors functioning as active loads and as current sources. This entire operational amplifier is available in a small chip, at a low price, and a user does not need to understand all the details of the actual circuit to use it properly. However, the basic understanding of the differential amplifier pair considered earlier is important as an aid in the overall operation and utilization of the IC chip.

Operational Amplifier Bias Connection

Most operational amplifiers require two dc bias supplies for full dynamic range operation. Normally, this is achieved with two separate dc power supplies. The most common arrangement is shown in Figure 13-14. One bias supply is established as a positive voltage of value V_{BB} with respect to ground, and one is established as a negative voltage $-V_{BB}$ with respect to ground. The most common values are $+15$ V and -15 V. The operational amplifier provides terminals for the two voltages. However, the ground terminal itself is normally not connected to the op-amp. Rather, all ground connections are referred to the common point between the two power supplies as shown in Figure 13-14. Referring back to Figure 13-13, the two bias points on this particular op-amp are identified as V+ and V−.

Operational Amplifier Symbols and Conventions

The most widely employed symbol for the operational amplifier, both with bias supplies connected and without bias supplies connected, is shown in Figure 13-15. In (a), the two bias connections are indicated as $+V_{BB}$ above and $-V_{BB}$ below the op-amp. Unless there is a particular reason for focusing on the power supply connections, it is customary in many op-amp circuit diagrams to omit the power supply connections. Thus they will be omitted from most subsequent diagrams, but their presence must be understood for operation.

The op-amp symbol with the power supply connections omitted and understood is shown in Figure 13-15(b). Most IC op-amps use a single-ended output, and it will be denoted here as v_o. Unlike the v_o variable used in the discrete-component

Figure 13-14 Dual power supply arrangement for biasing an op-amp.

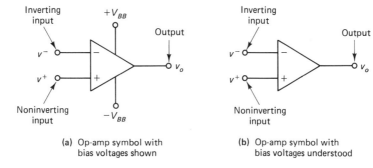

(a) Op-amp symbol with   (b) Op-amp symbol with
    bias voltages shown   bias voltages understood

Figure 13-15   Operational amplifier symbol with power supply
connections shown and with power supply connections understood.

circuit, however, this voltage is referred to ground. The earlier notation for inputs
(v_{i1} and v_{i2}) will be dropped at this point for reasons that will be clear later. Instead,
the two symbols v^+ and v^- will be used. (Do not confuse these lowercase instantaneous symbols with the op-amp power supply connections V+ and V− shown earlier on the representative op-amp schematic.)

The input labeled as (+) and denoted as v^+ is called the *noninverting input*.
The input labeled as (−) and denoted as v^- is called the *inverting input*. The differential input voltage v_d is defined as

$$v_d = v^+ - v^- \qquad\qquad (13\text{-}82)$$

The operational amplifier is characterized by a differential gain A_d, which is
typically 100,000 or more at dc and very low frequencies. The ideal output voltage
is related to the two input voltages as

$$v_o = A_d(v^+ - v^-) = A_d v_d \qquad\qquad (13\text{-}83)$$

Thus the op-amp output responds to the difference between the two inputs. This is
basically the same process as encountered for the simpler two transistor differential
amplifier. A major difference, however, is that A_d for the op-amp is considerably
larger than for the simpler differential amplifier circuit.

It should be stressed that the + and − symbols at the op-amp input terminals
have nothing to do with polarities of voltages at the inputs. Either input voltage may
be positive or negative. Rather, it is the sense of the amplification that is indicated.
A voltage of either polarity applied to the noninverting input will cause the output to
shift in the direction of the input. However, a voltage of either polarity applied to the
inverting input will cause the output to shift in the opposite direction of the input.

Transfer Characteristics

A typical transfer characteristic curve of an op-amp is shown in Figure 13-16. The
output voltage v_o versus the differential input voltage v_d is displayed. The slope of
the characteristic curve in the steep region is the differential gain A_d. Because of the

Figure 13-16 Typical transfer characteristic of an op-amp (different horizontal and vertical scales).

very large gain, the horizontal scale here is quite different from the vertical scale. If the scales were the same, the steep line would appear nearly vertical. Note that the ideal output voltage is zero when the differential input voltage is zero. This is true even though there are no coupling capacitors in the op-amp. This dc cancellation is achieved by level-shifting circuitry and the dual power supplies.

Saturation

The curve of Figure 13-16 was based on assumed power supply voltages of ± 15 V. It is observed that as the output voltage starts to approach either $+15$ V or -15 V, the characteristic curve starts to level off. Theoretically, the output could go as high as $+15$ V or as low as -15 V. In practice, however, certain transistors on the chip start to saturate before these levels are reached, and with most op-amps, the absolute power supply voltages are never quite reached.

The quantities $+V_{sat}$ and $-V_{sat}$ will be defined as the positive and negative saturation voltages, respectively. Typical values of V_{sat} range from 13 to 14 V, but they vary from one op-amp to another. Thus if $\pm V_{sat} = \pm 13$ V, *active-region* operation is from -13 to $+13$ V. As a good rule of thumb, when information is lacking, allow about 2 V margin or "backoff" from the two power supply limits. Thus for ± 15-V power supplies, the output should be able to vary from about -13 to $+13$ V.

Example 13-4

On the typical input/output characteristic curve of Figure 13-16, assume that the saturation level of 13 V is reached for a differential input voltage of 50 μV. Compute the differential gain A_d.

Solution

The differential gain is determined from the slope of the curve as

$$A_d = \frac{13 \text{ V}}{50 \times 10^{-6} \text{ V}} = 260,000 \qquad \text{(13-84)}$$

Example 13-5

A certain op-amp is characterized by a dc differential gain of $A_d = 120,000$. What is the value of differential input voltage that would just barely cause the output to go into saturation if $V_{\text{sat}} = 13.5$ V?

Solution

The output v_o is related to the differential input by

$$v_o = A_d v_d \qquad \text{(13-85)}$$

Solving for v_d and substituting $v_o = 13.5$ V and $A_d = 120,000$, we have

$$v_d = \frac{v_o}{A_d} = \frac{13.5 \text{ V}}{120,000} = 112.5 \ \mu\text{V} \qquad \text{(13-86)}$$

13-7 OPERATIONAL AMPLIFIER SPECIFICATIONS

We will now inspect some typical operational amplifier specifications to see how pertinent data concerning its operating characteristics are determined. The op-amp selected for this purpose is the widely used 741. This particular op-amp was the first to employ internal stabilization and was introduced by Fairchild Semiconductor in the late 1960s. Variations of the 741 are now produced by a number of different manufacturers.

Refer to the μA741 data sheets in Appendix D for the discussion that follows. (The prefix "μA" is one used by Fairchild. Different manufacturers utilize prefixes with specific device numbers.)

The first page of the specifications provides data on the various packages and pin diagrams. The two power supply terminals are denoted as V+ and V−, respectively. Again, do not confuse these symbols with the v^+ and v^- symbols that we are using in the text for the instantaneous input signal variables.

Some terminals that we have not considered thus far are the "+offset null" and "−offset null" terminals. These are used when it is necessary to provide fine adjustments on the dc output level with no signal applied. They need not be connected for noncritical applications. The use of these terminals will be discussed later in this section.

The "absolute maximum ratings" are also given on the first page of the specifications. The maximum values of the supply voltages are listed as ±22 V for three variations of the 741 and ± 18 V for the 741C. The maximum power dissipation depends on the type of package as noted.

The maximum values of the differential input voltage are ± 30 V. However, the maximum range of either input voltage is ± 15 V when power supplies of ± 15 V are employed. From the footnote it is seen that these voltages cannot extend beyond the range of the power supply voltages when the op-amp uses lower supply voltages.

Following the schematic on the second page, the next several pages provide various electrical characteristics of different versions of the 741 both at fixed temperature ($T_A = 25°C$) and over a temperature range. Not all of these data are appropriate to be considered at this time, but some entries will be considered. To simplify the discussion that follows, only the 741 data will be discussed. As can be verified, data for the 741A, 741C, and 741E differ somewhat.

Input Offset Voltage

Ideally, the op-amp should have a dc output voltage of zero when the differential input voltage is zero. In practice, a small offset (either positive or negative) will appear at the output. This offset as it appears at the output turns out to be a function of the external dc gain established for the composite circuit, and it is more meaningful to refer the voltage to the input in specifications. Hence the name *input offset voltage* is used to denote it. By referring the voltage to the input, it is easier to predict the actual offset at the output from the given specification and the composite circuit gain.

The circuit model that can be used to predict the effect of dc input offset voltage is shown in Figure 13-17. The input offset voltage is shown as a dc voltage of value V_{IO} referred to the noninverting input. This effect will be ignored except for certain applications where it is critical. Scanning through the electrical characteristics, the "input offset voltage" for the 741 at 25°C is observed to have a typical value of 1.0 mV and a maximum value of 5.0 mV.

Figure 13-17 Circuit model used to show effect of dc input offset voltage.

Input Bias and Offset Currents

The dc bias current used to establish the operating points of the op-amp input transistors must flow through any external circuit connected. For that reaon it is necessary that *there be a dc path to ground at both op-amp input terminals for proper circuit operation.*

Let I_B^+ represent the bias current for the noninverting input, and let I_B^- represent the input bias current for the inverting input. Let I_{IB} represent the average input bias current, which is defined as

$$I_{IB} = \frac{I_B^+ + I_B^-}{2} \qquad\qquad \text{(13-87)}$$

The value of this average bias current is denoted on specifications simply as "input bias current." For the 741 at 25°C, it has a typical value of 80 nA and a maximum value of 500 nA.

Another dc quantity of importance is the "input offset current," which will be denoted as I_{IO}, and it is defined as the magnitude of the difference between the two bias currents, that is,

$$I_{IO} = |I_B^+ - I_B^-| \qquad\qquad \text{(13-88)}$$

If I_B^+ and I_B^- were exactly equal, the input offset current would be zero. In practice, however, they differ somewhat, so the difference is not zero.

For the 741 at 25°C, the typical value of I_{IO} is 20 nA, and the maximum value is 200 nA.

Offset Null Circuit

As noted previously, there is usually a small dc output voltage at the op-amp even when the differential input voltage is zero. In some applications, this is of no consequence. However, when the op-amp is used to process small dc signals, this offset may be troublesome and it is desirable to null it out.

On the eighth page of the specifications is a circuit diagram showing the method recommended by the manufacturer for nulling out this voltage. A 10-kΩ potentiometer is connected between pins 1 and 5 and the negative power supply as shown. The potentiometer is adjusted for a null at the output.

The actual nulling procedure is performed with the op-amp connected to the external circuit resistances expected for the particular application since the actual offset is dependent on the external connections.

Bias-Compensating Resistor

Bias currents flowing in any resistances connected to the two op-amp inputs will produce small dc voltages, whose combined effects appear at the output as a dc offset. The dc offset resulting from bias currents is in addition to that produced by the input offset voltage and adds algebraically to the latter.

The offset produced by the bias currents can be minimized, whenever possible, by forcing the effective dc resistances to ground from the two op-amp inputs to be equal. Since the two bias currents are usually in the same range of values, this condition will result in two nearly equal dc voltages at the two inputs, and since their effects on the output voltage are in opposite directions, there will be a near-cancellation process. Theoretically, if the two bias currents were equal, and if the resistances to ground were also equal, perfect cancellation would result.

Any resistance connected from an input to the output of the op-amp is considered as being connected to ground for the purpose of determining a bias resistance. This is true because the op-amp output presents a very low effective resistance to ground. External inputs are considered as being connected to ground through any source resistance present.

To illustrate the process of bias compensation, consider the inverting amplifier circuit of Figure 13-18(a). Full operation of this circuit is considered in detail in Chapters 14 and 15. For the moment, we are concerned only with the process of bias compensation.

The basic form of Figure 13-18(a) will work fine for many applications where small dc levels are not encountered. However, in applications where offset effects are to be minimized, a bias compensation resistance R_c may be added as shown in Figure 13-18(b). Assuming an ideal source at the input and considering it momentarily replaced by a short, the net resistance to ground from the inverting input is the parallel combination of R_i and R_f (i.e., $R_i \parallel R_f$). A bias-compensation resistance R_c having this value is connected between the noninverting input and ground as shown. This resistor does not affect the primary operation of the circuit, but it does minimize the bias current offset effect.

Input Resistance

The input resistance is referred to by the more general term "input impedance" and is denoted as Z_i. The minimum value for the 741 is indicated as 0.3 MΩ, and the typical value is 2 MΩ.

Differential Gain

The differential gain of the 741 is found on the line indicated as "large-signal voltage gain." The minimum value for the 741 can be interpreted as 50,000, and a typical value is 200,000. These are the values of gain at dc.

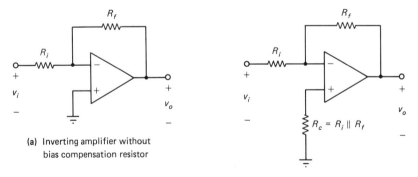

(a) Inverting amplifier without bias compensation resistor

(b) Addition of bias compensation resistor R_c

Figure 13-18 Manner in which bias compensation resistor is added to one amplifier configuration.

More information on gain is shown as curves on subsequent pages. On the sixth page of the specifications, a curve entitled "open-loop voltage gain as a function of supply voltage" is shown. As in the preceding listing, this is dc gain, and it is given in dB. The supply voltage is interpreted as both positive and negative (i.e., dual power supply voltages). For ±15 V, the gain is about 106 dB (or 200,000), which agrees with the typical value listed earlier.

On the ninth page of the specifications, a curve showing "open-loop frequency response" is given. The gain below 10 Hz is about 200,000, and this is in agreement with the typical value stated earlier. Observe, however, that this open-loop gain starts decreasing below 10 Hz and reaches a level of unity just above 1 MHz. The effect of this roll-off will be considered later.

Output Voltage Swing

On the seventh page of the specifications is a curve entitled "output voltage swing vs. supply voltage." The abscissa is interpreted as ± supply voltages. Note for bias supplies of ±15 V, the output voltage swing is given as a peak-to-peak value of 26 V. This corresponds to $+V_{\text{sat}} = 13$ V and $-V_{\text{sat}} = -13$ V, which is in agreement with our earlier rule of thumb.

Common-Mode Rejection Ratio

The common-mode rejection ratio (CMRR) was defined earlier, but the definition is repeated here. It is

$$\text{CMRR} = \frac{|A_d|}{|A_{\text{cm}}|} \tag{13-89}$$

where A_d is the differential gain and A_{cm} is the common-mode gain. On most specifications, CMRR is given in dB and this can be expressed as

$$\text{CMRR(dB)} = 20 \log_{10} \text{CMRR} \tag{13-90}$$

Values of common-mode rejection ratio are given on the third and fourth pages of the specifications. The minimum value is listed as 70 dB, and a typical value is listed as 90 dB. These correspond to absolute levels of 3162 and 31,623, respectively. Note how much larger these values are than the value computed in Example 13-1 for the single-stage differential amplifier.

It turns out that the common-mode rejection ratio varies with frequency. A curve displaying the variation with frequency is shown on the sixth page of the specifications.

DRILL PROBLEMS

13-1. Consider the circuit of Figure P13-1, and let $V_{BB} = 12$ V, $R_E = 11.3$ kΩ, and $R_C = 10$ kΩ. Initially assume that both base terminals are grounded.

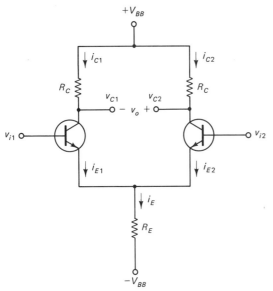

+V_{BB}

i_{C1}

i_{C2}

R_C

R_C

v_{C1}

v_{C2}

$- v_o +$

v_{i1}

v_{i2}

i_{E1}

i_{E2}

i_E

R_E

$-V_{BB}$

Figure P13-1

(a) Determine the dc emitter current I_E. (b) At the dc operating point, determine the values of i_{C1}, i_{C2}, v_{C1}, v_{C2}, and v_o.

13-2. Consider the circuit of Figure P13-1, and let $V_{BB} = 8$ V, $R_E = 1.825$ kΩ, and $R_C = 2$ kΩ. Initially assume that both base terminals are grounded. (a) Determine the dc emitter current I_E. (b) At the dc operating point, determine the values of i_{C1}, i_{C2}, v_{C1}, v_{C2}, and v_o.

13-3. For the circuit of Problem 13-1, determine the values of (a) A_d', (b) A_d, (c) A_{cm}, and (d) CMRR.

13-4. For the circuit of Problem 13-2, determine the values of (a) A_d', (b) A_d, (c) A_{cm}, and (d) CMRR.

13-5. For the circuit of Problems 13-1 and 13-3, assume that the two inputs are $v_{i1} = 12$ mV and $v_{i2} = 0$. Determine (a) the differential and common-mode input signals, (b) the differential and common-mode output signals, and (c) the emitter signal voltage.

13-6. For the circuit of Problems 13-2 and 13-4, assume that the two inputs are $v_{i1} = 10$ mV and $v_{i2} = 0$. Determine (a) the differential and common-mode input signals, (b) the differential and common-mode output signals, and (c) the emitter signal voltage.

13-7. Repeat the analysis of Problem 13-5 if $v_{i1} = 2.012$ V and $v_{i2} = 2$ V.

13-8. Repeat the analysis of Problem 13-6 if $v_{i1} = 1.5$ V and $v_{i2} = 1.49$ V.

13-9. Repeat the analysis of Problem 13-5 if $v_{i1} = -1.8$ V and $v_{i2} = -1.79$ V.

13-10. Repeat the analysis of Problem 13-6 if $v_{i1} = -2$ V and $v_{i2} = -2.01$ V.

13-11. In the circuit of Figure P13-11, let $V_{BB} = 12$ V and $R_B = 5$ kΩ. Determine I_C.

+V_{BB}

R_B

I_C

Figure P13-11

13-12. In the circuit of Figure P13-11, let V_{BB} = 24 V and R_B = 4 kΩ. Determine I_C.

13-13. In the circuit of Figure P13-11, let V_{BB} = 20 V. Determine the value of R_B required to establish a collector current I_C = 5 mA.

13-14. In the circuit of Figure P13-11, let V_{BB} = 15 V. Determine the value of R_B required to establish a collector current I_C = 10 mA.

13-15. The transfer characteristic of a certain op-amp, based on ±15-V power supplies, is shown in Figure P13-15. (Note that the horizontal and vertical scales are drastically different.) Determine **(a)** the saturation voltages and **(b)** the differential gain.

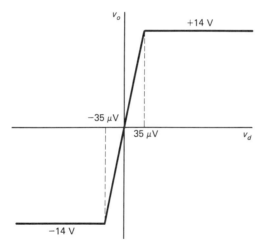

v_o

+14 V

−35 μV

35 μV

v_d

−14 V

Figure P13-15

13-16. The saturation voltages of a certain op-amp, based on ±15 V power supplies, are ±13.5 V. If saturation occurs for a differential input voltage of 50 μV, determine the differential gain.

13-17. A certain op-amp has a dc differential gain of 240,000. Determine the value of differential input voltage that would just barely cause the output to go into saturation if V_{sat} = 13.2 V.

13-18. A certain op-amp has a dc differential gain of 106 dB. Determine the value of differential input voltage that would just barely cause the output to go into saturation if $V_{\text{sat}} = 14$ V.

13-19. A certain op-amp has input bias currents $I_B^+ = 110$ nA and $I_B^- = 100$ nA. Determine the average input bias current I_{IB}.

13-20. A certain op-amp has bias currents $I_B^+ = 40$ nA and $I_B^- = 36$ nA. Determine the average input bias current I_{IB}.

13-21. Determine the input offset current for the op-amp of Problem 13-19.

13-22. Determine the input offset current for the op-amp of Problem 13-20.

13-23. A certain op-amp has $A_d = 200{,}000$ and $A_{\text{cm}} = 50$. Determine the values of CMRR and CMRR(dB).

13-24. A certain op-amp has $A_d = 80{,}000$ and $A_{\text{cm}} = 12$. Determine the values of CMRR and CMRR(dB).

13-25. A certain op-amp has a specified differential gain of 120,000 and a common-mode rejection ratio of 100 dB. Determine the value of the common-mode gain A_{cm}.

13-26. A certain op-amp has a specified differential gain of 100 dB and a common-mode rejection ratio of 80 dB. Determine the value of the common-mode gain A_{cm}.

DERIVATION PROBLEMS

13-27. Starting with (13-29) and (13-30), derive (13-31) and (13-32).

13-28. Assume that the output v_o of a differential amplifier can be expressed in terms of the two inputs v_1 and v_2 as

$$v_o = A_1 v_1 + A_2 v_2$$

where A_1 and A_2 are gain factors relating to the two separate inputs. The output v_o can also be expressed in terms of differential input v_d and common-mode input v_{cm} as

$$v_o = A_d v_d + A_{\text{cm}} v_{\text{cm}}$$

where A_d and A_{cm} are the differential and common-mode gains and v_d and v_{cm} are defined as

$$v_d = v_1 - v_2$$

and

$$v_{\text{cm}} = \frac{v_1 + v_2}{2}$$

Show that

$$A_d = \frac{A_1 - A_2}{2}$$

and

$$A_{cm} = A_1 + A_2$$

13-29. Referring to Problem 13-28, show that

$$A_1 = A_d + \frac{A_{cm}}{2}$$

and

$$A_2 = -A_d + \frac{A_{cm}}{2}$$

13-30. Referring to Problem 13-28, assume that $A_1 = A$ and $A_2 = -A$. Determine A_d and A_{cm}, and comment on the significance of the result.

COMPUTER PROBLEMS

13-31. Write a program to determine the various gain parameters of a differential amplifier of the form shown in Figure 13-1.

Input data: V_{EE}, R_E, R_C
Output data: A_d', A_d, A_{cm}, CMRR

13-32. Write a program to determine input and output signal quantities of a differential amplifier of the form shown in Figure 13-1 when excited by two input signals.

Input data: A_d, A_{cm}, v_{i1}, v_{i2}
Output data: v_{c1}, v_{c2}, v_o, v_{cd}, v_{ccm}, v_e

CHAPTER 14

Negative Feedback

OBJECTIVES

After completing this chapter, the reader should be able to:

1. Discuss the general concept of negative feedback and its advantages.
2. Draw a block diagram depicting a negative feedback system, and derive the equation relating closed-loop gain to open-loop gain and the feedback factor.
3. Show how the closed-loop gain is very nearly dependent only on the feedback factor and virtually independent of the forward gain when the loop gain is very large.
4. Show how the op-amp noninverting amplifier configuration fits the general feedback block diagram, and relate the various terms in the two cases.
5. Derive the closed-loop gain expression for the noninverting amplifier using the feedback equation.
6. Determine the gain, input resistance, and output resistance of the noninverting op-amp configuration.
7. Determine the gain, input resistance, and output resistance of the inverting op-amp configuration.

14-1 NEGATIVE FEEDBACK

One of the most important developments in the history of electronics was the concept of *negative feedback*. The use of negative feedback started in the vacuum-tube era and continued on through the transistor era. It remains as important as ever in the application of modern integrated circuits.

The development of the concept of negative feedback will utilize a block diagram form, from which the results may be adapted to specific devices and circuits. Consider the block diagram shown in Figure 14-1. The diagram contains three important elements required in a negative feedback system. They are (1) the *open-loop*

542

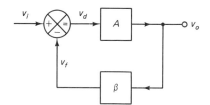

Figure 14-1 Block diagram illustrating negative feedback concept.

forward gain A, (2) the *feedback factor* β, and (3) the *input difference circuit* shown on the left.

The *open-loop forward gain A* represents the gain of the amplifier without feedback. The *feedback factor* represents the fraction of the output that is fed back to the input. The input *difference circuit* combines the feedback signal with the input. Specifically, the feedback signal is *subtracted* from the input signal in the model shown.

Assume that the input signal is v_i and the feedback signal is v_f. The output of the difference circuit is a voltage v_d given by

$$v_d = v_i - v_f \qquad \text{(14-1)}$$

This difference voltage is amplified by the gain A to produce the output voltage v_o according to

$$v_o = Av_d \qquad \text{(14-2)}$$

The feedback voltage v_f is proportional to the output voltage v_o and is

$$v_f = \beta v_o \qquad \text{(14-3)}$$

When (14-3) is substituted in (14-1), there results

$$v_d = v_i - \beta v_o \qquad \text{(14-4)}$$

Substitution of (14-4) in (14-2) results in

$$v_o = A(v_i - \beta v_o) = Av_i - A\beta v_o \qquad \text{(14-5)}$$

This last result may seem strange since v_o is expressed in terms of both the input v_i and itself. However, by regrouping all v_o terms on the left-hand side of the equation, v_o can be expressed in terms of the input. The regrouping procedure results in

$$v_o + A\beta v_o = Av_i \qquad \text{(14-6)}$$

Let A_{CL} represent the *closed-loop gain* (i.e., the *gain with feedback*). It is defined as

$$A_{\mathrm{CL}} = \frac{v_o}{v_i} \qquad \text{(14-7)}$$

Factoring out v_o on the left side of (14-6) and forming the ratio of (14-7), we obtain

$$\boxed{A_{\mathrm{CL}} = \frac{A}{1 + A\beta}} \qquad \text{(14-8)}$$

The quantity $A\beta$ in the denominator has a special significance and is defined as

$$\boxed{\text{loop gain} = A\beta} \qquad \textbf{(14-9)}$$

The *loop gain* is the product of all gain factors around the loop.

The result of (14-8) is a most important equation and should be emphasized. Stated in words, it says that *the closed-loop gain with feedback is the open-loop forward gain divided by one plus the loop gain.*

Normally, the closed-loop gain is considerably less than the open-loop gain. What is then accomplished by the process? The answer is that several important improvements can be achieved through this trade-off. Among these are stability, reduction of distortion and noise, and predictability of gain.

Example 14-1

A certain feedback amplifier can be modeled by the block diagram of Figure 14-1. The open-loop forward gain is $A = 100$, and the feedback is $\beta = 0.1$. Compute **(a)** the loop gain and **(b)** the closed-loop gain A_{CL} with feedback.

Solution

(a) The loop gain is

$$A\beta = 100 \times 0.1 = 10 \qquad \textbf{(14-10)}$$

(b) The closed-loop gain is

$$A_{\text{CL}} = \frac{100}{1 + 10} = 9.09 \qquad \textbf{(14-11)}$$

Limiting Case

Consider the case when $A\beta \gg 1$, that is, when the loop gain is very large. In this case the unity term in the denominator of (14-8) may be neglected, and the closed-loop gain may be closely approximated as

$$\boxed{A_{\text{CL}} \simeq \frac{A}{A\beta} = \frac{1}{\beta} \qquad \text{for } A\beta \gg 1} \qquad \textbf{(14-12)}$$

Simply stated, this result says that *when the loop gain is very large, the closed-loop gain is nearly equal to the reciprocal of the feedback factor.*

The implications of this result are most important and need to be emphasized. Usually, A is the gain of an active device that is difficult to control and hard to predict from one unit to another. On the other hand, β is usually the transfer ratio of a simple resistive network, which can be controlled to a high degree of accuracy. If the loop gain is made to be very large, the net gain will be virtually independent of the uncontrollable factor A and almost completely predictable from the controllable factor β.

What constitutes a large loop gain? There is no single answer for this question. However, as a general guideline, if $A\beta \geq 100$, the difference between the exact gain and the approximation of (14-12) is less than 1%.

Example 14-2

In a certain feedback amplifier system, $A = 20{,}000$ and $\beta = 0.01$. Compute **(a)** the exact closed-loop gain using (14-8) and **(b)** the approximate closed-loop gain using (14-12).

Solution

(a) The loop gain is $A\beta = 20{,}000 \times 0.01 = 200$. The exact gain is

$$A_{CL} = \frac{20{,}000}{1 + 200} = 99.5 \tag{14-13}$$

(b) The approximate gain is

$$A_{CL} \simeq \frac{1}{\beta} = \frac{1}{0.01} = 100 \tag{14-14}$$

The results differ by about 0.5%, a result of the large loop gain.

Example 14-3

To illustrate the stability of the closed-loop gain, consider the amplifier of Example 14-2. **(a)** Consider first that the amplifier gain drops to $A = 10{,}000$, and compute the closed-loop gain. **(b)** Next assume that the amplifier gain increases to $A = 40{,}000$, and compute the closed-loop gain. **(c)** Compare the preceding results, together with those of Example 14-2.

Solution

(a) For $A = 10{,}000$ and $\beta = 0.01$, we have

$$A_{CL} = \frac{10{,}000}{1 + 10^4 \times 0.01} = 99.01 \tag{14-15}$$

(b) For $A = 40{,}000$ and $\beta = 0.01$, we have

$$A_{CL} = \frac{40{,}000}{1 + 4 \times 10^4 \times 0.01} = 99.75 \tag{14-16}$$

(c) The various results of Examples 14-2 and 14-3 are tabulated below. The last case, $A = \infty$, corresponds to the approximation of (14-12).

A	A_{CL}
10,000	99.01
20,000	99.5
40,000	99.75
∞	100

The stability of A_{CL} as A varies over a very wide range is quite obvious.

Variation in Gain

We have seen that the variation of closed-loop gain is much smaller than the variation in open-loop gain when negative feedback is employed. A useful formula for predicting the approximate fractional variation in closed-loop gain is given in this section.

Let ΔA represent a change in open-loop gain, and let ΔA_{CL} represent the corresponding change in closed-loop gain. An approximate relationship between ΔA and ΔA_{CL} may be derived using calculus, and it is stated as follows:

$$\frac{\Delta A_{CL}}{A_{CL}} = \frac{1}{1 + A\beta}\frac{\Delta A}{A} \qquad (14\text{-}17)$$

The quantity $\Delta A/A$ may be considered as the fractional change in open-loop gain, and $\Delta A_{CL}/A_{CL}$ may be considered as the corresponding fractional change in closed-loop gain. This interpretation of (14-17) may be stated as

$$\left(\begin{array}{c}\text{fractional change} \\ \text{in } A_{CL}\end{array}\right) = \frac{1}{1 + A\beta} \times \left(\begin{array}{c}\text{fractional change} \\ \text{in } A\end{array}\right) \qquad (14\text{-}18)$$

These fractional changes may also be expressed as percentage changes if desired. Whenever it is actually desired to compute ΔA_{CL}, the nominal value of A is used in the equation. This result is most useful for dealing with fractional and percentage changes.

Although (14-17) and (14-18) are approximations, they provide close estimates of the relationship between open-loop and closed-loop fractional gain changes for small and moderate changes in A. For drastic changes in A, such as considered in Example 14-3, the formula is less accurate in predicting the exact changes. However, the sense of the result is correct even for large gain changes. Basically, the sense of the formula is that the variation in closed-loop gain decreases as the loop gain increases.

Example 14-4

A certain amplifier has a nominal open-loop gain of 20,000, and it is subject to a variation of ±10%. For $\beta = 0.01$, estimate the percentage variation in closed-loop gain.

Solution

The value of $\Delta A/A$ on a positive fractional basis is

$$\frac{\Delta A}{A} = 0.1 \tag{14-19}$$

The corresponding value of $\Delta A_{\text{CL}}/A_{\text{CL}}$ is estimated as

$$\frac{\Delta A_{\text{CL}}}{A_{\text{CL}}} = \frac{1}{1 + 20,000(0.01)} \times 0.1 = \frac{1}{201} \times 0.1 \simeq 0.0005 \tag{14-20}$$

where some realistic rounding was employed. Expressed as a percentage, we can say that the expected variation in the closed-loop gain is about ±0.05%.

14-2 NONINVERTING OPERATIONAL AMPLIFIER CIRCUIT

The closed-loop *noninverting amplifier circuit* employing an op-amp is considered in this section. This amplifier circuit readily fits the feedback block diagram form considered in the preceding section and may be analyzed by the concepts developed there.

The basic circuit form as it is more commonly drawn is shown in Figure 14-2(a). The input signal, referred to ground, is applied to the noninverting input. A feedback circuit consisting of R_i and R_f is arranged as a voltage divider between out-

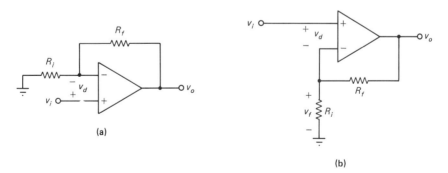

(a)

(b)

Figure 14-2 Noninverting amplifier drawn two ways.

put and input. The subscript i represents "input" and the subscript f represents "feedback."

An alternative way to draw the circuit to emphasize more clearly the feedback loop is shown in Figure 14-2(b). For this form, the inverting and noninverting inputs have been reversed in position.

The operational amplifier acts as both the difference circuit and the open-loop forward gain. We will momentarily change the notation of differential gain from A_d as defined in Chapter 13 to A, in order to conform to standard feedback notation. We have

$$v_o = Av_d = A(v_i - v_f) \tag{14-21}$$

The feedback voltage is the fraction of the output voltage appearing across R_i. At this point, any current flowing into or out of the op-amp input terminals will be neglected. With this assumption, the voltage v_f is determined by the voltage-divider rule as

$$v_f = \frac{R_i}{R_i + R_f} v_o \tag{14-22}$$

We immediately recognize that the feedback factor β is

$$\beta = \frac{R_i}{R_i + R_f} \tag{14-23}$$

Application of (14-8) yields for the closed-loop gain A_{CL},

$$A_{CL} = \frac{v_o}{v_i} = \frac{A}{1 + A\beta} = \frac{A}{1 + AR_i/(R_i + R_f)} \tag{14-24}$$

Equation (14-24) can be manipulated algebraically to the form

$$A_{CL} = \frac{(R_i + R_f)/R_i}{1 + (R_i + R_f)/AR_i} \tag{14-25}$$

Assume next that A is very large. In this case the second term in the denominator of (14-25) will be small compared with unity, and A_{CL} may be approximated as

$$\boxed{A_{CL} \simeq \frac{R_i + R_f}{R_i}} \tag{14-26}$$

Alternatively, this is equivalent to assuming that the loop gain is very large, in which case $A_{CL} \simeq 1/\beta$. The results are identical.

We see that the closed-loop gain of the noninverting amplifier circuit is essentially independent of the open-loop gain. However, the desired closed-loop gain is established by choosing appropriate values of R_i and R_f.

Example 14-5

A certain noninverting op-amp amplifier circuit has $R_i = 1$ kΩ, $R_f = 99$ kΩ, and $A = 200{,}000$. Determine **(a)** β, **(b)** the loop gain, **(c)** the exact closed-loop gain, and **(d)** the approximate closed-loop gain based on the assumption that $A \rightarrow \infty$.

Solution

(a) The value of β is

$$\beta = \frac{R_i}{R_i + R_f} = \frac{1\ \text{k}\Omega}{1\ \text{k}\Omega + 99\ \text{k}\Omega} = 0.01 \qquad \text{(14-27)}$$

(b) The value of the loop gain is

$$\text{loop gain} = A\beta = 2 \times 10^5 \times 0.01 = 2000 \qquad \text{(14-28)}$$

(c) The exact gain A_{CL} can be determined from either (14-8) or (14-25). Using the first form, we have

$$A_{CL} = \frac{200{,}000}{1 + 2000} = 99.95 \qquad \text{(14-29)}$$

(d) The approximate gain based on the assumption that $A \rightarrow \infty$ is determined from (14-26) as

$$A_{CL} = \frac{R_i + R_f}{R_i} = \frac{1\ \text{k}\Omega + 99\ \text{k}\Omega}{1\ \text{k}\Omega} = 100 \qquad \text{(14-30)}$$

Alternatively, A_{CL} is simply

$$A_{CL} = \frac{1}{\beta} = \frac{1}{0.01} = 100 \qquad \text{(14-31)}$$

Note that the gain as determined in (14-30) and (14-31) differs from (14-29) by about 0.05%.

14-3 EFFECT OF FEEDBACK ON INPUT AND OUTPUT RESISTANCES

In the preceding section, the input resistance of the op-amp was assumed to be infinite, and the output resistance was assumed to be zero. In this section the effects of a finite input resistance and a nonzero output resistance will be considered. Because the two effects are different, and the values differ by several orders of magnitude, it is possible to focus on each effect individually while simultaneously considering ideal conditions for the other effect. With this assumption, the mathematical expressions are greatly simplified with negligible loss in accuracy.

Effect on Input Resistance

The circuit model used to study the effect of feedback on finite input resistance in the noninverting amplifier is shown in Figure 14-3. The quantity r_d represents the *open-loop differential input resistance*. The differential gain is assumed again to be A. For this analysis, the output resistance is assumed to be zero since it contributes very little effect to the input resistance.

To determine the input resistance, a signal voltage v_i and an input current i_i are assumed. An expression must be determined for i_i in terms of v_i with all other variables eliminated.

From Figure 14-3 we can express i_i as

$$i_i = \frac{v_i - v_f}{r_d} \tag{14-32}$$

The feedback voltage v_f is

$$v_f = \beta v_o \tag{14-33}$$

However, from the work of Section 14-1, v_o can be expressed as

$$v_o = \frac{A}{1 + A\beta} v_i \tag{14-34}$$

Substitution of (14-34) in (14-33) yields

$$v_f = \frac{A\beta}{1 + A\beta} v_i \tag{14-35}$$

Next, v_f from (14-35) is substituted in (14-32).

$$i_i = \frac{1}{r_d}\left(v_i - \frac{A\beta}{1 + A\beta} v_i\right) = \frac{1}{r_d(1 + A\beta)} v_i \tag{14-36}$$

The input resistance R_{in} is then determined as

$$\boxed{R_{\text{in}} = \frac{v_i}{i_i} = (1 + A\beta)r_d} \tag{14-37}$$

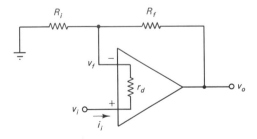

Figure 14-3 Circuit model used to study effect of feedback input resistance.

The actual circuit input resistance is thus $(1 + A\beta)$ times the open-loop differential input resistance.

Effect on Output Resistance

The circuit model used to study the effect of feedback on nonzero output resistance in the noninverting amplifier is shown in Figure 14-4. The quantity r_o represents the *open-loop op-amp output resistance,* and, as before, A represents the assumed finite differential gain. However, in this analysis, the input resistance is assumed to be infinite since it contributes very little effect to the output resistance.

To determine the output resistance, the input signal v_i has been deenergized (replaced by a short), and a fictitious voltage generator v_x is applied to the output. The current i_x that would flow into the output is then determined.

The current i_x is composed of two components i_o and i_1 as shown. However, the current i_1 is so small compared with i_o in a typical case that we can assume that $i_x \simeq i_o$. The value of i_o can then be determined as

$$i_o = \frac{v_x - (-Av_f)}{r_o} = \frac{v_x + Av_f}{r_o} \qquad \textbf{(14-38)}$$

The voltage v_f may be written as

$$v_f = \beta v_o = \beta v_x \qquad \textbf{(14-39)}$$

Substitution of (14-39) in (14-38) yields

$$i_o \simeq i_x = \frac{v_x + A\beta v_x}{r_o} = \frac{1 + A\beta}{r_o} v_x \qquad \textbf{(14-40)}$$

The output resistance R_{out} is then determined as

$$\boxed{R_{\text{out}} = \frac{v_x}{i_x} = \frac{r_o}{1 + A\beta}} \qquad \textbf{(14-41)}$$

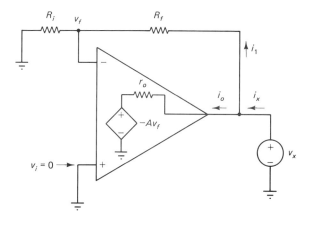

Figure 14-4 Circuit model used to study effect of feedback output resistance.

Sec. 14-3   Effect of Feedback on Input and Output Resistances   **551**

The actual circuit output resistance is thus the open-loop op-amp output resistance divided by $1 + A\beta$.

Voltage-Series Feedback

The noninverting amplifier circuit represents a type of feedback called voltage-series feedback. In this type of feedback, the variable sensed at the *output* is the *voltage* and the feedback is connected in *series* with the input. This is one of the most popular types of feedback arrangements, and its properties can be summarized as follows: *Voltage-series feedback causes the input resistance to increase by the factor $1 + A\beta$, and it causes the output resistance to decrease by the factor $1/(1 + A\beta$.)* These effects are desirable in an ideal voltage amplifier.

Voltage series feedback is, by no means, the only type of feedback used. Several other forms will be encountered in this book. Depending on the form, the input and output resistances may be increased or decreased.

Example 14-6

Consider the noninverting amplifier of Example 14-5, and assume that the open-loop op-amp input and output resistances are $r_d = 500 \text{ k}\Omega$ and $r_o = 100 \ \Omega$. Determine **(a)** the closed-loop input resistance and **(b)** the closed-loop output resistance.

Solution

It was determined in Example 14-5 that $A\beta = 2000$.

(a) The closed-loop input resistance is

$$R_{\text{in}} = (1 + A\beta)r_d = (1 + 2000)500{,}000 \simeq 1 \times 10^9 \ \Omega \qquad \textbf{(14-42)}$$

(b) The closed-loop output resistance is

$$R_{\text{out}} = \frac{r_o}{1 + A\beta} = \frac{100}{1 + 2000} \simeq 0.05 \ \Omega \qquad \textbf{(14-43)}$$

14-4 EFFECT OF FEEDBACK ON BANDWIDTH

In this section the effect of negative feedback on the bandwidth of an amplifier is discussed. Certain results are stated here without proof. The actual development of these results is deferred to Chapter 18. At that point in the book, steady-state phasor analysis will be used to facilitate a proof.

The closed-loop gain A_{CL} as a function of the open loop gain A and feedback ratio β has been shown to be

$$A_{\text{CL}} = \frac{A}{1 + A\beta} \qquad \textbf{(14-44)}$$

Thus far, A has been assumed to be a constant value, which results in a constant value for A_{CL} assuming, of course, that β is constant. In practice, the gain A will be frequency dependent. As we will see shortly, this results in a frequency-dependent closed-loop gain A_{CL}.

The high-frequency response of many amplifier circuits, particularly op-amps, is controlled over a wide frequency range by a dominating capacitance. (In op-amps, this dominating capacitive effect is established in the design process to achieve absolute stability.)

The Bode plot approximation of a typical open-loop gain response is shown by the upper curve in Figure 14-5. The value A_o represents the dc and very low frequency gain. The frequency f_o represents the 3-dB or "break" frequency in the open-loop gain response.

Assume that this amplifier is used as the forward gain block in a negative feedback circuit. Assume that a closed-loop low-frequency gain A_{CLO} is established. Obviously, $A_{CLO} < A_o$ as shown on Figure 14-5. The quality A_{CLO} is given by

$$A_{CLO} = \frac{A_o}{1 + A_o\beta} \tag{14-45}$$

It turns out that negative feedback causes the 3-dB closed-loop frequency to be greater than the corresponding open-loop frequency. Let f_{CL} represent the closed-loop 3-dB frequency. It will be shown in Chapter 18 with steady-state ac circuit analysis that f_{CL} and f_o are related by

$$\boxed{f_{CL} = (1 + A_o\beta)f_o} \tag{14-46}$$

Stated in words, *the closed-loop 3-dB frequency is $(1 + A_o\beta)$ times the open-loop 3-dB frequency.*

From Figure 14-5 it is noted that the Bode plot approximation of the closed-loop gain intersects the open-loop gain at the frequency f_{CL}. From this property, it can be inferred that *the bandwidth increases as the closed-loop gain decreases.*

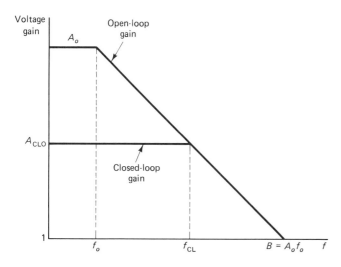

Figure 14-5 Open- and closed-loop amplitude response of amplifier with feedback.

Gain–Bandwidth Product

An interesting and meaningful property of the preceding development is the *gain–bandwidth* product. First, consider the product of the open-loop dc gain A_o and the open-loop 3-dB bandwidth f_o. This product is simply $A_o f_o$. Next, consider the product of the closed-loop dc gain A_{CLO} given by (14-45) and the closed-loop 3-dB bandwidth f_{CL} given by (14-46). This product is

$$A_{\text{CLO}} f_{\text{CL}} = \frac{A_o}{1 + A_o \beta} f_o (1 + A_o \beta) = A_o f_o \qquad \textbf{(14-47)}$$

The product thus remains the same. In effect, gain is traded for more bandwidth. This property appears in many electronic circuits and represents one of the inherent constraints in electronic circuits.

The *gain–bandwidth product* is also called the *unity-gain frequency*. The reason is that it is the bandwidth resulting from a closed-loop gain of unity. This is observed in Figure 14-5 by noting that the open-loop gain intersects the unity gain ordinate at a frequency $A_o f_o$. For convenience, a quantity B is defined as

$$\boxed{B = A_o f_o = A_{\text{CLO}} f_{\text{CL}}} \qquad \textbf{(14-48)}$$

Thus B is the *gain–bandwidth product* or the *unity-gain frequency*.

Example 14-7

Consider the noninverting op-amp amplifier circuit of Example 14-5. The gain will be interpreted as the dc gain (i.e., $A_o = 200{,}000$). Assume that $f_o = 5$ Hz. Determine **(a)** the gain–bandwidth product and **(b)** the closed-loop 3-dB frequency f_{CL}.

Solution

Since $\beta = 0.01$ in Example 14-5, $A_o \beta = 2000$.

(a) The gain–bandwidth product (or unity-gain frequency) is

$$B = A_o f_o = 200{,}000 \times 5 = 1 \times 10^6 \text{ Hz} = 1 \text{ MHz} \qquad \textbf{(14-49)}$$

(b) The closed-loop 3-dB frequency is

$$f_{\text{CL}} = f_o (1 + A\beta) = 5(1 + 2000) \approx 10 \text{ kHz} \qquad \textbf{(14-50)}$$

14-5 INVERTING OPERATIONAL AMPLIFIER CIRCUIT

An op-amp circuit of equal importance with the noninverting amplifier is the *inverting amplifier circuit*, and its basic form is shown in Figure 14-6. (The power supply connections are understood but are not shown.) In this case the noninverting termi-

Figure 14-6 Basic form of inverting amplifier.

nal is either grounded or connected through a bias compensation resistor to ground. For simplicity, it is shown here grounded. The signal is connected through a resistance R_i to the inverting input, and a feedback resistance R_f is connected between inverting input and output.

It turns out that β is the same in this circuit as in the noninverting amplifier. This can be seen by momentarily deenergizing the signal input point (v_i) to ground and noting that the feedback fraction β is simply

$$\beta = \frac{R_i}{R_i + R_f} \qquad (14\text{-}51)$$

However, the inverting amplifier differs from the noninverting amplifier in that the inverting amplifier employs *shunt* feedback at the input; that is, the feedback current is in parallel with the signal current. The result is that the circuit does not quite fit the standard block form given early in the chapter as well as the noninverting amplifier. Although the block form could be modified, we will employ a different approach here.

Assume that the voltage at the inverting input is v^- as shown in Figure 14-6. A KCL equation will be written at this node with currents expressed in terms of the voltages. It will be assumed that there is negligible current flowing into or out of the op-amp input terminals.

Summing the currents at the v^- mode, we have

$$\frac{v^- - v_i}{R_i} + \frac{v^- - v_o}{R_f} = 0 \qquad (14\text{-}52)$$

In addition, we have the op-amp equation that

$$v_o = Av_d = A(0 - v^-) = -Av^- \qquad (14\text{-}53)$$

Solving for v^- from (14-53) and substituting in (14-52), we obtain

$$v_o \left[\frac{1}{R_f} + \frac{1}{A} \left(\frac{1}{R_i} + \frac{1}{R_f} \right) \right] = \frac{-v_i}{R_i} \qquad (14\text{-}54)$$

After some manipulation, the closed-loop gain can be expressed as

$$A_{\text{CL}} = \frac{v_o}{v_i} = \frac{-R_f/R_i}{1 + 1/A\beta} \qquad (14\text{-}55)$$

where β was defined in (14-51).

Next, assume that $A\beta \gg 1$. In this case (14-55) reduces to

$$\boxed{A_{\text{CL}} \simeq -\frac{R_f}{R_i}} \qquad \textbf{(14-56)}$$

The closed-loop gain of the inverting amplifier under large loop gain conditions is a simple resistance ratio. The negative sign indicates the inherent sign inversion in the process. Once again, we see the beauty of the feedback process.

Input Resistance

Because of the shunt form of the feedback in the inverting amplifier, the input resistance *at the inverting input* approaches zero. This can be seen from (14-53) by first solving for v^-. We have

$$v^- = \frac{-v_o}{A} \qquad \textbf{(14-57)}$$

For any finite v_o, as A becomes very large, v^- approaches zero. This property is referred to as a *virtual ground;* that is, the voltage approaches zero, but it is not connected to the circuit ground. In fact, this point *must not be* connected to the actual circuit ground since it is necessary for a small voltage to exist across the input terminals for the circuit to function properly.

The external circuit input resistance R_{in} is the most meaningful property, and this resistance will be defined as the resistance seen at the point where v_i is connected. Since $v^- \approx 0$, the resistance seen by v_i is simply

$$\boxed{R_{\text{in}} = R_i} \qquad \textbf{(14-58)}$$

Output Resistance

As far as the output resistance is concerned, the inverting amplifier functions in the same manner as the noninverting circuit. This is readily observed by deenergizing the input signal and noting that the circuit structure is then the same as for the noninverting configuration. Thus the output resistance is

$$\boxed{R_{\text{out}} = \frac{r_o}{1 + A\beta}} \qquad \textbf{(14-59)}$$

Example 14-8

A certain inverting op-amp amplifier circuit has $R_i = 1$ kΩ, $R_f = 24$ kΩ, and $A = 100{,}000$. Determine **(a)** β, **(b)** the loop gain, **(c)** the exact closed-loop gain, and **(d)** the approximate closed-loop gain based on the assumption that $A \to \infty$.

Solution

(a) The value of β is

$$\beta = \frac{R_i}{R_i + R_f} = \frac{1 \text{ k}\Omega}{1 \text{ k}\Omega + 24 \text{ k}\Omega} = 0.04 \qquad \textbf{(14-60)}$$

(b) The value of the loop gain is

$$\text{loop gain} = A\beta = 10^5 \times 0.04 = 4000 \qquad \textbf{(14-61)}$$

(c) The exact gain can be determined from (14-55) as

$$A_{\text{CL}} = \frac{-24 \text{ k}\Omega/1 \text{ k}\Omega}{1 + 1/4000} = -23.994 \qquad \textbf{(14-62)}$$

(d) The approximate gain based on the assumption that $A \rightarrow \infty$ is determined from (14-56) as

$$A_{\text{CL}} = \frac{-24 \text{ k}\Omega}{1 \text{ k}\Omega} = -24$$

The approximate gain differs from the exact gain by about 0.025%.

DRILL PROBLEMS

14-1. A certain feedback amplifier can be modeled by the block diagram of Figure 14-1. The open-loop forward gain is $A = 250$, and the feedback factor is $\beta = 0.08$. Compute **(a)** the loop gain and **(b)** the closed-loop gain A_{CL} with feedback.

14-2. Repeat the computations of Problem 14-1 if $A = 500$ and $\beta = 0.05$.

14-3. In a certain feedback amplifier system, $A = 7500$ and $\beta = 0.02$. Compute **(a)** the exact closed-loop gain and **(b)** the approximate closed-loop gain based on the assumption that $A \rightarrow \infty$.

14-4. Repeat the computations of Problem 14-3 if $A = 25,000$ and $\beta = 0.04$.

14-5. A certain feedback amplifier has a fixed feedback ratio $\beta = 0.05$. Determine the closed-loop gain A_{CL} for each of the following values of open-loop gain A: **(a)** 100; **(b)** 1000; **(c)** 10^4; **(d)** ∞.

14-6. A certain feedback amplifier has a fixed feedback ratio $\beta = 0.02$. Determine the closed-loop gain A_{CL} for each of the following values of open-loop gain A: **(a)** 100; **(b)** 1000; **(c)** 10^4; **(d)** ∞.

14-7. A certain amplifier has a nominal open-loop gain of 5000, and it is subject to a variation of $\pm 12\%$. For $\beta = 0.02$, estimate the percentage variation in closed-loop gain.

14-8. A certain amplifier has a nominal open-loop gain of 12,000, and it is subject to a variation of $\pm 20\%$. For $\beta = 0.02$, estimate the percentage variation in closed-loop gain.

14-9. A certain noninverting op-amp amplifier circuit has $R_i = 2$ kΩ, $R_f = 98$ kΩ, and $A = 120,000$. Determine **(a)** β, **(b)** the loop gain, **(c)** the exact closed-loop gain, and **(d)** the approximate closed-loop gain based on the assumption that $A \to \infty$.

14-10. Repeat the computation of Problem 14-9 if $R_i = 1$ kΩ, $R_f = 19$ kΩ, and $A = 50,000$.

14-11. Based on the assumption that $A \to \infty$ and $R_i = 1$ kΩ, compute the values of noninverting gain for each of the following values of R_f: **(a)** 9 kΩ; **(b)** 19 kΩ; **(c)** 49 kΩ; **(d)** 199 kΩ.

14-12. Based on the assumption that $A \to \infty$ and $R_f = 100$ kΩ, compute the values of noninverting gain for each of the following values of R_i: **(a)** 100 Ω; **(b)** 1 kΩ; **(c)** 10 kΩ; **(d)** 50 kΩ.

14-13. Consider the amplifier of Problem 14-9, and assume that the open-loop op-amp input and output resistances are $r_d = 400$ kΩ and $r_o = 60$ Ω. Determine **(a)** the closed-loop input resistance and **(b)** the closed-loop output resistance.

14-14. Repeat the computations of Problem 14-13 for the amplifier of Problem 14-10 if $r_d = 1$ MΩ and $r_o = 50$ Ω.

14-15. Consider the noninverting op-amp circuit of Problem 14-9, and assume that the given gain is the dc gain (i.e., $A_o = 120,000$). Assume that $f_o = 20$ Hz. Determine **(a)** the gain–bandwidth product and **(b)** the closed-loop 3-dB frequency f_{CL}.

14-16. Consider the noninverting op-amp circuit of Problem 14-10, and assume that the given gain is the dc gain (i.e., $A_o = 50,000$). Assume that $f_o = 10$ Hz. Determine **(a)** the gain–bandwidth product and **(b)** the closed-loop 3-dB frequency f_{CL}.

14-17. A certain op-amp has a gain–bandwidth product $B = 2 \times 10^6$. Determine the 3-dB closed-loop bandwidth for each of the following noninverting closed-loop values of dc gain: **(a)** 10^3; **(b)** 100; **(c)** 10; **(d)** 1.

14-18. A certain op-amp connected as a noninverting amplifier has a closed-loop gain of 50 and a closed-loop 3-dB bandwidth of 30 kHz. **(a)** Determine the gain–bandwidth product. **(b)** If the open-loop 3-dB bandwidth is 7.5 Hz, determine the open-loop dc gain.

DERIVATION PROBLEMS

14-19. (Requires calculus.) Starting with (14-8), show that

$$\frac{dA_{CL}}{dA} = \frac{1}{(1 + A\beta)^2}$$

14-20. (Requires calculus.) Using the results of Problem 14-19 and employing differentials, derive (14-17).

DESIGN PROBLEMS

14-21. It is desired to add feedback to a certain amplifier to improve the stability. The nominal value of the open-loop gain is 1000. Determine the value of the feedback factor β required to establish a closed-loop gain of 100.

14-22. It is desired to add feedback to a certain amplifier to improve the stability. The loop gain established for the amplifier is $A\beta = 20$. Determine the values of open-loop gain and feedback factor required to achieve a closed-loop gain of 12.

14-23. The open-loop gains of amplifiers in a certain system are subject to a $\pm 16\%$ variation. It is desired to reduce this uncertainty with feedback. Determine the value of loop gain required to reduce the variation to about $\pm 0.5\%$.

14-24. It is desired to achieve a net gain of 10,000 by connecting a number of amplifiers in cascade. Each amplifier has a gain of 200 ± 20. Feedback is to be employed in each stage such as to reduce the percentage variation to a level of about $\pm 0.5\%$. Determine **(a)** the required closed-loop gain of each stage if all stages are to have equal gains and **(b)** the number of stages required.

COMPUTER PROBLEMS

14-25. Write a computer program to determine the exact and approximate closed-loop gain values of a feedback system (as modeled in Figure 14-1) in terms of the open-loop forward gain and the feedback factor.

> Input data: A, β
> Output data: A_{CL} (exact value), A_{CL} (approximate value based on $A = \infty$)

14-26. Write a computer program to determine the exact and approximate closed-loop gain and resistance parameters of a noninverting op-amp amplifier circuit in terms of the open-loop parameters and the input and feedback resistances.

Input data: A, R_i, R_f, r_d, r_o

Output data: A_{CL} (exact value), A_{CL} (approximate value based on $A = \infty$), R_{in}, R_{out}

14-27. Write a computer program to determine the closed-loop 3-dB bandwidth of a noninverting op-amp amplifier in terms of the open-loop dc gain, 3-dB frequency, and the input and feedback resistances.

Input data: A_o, f_o, R_i, R_f

Output data: f_{CL}

14-28. Write a computer program to determine the exact and approximate closed-loop gain values of an inverting op-amp amplifier circuit in terms of the open-loop gain and the input and feedback resistances.

Input data: A, R_i, R_f

Output data: A_{CL} (exact value), A_{CL} (approximate value based on $A = \infty$)

CHAPTER 15

Operational Amplifier Linear Circuits

OBJECTIVES

After completing this chapter, the reader should be able to:

1. State the three assumptions and the associated implications used in the idealized analysis of op-amp circuits.
2. Analyze an inverting amplifier circuit by employing the idealized assumptions.
3. Design an inverting amplifier circuit to meet prescribed specifications.
4. Analyze a noninverting amplifier circuit by employing the idealized assumptions.
5. Design a noninverting amplifier circuit to meet prescribed specifications.
6. Compare advantages and disadvantages of inverting and noninverting configurations.
7. Analyze or design an op-amp summing circuit.
8. Analyze or design an op-amp difference amplifier circuit.
9. Analyze or design a current-controlled voltage source.
10. Analyze or design a voltage-controlled current source.
11. Analyze or design a current-controlled current source.
12. Discuss the operation of an active filter and its advantages.
13. Design an active Butterworth filter to meet specified requirements.

15-1 SIMPLIFIED ANALYSIS OF OP-AMP CIRCUITS

The basis for a very simplified approach to the analysis of op-amp circuits will be established in this section. This simplified analysis is based on certain assumptions which may be made in a wide variety of applications. Designers select op-amps with specifications that justify these assumptions whenever possible. A wide variety of application circuits may be readily analyzed and designed around this approach.

As a quick review of material previously considered, the symbol of an op-amp is shown in Figure 15-1(a). The two input terminals are known as the *noninverting*

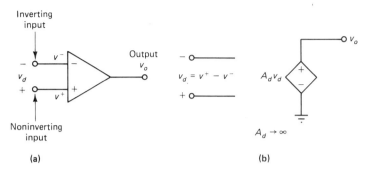

Inverting
input

Output
v_o

$v_d = v^+ - v^-$

$A_d v_d$

v_o

$A_d \to \infty$

Noninverting
input

(a)   (b)

Figure 15-1 Op-amp symbol and ideal circuit model.

input (indicated by +) and the *inverting input* (indicated by −). The voltage at the noninverting input with respect to ground is indicated as v^+, and the voltage at the inverting input with respect to ground is indicated as v^-. The differential input voltage v_d is defined as the difference between these voltages in the following sense:

$$v_d = v^+ - v^- \tag{15-1}$$

The output voltage with respect to ground is denoted as v_o. This voltage is related to the differential input voltage and the differential gain A_d by

$$v_o = A_d v_d = A_d(v^+ - v^-) \tag{15-2}$$

It is assumed that the op-amp is powered by dual power supplies with a common ground established at the midpoint as discussed in Chapter 13. However, the power supply connections are not shown here.

A model depicting the ideal assumptions is shown in Figure 15-1(b). The ideal assumptions are as follows:

1. The input resistance of the op-amp is infinite. This is indicated on the model by the presence of an open circuit between the two input terminals.
2. The output resistance of the op-amp is zero. This is indicated on the model by the absence of a series resistance in the output circuit.
3. The open-loop differential gain A_d becomes infinite in the ideal case. This is indicated below the dependent source since it would be awkward to try to show an infinite value directly on the model schematic.

The implications of these three assumptions are the key to the simplified analysis and will now be discussed.

1. The assumed infinite input resistance means that the op-amp input current can be assumed to be zero.
2. The assumed zero output resistance means that the op-amp output voltage can be assumed to be independent of the load.
3. The implication of the assumed infinite gain is best seen by first solving for v_d in (15-2), in which case we have

$$v_d = \frac{v_o}{A_d} \qquad (15\text{-}3)$$

Assume that negative feedback forces v_o to be finite and in the linear operating region. If A_d is allowed to increase without limit, we have

$$\lim_{A_d \to \infty} v_d = \lim_{A_d \to \infty} \frac{v_o}{A} = 0 \qquad (15\text{-}4)$$

The differential voltage thus approaches zero and may be stated in either of the following ways:

$$\boxed{v_d = v^+ - v^- = 0} \qquad (15\text{-}5)$$

or

$$\boxed{v^+ = v^-} \qquad (15\text{-}6)$$

Stated in words, *the voltages at the two op-amp inputs are forced to be essentially equal under closed-loop stable conditions.*

A major qualification about the last implication is in order. Since the gain is never infinite, there must always be a small potential difference between these terminals to make the op-amp work. Thus *the two input terminals must not be connected together*. However, in a variety of applications, it is acceptable to assume equal voltages at the two terminals since the difference voltage will be very small compared with external circuit voltages.

Example 15-1

Assume that in a certain op-amp closed-loop feedback circuit, the output is forced through feedback to be $v_o = 10$ V. Calculate the differential input voltage v_d for each of the following values of open-loop differential gain A_d: **(a)** 200,000; **(b)** 50,000; **(c)** 10,000.

Solution

This problem may seem backward in that we are given the output and gain and wish to determine the input. In a sense, however, this is the way in which many feedback circuits appear to work. Through a large loop gain, the output is forced to assume a value nearly independent of active-device characteristics. The input to the op-amp then assumes the value necessary to force the output to be the required value. Do not confuse the op-amp differential input voltage with the external input signal voltage, which is not given, but which is assumed to remain constant in this example.

The differential input voltage for each value of A_d is determined from (15-3) as

$$v_d = \frac{v_o}{A_d} = \frac{10}{A_d} \qquad \textbf{(15-7)}$$

Each of the three values of A_d may be substituted in (15-7), and the results are summarized as follows:

	A_d	v_d
(a)	200,000	50 μV
(b)	50,000	200 μV
(c)	10,000	1 mV

Note that as the gain drops, the differential input voltage increases to maintain the assumed output voltage, and the assumption that $v_d = 0$ becomes less valid. Depending on the circuit, the assumption that $v_o = 10$ V will similarly become less valid, and the output will begin to depend somewhat on the open-loop gain.

15-2 INVERTING AMPLIFIER REVISITED

We begin the simplified analysis of op-amp circuits by considering again the inverting amplifier, of which a more exact analysis was given in Chapter 14. The analysis here employs the ideal assumptions of the preceding section.

The inverting amplifier with certain variables identified is shown in Figure 15-2. Since the noninverting input terminal is grounded, $v^+ = 0$. Based on the assumption that the differential input voltage is zero, we can therefore assume that $v^- = 0$. As discussed in Chapter 14, this situation is referred to as a *virtual ground;* that is, the inverting input is forced to assume ground potential even though it is not connected (and *should not be connected*) to ground.

The existence of the virtual ground at the inverting input means that the input signal voltage v_i appears across R_i. The input current i_i is

$$i_i = \frac{v_i}{R_i} \qquad \textbf{(15-8)}$$

The current i_i cannot flow into the op-amp because of the assumed infinite in-

Figure 15-2 Inverting op-amp amplifier circuit with certain variables labeled.

put resistance, so it flows through R_f. A voltage v_f is produced across R_f given by

$$v_f = R_f i_i \tag{15-9}$$

Substitution of (15-8) in (15-9) yields

$$v_f = R_f \frac{v_i}{R_i} = \frac{R_f}{R_i} v_i \tag{15-10}$$

Since the assumed positive reference terminal of v_f is at ground potential, the output voltage v_o is simply

$$v_o = -v_f = -\frac{R_f}{R_i} v_i \tag{15-11}$$

Let A_{CL} represent the closed-loop gain. It is determined as

$$\boxed{A_{\mathrm{CL}} = \frac{v_o}{v_i} = -\frac{R_f}{R_i}} \tag{15-12}$$

This result was, of course, established in Chapter 14 by a more elaborate procedure after limiting conditions had been made. The procedure just given was much simpler and more intuitive.

The input resistance R_{in} of the noninverting amplifier is readily determined from (15-8) as

$$\boxed{R_{\mathrm{in}} = \frac{v_i}{i_i} = R_i} \tag{15-13}$$

Do not confuse this input resistance of the composite circuit with the input resistance of the op-amp alone, which has been assumed to be infinite.

The output resistance R_{out} in the ideal case is simply

$$\boxed{R_{\mathrm{out}} = 0} \tag{15-14}$$

Saturation

It must be understood that the gain equation of (15-12) is valid only for operation in the linear active region. For the common power supply voltages of $+15$ V and -15 V, the output saturation voltages are typically about $\pm V_{\mathrm{sat}} = \pm 13$ V. Consequently, if the input should be large enough to cause the output to try to exceed these limits, the output will be clipped, and distortion will result.

Let V_{ip} represent the magnitude of the input at which saturation would start to occur. This value is

$$\boxed{V_{ip} = \frac{V_{\mathrm{sat}}}{|A_{\mathrm{CL}}|}} \tag{15-15}$$

based on symmetry in the two saturation voltages. The linear operating region for the input signal is $-V_{ip} < v_i < +V_{ip}$.

Simplified Design

It is possible to design useful amplifier circuits with the material just provided. A typical design would be based on a specification of the closed-loop gain A_{CL} and the minimum value of the input resistance R_{in}. The required gain constrains the ratio R_f/R_i as given by (15-12). The minimum input resistance constrains the minimum value of R_i.

In general, there may be a number of combinations of R_f and R_i that satisfy all the requirements. In deciding which values to use, the following points should be kept in mind: (1) Smaller values of resistance reduce the input resistance to the circuit and force the op-amp to supply more current to the feedback circuit. (2) Larger values of resistance tend to generate more thermal noise and cause possible offset effects due to bias currents.

As a general guideline, resistance values between a few thousand ohms and up to a few hundred thousand ohms tend to be best for most applications. However, this should be considered as a rough guideline rather than as an exact constraint.

Example 15-2

Consider the inverting amplifier circuit of Figure 15-3. **(a)** Determine the closed-loop gain A_{CL}. **(b)** Determine the input resistance R_{in}. **(c)** For $\pm V_{sat} = \pm 13$ V, determine the input voltage level at which saturation occurs. **(d)** Determine the output voltage for each of the following input voltages: $v_i = 0, 1$ V, -1 V, 2 V, and 4 V.

Figure 15-3 Amplifier circuit of Example 15-2.

Solution

(a) The closed-loop gain is

$$A_{CL} = -\frac{R_f}{R_i} = -\frac{100,000}{20,000} = -5 \tag{15-16}$$

(b) The input resistance is

$$R_{in} = R_i = 20 \text{ k}\Omega \tag{15-17}$$

(c) For $\pm V_{sat} = \pm 13$ V, the input voltage magnitude at which saturation starts to occur is

$$V_{ip} = \frac{13}{5} = 2.6 \text{ V} \qquad (15\text{-}18)$$

The linear range in volts for the input is $-2.6 < v_i < 2.6$.

(d) For any value of input voltage in the linear region, the output voltage is

$$v_o = -5v_i \qquad (15\text{-}19)$$

The various output voltages for the first four cases are summarized as follows:

v_i (V)	v_o (V)
0	0
1	−5
−1	5
2	−10

Note that the output voltage is zero when the input is zero. For the other cases, the output magnitude is five times as large as the input, but is inverted in polarity. Since the op-amp is a dc-coupled device, the input and output could be dc levels, or they could represent certain properties (e.g., peak values) of ac signals.

Next, consider the last voltage (i.e., $v_i = 4$ V). If one simply substituted in (15-19), the output would appear to be -20 V. However, this is impossible since the output cannot extend outside the range from -13 to $+13$ V. The problem is that the peak input voltage as computed in (15-18) has been exceeded (i.e., saturation has occurred). For this condition, we can only say that $v_o = -13$ V. Note that negative saturation occurs for a positive input voltage, and vice versa.

Example 15-3

Design an inverting op-amp amplifier circuit to meet the following specifications:

$$A_{CL} = -15$$

$$R_{in} \geq 10 \text{ k}\Omega$$

Use standard 5% resistance values, and keep all values below 200 kΩ to minimize noise and bias offset effects.

Solution

The gain specification requires that

$$\frac{R_f}{R_i} = 15 \qquad\qquad (15\text{-}20)$$

or

$$R_f = 15R_i \qquad\qquad (15\text{-}21)$$

Since $R_{in} = R_i$, the second specification means that the smallest acceptable resistance would be 10 kΩ, which is a standard value. It turns out that 15 × 10 kΩ = 150 kΩ is also a standard value. Thus one solution would be R_i = 10 kΩ and R_f = 150 kΩ. However, to allow some extra leeway for tolerances on the input resistance, a better solution with standard values might be R_i = 12 kΩ and R_f = 180 kΩ.

15-3 NONINVERTING AMPLIFIER REVISITED

Let us now analyze the noninverting amplifier configuration using the idealized assumptions of Section 15-1. The noninverting amplifier with certain variables identified is shown in Figure 15-4. The input voltage v_i is applied to the noninverting input, and this results in $v^+ = v_i$. Based on the assumption that the voltages at the two input terminals are forced to be the same, we can say that

$$v^- = v^+ = v_i \qquad\qquad (15\text{-}22)$$

The voltage across the resistance R_i is thus v_i, and the current i_i is

$$i_i = \frac{v_i}{R_i} \qquad\qquad (15\text{-}23)$$

The current i_i cannot flow out of the op-amp input, so it must be supplied by the op-amp output through R_f. A voltage v_f is produced across R_f given by

$$v_f = R_f i_i \qquad\qquad (15\text{-}24)$$

Substitution of (15-23) in (15-24) yields

$$v_f = R_f \frac{v_i}{R_i} = \frac{R_f}{R_i} v_i \qquad\qquad (15\text{-}25)$$

Figure 15-4 Noninverting op-amp amplifier circuit with certain variables labeled.

A KVL loop is now formed from ground to the inverting input, over to the output, and back to ground. This equation reads

$$-v_i - v_f + v_o = 0 \qquad (15\text{-}26)$$

Solution for v_o and substitution of (15-25) for v_f result in

$$v_o = v_i + \frac{R_f}{R_i} v_i \qquad (15\text{-}27)$$

The closed-loop voltage gain A_{CL} is then determined as

$$\boxed{A_{\mathrm{CL}} = \frac{v_o}{v_i} = 1 + \frac{R_f}{R_i} = \frac{R_i + R_f}{R_i}} \qquad (15\text{-}28)$$

where two equivalent forms have been given. As in the case of the inverting amplifier, the closed-loop gain expression agrees with the result obtained in Chapter 14.

Since the signal source looks directly into an assumed open circuit at the noninverting input, the circuit input resistance R_{in} in this case is assumed to be

$$\boxed{R_{\mathrm{in}} = \infty} \qquad (15\text{-}29)$$

The output resistance in this ideal case is simply

$$\boxed{R_{\mathrm{out}} = 0} \qquad (15\text{-}30)$$

Saturation

The discussion concerning saturation as given in Section 15-2 for the inverting amplifier applies to the noninverting amplifier as well. The appropriate noninverting gain is used in (15-15) to predict the input at which saturation starts to occur. In the case of the noninverting amplifier, positive saturation would occur for a positive input voltage, and vice versa.

Voltage Follower

A special case of a noninverting amplifier is the voltage follower, whose basic form is shown in Figure 15-5. The result of (15-28) may be adapted to this case by setting $R_f = 0$ and $R_i = \infty$. The result is

$$\boxed{A_{\mathrm{CL}} = 1} \qquad (15\text{-}31)$$

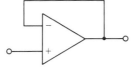

Figure 15-5 Voltage-follower circuit.

The voltage follower thus has unity gain. However, since the ideal input resistance is infinite and the ideal output resistance is zero, it is very useful as a buffer. It is used to isolate a high-resistance source, in which minimal loading can be tolerated, from a load.

Example 15-4

Consider the noninverting amplifier circuit of Figure 15-6. **(a)** Determine the closed-loop gain A_{CL}. **(b)** For $\pm V_{sat} = \pm 13$ V, determine the input voltage level at which saturation occurs. **(c)** Determine the output voltage for each of the following input voltages: $v_i = 0, 0.2$ V, -0.2 V, 0.5 V, 2 V.

Figure 15-6 Amplifier circuit of Example 15-4.

Solution

(a) The closed-loop voltage gain is

$$A_{CL} = 1 + \frac{R_f}{R_i} = 1 + \frac{18 \times 10^3}{2 \times 10^3} = 1 + 9 = 10 \qquad \textbf{(15-32)}$$

(b) For $\pm V_{sat} = \pm 13$ V, the peak input magnitude is

$$V_{ip} = \frac{13}{10} = 1.3 \text{ V} \qquad \textbf{(15-33)}$$

The linear range for the input is $-1.3 < v_i < +1.3$.

(c) For any value of input voltage in the linear region, the output voltage is

$$v_o = 10v_i \qquad \textbf{(15-34)}$$

The various output voltages for the first four cases are summarized as follows:

v_i (V)	v_o (V)
0	0
0.2	2
−0.2	−2
0.5	5

Note that the output voltages have the same sign as the input voltages in this case. The last voltage, $v_i = 2$ V, exceeds the peak input permitted for linear operation, so saturation will occur. In this case, $v_o = 13$ V.

Example 15-5

Design a noninverting op-amp amplifier circuit to achieve a gain $A_{CL} = 12$. Use standard 5% resistance values and keep all values below 100 kΩ in this particular design.

Solution

The gain specification requires that

$$1 + \frac{R_f}{R_i} = 12 \qquad \textbf{(15-35)}$$

which simplifies to

$$\frac{R_f}{R_i} = 11 \qquad \textbf{(15-36)}$$

or

$$R_f = 11R_i \qquad \textbf{(15-37)}$$

A search through Appendix A results in the following acceptable combination:

$$R_i = 3 \text{ k}\Omega \qquad \text{and} \qquad R_f = 33 \text{ k}\Omega$$

15-4 COMPARISON OF AMPLIFIER CIRCUITS

A comparison of the inverting and noninverting amplifiers will now be made. The formulas for voltage gain, input resistance, and output resistance for the ideal cases are summarized in Table 15-1. For a given R_f/R_i ratio, the gain magnitude of the noninverting circuit is slightly higher than that of the inverting amplifier because of the additional unity term. However, this is neither an advantage nor a disadvantage since resistance ratios can always be adjusted easily.

The major advantage of the noninverting amplifier as compared to the invert-

TABLE 15-1 SUMMARY OF IDEAL RELATIONSHIPS FOR INVERTING AND NONINVERTING AMPLIFIER CIRCUITS

	Inverting	Noninverting
A_{CL}	$-\dfrac{R_f}{R_i}$	$1 + \dfrac{R_f}{R_i}$
R_{in}	R_i	∞
R_{out}	0	0

ing amplifier is the very high input resistance (assumed here as infinite) of the former. In designing inverting amplifiers, one must always select a value of R_i no smaller than the minimum acceptable input resistance, and this sometimes limits the use of an inverting amplifier.

On the other hand, the inverting amplifier has one advantage over the noninverting amplifier in the following sense: Inverting amplifiers can be cascaded to achieve noninverting gains if desired, provided that an even number of stages is used. However, one can never achieve inversion with a noninverting amplifier. Since many applications require inverting gains, the inverting amplifier circuit is essential.

Certain types of signal processing operations (e.g., summing) are more cumbersome to achieve with the noninverting amplifier. In fact, some operations (e.g., ideal integration) are theoretically impossible to achieve with the noninverting amplifier. The inverting amplifier is therefore more flexible in its possible range of applications.

The conclusion is that *both* the *noninverting* and *inverting* amplifiers are very important, and both have numerous applications in modern electronic circuits. Both the noninverting and inverting amplifier configurations may be considered in their basic forms as *voltage-controlled voltage sources*.

15-5 SUMMING AMPLIFIER CIRCUIT

The operational amplifier may be used to combine several signal voltages together while maintaining a common ground for each signal. The simplest designation for such a circuit is a *summing amplifier*. However, the circuit allows the relative weighting of each signal to be adjusted, so the circuit is also called a *linear combination circuit*. The first term is used in this book.

The basic form of a summing amplifier circuit is shown in Figure 15-7. The voltages v_1, v_2, . . . , v_n represent an arbitrary number of input voltages, each referred to circuit ground.

The inverting input terminal is forced to assume a virtual ground condition. Thus each input voltage appears across its particular input resistance. The various currents i_1, i_2, . . . , i_n may then be expressed as follows:

$$i_1 = \frac{v_1}{R_1} \tag{15-38}$$

Figure 15-7 Summing amplifier circuit.

$$i_2 = \frac{v_2}{R_2} \tag{15-39}$$

$$\vdots$$

$$i_n = \frac{v_n}{R_n} \tag{15-40}$$

By KCL, the current i_f may be expressed as

$$i_f = i_1 + i_2 + \cdots + i_n \tag{15-41a}$$

$$= \frac{v_1}{R_1} + \frac{v_2}{R_2} + \cdots + \frac{v_n}{R_n} \tag{15-41b}$$

where the various expressions for the currents were substituted in (15-41b).

The output voltage v_o may be expressed as

$$v_o = -R_f i_f \tag{15-42}$$

Substituting (15-41b) in (15-42), the output voltage is

$$v_o = -\frac{R_f}{R_1}v_1 - \frac{R_f}{R_2}v_2 - \cdots - \frac{R_f}{R_n}v_n \tag{15-43}$$

This last equation may be expressed as

$$\boxed{v_o = A_1 v_1 + A_2 v_2 + \cdots + A_n v_n} \tag{15-44}$$

where

$$A_1 = -\frac{R_f}{R_1} \tag{15-45}$$

$$A_2 = -\frac{R_f}{R_2} \tag{15-46}$$

$$\vdots$$

$$A_n = -\frac{R_f}{R_n} \tag{15-47}$$

The output voltage v_o is thus an algebraic sum of the n input voltages, with each input voltage weighted by a gain factor determined from a resistance ratio. The resistance R_f is common to all the gain factors, but each gain has the input resistance for the particular signal in the denominator, allowing individual adjustment for that gain.

As a by-product of this circuit, the gain factors of all the input signals are inverted. This may pose a minor inconvenience in some applications. However, it is readily circumvented by following this circuit with a simple inverting amplifier with a gain of -1. This causes the resulting sense of all the gain factors to be positive.

Because of the inversion property, a logical question is whether or not the non-inverting amplifier may be used as a summing amplifier. The answer is that it can be used, but the circuit is not nearly as convenient to use. The beauty of the inverting amplifier is that the input currents sum at a virtual ground point, and each is independent of the others. With noninverting summing circuits, each input circuit affects all the other inputs, and the circuit design is somewhat awkward. We will restrict our consideration here to the inverting case since it is the most popular circuit. An additional inversion, when required, is a small price to pay for the convenience of this circuit.

In some cases it is desired that all combined signals be amplified by the same level. Assume that the desired output is

$$v_o = -A(v_1 + v_2 + \cdots + v_n) \qquad (15\text{-}48)$$

where $-A$ is the desired amplification. This is achieved by selecting $R_1 = R_2 = \cdots$ $R_n = R$ and $R_f = AR$. The resulting circuit is shown in Figure 15-8.

Figure 15-8 Summing amplifier circuit with equal amplification for all signals.

Example 15-6

(a) For the circuit of Figure 15-9, determine an expression for the output voltage in terms of the three input voltages. **(b)** Determine the output voltage for the following input condition: $v_1 = 0.5$ V, $v_2 = 0.8$ V, $v_3 = -3$ V.

Figure 15-9 Circuit of Example 15-6.

Solution

(a) The gain factors for the three input signals are determined as follows:

$$A_1 = -\frac{R_f}{R_1} = -\frac{100 \text{ k}\Omega}{10 \text{ k}\Omega} = -10 \qquad (15\text{-}49)$$

$$A_2 = -\frac{R_f}{R_2} = -\frac{100 \text{ k}\Omega}{20 \text{ k}\Omega} = -5 \qquad \textbf{(15-50)}$$

$$A_3 = -\frac{R_f}{R_3} = -\frac{100 \text{ k}\Omega}{100 \text{ k}\Omega} = -1 \qquad \textbf{(15-51)}$$

The output voltage is then expressed as

$$v_o = -10v_1 - 5v_2 - v_3 \qquad \textbf{(15-52)}$$

(b) For the three values of input voltages given in the problem statement, we have

$$v_o = -10(0.5) - 5(0.8) - 1(-3) = -5 - 4 + 3 = -6 \text{ V} \qquad \textbf{(15-53)}$$

Note that the net output effects of the first two voltages are in a negative sense, while the net effect of the third voltage on the output is in a positive sense. In the last case, a negative input multiplied by a negative gain factor produces a positive output sense.

Example 15-7

It is desired to combine two signals v_1 and v_2 and produce an output v_o according to the following requirement:

$$v_o = -12v_1 - 6v_2$$

The input resistances at the two inputs must not be less than 10 kΩ. Design a circuit to achieve the objective. Use resistor values less than 200 kΩ.

Solution

The two gain specifications force the following constraints:

$$\frac{R_f}{R_1} = 12 \qquad \textbf{(15-54)}$$

and

$$\frac{R_f}{R_2} = 6 \qquad \textbf{(15-55)}$$

There are two gain constraints and three unknowns. This means that one resistance may be selected somewhat arbitrarily. However, the minimum input resistance of 10 kΩ establishes a lower bound at the input. Note that the *smallest* input resistance corresponds to the *highest* gain. One solution would be to select $R_1 = 10$ kΩ. The other two resistances would then be $R_f = 120$ kΩ and $R_2 = 20$ kΩ, all of which are standard values. In searching through Appendix A, however, a second possible solution is $R_1 = 15$ kΩ, $R_f = 180$ kΩ, and $R_2 = 30$ kΩ. Since the second solution provides a higher input resistance, it will be selected in this case. The circuit is shown in Figure 15-10.

180 kΩ

15 kΩ

v_1

30 kΩ

v_2

v_o

Figure 15-10 Circuit design of Example 15-7.

15-6 DIFFERENCE AMPLIFIER CIRCUIT

The operational amplifier may be used to form the difference between two signals and, if desired, amplify this difference at the same time. This circuit is called the *difference amplifier*.

The form of a difference amplifier is shown in Figure 15-11. Note that in this form, two resistances have the value R, while the other two have the value AR. It is possible to create different weightings of the two signals by changing one of the resistances to a different value, but that situation will not be considered here. The circuit shown is the most common form.

The difference circuit with certain variables labeled to facilitate analysis is shown in Figure 15-12. The starting point for the analysis is to recognize that the voltage v^+ at the noninverting input is determined solely by v_i since no other part of the circuit is connected there. By the voltage-divider rule, we have

$$v^+ = \frac{AR}{AR + R} v_1 = \frac{A}{A + 1} v_1 \qquad (15\text{-}56)$$

where the common factor R was canceled in the last form.

Figure 15-11 Difference amplifier circuit.

Figure 15-12 Difference amplifier labeled for analysis.

Assuming linear operation, the voltage v^- must be the same as v^+, that is,

$$v^- = v^+ = \frac{Av_1}{A + 1} \tag{15-57}$$

A current i must flow between v^- and v_2. Assuming a positive reference for v^-, this current is

$$i = \frac{v^- - v_2}{R} \tag{15-58}$$

Substitution of (15-57) in (15-58) results in

$$i = \frac{Av_1}{(A + 1)R} - \frac{v_2}{R} \tag{15-59}$$

A voltage v_f is produced across the feedback resistance and is given by

$$v_f = ARi = \frac{A^2}{A + 1}v_1 - Av_2 \tag{15-60}$$

Finally, the output voltage v_o is

$$v_o = v_f + v^- = \frac{A^2}{A + 1}v_1 - Av_2 + \frac{Av_1}{A + 1} \tag{15-61}$$

$$= \frac{A(A + 1)}{A + 1}v_1 - Av_2 \tag{15-62}$$

Simplification yields

$$\boxed{v_o = A(v_1 - v_2)} \tag{15-63}$$

The output voltage is thus A times the difference between the input voltages.

A special case for this circuit is when all four resistances are equal. In that case, the output voltage is simply the difference between the two input voltages.

Example 15-8

(a) For the circuit of Figure 15-13, determine an expression for the output voltage in terms of the two input voltages. **(b)** Determine the output voltage when $v_1 = 2$ V and $v_2 = 1.9$ V

Figure 15-13 Circuit of Example 15-8.

Solution

(a) The gain A is determined as

$$A = \frac{50 \text{ k}\Omega}{10 \text{ k}\Omega} = 5 \tag{15-64}$$

The output voltage can be expressed as

$$v_o = 5(v_1 - v_2) \tag{15-65}$$

(b) For $v_1 = 2$ V and $v_2 = 1.9$ V, we have

$$v_o = 5(2 - 1.9) = 5 \times 0.1 = 0.5 \text{ V} \tag{15-66}$$

15-7 CONTROLLED SOURCES

The noninverting and inverting amplifier circuits considered earlier in the chapter may be both considered as *voltage-controlled voltage sources*. In each case, the controlling variable was the input voltage, and the controlled variable was the output voltage. In this section it will be shown how each of the three other types of dependent sources may be implemented with an op-amp.

Current-Controlled Voltage Source

The schematic diagram of the simplest type of current-controlled voltage source (ICVS) is shown in Figure 15-14. The circuit has essentially the form of an inverting amplifier. However, no input resistance is shown. Instead, the input variable is considered as the current i_i. Since the inverting input assumes a virtual ground level, the input current sees a very low input resistance. This is the ideal condition for a current-driven amplifier, which is in direct contrast to a voltage-driven amplifier.

The current i_i flows through the resistance R and produces a voltage across it of value Ri_i, with the positive terminal on the left. The output voltage is

$$\boxed{v_o = -Ri_i} \tag{15-67}$$

The output voltage is thus a constant times the input current, the desired situation for a current-controlled voltage source.

Figure 15-14 Current-controlled voltage source.

A constant that relates a dependent output voltage to an independent input current has the dimensions of ohms and is called the *transresistance* R_m. The transresistance for this circuit is simply

$$\boxed{R_m = R} \tag{15-68}$$

Voltage-Controlled Current Source

The schematic diagram of the simplest form of a *voltage-controlled current source* (VCIS) is shown in Figure 15-15. The circuit has the same form as an inverting amplifier. However, it differs from the inverting amplifier in that the output variable is the desired current i_L through the load resistance R_L.

The current i_L is the same as the current established through the resistance R on the left and is

$$\boxed{i_L = \frac{v_i}{R}} \tag{15-69}$$

The current is thus a constant times the input voltage and is independent of the resistance R_L. These are the requirements for a voltage-controlled current source.

A constant that relates a dependent load current to an independent input voltage has the dimensions of siemens and is the *transconductance* g_m. The transconductance for the circuit is simply

$$\boxed{g_m = \frac{1}{R}} \tag{15-70}$$

Since the op-amp output voltage is not of prime interest in this circuit, it is easy to overlook the fact that this output voltage must be maintained in the linear region of the op-amp. This limits the maximum value of the output resistance. The reader is invited to show (Problem 15-27) that linear operation requires that the following constraint be satisfied:

$$\boxed{R_L i_L < V_{\text{sat}}} \tag{15-71}$$

Figure 15-15 Voltage-controlled current source for a floating load.

Howland Current Source

The preceding current source requires that the load be floating with respect to ground. Whenever the terminal of the load must be connected to actual ground, the Howland current source can be used. This circuit is shown in Figure 15-16. A guided exercise in the derivation of the input-output relationship is given as Problem 15-28. The basic relationship, which is certainly not obvious, is

$$\boxed{i_L = \frac{v_i}{R}} \tag{15-72}$$

The transconductance is

$$\boxed{g_m = \frac{1}{R}} \tag{15-73}$$

The reader is invited to show (Problem 15-29) that linear operation requires that the following constraint be satisfied:

$$R_L i_L < \frac{V_{\text{sat}}}{2} \tag{15-74}$$

Figure 15-16 Howland voltage-controlled current source for grounded load.

Current-Controlled Current Source

The schematic diagram of a *current-controlled current source* (ICIS) is shown in Figure 15-17. As in the case of the ICVS, the input resistance is very low as required. A guided exercise in the derivation of the input/output relationship for this circuit is given as Problem 15-30. The basic relationship is

$$\boxed{i_L = \left(1 + \frac{R_2}{R_1}\right) i_i} \tag{15-75}$$

Figure 15-17 Current-controlled current source.

The current gain is denoted as β, and for this circuit it is

$$\boxed{\beta = 1 + \frac{R_2}{R_1}} \qquad \text{(15-76)}$$

For linear operation, it can be shown that the following requirement must be met:

$$\left[R_2 + R_L\left(1 + \frac{R_2}{R_1}\right) \right]|i_i| < V_{\text{sat}} \qquad \text{(15-77)}$$

Example 15-9

It is desired to establish a dc current of 5 mA through a certain nonlinear floating resistive load. The circuit of Figure 15-18 is proposed. Assume that the nonlinear resistance is sufficiently small that saturation will not occur. Determine the value of R required.

Figure 15-18 Circuit of Example 15-9.

Solution

The circuit is a voltage-controlled current source with a 15-V dc supply used as the controlling voltage. Since $i_L = i_i = v_i/R$, the value of R is

$$R = \frac{v_i}{i_i} = \frac{v_i}{i_L} = \frac{15\text{ V}}{5 \times 10^{-3}\text{ A}} = 3\text{ k}\Omega \qquad \text{(15-78)}$$

Some readers may question the need for using an op-amp since one could readily establish a fixed current through a simple series circuit consisting of a variable resistance and the nonlinear load. However, that type of circuit would require a readjustment after each change of the load resistance. The circuit shown will maintain a fixed load current of 5 mA for any value of load provided that linear operation is maintained.

Example 15-10

Consider the Howland current source of Figure 15-19. **(a)** Determine the transconductance. **(b)** Determine the load current for $v_i = 2$ V and $v_i = 4$ V, assuming linear operation. **(c)** For $v_i = 4$ V, determine the range of R_L that will ensure linear operation. Assume that $\pm V_{\text{sat}} = \pm 13$ V.

Figure 15-19   Circuit of Example 15-10.

Solution

(a) The transconductance g_m is

$$g_m = \frac{1}{R} = \frac{1}{2 \times 10^3} = 0.5 \times 10^{-3} \text{ S} \qquad \textbf{(15-79)}$$

(b) For any input voltage v_i, the load current i_L in the linear region is

$$i_L = g_m v_i = 0.5 \times 10^{-3} v_i \qquad \textbf{(15-80)}$$

For $v_i = 2$ V, i_L is

$$i_L = 0.5 \times 10^{-3} \times 2 = 1 \text{ mA} \qquad \textbf{(15-81)}$$

For $v_i = 4$ V, i_L is

$$i_L = 0.5 \times 10^{-3} \times 4 = 2 \text{ mA} \qquad \textbf{(15-82)}$$

(c) Linear operation is based on the constraint of (15-74). For $i_L = 2$ mA, we have

$$R_L \times 2 \times 10^{-3} < \frac{13}{2} \qquad \textbf{(15-83)}$$

This leads to the requirement that

$$R_L < 3250 \ \Omega \qquad\qquad \textbf{(15-84)}$$

Thus the current through the load should remain fixed at 2 mA for all values of R_L less than 3250 Ω.

15-8 ACTIVE FILTERS

A *filter* is a circuit that passes a certain range of frequencies while rejecting or strongly attenuating frequencies outside this range. The range of frequencies experiencing little or no attenuation is called the *passband,* and the range of frequencies in which a large reduction in amplitude occurs is called the *stopband.* Practical filters have a *transition band* between the passband and stopband. The boundaries between these ranges are somewhat arbitrary and depend on the filter type as well as the specifications.

The subject of electrical filters is a vast subject area for which many complete books have been devoted. A brief introduction to a few so-called *active filters* is given in this section. However, a sufficient amount of material is given to permit the reader to design and construct workable filters.

Passive versus Active Filters

A *passive* filter is one utilizing resistance, capacitance, and inductance, but with no active elements required. The only power requirement for a passive filter, therefore, is the power dissipated by the input signal source. Passive filters dominated the electronics industry in the early phase of its evolution. Passive filters are very difficult to implement at low frequencies, however, because of the large values of inductance required and the lossy nature of such large inductance.

An *active filter* is one utilizing resistance, capacitance, and one or more active devices (such as an op-amp). Through feedback, the active device eliminates the necessity for inductors. Consequently, it is possible to achieve the same degree of filtering with active *RC* filters as with passive *RLC* filters. At low frequencies (typically a few hundred hertz and below), active filters are far superior to passive filters. Active filters require one or more power supplies to function properly.

Active filters first appeared on the scene in the 1950s, and their use has increased markedly through the past several decades. Today, active filters are widely used for many applications in electronics, instrumentation, control, and related areas.

Filter Types

Filters can be classified as *low-pass*, *high-pass*, *bandpass*, and *band-rejection*. The general characteristics are as follows:

1. *Low-pass.* A low-pass filter passes frequencies *below* a certain reference frequency while strongly attenuating frequencies above this range.

Sec. 15-8   Active Filters

2. *High-pass.* A high-pass filter passes frequencies *above* a certain reference frequency while strongly attenuating frequencies below this range.

3. *Bandpass.* A bandpass filter passes frequencies within a certain limited range while strongly attenuating frequencies both above and below this range.

4. *Band-stop.* A band-stop filter strongly attenuates frequencies within a certain limited range while passing frequencies both above and below this range.

The most commonly used active filters are the *low-pass* types.

Filters can be classified by the types of passband and/or stopband shape. The most common type is the *Butterworth* characteristic. The Butterworth amplitude characteristic is referred to as a *maximally flat amplitude response.* Within the passband, the response is as flat as possible for the complexity of the filter.

Filters can also be classified by the *number of poles.* This term relates to certain advanced mathematical functions used in establishing the response form. For the active filters considered here, the *number of poles is the number of capacitors in the basic circuit form.*

The treatment here will be limited to low-pass, active Butterworth filters with two poles, three poles, and four poles. These particular filters can be used for a variety of applications with moderate filtering requirements at relatively low and very low frequencies (typically below 10 kHz).

Butterworth Filter Curves

Curves displaying the low-pass amplitude response characteristics for two-pole, three-pole, and four-pole Butterworth filters are shown in Figure 15-20. The frequency f_c is defined as the *cutoff frequency,* and it is the frequency at which the response is down 3 dB. The frequency range to the left is considered as the passband, and the range to the right is considered as the stopband. Note that all curves pass through the point at which the response is down 3 dB.

As the number of poles increases, two trends can be observed: (1) the passband response is flatter, and (2) the slope of the curve in the stopband is steeper. Thus more ideal filtering characteristics are achieved as the number of poles is increased.

An *octave* of frequency change corresponds to a doubling of the frequency. Let N represent the number of poles. It can be shown both theoretically and from the curves of Figure 15-20 that the slope of a given amplitude response well above cutoff is very close to

$$\text{slope} = -6N \text{ dB/octave} \qquad \textbf{(15-85)}$$

Note that the frequency scale provided ranges from $0.1 f_c$ to $10 f_c$. The curves then apply to any filter of the given types irrespective of the actual frequency range provided that the scale is interpreted correctly. For example, if $f_c = 500$ Hz, the frequency scale may be considered to extend from 50 Hz to 5 kHz.

Curves such as given in Figure 15-20 are useful not only to predict the amplitude response of the pertinent filters, but they can also be used to determine the required number of poles to meet a prescribed specification. This process is illustrated in Example 15-11.

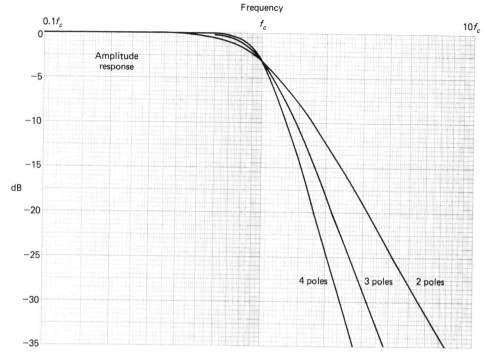

Figure 15-20 Butterworth low-pass amplitude response curves.

Example 15-11

A low-pass filter is desired to meet the following specifications:

3-dB cutoff frequency = 1 kHz
Response to be down by at least 15 dB at 2 kHz

Determine if any of the Butterworth filter characteristics of Figure 15-20 will meet the specifications, and if so, determine the minimum number of poles.

Solution

The cutoff frequency is $f_c = 1$ kHz. The other frequency of interest is 2 kHz, which is twice the cutoff frequency. This means that $2f_c$ is the pertinent frequency at which to focus on Figure 15-20. At $2f_c$, a two-pole filter is down by slightly more than 12 dB, so the specification is not met. However, the response of a three-pole filter is down by about 18 dB at $2f_c$. Thus the three-pole filter meets the specification with about 3 dB of reserve. Obviously, a four-pole filter would meet the specification with much reserve since it is down by about 24 dB at $2f_c$. The minimum number of poles required is *three*. The filter involved will be designed in Example 15-13.

Filter Circuits

The circuit diagrams for two-pole, three-pole, and four-pole active low-pass Butterworth filters are shown in Figures 15-21, 15-22, and 15-23, respectively. The two-pole and three-pole circuits use one op-amp, while the four-pole circuit uses two op-amps. All op-amps in these circuits are connected as voltage followers. There are other active filter forms in which the op-amps are connected to achieve gains greater than unity, but the voltage follower forms are among the most common active filters.

Formulas for determining element values are given beside the appropriate components. The quantity ω_c is the radian 3-dB cutoff frequency, and it is related to the cyclic cutoff frequency f_c in hertz by

$$\omega_c = 2\pi f_c \tag{15-86}$$

In each circuit diagram, there is at least one parameter value that may be selected arbitrarily. This gives the designer some flexibility in selecting components around available stock values.

*A bias compensation resistor may be inserted here if desired

Figure 15-21 Two-pole active low-pass Butterworth filter.

*A bias compensation resistor may be inserted here if desired

Figure 15-22 Three-pole active low-pass Butterworth filter.

$C_1 = \dfrac{1.082}{\omega_c R_1}$

$C_3 = \dfrac{2.613}{\omega_c R_2}$

$C_2 = \dfrac{0.9241}{\omega_c R_1}$

$C_4 = \dfrac{0.3825}{\omega_c R_2}$

*A bias compensation resistor may be inserted here if desired

Figure 15-23   Four-pole active low-pass Butterworth filter.

Determining Element Values

First, consider the two-pole and three-pole circuits since only one value of resistance is required. Assume that the 3-db frequency f_c is known and that ω_c is calculated. There are two approaches that may be used, which will be summarized in the next two paragraphs.

Select R.   The first approach is to select a value of R, which will then permit the computation of all capacitor values. If the resulting capacitor values are not close to stock values, some trial and error may be required. If the resulting values of C are too large, increase the value of R and repeat the process. If the resulting values of C are too small, decrease the value of R and repeat the process. With some experience, the process will become somewhat intuitive.

Select C.   The second approach is to force one of the capacitor values (often the largest) to assume a standard stock value and to determine the value of R required. The remaining values of capacitance are then calculated as before. It is usually easier to implement "odd" values of resistance than capacitance, so capacitance values are often more satisfactory with this approach than the first procedure. This procedure is illustrated in Example 15-13.

The preceding discussion is applicable to the four-pole filter of Figure 15-23 with the following modification: Since the four-pole filter has two sections, the procedure is applied separately to each section. Note that each section has a separate resistance. As one possibility, they could be selected to be the same (i.e., $R_1 = R_2$). However, by allowing them to be different, one capacitance in each section could be forced to assume a standard value.

It is important that the element values of active filters be as close as possible to the ideal values. Capacitor types should be polystyrene when possible, but Mylar and high-quality mica types may also be used. Do not use electrolytic or other polarized capacitors. Avoid disk ceramic capacitors because of the wide tolerances.

Example 15-12

Design a two-pole, low-pass, active Butterworth filter with a 3-dB cutoff frequency of 1 kHz. Select $R = 10$ kΩ.

Solution

The pertinent circuit is shown in Figure 15-21. The value of ω_c is

$$\omega_c = 2\pi f_c = 2\pi \times 1 \times 10^3 = 6283.19 \text{ rad/s} \qquad \textbf{(15-87)}$$

With $R = 10^4$ Ω, the values of C_1 and C_2 are calculated as

$$C_1 = \frac{1.414}{\omega_c R} = \frac{1.414}{6283.19 \times 10^4} = 0.0225 \ \mu\text{F} \qquad \textbf{(15-88)}$$

and

$$C_2 = \frac{0.7071}{\omega_c R} = \frac{0.7071}{6283.19 \times 10^4} = 0.0113 \ \mu\text{F} \qquad \textbf{(15-89)}$$

The circuit diagrams with the ideal element values is shown in Figure 15-24. Although C_1 and C_2 are not exactly standard values, they are very close to the standard values of 0.022 μF and 0.011 μF, respectively. For most applications, these stock values should suffice.

Figure 15-24 Active filter design of Example 15-12.

Example 15-13

Design an active filter to meet the specifications of Example 15-11. Perform the design so that the largest value of capacitance is the stock value of 0.047 μF.

Solution

It was determined in Example 15-11 that a three-pole filter will meet the requirements, so the pertinent schematic diagram is shown in Figure 15-22. Inspection of the formulas for determining capacitance reveals that C_1 is the largest capacitance. Since the 3-dB frequency is $f_c = 1$ kHz, $\omega_c = 6283.19$ rad/s as in Example 15-12. We then force C_1 to be 0.047 μF by equating the expression for C_1 to this value; that is,

$$C = \frac{3.546}{\omega_c R} = \frac{3.546}{6283.19R} = 0.047 \times 10^{-6} \qquad \textbf{(15-90)}$$

Solving for R, we obtain

$$R = \frac{3.546}{6283.19 \times 0.047 \times 10^{-6}} = 12.01 \text{ k}\Omega \qquad \textbf{(15-91)}$$

We are lucky because the calculated value of R is very close to the stock value of 12 kΩ.

Calculations of C_2 and C_3 now proceed as follows:

$$C_2 = \frac{1.392}{\omega_c R} = \frac{1.392}{6283.19 \times 12.01 \times 10^3} = 0.01845 \ \mu\text{F} \qquad \textbf{(15-92)}$$

and

$$C_3 = \frac{0.2024}{\omega_c R} = \frac{0.2024}{6283.19 \times 12.01 \times 10^3} = 0.002682 \ \mu\text{F} \qquad \textbf{(15-93)}$$

The circuit diagram is shown in Figure 15-25. The resistances have been rounded to 12 kΩ, but the capacitance values shown are as calculated. Although C_2 and C_3 are not exactly standard values, they are very close to the standard values of 0.018 μF and 0.0027 μF, respectively.

Figure 15-25 Active filter design of Example 15-13.

15-1. Assume that in a certain op-amp closed-loop application, the output is forced through feedback to be $v_o = 12$ V. Calculate the differential input voltage v_d for each of the following values of open-loop differential gain A_d: **(a)** 10^5; **(b)** 20,000; **(c)** 5000.

15-2. Assume that in a certain op-amp closed-loop application, the output is forced through feedback to be $v_o = -10$ V. Calculate the differential input voltage v_d for each of the following values of open-loop differential gain A_d: **(a)** 120,000; **(b)** 40,000; **(c)** 8000.

15-3. The two input signal voltages for a certain op-amp circuit are $v^+ = 6.002$ V and $v^- = 6.000$ V. Determine the output voltage v_o if the differential gain is $A_d = 4000$.

15-4. The two input signal voltages for a certain op-amp circuit are $v^+ = 4.000$ V and $v^- = 4.003$ V. Determine the output voltage v_o if the differential gain is $A_d = 2500$.

15-5. Consider the amplifier circuit of Figure P15-5. **(a)** Determine the closed-loop gain A_{CL}. **(b)** Determine the input resistance R_{in}. **(c)** For $\pm V_{\text{sat}} = \pm 13$V, determine the input voltage level at which saturation occurs. **(d)** Determine the output voltage for each of the following input voltages: $v_i = 0, 0.2$ V, -0.2 V, 0.5 V, 1 V.

Figure P15-5

15-6. For the amplifier circuit of Figure P15-6, repeat the computations of Problem 15-5.

Figure P15-6

15-7. For the circuit of Figure P15-7, determine an expression for the output voltage in terms of the two input voltages.

Figure P15-7

15-8. For the circuit of Figure P15-8, determine an expression for the output voltage in terms of the three input voltages.

Figure P15-8

15-9. For the circuit of Problem 15-7 (Figure P15-7), determine the output voltage for each of the following combinations of input voltages: **(a)** $v_1 = 3$ V, $v_2 = 2$ V; **(b)** $v_1 = 3$ V, $v_2 = -4$ V; **(c)** $v_1 = 5$ V, $v_2 = -8$ V; **(d)** $v_1 = -5$ V, $v_2 = -8$ V. Assume that $\pm V_{sat} = \pm 13$ V.

15-10. For the circuit of Problem 15-8 (Figure P15-8), determine the output voltage for each of the following combinations of input voltages: **(a)** $v_1 = 4$ V, $v_2 = -4$ V, $v_3 = 2$ V; **(b)** $v_1 = 2$ V, $v_2 = 3$ V, $v_3 = -2$ V; **(c)** $v_1 = -1$ V, $v_2 = 2$ V, $v_3 = 1$ V. Assume that $\pm V_{sat} = \pm 13$ V.

15-11. For the circuit of Figure P15-11, determine an expression for the output voltage in terms of the two input voltages.

Figure P15-11

15-12. For the circuit of Figure P15-12, determine an expression for the output voltage in terms of two input voltages.

Figure P15-12

15-13. For the circuit of Problem 15-11 (Figure P15-11), determine the output voltage for each of the following combinations of input voltages: **(a)** $v_1 = 2$ V, $v_2 = 2$ V; **(b)** $v_1 = 2$ V, $v_2 = -2$ V; **(c)** $v_1 = -2$ V, $v_2 = 2$ V; **(d)** $v_1 = 5$ V, $v_2 = -5$ V. Assume that $\pm V_{sat} = \pm 13$ V.

15-14. For the circuit of Problem 15-12 (Figure P15-12), determine the output voltage for each of the following combinations of input voltages: **(a)** $v_1 = 2$ V, $v_2 = 2$ V; **(b)** $v_1 = 0.1$ V, $v_2 = -0.1$ V; **(c)** $v_1 = -0.1$ V, $v_2 = 0.1$ V; **(d)** $v_1 = 5$ V, $v_2 = -5$ V. Assume that $\pm V_{sat} = \pm 13$ V.

15-15. Consider the current-controlled voltage source of Figure P15-15. **(a)** Determine the transresistance. **(b)** Determine the output voltage for each of the following values of input current i_i: 0, 2 mA, -2 mA, 10 mA. Assume that $\pm V_{sat} = \pm 13$

Figure P15-15

15-16. Consider the current-controlled voltage source of Figure P15-16. **(a)** Determine the transresistance. **(b)** Determine the output voltage for each of the following values of input curent i_i: 0, 50 μA, 0.6 mA, -1 mA, -2 mA. Assume that $\pm V_{sat} = \pm 13$ V.

Figure P15-16

15-17. Consider the voltage-controlled current source of Figure P15-17. Assume that $\pm V_{sat} = \pm 13$ V. **(a)** Determine the transconductance. **(b)** For $v_i = 2$ V, determine the maximum value of R_L for linear operation. **(c)** For $R_L = 10$ kΩ, determine the maximum value of v_i for linear operation.

Figure P15-17

15-18. Consider the voltage-controlled current source of Figure P15-18. Assume that $\pm V_{\text{sat}} = \pm 13$ V. **(a)** Determine the transconductance. **(b)** For $v_i = 5$ V, determine the maximum value of R_L for linear operation. **(c)** For $R_L = 40$ kΩ, determine the maximum value of v_i for linear operation.

Figure P15-18

15-19. Consider a Howland current source of the form shown in Figure 15-16 with $R = 2$ kΩ. Assume that $\pm V_{\text{sat}} = \pm 13$ V. **(a)** Determine the transconductance. **(b)** For $v_i = 2$ V, determine the maximum value of R_L for linear operation. **(c)** For $R_L = 10$ kΩ, determine the maximum value of v_i for linear operation.

15-20. Consider a Howland current source of the form shown in Figure 15-16 with $R = 16$ kΩ. Assume that $\pm V_{\text{sat}} = \pm 13$ V. **(a)** Determine the transconductance. **(b)** For $v_i = 5$ V, determine the maximum value of R_L for linear operation. **(c)** For $R_L = 40$ kΩ, determine the maximum value of v_i for linear operation.

15-21. Consider the current-controlled current source of Figure P15-21. **(a)** Determine the current gain β. **(b)** Determine the load current i_L if $i_i = 200$ μA, $R_L = 5$ kΩ, and $\pm V_{\text{sat}} = \pm 13$ V.

Figure P15-21

15-22. Consider the current-controlled current source of Figure P15-22. **(a)** Determine the current gain β. **(b)** Determine the load current i_L if $i_i = 50 \mu$A. **(c)** Determine the load current i_L if $i_i = 100 \mu$A, $R_L = 1$ kΩ, and $\pm V_{\text{sat}} = \pm 13$ V.

Figure P15-22

15-23. A certain three-pole, low-pass Butterworth filter has a 3-dB cutoff frequency of 1.5 kHz. Using Figure 15-20, determine the approximate attenuation in decibels at 4.5 kHz.

15-24. A certain four-pole, low-pass Butterworth filter has a 3-dB cutoff frequency of 800 Hz. Using Figure 15-20, determine the approximate attenuation in decibels at 2 kHz.

DERIVATION PROBLEMS

15-25. Show that the differential input voltage for an inverting amplifier configuration is given by

$$v_d = -\frac{R_f}{R_i}\frac{v_i}{A_d}$$

15-26. Show that the differential input voltage for a noninverting amplifier configuration is given by

$$v_d = \left(1 + \frac{R_f}{R_i}\right)\frac{v_i}{A_d}$$

15-27. For the voltage-controlled current source of Figure 15-15, derive the constraint of (15-71).

15-28. The objective of this exercise is to derive the current relationship of the Howland current source as given by (15-72). On the circuit diagram of Figure 15-16, label the voltages v_o, v^+, and v^- at the appropriate points. Note that $v^- = [R/(R + R)]v_o = v_o/2$ and $v^+ = v^-$. Next, sum the currents leaving the noninverting terminal, and after appropriate substitution, derive (15-72).

15-29. For the Howland current source of Figure 15-16, derive the constraint of (15-74).

15-30. The objective of this exercise is to derive the current-gain relationship of a current-controlled current source as given by (15-75). On the circuit diagram of Figure 15-17, label a current i_2 flowing up through R_2, a current i_1 flowing from ground through R_1, a voltage v_2 across R_2 (+ at bottom), and a voltage v_1 across R_1 (+ on left). The current i_2 flows through R_2 producing the voltage v_2. Since the inverting input is at virtual ground, $v_1 = v_2$. The voltage across R_1 results in the current i_1. Finally, i_L is determined from i_1 and i_2. Using this approach, derive (15-75).

15-31. For the circuit of Figure P15-31, show that the output voltage is given by

$$v_o = \frac{R_3 + R_4}{R_3}\left(\frac{R_2}{R_1 + R_2}v_1 + \frac{R_1}{R_1 + R_2}v_2\right)$$

Figure P15-31

15-32. For the circuit of Figure P15-32, show that the output voltage is given by

$$v_o = \frac{R_2}{R_1 + R_2} \frac{R_3 + R_4}{R_3} v_1 - \frac{R_4}{R_3} v_2$$

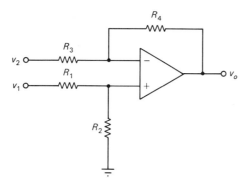

Figure P15-32

DESIGN PROBLEMS

15-33. Design an inverting op-amp amplifier circuit to meet the following specifications:

$$A_{CL} = -12$$
$$R_{in} \geq 10 \text{ k}\Omega$$

Use standard 5% resistance values, and keep all values below 200 kΩ to minimize noise and bias offset effects.

15-34. Design an inverting op-amp amplifier circuit to meet the following specifications:

$$A_{CL} = -25$$
$$R_{in} \geq 20 \text{ k}\Omega$$

Use standard 5% resistance values, and keep all values below 1 MΩ to minimize noise and bias offset effects.

15-35. Design a noninverting op-amp amplifier circuit to achieve a gain $A_{CL} = 25$. Use standard 5% resistance values, and keep all values below 500 kΩ.

15-36. Design a noninverting op-amp amplifier circuit to achieve a gain $A_{CL} = 48$. Use standard 5% resistance values, and keep all values below 500 kΩ.

15-37. It is desired to combine two signals v_1 and v_2 to produce an output v_o according to the following requirement:

$$v_o = -5V_1 - 12v_2$$

The input resistance at the two inputs must not be less than 5 kΩ. Design a circuit to achieve the objective. Use standard 5% resistance values less than 200 kΩ.

15-38. It is desired to combine three signals v_1, v_2, and v_3 to produce an output v_o according to the following requirement:

$$v_o = -v_1 - 4v_2 - 12v_3$$

The input resistance at the two inputs must not be less than 10 KΩ. Design a circuit to achieve the objective. Use standard 5% resistance values less than 200 kΩ.

15-39. Design a current-controlled voltage source to achieve a transresistance of 1.5 kΩ.

15-40. Design a voltage-controlled current source for a floating load to achieve a transconductance of 1 mS.

15-41. Design a Howland current source for a grounded load to achieve a transconductance of 1 mS.

15-42. Design a current-controlled current source to achieve a current gain of 28. Use standard 5% resistance values.

15-43. A low-pass filter is desired to meet the following specifications:

3-dB cutoff frequency = 500 Hz
Response to be down by at least 26 dB at 1500 Hz

Determine the minimum number of poles required in a Butterworth filter.

15-44. A low-pass filter is desired to meet the following specifications:

3-dB cutoff frequency = 500 Hz
Response to be down by at least 30 dB at 1250 Hz

Determine the minimum number of poles required in a Butterworth filter.

15-45. Design a three-pole, low-pass, active Butterworth filter with a 3-dB frequency of 500 Hz. Select $R = 20$ kΩ.

15-46. Design a four-pole, low-pass, active Butterworth filter with a 3-dB frequency of 500 Hz. Select $R_1 = R_2 = 20$ kΩ.

15-47. Repeat the design of Problem 10-45 by selecting the largest capacitor to be 0.068 μF.

15-48. Repeat the design of Problem 10-46 by selecting the largest capacitor in each section to be 0.068 μF.

TROUBLESHOOTING PROBLEMS

15-49. You are troubleshooting an inverting amplifier circuit of the form shown in Figure 15-4. Specifications call for $v_o = -12$ V when $v_i = 1$ V. However, when v_i is a sine wave with a peak value of 1 V, the actual output is observed to be a square-wave oscillating between about -13 and $+13$ V. The frequency of the square-wave is the same as that of the sine wave. Which *two* of the following *could* be the difficulty? **(a)** R_i too small; **(b)** R_i too large; **(c)** R_f too small; **(d)** R_f too large; **(e)** defective power supply; **(f)** defective op-amp.

15-50. You are troubleshooting a noninverting amplifier circuit of the form shown in Figure 15-4. Specifications call for $v_o = 12$ when $v_i = 1$ V. However, when a 1-V dc level is established at v_i, the following voltages are measured: $v_o = 5.5$ V, $v^- = 0.5$ V. The power supply voltages are correct at ± 15 V. Which *one* of the following *could* be the difficulty? **(a)** R_i too small; **(b)** R_i too large; **(c)** R_f too small; **(d)** R_f too large; **(e)** defective op-amp.

COMPUTER PROBLEMS

15-51. Write a program to design two-pole, low-pass, active Butterworth filters.

Input data: f_c, R
Output data: C_1, C_2

15-52. Write a program to design three-pole, low-pass, active Butterworth filters.

Input data: f_c, R
Output data: C_1, C_2, C_3

15-53. Write a program to design four-pole, low-pass, active Butterworth filters.

Input data: f_c, R_1, R_2
Output data: C_1, C_2, C_3, C_4

CHAPTER 16

Nonlinear Electronic Circuits

OBJECTIVES

After completing this chapter, the reader should be able to:

1. Discuss the major differences between linear and nonlinear electronic circuits.
2. Analyze the operation of active rectifier circuits.
3. Draw waveforms and write equations to describe capacitor charge and discharge.
4. Analyze the operation of passive and active peak detector circuits.
5. Analyze the operation of an envelope detector circuit.
6. Analyze the operation of passive and active clamping circuits.
7. Analyze the operation of comparator circuits implemented with a general-purpose op-amp.

16-1 OVERVIEW

For the past several chapters, most of our attention was directed toward the analysis, modeling, and design of various amplifier circuits. Ideal amplifiers are examples of *linear* circuits. Ideal linear amplifiers are characterized by the fact that the output waveform has the same form as the input waveform (i.e., no distortion is introduced).

In this chapter we observe some of the most common *nonlinear* electronic circuits. For this purpose a nonlinear circuit will be defined as one in which the output waveform does not have the same shape as the input waveform. Obviously, if an amplifier is nonlinear, it is an undesirable situation. However, there are numerous applications in which it is desired to reshape or change the form of a waveform or a signal in some predetermined manner. Thus the nonlinear circuits to be considered here can be used to advantage in various waveshaping applications.

Along with passive nonlinear circuits, the op-amp will be shown to be quite useful in realizing precision nonlinear operations. When op-amps appear in applications, bear in mind that the feedback loop may be closed in some applications, while in others, the op-amp may be allowed to assume a saturation state. In the first case, the assumption of a zero differential input voltage will normally still be valid. In the latter case, however, the loop will often be broken, and the differential input voltage will no longer be forced to assume a nearly zero value. In designing such circuits, care must be taken that the maximum input differential voltage not be exceeded.

The nonlinear circuits of this chapter utilizing op-amps may be classified as either *saturating* or *nonsaturating*. A *saturating circuit* is one in which the op-amp output becomes saturated for some portion of the normal range of the input signal. A *nonsaturating circuit* is one in which the op-amp output is prevented from becoming saturated for the normal range of the input signal.

It takes an appreciable amount of time for an op-amp to move into and out of saturation. Consequently, saturating circuits are considerably slower than nonsaturating circuits and may be used only at very low frequencies. However, their simplicity justifies their use in applications in which the speed of response is not critical.

16-2 ACTIVE RECTIFIER CIRCUITS

Several passive small-signal rectifier circuits were considered in Chapter 3 to illustrate simple applications of diodes. These circuits were all characterized by certain errors between the input and the desired output as a result of forward-biased diode voltage drops. In this section several rectifier circuits utilizing one or more op-amp circuits are considered. The use of an op-amp with feedback can virtually eliminate the forward diode drop in the signal path. Rectifiers employing op-amps to eliminate the diode drop will be referred to as *active rectifier circuits*. Such circuits are also denoted elsewhere in the literature as *precision rectifier circuits*.

Noninverting Half-Wave Circuit

The simplest form of an active rectifier circuit is the noninverting half-wave rectifier circuit of Figure 16-1. The circuit has essentially the same form as a voltage follower except for the presence of the diode in the forward block. When v_i is positive, v_a will increase in a positive direction, and this will forward bias the diode. The feedback loop will thus be closed, and this will force the noninverting output voltage to be the same as the input voltage v_i. However, the noninverting input is the same as the output, so $v_o = v_i$ for this condition. The voltage v_a will be clamped at about 0.7 V higher than v_i to maintain the feedback loop.

When v_i is negative, v_a will become negative, and this will open the loop. Consequently, v_a will assume the value of negative saturation (i.e., $v_a = -V_{\text{sat}}$). The voltage v_o is thus zero in this case.

A summary of the preceding conditions is as follows:

$$v_o = \begin{cases} v_i & \text{for } v_i > 0 & \text{(16-1a)} \\ 0 & \text{for } v_i < 0 & \text{(16-1b)} \end{cases}$$

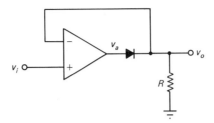

Figure 16-1 Noninverting active half-wave rectifier circuit.

Strictly speaking, v_i must be large enough to cause v_a to exceed the diode drop for (16-1a) to be valid, but since this value for v_i is of the order of microvolts, the condition of (16-1a) is quite accurate. The transfer characteristic for this rectifier is shown in Figure 16-2. The voltage v_a is given by

$$v_a = \begin{cases} v_i + 0.7 & \text{for } v_i > 0 \quad \textbf{(16-2a)} \\ -V_{\text{sat}} & \text{for } v_1 < 0 \quad \textbf{(16-2b)} \end{cases}$$

The primary disadvantage of this rectifier is that it is a *saturating circuit*, so its speed is somewhat limited.

Inverting Half-Wave Circuit

A nonsaturating rectifier can be achieved by employing two diodes in an inverting half-wave circuit. This circuit as arranged for active operation in the second quardrant is shown in Figure 16-3.

Refer to Figure 16-4(a) for the discussion that follows. When v_i is positive, v_a will become negative, and this will forward bias D1. However, D2 will be reverse biased, which uncouples v_o from v_a. The feedback loop is thus closed around D1,

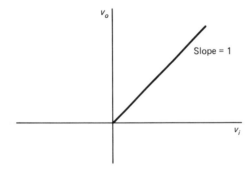

Figure 16-2 Transfer characteristic curve of active noninverting rectifier circuit.

Figure 16-3 Inverting active half-wave rectifier circuit.

(a) $v_i > 0$

$v_a = v_o + 0.7 = -v_i + 0.7$

(b) $v_i < 0$

Figure 16-4 Two states in analyzing active rectifier circuit of Figure 16-3.

and since the inverting input must assume a virtual ground for this condition, $v_o = 0$. The voltage v_a is clamped at 0.7 V below ground (i.e., $v_a = -0.7$ V).

Refer now to Figure 16-4(b). When v_i is negative, v_a will become positive, and this will reverse bias D1. However, D2 will be forward biased for this condition. The feedback loop is thus closed around D2 with D1 open. Under this condition, the circuit is operating as an inverting amplifier with a diode added to the forward gain block. The output v_o is thus forced to be $v_o = -(R/R)v_i = -v_i$. The voltage v_a assumes a voltage of 0.7 V above the level of v_o.

The preceding results for v_o are summarized as follows:

$$v_o = \begin{cases} 0 & \text{for } v_i > 0 \\ -v_i & \text{for } v_i < 0 \end{cases}$$

(16-3a)

(16-3b)

The transfer characteristic for this rectifier is shown in Figure 16-5.

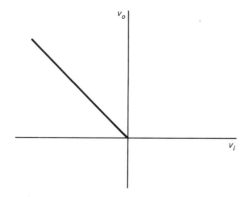

Figure 16-5 Transfer characteristic curve of active inverting rectifier circuit.

The two states for v_a are summarized as follows:

$$v_a = \begin{cases} -0.7 \text{ V} & \text{for } v_i > 0 \qquad \text{(16-4a)} \\ -v_i + 0.7 \text{ V} & \text{for } v_i < 0 \qquad \text{(16-4b)} \end{cases}$$

Although v_a switches abruptly between two levels as the input changes sign, it will remain out of saturation (unless v_i comes within 0.7 V of saturation). Consequently, this rectifier circuit is much faster than the noninverting circuit considered earlier.

Example 16-1

For the rectifier circuit of Figure 16-1, determine v_o and v_a for $v_i =$ **(a)** 3 V and **(b)** −3 V. Assume that $\pm V_{\text{sat}} = \pm 13$ V.

Solution

(a) When $v_i = 3$ V, the loop is closed through the diode. This corresponds to the first quardrant of Figure 16-2 and

$$v_o = 3 \text{ V} \qquad \text{(16-5)}$$

The voltage v_a must be 0.7 V higher; that is,

$$v_a = 3.7 \text{ V} \qquad \text{(16-6)}$$

(b) When $v_i = -3$ V, the loop is open and

$$v_o = 0 \qquad \text{(16-7)}$$

The op-amp output assumes negative saturation; that is,

$$v_a = -13 \text{ V} \qquad \text{(16-8)}$$

Example 16-2

For the rectifier circuit of Figure 16-3, determine v_o and v_a for $v_i =$ **(a)** 3 V and **(b)** −3 V.

Solution

(a) When $v_i = 3$ V, D1 is forward biased and D2 is reverse biased. The output is

$$v_o = 0 \qquad \text{(16-9)}$$

The voltage v_a is

$$v_a = -0.7 \text{ V} \qquad \text{(16-10)}$$

(b) When $v_i = -3$ V, D1 is reverse biased and D2 is forward biased. In this case

$$v_o = -v_i = -(-3) = 3 \text{ V} \qquad \textbf{(16-11)}$$

The voltage v_a is

$$v_a = -v_i + 0.7 = -(-3) + 0.7 = 3.7 \text{ V} \qquad \textbf{(16-12)}$$

16-3 CAPACITOR CHARGE AND DISCHARGE

Many nonlinear electronic circuits and timers utilize the charging and/or discharging of a capacitor to control the timing. It is, therefore, necessary to understand this process before undertaking a study of circuits utilizing the effect.

Once a voltage is established across an ideal capacitor, the capacitor will hold or store the voltage until it is discharged. If the capacitor were ideal and had no losses, the voltage would remain indefinitely. Any practical capacitor will, of course, have some leakage, so it will eventually discharge. However, many capacitors are capable of holding a charge for a long period of time.

Capacitor Discharge

Although the order of presentation here may seem backward, it is easier mathematically to begin the discussion by considering the discharge of a capacitor that is already charged to some initial value. Consider the capacitor shown in Figure 16-6(a), which has been previously charged to a voltage V_o. The quantity v_C represents the instantaneous voltage, which will change with time once discharge is initiated. The capacitor is assumed to be lossless, so $v_C = V_o$ initially.

At some time, which we will conveniently designate as $t = 0$, assume that the resistor is connected across the capacitor. Current starts to flow through the resistor, and the initial stored energy starts to dissipate in the resistor.

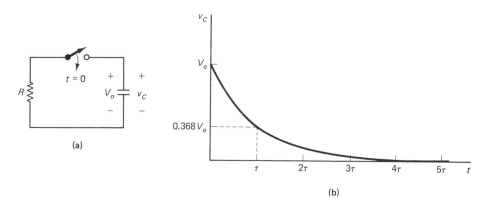

Figure 16-6 Initially charged capacitor and the discharge voltage waveform.

The exact mathematical form of the voltage v_C is given by

$$v_C = V_o e^{-t/\tau} = V_o e^{-t/RC} \qquad \text{(16-13)}$$

The quantity e represents the base of the natural system of logarithms. The value of e to four significant figures is 2.718. The quantity $-t/\tau = -t/RC$ is the exponent of e. The function e^x is available on all modern scientific calculators.

The form of the voltage discharge is shown in Figure 16-6(b). The initial value of the exponential function is unity, so the initial value of v_C is V_o, as required.

The quantity τ is defined as the *time constant* and is

$$\boxed{\tau = RC} \qquad \text{(16-14)}$$

After a time $t = \tau$, the voltage has discharged to about 36.8% of its initial value. In a time $t = 2\tau$, the voltage has discharged to about 13.5% of its initial value. Other values are readily determined with a calculator. In a time $t = 5\tau$, the voltage has discharged to a level less than 1% of its initial value.

Charging a Capacitor

Consider the circuit of Figure 16-7(a) and assume that the capacitor is initially uncharged. At $t = 0$, the switch is closed, and the capacitor is connected through a resistance to a dc voltage. Current begins to flow, and the voltage v_C across the capacitor starts to change. The exact mathematical form of the voltage v_C is given by

$$\boxed{v_C = V_B(1 - e^{-t/\tau})} \qquad \text{(16-15)}$$

where $\tau = RC$ was defined in (16-14).

The form of the voltage across the capacitor is shown in Figure 16-7(b). After a time $t = \tau$, the voltage will reach 63.2% of its final voltage. After a time $t = 2\tau$, the capacitor voltage will reach about 86.5% of its final value. In a time $t = 5\tau$, the voltage is over 99.3% of the final value. For most practical purposes, the capacitor can be thought of as reaching the final value in a time $t = 5\tau$.

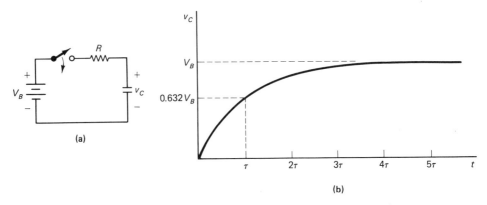

Figure 16-7 Initially charged capacitor and the charging voltage waveform.

Changing Levels

Thus far we have considered the complete discharge of a capacitor and the process of charging an initially uncharged capacitor. Next we consider the process of charging or discharging a capacitor from one level to a second level. Two possibilities must be considered: (1) discharging from a higher level to a lower level, and (2) charging from a lower level to a higher level.

The circuit to be considered is shown in Figure 16-8(a). The capacitor is assumed to be charged to a level V_o, and a dc voltage of value V_B is connected at $t = 0$.

$V_B < V_o$. If the applied voltage is smaller than the initial capacitor voltage, the capacitor will discharge toward the level of V_B. This situation is shown in Figure 16-8(b).

$V_B > V_o$. If the applied external voltage is larger than the initial capacitor voltage, the capacitor will charge toward the level of V_B. This situation is shown in Figure 16-8(c).

It turns out that the same equation can be used to describe both situations. The voltage v_C can be expressed as

$$v_C = V_B + (V_o - V_B)e^{-t/\tau} \qquad (16\text{-}16)$$

Depending on which of the two cases is being considered, the factor of the exponential in (16-16) can be either positive or negative.

(a)

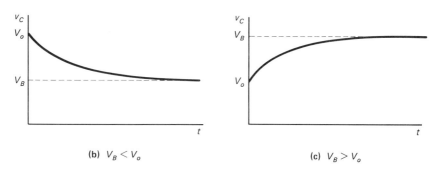

(b) $V_B < V_o$ (c) $V_B > V_o$

Figure 16-8 Initially charged capacitor, partial discharge to a lower level, and charging to a higher level.

Example 16-3

A portion of a certain electronic switching ciruit is shown in Figure 16-9. Initially both switches are open. First, S1 is closed and the capacitor is charged from some external source not shown through the 100 Ω resistor. S1 is then opened and the voltage remains on the capacitor. At a later time, S2 is closed, and the capacitor discharges through the 100-kΩ resistor. **(a)** Determine the charging and discharging time constants. **(b)** Based on the 5τ rule, determine the minimum times required for charging and discharging of the capacitor.

Figure 16-9   Circuit of Example 16-3.

Solution

(a) Let τ_c and τ_d represent the charging and discharging time constants, respectively. For charging, the value of R is 100 Ω, and for discharging the value of R is 100 kΩ. The value of C is, of course, common for both. We have

$$\tau_c = 100 \times 0.01 \times 10^{-6} = 10^{-6} \text{ s} = 1 \ \mu\text{s} \qquad \textbf{(16-17)}$$

and

$$\tau_d = 10^5 \times 0.01 \times 10^{-6} = 10^{-3} \text{ s} = 1 \text{ ms} \qquad \textbf{(16-18)}$$

(b) Let T_c and T_d represent the minimum times for charging and discharging, respectively, based on the 5τ rule. These values are

$$T_c = 5\tau_c = 5 \times 1 \ \mu\text{s} = 5 \ \mu\text{s} \qquad \textbf{(16-19)}$$

$$T_d = 5\tau_d = 5 \times 1 \text{ ms} = 5 \text{ ms} \qquad \textbf{(16-20)}$$

It will thus take about 1000 times as long to discharge the capacitor as to charge it because of the drastically different time constants. Understand that the "5τ rule" is simply a convenient approximation.

16-4 PEAK DETECTOR CIRCUITS

An ideal *peak detector circuit* is one that extracts the peak value of a waveform and holds that value for some interval of time. A peak detector may be classified as either a *positive* peak detector or a *negative* peak detector, depending on the side of the waveform to which it responds. It may be further classified as a *passive* or an *active* peak detector, depending on whether or not an active device is used.

The action of an ideal positive peak detector is illustrated in Figure 16-10. Assume that the input waveform v_i of (a) is first turned on at $t = 0$. The output v_o shown in (b) follows the input up to the peak level V_{ip}. However, when the input drops below that level, the output remains at the value V_{ip} as shown. A negative peak detector would produce an output at the level $-V_{ip}$.

Passive Peak Detectors

The passive peak detector is the simplest circuit of this type, although it does introduce some error in the output level. The form of a passive positive peak detector is shown in Figure 16-11(a), and the corresponding negative peak detector is shown in (b).

A brief analysis will now be made of the positive peak detector of Figure 16-11(a). On the positive half-cycle of the input, the capacitor charges through the diode resistance and any source resistance (not shown) to a voltage equal to the peak positive voltage minus the diode drop. It is assumed that the charging time constant is very small compared with the waveform period, which is usually the case. Assuming a drop of 0.7 V and a peak input voltage level of V_{ip}, the voltage v_o reaches a level of

$$v_o \simeq V_{ip} - 0.7 \qquad\qquad (16\text{-}21)$$

(a)

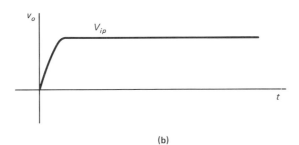

(b)

Figure 16-10 Waveforms of ideal peak detector circuit.

(a) Positive peak detector

(b) Negative peak detector

Figure 16-11 Passive peak detector circuits.

The output voltage is thus lower than the input peak by one forward-biased diode drop.

After the capacitor has been charged to its peak value and the input voltage starts to drop, the diode becomes reverse based. If there were no shunting resistance across the capacitor, the capacitor would remain at the voltage level of (16-21) indefinitely as long as the peak value of the input did not increase.

In practice, there is almost always some type of load across the capacitor. This is indicated by the resistance R in Figure 16-11. After all, if the output voltage is to serve any purpose, it must be connected to some additional circuit. In the normal type of application, the shunting resistance is much larger than the resistance of the charging part of the circuit. Consequently, the discharge time constant is very large compared to a period of the input voltage, and the capacitor will discharge only slightly during a cycle. At a point near the peak of the next cycle, the diode will turn on briefly and recharge the capacitor to the peak input level.

Even if there were no load at all on the capacitor, it might be desirable in some applications to provide a means to discharge the capacitor slowly. Consider the situation where it is desired to monitor the peak value of a waveform, but in which the peak might be expected to change over a period of time. If the peak value were to increase, the capacitor voltage will adjust to the new level quickly. However, if the peak value decreased, the output voltage could never readjust unless some means for discharge is provided.

The passive peak detector circuits are widely used in noncritical applications where some error in the exact peak level can be tolerated. When this circuit is used in instrumentation, it is possible to calibrate the output level to compensate for much of the error in the diode drop.

Active Peak Detector

In applications requiring high accuracy, an active peak detector employing an op-amp with negative feedback may be used. Op-amp active detectors may be classified as either *saturating* or *nonsaturating*. A saturating detector moves into and out of saturation once each cycle, while a nonsaturating detector always remains in the active region. Consequently, a nonsaturating peak detector is capable of operating at a much higher frequency. The treatment here will be limited to a nonsaturating circuit. The circuit arranged as a positive peak detector is shown in Figure 16-12. The analysis of this circuit is a little tricky and must be carefully studied.

Assume first that the capacitor is uncharged and that the input v_i starts increasing from an initial zero level. The positive value of v_i causes the output v_{o1} of OA-1

Figure 16-12 Active precision nonsaturating positive peak detector.

to become positive, and this forward biases D2 and starts to charge the capacitor C. OA-2 is connected as a voltage follower, and v_o thus follows the capacitor voltage. The voltage v_{o1} is held at 0.7 V above this voltage by D2, and this causes D1 to remain reverse biased.

Refer now to Figure 16-13 in which possible states of the circuit are shown. When v_i reaches its peak value (i.e., $v_i = V_{ip}$), the situation is shown in Figure 16-

(a) $v_i = V_{ip}$

(b) $v_i < V_{ip}$ (after capacitor is charged to V_{ip})

Figure 16-13 Two states in analyzing active peak detector circuit of Figure 16-12.

13(a). It is assumed here that the charging time constant is sufficiently small that the capacitor reaches its peak level at essentially the same time as the voltage reaches its peak value. The voltage v_{o1} is forced to be $v_{o1} = V_{ip} + 0.7$. The output is $v_o = V_{ip}$, and diode D1 is still reverse biased.

Assume next that v_i decreases after the capacitor has been charged to V_{ip}. The voltage v_{o1} now starts to drop and diode D2 becomes reverse biased. However, this drop causes D1 to turn on. The two stages each have individual feedback loops now, and the inverting input of OA-1 is forced to assume a value v_i. The output v_{o1} will thus be clamped at 0.7 V lower than the input. This situation is shown in Figure 16-13(b).

As the circuit has been shown, the capacitor can only discharge through its internal leakage or through the noninverting input of OA-2. When it is desired to hold a peak voltage for a very long time, the circuit should employ an op-amp for OA-2 with a FET input and a capacitor with very low losses. In an application where it is desired to provide a nominal discharge path, a large value of resistance may be connected in parallel with the capacitor.

Example 16-4

Consider an active peak detector of the form shown in Figure 16-12. Assume that a random signal varying from -6 V to 6 V is applied at the input. Determine the values of v_o and v_{o1} for each of the following conditions: **(a)** $v_i = 6$ V on the first positive peak; **(b)** $v_i = 2$ V after the first positive peak. Assume that the charging time constant is negligible compared with the time scale of observation.

Solution

(a) When $v_i = 6$ V, the capacitor charges quickly to 6 V, and we have

$$v_o = 6 \text{ V} \tag{16-22}$$

The voltage v_{o1} assumes the value

$$v_{o1} = 6 + 0.7 = 6.7 \text{ V} \tag{16-23}$$

(b) Once the capacitor is charged, it remains at the previous peak level and

$$v_o = 6 \text{ V} \tag{16-24}$$

For $v_i = 2$ V, the loop around OA-1 forces the inverting input to have the same value (i.e., 2 V). The voltage v_{o1} is one diode drop below this value; that is,

$$v_{o1} = 2 - 0.7 = 1.3 \text{ V} \tag{16-25}$$

16-5 ENVELOPE DETECTOR

. An important circuit widely employed in radio receivers is the *envelope detector*. A passive positive envelope detector circuit is shown in Figure 16-14(a). The circuit has the same form as that of a passive peak detector, but the discharge-time constant established for the circuit is what distinguishes the different type of operation.

To discuss the operation of the envelope detector, it is necessary to inspect the waveform of an *amplitude modulated* (AM) signal. A simplified AM signal is shown

(a)

(b)

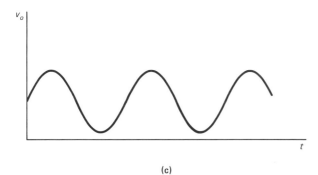

(c)

Figure 16-14 Passive envelope detector and waveforms.

in Figure 16-14(b). In an AM signal, the amplitude of a high-frequency sinusoid (called a *carrier*) is varied (or *modulated*) with a lower-frequency signal. Both the positive and negative sides of the carrier then vary up and down in accordance with the lower-frequency signal. To enhance the illustration, only a few cycles of the higher-frequency signal are shown here. In practice, the frequency of the carrier may be 100 or more times as great as the lower-frequency modulating signal.

Let T_c represent the period corresponding to the carrier, and let T_m represent the period of the modulating signal. The time constant $\tau = RC$ of the envelope detector is chosen to satisfy the following two inequities:

$$\tau \gg T_c \tag{16-26}$$

$$\tau \ll T_m \tag{16-27}$$

It is possible to satisfy both inequalities since a typical value of T_m will be 100 or more times as great as T_c.

The inequality of (16-26) means that the voltage v_o cannot follow the high-frequency variations of the carrier. However, the inequality of (16-27) allows v_o to follow the envelope. The result is that the voltage v_o follows the positive envelope of the waveform reasonably well. There will be some error due to the diode drop and some slight discharge between "spurts" of the carrier. However, this basic circuit form has been used in millions of AM radios over the years.

The somewhat idealized form of the output is indicated as v_o in Figure 16-14(c). This signal contains a superfluous dc level, but it can be readily removed with an RC coupling circuit (not shown).

Example 16-5

Consider a passive envelope detector of the form shown in Figure 16-14(a). The carrier frequency in a certain application is 500 kHz. Assume that the time constant τ is arbitrarily selected to satisfy the criterion $\tau = 20T_c$, where T_c is the carrier period. (a) If $C = 0.01 \ \mu F$, determine the value of R required. (b) For a modulating frequency of 2 kHz, show that $T_m \gg T_c$, where T_m is the modulating period.

Solution

(a) The carrier period is

$$T_c = \frac{1}{500 \times 10^3} = 2 \ \mu s \tag{16-28}$$

Based on the criterion employed,

$$\tau = 20T_c = 20 \times 2 \ \mu s = 40 \ \mu s \tag{16-29}$$

Since $\tau = RC$ and $C = 0.01 \ \mu F$, we have

$$R = \frac{\tau}{C} = \frac{40 \times 10^{-6}}{0.01 \times 10^{-6}} = 4 \text{ k}\Omega \qquad (16\text{-}30)$$

(b) The modulating signal period is

$$T_m = \frac{1}{2 \times 10^3} = 0.5 \text{ ms} = 500 \ \mu s \qquad (16\text{-}31)$$

Comparing the modulating signal period of 500 μs with the carrier period of 2 μs, the former is 250 times as large as the latter. Thus $T_m \gg T_c$.

16-6 CLAMPING CIRCUITS

A clamping circuit is one that shifts a signal either up or down about a specified reference level. A common application of a clamping circuit is that of a *dc restorer*. This application involves shifting the level of a signal that has been ac coupled in order to restore a dc level. The dc restorer circuit has been widely used in the video amplifiers of television receivers.

Passive Clamping Circuit

The simplest version of a clamping circuit is the passive version shown in Figure 16-15(a). On the first negative portion of the input signal, the diode conducts and charges the capacitor to the negative peak value minus one diode drop. The positive terminal of this voltage will be on the right. The diode then remains reverse biased as the input voltage changes in a positive direction. The voltage across the capacitor then adds to the input voltage and shifts the net output voltage upward.

Waveforms illustrating this clamping action are shown in Figure 16-15(b) and (c). These waveforms apply *after* the initial clamping action has been established. The input waveform v_i of (b) has no dc level. However, the output waveform v_o of (c) is shifted upward. If the diode were perfect, the bottom level of the output would be exactly at a voltage level of zero. The diode drop causes the output to have a slight negative excursion as shown.

Active Clamping Circuit

The error resulting from the diode drop can be virtually eliminated with the active clamping circuit of Figure 16-16(a) utilizing an op-amp. On the first negative half-cycle of the input waveform, the op-amp loop is closed though the diode, and the inverting input is forced to assume a virtual ground level. This forces the capacitor to charge to the negative peak of the input with the positive terminal on the right. When the input becomes more positive, the loop is broken, and the output of the op-amp goes into negative saturation. However, the capacitor voltage adds to the input voltage to produce an output clamped at zero volts.

(a)

(a)

(b)

(b)

≈ −0.7 V

(c)

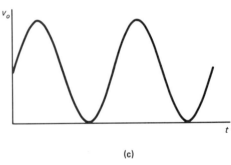

(c)

Figure 16-15 Passive clamping circuit and typical waveforms.

Figure 16-16 Active clamping circuit and typical waveforms.

Representative waveforms are shown in Figure 16-16(b) and (c). For the input v_i of (b), the output v_o will be of the form shown in (c).

Example 16-6

Consider an ac signal of the form of (b) in either Figure 16-15 or 16-16, and assume that it oscillates between −6 V and 6 V. Determine the peak value V_{op} of the output for the following two circuits: **(a)** passive clamping circuit of Figure 16-15(a); **(b)** active clamping circuit of Figure 16-16(a).

Solution

(a) For the circuit of Figure 16-15(a), the peak value of the output is

$$V_{op} = 2 \times 6 - 0.7 = 12 - 0.7 = 11.3 \text{ V} \qquad (16\text{-}32)$$

(b) For the circuit of Figure 16-16(a), the peak value of the output is

$$V_{op} = 2 \times 6 = 12 \text{ V} \qquad (16\text{-}33)$$

16-7 COMPARATOR CIRCUITS

A comparator circuit is one that provides one of two possible output levels in accordance with the range of an input signal. A variety of possible comparator types may be implemented with general-purpose op-amps, although their switching speeds are limited by the op-amp characteristics. Dedicated comparators are available in which the switching speed is optimized. The treatment here will be limited to the use of general-purpose op-amps as comparators since this approach will most clearly illustrate the principle of operation.

Noninverting Comparator

The simplest form of a noninverting op-amp comparator is shown in Figure 16-17(a). The op-amp is operated in an open-loop arrangement, and the inverting input is grounded. When the input voltage v_i exceeds V_{sat}/A_d (typically in the microvolt range), the output v_o moves into positive saturation and remains there for subsequent increases of the input.

The transfer characteristic curve is shown in Figure 16-17(b). Note that there is a small transition inverval on either side of $v_i = 0$.

A comparator circuit may be used to convert a general time-varying signal into a sequence of positive and negative pulses. For example, consider the sinusoidal waveform of Figure 16-18(a). When $v_i > 0$ (by more than a few microvolts), the output is $v_o = V_{sat}$. The result is a square wave, as shown in Figure 16-18(b).

Once again, it should be stressed that the switching speed is limited in general-purpose op-amps, and a square wave generated by this process can only switch at a moderate rate. The square wave of Figure 16-18(b) is assumed to be at a sufficiently low frequency that the transition times are not visible.

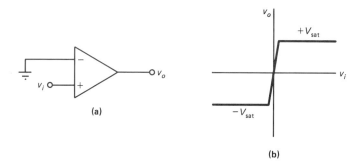

(a)

(b)

Figure 16-17 Op-amp noninverting comparator and its transfer characteristic.

(a)

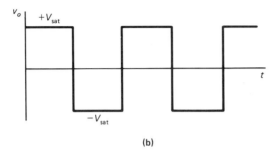

(b)

Figure 16-18 Sinusoidal input to noninverting comparator and form of the output (at low frequencies).

Inverting Comparator

An inverting comparator can be implemented by grounding the noninverting input and applying the input signal to the inverting input as shown in Figure 16-19(a). In this case, a positive input causes the output to move into negative saturation, and vice versa. The input/output characteristic is shown in Figure 16-19(b).

A sinusoidal input of the form of Figure 16-20(a) now results in an inverted square wave as shown in Figure 16-20(b). The discussion concerning switching speed for the noninverting comparator applies to the inverting comparator as well.

Example 16-7

Assume for a given op-amp that $\pm V_{sat} = \pm 13$ V. If the differential gain is $A_d = 2 \times 10^5$, calculate the positive and negative input voltages at which saturation will be reached.

Solution

For $v_o = 13$ V, the value of v_i is

$$v_i = \frac{13}{A_d} = \frac{13}{2 \times 10^5} = 65 \ \mu V \qquad \textbf{(16-34)}$$

From symmetry, the corresponding negative voltage required to reach $v_o = -13$ V is

$$v_i = -65 \ \mu V \qquad \textbf{(16-35)}$$

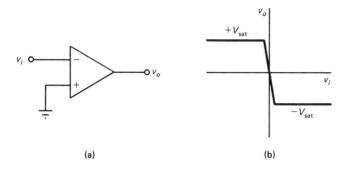

Figure 16-19 Op-amp inverting comparator and transfer characteristic.

(a)   (b)

(a)

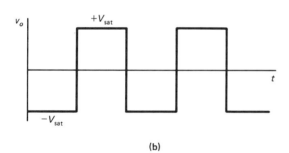

(b)

Figure 16-20 Sinusoidal input to inverting comparator and form of the output (at low frequencies).

DRILL PROBLEMS

16-1. For the rectifier circuit of Figure 16-1, determine v_a and v_o for $v_i =$ **(a)** 8 V and **(b)** -8 V. Assume that $\pm V_{sat} = \pm 13.5$ V.

16-2. For the rectifier circuit of Figure 16-1, determine v_a and v_o for $v_i =$ **(a)** 12 V and **(b)** -12 V. Assume that $\pm V_{sat} = \pm 13$ V.

16-3. For the rectifier circuit of Figure 16-3, determine v_a and v_o for $v_i =$ **(a)** 8 V and **(b)** -8 V. Assume that $\pm V_{sat} = \pm 13.5$ V.

16-4. For the rectifier circuit of Figure 16-3, determine v_a and v_o for $v_i =$ **(a)** 12 V and **(b)** -12 V. Assume that $\pm V_{sat} = \pm 13$ V.

16-5. A certain discharge circuit is of the form shown in Figure 16-6(a) with $R = 220 \ \Omega$ and $C = 0.001 \ \mu$F. Using the 5τ rule, determine the approximate time required to discharge the capacitor.

16-6. Repeat the analysis of Problem 16-5 if $R = 220$ kΩ and $C = 0.1$ μF.

16-7. A certain charging circuit is of the form shown in Figure 16-7(a) with $R = 2$ MΩ and $C = 12$ μF. Using the 5τ rule, determine the approximate time required to charge the capacitor.

16-8. Repeat the analysis of Problem 16-7 if $R = 470$ Ω and $C = 220$ pF.

16-9. A certain discharge circuit is of the form shown in Figure 16-6(a) with $V_o = 100$ V and $\tau = 1$ ms. Use the exponential function on your calculator to determine the capacitor voltage at the following values of t: **(a)** 0; **(b)** 1 ms; **(c)** 2 ms; **(d)** 5 ms; **(e)** 10 ms.

16-10. A certain charging circuit is of the form shown in Figure 16-7(a) with $V_B = 100$ V and $\tau = 1$ ms. Use the exponential function on your calculator to determine the capacitor voltage at the following values of t: **(a)** 0; **(b)** 1 ms; **(c)** 2 ms; **(d)** 5 ms; **(e)** 10 ms.

16-11. Assume a passive positive peak detector of the form shown in Figure 16-11(a). Assume that a random signal varying from -6 to 6 V is applied to the input. **(a)** Determine the value of v_o after the first positive peak of the input voltage has occurred. **(b)** Determine the peak reverse voltage V_{rp} that will appear across the diode.

16-12. Assume a passive negative peak detector of the form shown in Figure 16-11(b). Assume that a random signal varying from -6 to 6 V is applied to the input. **(a)** Determine the value of v_o after the first negative peak of the input voltage has occurred. **(b)** Determine the peak reverse voltage V_{rp} that will appear across the diode.

16-13. Assume an active positive peak detector of the form shown in Figure 16-12. Assume that a random signal varying from -4 to 4 V is applied to the input. Determine the values of v_o and v_{o1} for each of the following conditions: **(a)** $v_i = 4$ V on the first positive peak; **(b)** $v_i = -1$ V after the first positive peak.

16-14. Assume an active negative peak detector based on the circuit of Figure 16-12, but with the two diodes reversed. Determine the values of v_o and v_{o1} for each of the following conditions: **(a)** $v_i = -4$ V on the first negative peak; **(b)** $v_i = 4$ V after the first negative peak.

16-15. Consider a passive clamping circuit of the form shown in Figure 16-15(a), and assume an ac signal of the form shown in Figure 16-16(b) that oscillates between -8 and 8 V. **(a)** Determine the peak value V_{op} of the output voltage. **(b)** Determine the peak reverse voltage V_{rp} that will appear across the diode.

16-16. Consider a passive clamping circuit based on the form shown in Figure 16-15(a), but with the diode reversed. Assume an ac signal of the form shown in Figure 16-16(b) that oscillates between -8 and 8 V. Determine the upper and lower limits of the output voltage.

16-17. Consider an active clamping circuit of the form shown in Figure 16-16(a), and assume an ac signal of the form shown in Figure 16-16(b) that oscillates between -8 and 8 V. **(a)** Determine the peak value V_{op} of the output voltage. **(b)** Determine the peak reverse voltage V_{rp} that will appear across the diode. Assume that $\pm V_{\text{sat}} = \pm 13$ V.

16-18. Consider an active clamping circuit based on the form shown in Figure 16-16(a) but with the diode reversed. Assume an ac signal of the form shown in Figure 16-16(b) that oscillates between -8 and 8 V. Determine the upper and lower limits of the output voltage.

16-19. Consider an op-amp noninverting comparator circuit of the form shown in Figure 16-17(a), and assume that $\pm V_{\text{sat}} = \pm 13$ V. Determine v_o when $v_i = $ **(a)** 2 V and **(b)** -2 V.

16-20. Consider an op-amp inverting comparator circuit of the form shown in Figure 16-19(a), and assume that $\pm V_{\text{sat}} = \pm 13$ V. Determine v_o when $v_i = $ **(a)** 2 V and **(b)** -2 V.

DERIVATION PROBLEMS

16-21. Consider the circuit of Figure P16-21. **(a)** Determine equations for v_a and v_o in terms of v_i. **(b)** Sketch the transfer characteristic relating v_o to v_i.

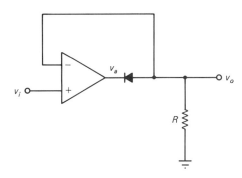

Figure P16-21

16-22. Consider the circuit of Figure P16-22. **(a)** Determine equations for v_a and v_o in terms of v_i. **(b)** Sketch the transfer characteristic relating v_o to v_i.

Figure P16-22

16-23. When $t \ll \tau$, the exponential function may be approximated as

$$e^{-t/\tau} \simeq 1 - \frac{t}{\tau}$$

Consider a capacitor discharge circuit as shown in Figure 16-6(a). Show that for $t \ll \tau$, the change in voltage Δv_C (from the initial value V_o) is

$$\Delta v_C \simeq -\frac{V_o t}{\tau}$$

16-24. Consider a capacitor charging circuit as shown in Figure 16-7(a). Show that for $t \ll \tau$, the voltage v_C can be approximated as

$$v_c \simeq \frac{V_B t}{\tau}$$

(See the approximation of Problem 16-23.)

DESIGN PROBLEMS

16-25. A passive envelope detector circuit is to be designed to detect modulation on a 2-MHz carrier. Assume that the time constant is selected to satisfy the criterion $\tau = 40T_c$, where T_c is the carrier period. **(a)** If $C = 0.002\ \mu F$, determine the value of R required. **(b)** For a modulating frequency of 5 kHz, show that $T_m \gg T_c$.

16-26. For the envelope detector of Example 16-5, assume that the time constant is to be selected as the geometric mean between the carrier period and the modulating period; that is, $\tau = \sqrt{T_c T_m}$. Selecting $C = 0.01\ \mu F$ as in Example 16-5, determine the value of R required.

TROUBLESHOOTING PROBLEMS

16-27. You are troubleshooting an active rectifier circuit of the form shown in Figure 16-3, which is not working properly. When $v_i = -2$ V, $v_o = 2$ V, but when $v_i = 2$ V, $v_o = 2$ V. Which one of the following *could be* the difficulty? **(a)** D1 shorted; **(b)** D1 open; **(c)** D2 shorted; **(d)** D2 open.

16-28. If instead of the voltage indicated in Problem 16-27, suppose that when $v_i = 2$ V, $v_o = 0$ V, but when $v_i = -2$ V, $v_o = 0$ V. Which one of the following *could be* the difficulty? **(a)** D1 shorted; **(b)** D1 open; **(c)** D2 shorted; **(d)** D2 open.

COMPUTER PROBLEMS

16-29. Write a program to determine the discharge voltage across a capacitor at any arbitrary time.

Input data: V_o, τ (or R and C), t

Output data: v_C

16-30. Write a program to determine the charging voltage across a capacitor at any arbitrary time.

Input data: V_B, τ (or R and C), t

Output data: v_C

CHAPTER 17

Oscillator Circuits

OBJECTIVES

After completing this chapter, the reader should be able to:

1. Discuss the general properties of oscillator circuits.
2. Define the Barkhausen criterion.
3. Recognize the form and analyze the following *RC* oscillators: Wien bridge oscillator and phase shift oscillator.
4. Recognize the form and analyze the following *LC* oscillators: Colpitts oscillator, crystal oscillator, Hartley oscillator, and Clapp oscillator.
5. Distinguish between the three types of multivibrators: (a) bistable, (b) monostable, and (c) astable.
6. Recognize the form and analyze the op-amp square-wave generator.
7. Analyze the operation of the 555 timer in both the astable and monostable modes.
8. Discuss the operation of a unijunction transistor and its application as an oscillator.

17-1 OVERVIEW

An *oscillator* is a circuit that generates a periodic waveform of a predetermined form and frequency. All electronic circuits containing feedback are subject to generate oscillations if certain gain and phase shift conditions exist. In an amplifier circuit, these oscillations are undesirable. In an oscillator circuit, however, the design goal is to generate oscillations in a controlled manner.

Oscillators may be designed to produce a fixed single frequency, or they may be tunable over a range of frequencies. An oscillator may produce a continuous output, or it may be triggered off and on from an external signal. Output waveforms of oscillators include sinusoids, square waves, pulse waveforms, triangular waveforms, and many others.

Oscillations in a circuit are produced by *positive feedback*. This is in contrast to *negative feedback,* which is used to stabilize amplifiers and other circuits, as we have seen for the past several chapters. However, what is intended as negative feedback may turn out to be positive feedback due to unwanted phase shifts in the circuit. Sooner or later, everyone working with electronic circuits will encounter spurious oscillations occurring because of unintended positive feedback. Our focus throughout this chapter is on the *planned* use of positive feedback to establish and maintain oscillations of a specified type.

Barkhausen Criterion

The concept of the *Barkhausen criterion* as the basis for producing steady-state oscillations is developed in this section. The development is based on a simplified intuitive approach as opposed to a more rigorous mathematical basis.

Consider the block diagram of Figure 17-1 containing a forward gain block of value A and a feedback block of value β. This configuration differs from the negative feedback system of earlier chapters in two respects: (1) there is no subtraction or inversion of the input as was the case for negative feedback, and (2) there is no input provided for an external signal.

The diagram here, then, describes a *positive feedback loop,* for which self-sustained oscillations *may be* generated. Understand that we are observing only signal conditions in the loop. The energy required to sustain the oscillations is provided by the power supplies for the active components.

Assume that somehow a signal e appears at the input to the A block. This signal is amplified and appears at the output as Ae. This output is multiplied by β, and the output of the β-block is then $A\beta e$. To sustain the oscillation, this output must equal the value assumed at the beginning. Thus

$$A\beta e = e \qquad (17\text{-}1)$$

We now define the *loop gain* G_L as

$$\boxed{G_L = A\beta} \qquad (17\text{-}2)$$

Substituting this definition in (17-1) and canceling the common e factors, we determine the criterion for sustaining oscillations as

$$\boxed{G_L = 1} \qquad (17\text{-}3)$$

This results in an expression for the Barkhausen criterion. Stated in words, it says that *to sustain oscillations in a loop, the net loop gain must be equal to unity*.

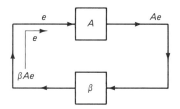

Figure 17-1 Block diagram of positive feedback loop.

From the point of view of ac circuit theory, the Barkhausen criterion may be restated with the following two separate conditions:

$$\text{Loop gain magnitude} = 1 \qquad \textbf{(17-4a)}$$

$$\text{Loop phase shift} = n \times 360° \qquad \textbf{(17-4b)}$$

where n is any integer. In practice, these two criteria are usually employed in analyzing the conditions for oscillation in practical circuits. Note that (17-4b) applies for any integer multiple of 360°, including 0°. Thus if the loop gain magnitude is unity, oscillations will occur if the phase shift around the loop is 0°, 360°, $-360°$, 720°, and so on.

The preceding discussion justifies the process of sustaining an oscillation, but the skeptical reader may wonder how the process was started. From where did the assumed signal e arise?

Actually, it turns out that noise in the circuit is sufficient to start the process. In practical oscillators, the loop gain will be slightly greater than unity at very small-signal levels so that the oscillation, once started, will grow in amplitude. However, as the level grows, the loop gain drops to a level just sufficient to maintain oscillations.

It may be difficult to maintain the loop gain at exactly unity without some type of compensation. Some of the best oscillators employ both positive and negative feedback. The oscillations are established and maintained by the positive feedback. The negative feedback is used to control the amplitude of oscillations and to keep the loop gain maintained at exactly unity.

RC and LC Oscillators

Many of the standard oscillator circuits are classified as either *RC* or *LC* oscillators. The distinction lies in the types of components used to generate necessary phase shifts in the β-block of the loop. *RC* oscillators use resistance and capacitance to create the phase shift, while *LC* oscillators use inductance and capacitance.

As a general rule, *RC* oscillators are easier to design and have superior operating characteristics at relatively lower frequencies, while *LC* oscillators are superior at higher frequencies. There is no rigid boundary, and there are frequencies where either type would work as well as the other. As a rough guideline, *RC* oscillators work best below about 1 MHz or so.

Example 17-1

A certain loop consists of an attenuator, an amplifier, and a phase shift circuit as shown in Figure 17-2. The output/input ratio of the attenuator is 0.04. The amplifier has a gain magnitude A and a phase shift θ at a certain frequency. At this reference frequency, the phase shift of the phase shift circuit is $-60°$, and its amplitude response is unity. What values of A and θ will produce oscillations in the circuit?

Figure 17-2 Loop of Example 17-1.

Solution

The magnitude-angle (or polar) form of the attenuator output/input ratio is denoted as $0.04\underline{/0°}$, and a similar form for the phase shift circuit is $1\underline{/-60°}$. The net loop gain is determined by multiplying the magnitudes and adding the angles; that is,

$$G_L = (0.04\underline{/0°})(A\underline{/\theta})(1\underline{/-60°})$$
$$= 0.04A\underline{/\theta - 60°} \qquad \textbf{(17-5)}$$

Applying the Barkhausen criteria to the magnitude and angle, we have

$$0.04A = 1 \qquad \textbf{(17-6)}$$

and

$$\theta - 60° = 0° \qquad \textbf{(17-7)}$$

where $n = 0$ was chosen as the simplest case. These equations result in

$$A = 25 \qquad \textbf{(17-8)}$$

and

$$\theta = 60° \qquad \textbf{(17-9)}$$

Thus oscillations will occur at the reference frequency if the amplifier gain is 25, and the amplifier phase shift is 60°.

The reader may be puzzled as to why we chose $n = 0$ in the phase constraint. It turns out that it does not really matter which *integer* value of n is selected since all the results are identical when the proper interpretation of the angle is made. For example, if 360° had been chosen in (17-7), the result would be 420°. However, an angle of 420° is equivalent to an angle of 60° with respect to the phasor rotation. Thus the simplest form of the angle is usually chosen.

17-2 RC OSCILLATORS

Two of the most common types of *RC* oscillators are discussed in this section. As indicated earlier, *RC* oscillators are usually best for applications at relatively low frequencies.

The desired output waveforms from the circuits to be considered are sine waves. In practice, sine-wave oscillators will contain some distortion. The optimum design is one that minimizes the distortion and produces a nearly perfect sine wave.

Wien Bridge Oscillator

The most popular *RC* oscillator circuit employing the unity-gain feedback concept is the *Wien bridge oscillator,* whose basic form is shown in Figure 17-3. Any active device capable of producing a noninverting voltage gain $A = 3$ and negligible loading on the feedback circuit may be used. This gain requirement arises from the fact that the output voltage of the feedback circuit (on the left) is one-third of the input voltage (on the right) at the frequency of oscillation. At that frequency, the phase shift through the feedback circuit is zero. Thus the loop gain is $3 \times \frac{1}{3} = 1$ at the frequency of oscillation, and the Barkhausen criterion is met. Development of these results will be left as guided exercises (Problems 17-17 and 17-18).

The frequency of oscillation f_o is related to the circuit components by

$$f_o = \frac{1}{2\pi RC} \qquad (17\text{-}10)$$

Note that two resistors have the same value R, and two capacitors have the same value C. If the R's and C's were different, the gain requirement and frequency of oscillation would be different than stated previously. Virtually all circuits built utilize equal values for the resistances and equal values for the capacitances.

One of the outstanding virtues of the Wien bridge oscillator is its ease of tuning. The two values of R can be realized by a ganged potentiometer, or the two values of C can be realized by a ganged capacitor. In either case, the particular parameter is changed by turning one control, and this provides a simple way to vary the frequency. Many general-purpose laboratory oscillators utilize the Wien bridge oscillator. If designed properly, the output waveform is nearly sinusoidal and contains very little distortion.

In practice, some means of stabilizing the gain to the exact required level is

Figure 17-3 Basic form of a Wien bridge oscillator.

necessary. A number of ways utilizing such varied means as FET voltage-controlled resistors, diodes, thermistors, and nonlinear resistances have been employed. The process will be illustrated here with an incandescent lamp.

Consider the circuit of Figure 17-4. The required positive gain is achieved with a noninverting op-amp amplifier circuit. The gain of this amplifier is

$$A_{CL} = 1 + \frac{R_f}{R_i} \tag{17-11}$$

To achieve $A_{CL} = 3$, it is necessary that $R_f = 2R_i$.

The gain can be stabilized by employing a nonlinear component as a portion of or all of either R_f or R_i. If the resistance *increases* with increasing output level, the nonlinear element should be part of or all of R_i. The gain will then be reduced if the output signal level tends to increase, since an increase in R_i corresponds to a lower gain. If the resistance *decreases* with increasing output level, the nonlinear element should be part of or all of R_f. In the circuit of Figure 17-4, it is assumed that the resistance of the incandescent lamp increases with the output level. This particular scheme is a very common one for Wien bridge oscillators.

Phase Shift Oscillator

Another popular *RC* oscillator is the *phase shift oscillator* shown in Figure 17-5. It turns out that the phase shift through this circuit at the desired frequency of oscilla-

Figure 17-4 Wein bridge oscillator with stabilization using an incandescent lamp.

Figure 17-4 Wien bridge oscillator with stabilization using an incandescent lamp.

627

tion is 180° so an inverting gain is required for this circuit. If there is negligible loading on the RC phase shift circuit; that is, if the input resistance to the amplifier can be assumed to be infinite, the magnitude of the output voltage of the phase shift circuit is $\frac{1}{29}$ of the input at the frequency of 180° phase shift. This establishes the requirement for the gain to be

$$\boxed{-|A| = -29} \tag{17-12}$$

where the negative sign implies an inverting gain. The frequency f_o at which oscillations occur is

$$\boxed{f_o = \frac{1}{2\pi\sqrt{6}\,RC}} \tag{17-13}$$

Note that the three values of R are equal and the three values of C are equal.

Example 17-2

A certain Wien bridge oscillator employs the circuit of Figure 17-4 with $R = 10 \text{ k}\Omega$ and $C = 0.01 \ \mu\text{F}$ (two values of each as shown). The feedback resistor is $R_f = 2 \text{ k}\Omega$, and a nonlinear lamp is used. Determine **(a)** the frequency of oscillation and **(b)** the resistance of the lamp when stable oscillations are produced.

Solution

(a) The oscillation frequency is

$$f_o = \frac{1}{2\pi RC} = \frac{1}{2\pi \times 10^4 \times 0.01 \times 10^{-6}} = 1592 \text{ Hz} \tag{17-14}$$

(b) For stable oscillations, $R_f = 2R_i$, or

$$R_i = \frac{R_f}{2} = \frac{2 \times 10^3}{2} = 1 \text{ k}\Omega \tag{17-15}$$

This resistance will generally be greater than the "cold" resistance of the bulb as measured by an ohmmeter.

Example 17-3

A certain Wien bridge oscillator is being designed to be tunable from 50 to 500 Hz with a dual potentiometer. Fixed values of 0.1 μF are employed for the two capacitors. Determine the range of resistances required to tune the specified range.

Solution

From (17-10), the resistance value is

$$R = \frac{1}{2\pi f_o C} = \frac{1}{2\pi \times 0.1 \times 10^{-6} f_o} \qquad \textbf{(17-16)}$$

Let R_{min} and R_{max} represent the minimum and maximum values of the resistance required. Since f_o appears in the denominator of (17-16), the *highest* frequency results in the *minimum* value of resistance, and vice versa. The resistance values are then calculated as follows:

$$R_{min} = \frac{1}{2\pi \times 0.1 \times 10^{-6} \times 500} = 3183 \ \Omega \qquad \textbf{(17-17)}$$

$$R_{max} = \frac{1}{2\pi \times 0.1 \times 10^{-6} \times 50} = 31.83 \ k\Omega \qquad \textbf{(17-18)}$$

The two resistances must each be adjustable from 3183 Ω to 31.83 kΩ in order to tune the oscillator for the required frequency range. This can be achieved by two fixed resistances, each connected in series with a section of a dual potentiometer.

Example 17-4

A certain phase shift oscillator is of the form shown in Figure 17-5. The resistance and capacitance values are $R = 10$ kΩ and $C = 0.01$ μF (three values of each). Determine the frequency of oscillation.

Solution

The frequency is

$$f_o = \frac{1}{2\pi \sqrt{6} \ RC} = \frac{1}{2\pi \sqrt{6} \times 10^4 \times 0.01 \times 10^{-6}} = 650 \ \text{Hz} \qquad \textbf{(17-19)}$$

17-3 LC OSCILLATORS

In the frequency range from about 1 MHz or so up to several hundred megahertz, *LC* oscillators are more practical than *RC* oscillators. Over the years, quite a number of *LC* oscillators have been developed, and a complete treatment of this subject is best relegated to a textbook dealing with communications electronics. However, a few of the more popular *LC* oscillators are surveyed in this section.

Although it is possible to implement *LC* oscillators with op-amps in a limited frequency range, it turns out that the higher-frequency range where *LC* oscillators are best suited is the frequency range where op-amp usage is more limited. Consequently, most *LC* oscillators are implemented with BJTs and FETs.

Colpitts Oscillator

One of the most popular of the LC oscillators in the Colpitts oscillator, for which one implementation utilizing a BJT is shown in Figure 17-6. The resistors R_1, R_2, and R_E are used to establish the bias operating point, and C_E is used to bypass the emitter resistor as in an amplifier. The RF choke is an inductance whose reactance is large at the frequency of oscillation and would be an open circuit if it were not for losses. However, its losses create an effective resistance which, when coupled with any reflected resistance from the feedback circuit, acts as a dynamic load for the BJT. The reactance of C_3 is assumed to be small at the oscillation frequency.

The feedback circuit consists of L, C_1, and C_2. It turns out that the oscillation frequency f_o is given by

$$f_o = \frac{1}{2\pi\sqrt{LC_{eq}}} \qquad \text{(17-20)}$$

where C_{eq} is the series equivalent capacitance of C_1 and C_2 as given by

$$C_{eq} = \frac{C_1 C_2}{C_1 + C_2} \qquad \text{(17-21)}$$

Example 17-5

A certain Colpitts oscillator of the form shown in Figure 17-6 has $C_1 = 100$ pF, $C_2 = 500$ pF, and $L = 60$ μH. Determine the frequency of oscillation.

Figure 17-6 Colpitts oscillator employing a BJT.

Solution

The equivalent capacitance from (17-21) is

$$C_{eq} = \frac{C_1 C_2}{C_1 + C_2} = \frac{100 \times 10^{-12} \times 500 \times 10^{-12}}{600 \times 10^{-12}} = 83.33 \text{ pF} \qquad \textbf{(17-22)}$$

The oscillation frequency from (17-20) is

$$f_o = \frac{1}{2\pi \sqrt{LC_{eq}}}$$

$$= \frac{1}{2\pi \sqrt{60 \times 10^{-6} \times 83.3 \times 10^{-12}}} = 2.251 \text{ MHz} \qquad \textbf{(17-23)}$$

Crystal Oscillator

A crystal oscillator is a circuit that employs a *crystal* to establish the oscillation frequency. The most common type of crystal is the *quartz crystal*. It has been included with *LC* oscillators because the equivalent circuit has the form of an *LC* circuit.

Quartz is a substance that exhibits a property called the *piezoelectric effect*. When a mechanical stress is applied across the crystal, a voltage is developed at the frequency of the mechanical vibration. Conversely, when a voltage is applied across the crystal, it vibrates at the frequency of the voltage.

The schematic symbol of a crystal is shown in Figure 17-7(a), and an equivalent-circuit model is shown in Figure 17-7(b). The crystal exhibits both a series resonant frequency and a parallel resonant frequency. The parallel resonant frequency is higher than the series resonant frequency.

An oscillator circuit employing a crystal is shown in Figure 17-8. This particular circuit is a modified Colpitts oscillator, and the parallel resonant mode of the crystal is utilized.

The primary advantage of a crystal oscillator is the excellent frequency stability provided. Commercial broadcast transmitters utilize crystal oscillators with the crystal kept in a temperature stabilized oven to provide additional stability in the operating frequency.

(a)

(b)

Figure 17-7 Schematic symbol for a quartz crystal and an equivalent-circuit model.

Figure 17-8 Crystal oscillator based on Colpitts circuit.

Hartley Oscillator

A *Hartley oscillator* is shown in Figure 17-9. The primary difference between this circuit and the Colpitts oscillator is that the phase shift circuit of the Hartley oscillator consists of two inductances, denoted by L_1 and L_2, and a single capacitor C.

The frequency of oscillation f_o of the Hartley oscillator is given by

$$ f_o = \frac{1}{2\pi \sqrt{L_{eq} C}} \tag{17-24} $$

Figure 17-9 Hartley oscillator employing a BJT.

Oscillator Circuits   Chap. 17

where L_{eq} is the equivalent inductance of the series combination of L_1 and L_2 as given by

$$\boxed{L_{\text{eq}} = L_1 + L_2}$$ (17-25)

Example 17-6

A certain Hartley oscillator of the form shown in Figure 17-9 has $L_1 = 80 \ \mu\text{H}$, $L_2 = 10 \ \mu\text{H}$, and $C = 100$ pF. Determine the frequency of oscillation.

Solution

The equivalent inductance from (17-25) is

$$L_{\text{eq}} = L_1 + L_2 = 80 + 10 = 90 \ \mu\text{H}$$ (17-26)

The oscillation frequency from (17-24) is

$$f_o = \frac{1}{2\pi\sqrt{L_{\text{eq}}C}}$$

$$= \frac{1}{2\pi\sqrt{90 \times 10^{-6} \times 100 \times 10^{-12}}} = 1.678 \ \text{MHz}$$ (17-27)

Clapp Oscillator

An *LC* oscillator based on the Colpitts circuit, but with some improvement in performance, is the Clapp oscillator shown in Figure 17-10. A capacitor C_3 is placed in

Figure 17-10   Clapp oscillator employing a BJT.

series with L, and it is chosen to be somewhat smaller than either C_1 or C_2. The frequency of oscillation is given by an expression of the form of (17-20), but C_{eq} in this case is determined from

$$\frac{1}{C_{eq}} = \frac{1}{C_1} + \frac{1}{C_2} + \frac{1}{C_3} \qquad \text{(17-28)}$$

If $C_3 \ll C_1$ and $C_3 \ll C_2$, then to a first approximation

$$C_{eq} \simeq C_3 \qquad \text{(17-29)}$$

The capacitance C_3 is affected much less by transistor input and output capacitances than either C_1 or C_2. Consequently, the frequency of oscillation is more stable with respect to uncontrollable circuit capacitances than with the Colpitts oscillator.

Example 17-7

A certain Clapp oscillator of the form shown in Figure 17-10 has $C_1 = 220$ pF, $C_2 = 1000$ pF, $C_3 = 20$ pF, and $L = 50$ μH. Determine the frequency of oscillation.

Solution

The equivalent capacitance is determined from (17-28) with the following steps:

$$\frac{1}{C_{eq}} = \frac{1}{220 \times 10^{-12}} + \frac{1}{1000 \times 10^{-12}} + \frac{1}{20 \times 10^{-12}} \qquad \text{(17-30)}$$

This leads to

$$C_{eq} = 18.00 \text{ pF} \qquad \text{(17-31)}$$

The frequency of oscillation is then

$$f_o = \frac{1}{2\pi \sqrt{LC_{eq}}}$$

$$= \frac{1}{2\pi \sqrt{50 \times 10^{-6} \times 18 \times 10^{-12}}} = 5.305 \text{ MHz} \qquad \text{(17-32)}$$

17-4 SQUARE-WAVE OSCILLATORS

All the oscillator circuits considered up to this point have as their major objective a sinusoidal output waveform. However, there are many applications in which a nonsinusoidal waveform is desired. The most common nonsinusoidal signal is the *square wave*. A *bipolar* square wave is one that oscillates between positive and negative levels, and a *unipolar* square wave is one that oscillates between zero and a level of one polarity (either positive or negative).

Multivibrators

The term *multivibrator* is one that has long-standing usage in the electronics industry and dates back to the vacuum-tube era. It refers to certain circuits used for pulse and/or square-wave generation. There are three types of multivibrators: (1) *bistable* (or *flip-flop*), (2) *monostable* (or *one-shot*), and (3) *astable* (or *free-running*).

The *bistable multivibrator* has two stable states and will remain in either state until it receives a proper signal, in which case it will change states. Bistable multivibrators are used as memory devices in digital systems.

The *monostable multivibrator* has one stable state. Upon receiving a proper trigger, it will change states for a specified interval of time, but it will then return to the original stable state.

The *astable* multivibrator has no stable states. Consequently, it oscillates back and forth between states at a rate determined by the circuit parameters. It is, therefore, capable of generating a square wave in the process.

Op-Amp Square-Wave Generator

As a simple example of an astable multivibrator, the op-amp square-wave generator of Figure 17-11 will be considered. As shown, the circuit produces an output square wave that oscillates between the positive and negative saturation levels. By the use of zener diodes, lower output magnitude levels can be established. However, the analysis here will be based on the two levels $+V_{sat}$ and $-V_{sat}$.

To analyze a multivibrator, it is usually necessary first to assume that steady-state conditions exist. The initial assumptions on levels may seem strange, but by moving through a complete cycle, these initial conditions may be verified.

In addition to the circuit diagram of Figure 17-11, refer to the waveforms of Figure 17-12 in the discussion that follows. Assume that a transition occurs at $t = 0$ with v_o assuming the positive state (i.e., $v_o = +V_{sat}$). By the voltage-divider rule, the voltage v^+ will be a fraction of v_o. We will define a feedback voltage V_f as

$$V_f = \frac{R_2}{R_1 + R_2} V_{sat} \tag{17-33}$$

Thus, $v^+ = V_f$ when $v_o = V_{sat}$ as shown in Figure 17-12(a) and (b), respectively. For reasons that will be clear later, assume that $v^- = -V_f$ at $t = 0$ as indicated in Figure 17-12(c). Note that v^- is the voltage across the capacitor at any given time.

Figure 17-11 Square-wave generator using a single op-amp.

(a)

(b)

(c)

Figure 17-12 Waveforms of op-amp square-wave generator.

The top portion of the circuit consists of a capacitance initially changed to a level $-V_f$, a resistance R, and a positive voltage $+V_{sat}$. The voltage across the capacitance will start climbing toward the level $+V_{sat}$ as indicated on Figure 17-12(c). However, when it reaches and slightly exceeds the level $+V_f$, the differential voltage $v_d = v^+ - v^-$ becomes negative. This will cause the output v_o to abruptly change states, in which case $v_o = -V_{sat}$. In turn, the voltage v^+ changes to $v^+ = -V_f$. The capacitor voltage now starts discharging toward a level $-V_{sat}$ as shown in Figure 17-12(c). When it reaches the level $-V_f$, the circuit changes state again, and a full cycle is complete.

The derivation of the expression for a period is left as a guided exercise at the end of the chapter (Problem 17-19). The period T can be determined from the equation

$$T = 2RC \ln \left(1 + \frac{2R_2}{R_1} \right)$$

(17-34)

Oscillator Circuits   Chap. 17

where "ln" refers to the "natural logarithm" of the quantity on the right. Note that the result is independent of the actual value of V_{sat}. The frequency of oscillation f_o is

$$f_o = \frac{1}{T} \qquad (17\text{-}35)$$

Example 17-8

A certain op-amp multivibrator of the form shown in Figure 17-11 has $R = 200$ kΩ, $C = 0.01$ μF, $R_1 = 3$ kΩ, and $R_2 = 1$ kΩ. Determine the period and frequency of oscillation.

Solution

The period is determined from (17-34) as

$$T = 2 \times 2 \times 10^5 \times 0.01 \times 10^{-6} \times \ln\left(1 + \frac{2 \times 1 \times 10^3}{3 \times 10^3}\right)$$

$$= 4 \times 10^{-3} \times \ln 1.667 \qquad (17\text{-}36)$$

$$= 4 \times 10^{-3} \times 0.5110 = 2.044 \times 10^{-3} \text{ s} = 2.044 \text{ ms}$$

The frequency is

$$f_o = \frac{1}{2.044 \times 10^{-3}} = 489.2 \text{ Hz} \qquad (17\text{-}37)$$

Example 17-9

Design an op-amp square-wave generator of the form shown in Figure 17-11 to generate a 1-kHz square-wave signal.

Solution

The required period T is

$$T = \frac{1}{f_o} = \frac{1}{10^3} = 10^{-3} \text{ s} \qquad (17\text{-}38)$$

We thus have the requirement that

$$2RC \ln\left(1 + \frac{2R_2}{R_1}\right) = 10^{-3} \qquad (17\text{-}39)$$

There are four parameter values to select and only one equation to satisfy. Consequently, three of the four values can be selected arbitrarily.

As a reasonable approach, first consider R_1 and R_2 since the ratio R_2/R_1 appears in the logarithmic function. Refer back to Figure 17-12 and equation (17-33). If $R_2 \ll R_1$, then $V_f \ll V_{sat}$, and the timing waveform barely crosses zero in both directions. This causes the period to be more sensitive to the op-amp open-loop gain and offset effects. At the other extreme, if $R_2 \gg R_1$, V_f is nearly as large as V_{sat}, and this results in a poorly defined point of triggering on the upper part of the exponential curve. Thus these extreme cases should be avoided. Although there is no optimum point, a ratio of R_2/R_1 between, roughly, about $\frac{1}{9}$ and 1 should suffice. These two limits correspond to $V_f = 0.1V_{sat}$ and $0.5V_{sat}$.

As an arbitrary choice in this range, we will select $R_2/R_1 = \frac{1}{2}$, which can be achieved with $R_1 = 20 \text{ k}\Omega$ and $R_2 = 10 \text{ k}\Omega$ (or other values with the same ratio). This corresponds to $V_f = V_{sat}/3$. When $R_2/R_1 = \frac{1}{2}$ is substituted in (17-39), the equation reduces to

$$2RC \ln (1 + 2 \times \tfrac{1}{2}) = 10^{-3} \qquad \textbf{(17-40)}$$

This simplifies to

$$RC = \frac{10^{-3}}{2 \ln 2} = \frac{10^{-3}}{2(0.6931)} = 0.7213 \times 10^{-3} \text{ s} \qquad \textbf{(17-41)}$$

We have now reduced the number of unknowns to two. Any RC product satisfying (17-41) will theoretically work. However, in the capacitance range involved, it is easier to adjust R than C. Consequently, the best approach is to select C and calculate the required value of R.

Let's try $C = 0.01 \ \mu\text{F}$, which is a common standard value. With this choice, R can be calculated as

$$R = \frac{0.7213 \times 10^{-3}}{0.01 \times 10^{-6}} = 72.13 \text{ k}\Omega \qquad \textbf{(17-42)}$$

While this value of R is not a standard one, it can easily be achieved with a potentiometer alone or with a fixed resistance plus a potentiometer. In fact, the final design of a circuit such as this is best adjusted in the lab.

The resulting design based on Figure 17-11 is summarized as follows:

$$R_1 = 20 \text{ k}\Omega$$

$$R_2 = 10 \text{ k}\Omega$$

$$R = 72.13 \text{ k}\Omega$$

$$C = 0.01 \ \mu\text{F}$$

17-5 555 TIMER

Integrated circuits that generate square-waves and perform other multivibrator functions are readily available. Probably the most popular of such circuits is the widely used 555 timer, which was first introduced by Signetics, Inc.

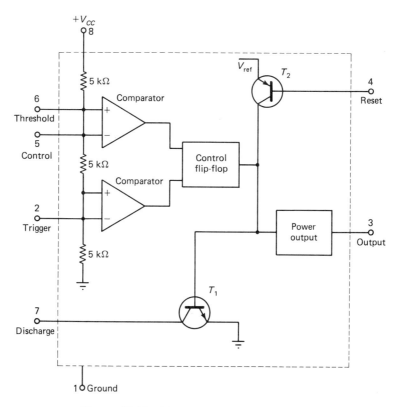

Figure 17-13 Block diagram of 555 timer IC.

A block diagram of the 555 timer is shown in Figure 17-13. It consists of two comparators, a control flip-flop (bistable multivibrator), a discharge transistor (T_1), a reset transistor (T_2), and an output stage. It can be operated with a dc voltage ($+V_{CC}$) between $+5$ and $+18$ V.

A simplified description of the logic is as follows: (1) If the voltage at the threshold terminal (pin 6) slightly exceeds $2\,V_{CC}/3$, the output (pin 3) assumes a low state (close to zero); and (2) if the trigger input (pin 2) drops slightly below $V_{CC}/3$, the output will assume a high state (close to V_{CC}). When the output is in a high state, the discharge transistor is at cutoff, and when the output is in a low state, the discharge transistor is saturated.

Astable Operation

The 555 connected for basic astable operation is shown in Figure 17-14. The only external components required for operation are R_A, R_B, C, and the 0.01-μF bypass capacitor.

Equivalent circuits for the timing process are shown in Figure 17-15, and pertinent waveforms are shown in Figure 17-16. Assume just prior to $t = 0$ that the capacitor voltage has dropped to a level slightly less than $V_{CC}/3$. The output v_o then assumes a high state (i.e., $v_o = V_H$). When the output is in the high state, the

Figure 17-14 Astable multivibrator connection for 555.

(a) Output high state   (b) Output low state

Figure 17-15 Timing equivalent circuits in 555 astable mode.

(a)

(b)

Figure 17-16 Astable 555 waveforms.

discharge transistor (pin 7) is off. The equivalent circuit is shown in Figure 17-15(a). Based on this equivalent circuit, the capacitor voltage will rise exponentially toward V_{CC} (to which level it never reaches).

When v_C slightly exceeds 2 $V_{CC}/3$, the output v_o will assume the low state. When the output is in the low state, the discharge transistor becomes saturated, which effectively brings pin 7 to a ground level. The equivalent circuit then assumes the form shown in Figure 17-15(b). Based on this equivalent circuit, the capacitor voltage will discharge exponentially toward zero (which it never reaches).

When v_C drops slightly below $V_{CC}/3$, the output again assumes a high state. One complete cycle has now been completed.

Observe that the time constant during charging (output high level) is different than the time constant during the discharge process (output low level). Let τ_H represent the charging time constant and let τ_L represent the discharge time constant. We have

$$\tau_H = (R_A + R_B)C \qquad \text{(17-43)}$$

and

$$\tau_L = R_B C \qquad \text{(17-44)}$$

The total period T is composed of two intervals T_H and T_L as shown in Figure 17-16. It can be shown that

$$\boxed{T_H = \tau_H \ln 2 = 0.693(R_A + R_B)C} \qquad \text{(17-45)}$$

and

$$\boxed{T_L = \tau_L \ln 2 = 0.693 R_B C} \qquad \text{(17-46)}$$

The total period is then

$$\boxed{T = T_H + T_L = 0.693(R_A + 2R_B)C} \qquad \text{(17-47)}$$

The frequency $f_o = 1/T$ is

$$\boxed{f_o = \frac{1.443}{(R_A + 2R_B)C}} \qquad \text{(17-48)}$$

Note from (17-45) and (17-46) that $T_H > T_L$; that is, the high interval is always longer than the low interval. Depending on the application, this property may or may not be significant. By choosing R_A to be small compared with R_B, the difference in the two intervals may be made to be very small. However, there is a limit to how small R_A can be made. During the low output interval, pin 7 assumes ground potential, and the power supply voltage is directly across R_A. The power dissipated in R_A during this interval is V_{CC}^2/R_A, and the power rating of the resistor will limit the minimum value of R_A.

A convenient parameter in characterizing the period is the *duty cycle d*. It will be defined as

$$d = \frac{T_H}{T} \times 100\% = \frac{T_H}{T_H + T_L} \times 100\% \qquad \textbf{(17-49)}$$

Stated in words, the duty cycle is the percentage of the period in which the output assumes the high state.

By substituting (17-45) and (17-46) in (17-49), the duty cycle can be expressed as

$$d = \frac{R_A + R_B}{R_A + 2R_B} \times 100\% \qquad \textbf{(17-50)}$$

Monostable Operation

The 555 connected for monostable operation is shown in Figure 17-17. The stable state is with the output voltage low. The trigger circuit consisting of a diode, a resistance R_t, and a capacitance C_t, is connected to the trigger input terminal (pin 2). The timing circuit consists of a single resistance R and a capacitance C. The time constant τ is

$$\tau = RC \qquad \textbf{(17-51)}$$

An equivalent circuit for the timing process is shown in Figure 17-18, and pertinent waveforms are shown in Figure 17-19. Assume that the output v_o is initially in its low state. The discharge transistor is thus on, and this causes v_C to be at ground potential.

At $t = 0$, a negative pulse of sufficient magnitude to drop the voltage at the trigger input below $V_{CC}/3$ is received. The output then assumes the high state (i.e., $v_C = V_H$), as shown in Figure 17-19. The discharge transistor now assumes an open-circuit condition, and the timing circuit of Figure 17-18 is applicable. The capacitor

Figure 17-17 Monostable multivibrator connection for 555.

Oscillator Circuits   Chap. 17

(a)

Figure 17-18 Timing equivalent
circuit in 555 monostable mode.

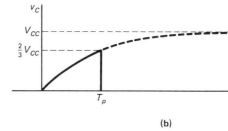

(b)

Figure 17-19 Monostable 555 waveforms.

voltage charges exponentially toward a final value of V_{CC} (which it will never reach). Note that v_C is also the voltage applied to the upper comparator. When this voltage reaches a level slightly greater than $2V_{CC}/3$, the output drops to the low state, and the discharge transistor is turned on. The capacitor is quickly discharged through the transistor, and the circuit is returned to its stable state.

The pulse width T_p is the time required for the voltage across the capacitor to rise from zero to $2V_{CC}/3$. It can be shown that T_p is

$$\boxed{T_p = \tau \ln 3 = RC \ln 3 = 1.1RC} \qquad (17\text{-}52)$$

The exact value of the trigger circuit time constant $R_t C_t$ is not critical, but it should be small compared with the pulse width T_p.

Example 17-10

A 555 timer connected for astable operation as shown in Figure 17-14 has $R_A = 1\ \text{k}\Omega$, $R_B = 4\ \text{k}\Omega$, and $C = 0.02\ \mu\text{F}$. Determine the high and low timing intervals, the total period, the frequency, and the duty cycle.

Solution

The high and low timing intervals are determined from (17-45) and (17-46) as

$$T_H = 0.693(R_A + R_B)C$$

$$= 0.693(10^3 + 4 \times 10^3) \times 0.02 \times 10^{-6} = 69.3\ \mu\text{s} \qquad (17\text{-}53)$$

and

$$T_L = 0.693R_B C$$

$$= 0.693 \times 4 \times 10^3 \times 0.02 \times 10^{-6} = 55.4 \ \mu s \qquad \textbf{(17-54)}$$

The total period T is

$$T = 69.3 + 55.4 = 124.7 \ \mu s \qquad \textbf{(17-55)}$$

The frequency f_o is

$$f_o = \frac{1}{124.7 \times 10^{-6}} = 8.019 \ \text{kHz} \qquad \textbf{(17-56)}$$

The duty cycle d is

$$d = \frac{T_H}{T} \times 100\% = \frac{69.3 \ \mu s}{124.7 \ \mu s} \times 100\% = 55.6\% \qquad \textbf{(17-57)}$$

Example 17-11

A 555 timer connected for monostable operation as shown in Figure 17-17 has $R = 10 \ \text{k}\Omega$ and $C = 0.1 \ \mu\text{F}$. Determine the pulse width.

Solution

From (17-52), the pulse width T_p is

$$T_p = 1.1RC = 1.1 \times 10^4 \times 0.1 \times 10^{-6} = 1.1 \ \text{ms} \qquad \textbf{(17-58)}$$

Example 17-12

Design a 555 astable circuit to produce a 500-Hz square wave with a duty cycle of 60%.

Solution

The pertinent design equations involve T_H and T_L as given by (17-45) and (17-46). Since T_H involves both R_A and R_B, while T_L involves only R_B, it is easier to work with the latter timing interval first. The period T is given by

$$T = \frac{1}{f_o} = \frac{1}{500} = 2 \ \text{ms} \qquad \textbf{(17-59)}$$

The values of T_H and T_L are given by

$$T_H = 0.6T = 0.6 \times 2 = 1.2 \ \text{ms} \qquad \textbf{(17-60)}$$

$$T_L = T - T_H = 2 - 1.2 = 0.8 \ \text{ms} \qquad \textbf{(17-61)}$$

From (17-46), we have the requirement that

$$0.693 R_B C = 0.8 \times 10^{-3} \qquad \text{(17-62)}$$

or

$$R_B C = 1.154 \times 10^{-3} \qquad \text{(17-63)}$$

The value of C will be arbitrarily selected as $C = 0.1\ \mu\text{F}$. The value of R_B is then determined as

$$R_B = \frac{1.154 \times 10^{-3}}{0.1 \times 10^{-6}} = 11.54\ \text{k}\Omega \qquad \text{(17-64)}$$

From (17-45), we have the requirement that

$$0.693(R_A + R_B)C = 1.2 \times 10^{-3} \qquad \text{(17-65)}$$

or

$$(R_A + R_B)C = 1.732 \times 10^{-3} \qquad \text{(17-66)}$$

Substituting $C = 0.1\ \mu\text{F}$ leads to

$$R_A + R_B = \frac{1.732 \times 10^{-3}}{0.1 \times 10^{-6}} = 17.32\ \text{k}\Omega \qquad \text{(17-67)}$$

Finally, $R_B = 11.54\ \text{k}\Omega$ is substituted in (17-67), and this leads to

$$R_A = 17.32 - 11.54 = 5.78\ \text{k}\Omega \qquad \text{(17-68)}$$

To summarize, the circuit will be of the form shown in Figure 17-14 with the following parameter values:

$$R_A = 5.78\ \text{k}\Omega$$

$$R_B = 11.54\ \text{k}\Omega$$

$$C = 0.1\ \mu\text{F}$$

17-6 UNIJUNCTION TRANSISTOR AND OSCILLATOR

Because of its importance in generating nonsinusoidal waveforms, the operation of a unijunction transistor has been delayed to the present chapter. Along with the principle of operation, application of this device to a representative oscillator circuit is considered.

Unijunction Transistor

A diagram suggesting the layout of a *unijunction transistor* (*UJT*) is shown in Figure 17-20(a). The UJT has two semiconductor regions and three external terminals. The three external terminals consist of one emitter terminal and two base terminals. The

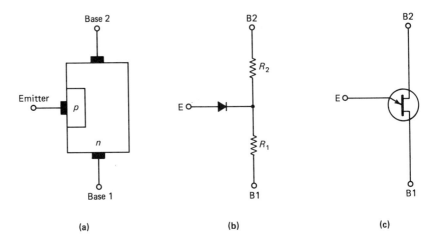

Figure 17-20 Unijunction transistor and schematic symbol.

p region is heavily doped, but the *n* region is lightly doped. Because of the light doping in the emitter region, the resistance between the two bases is relatively high when the emitter is open.

An equivalent circuit when the emitter is open is shown in Figure 17-20(b) and the schematic symbol is shown in (c). The net resistance between the two bases when the emitter is open is called the *interbase resistance* and is denoted as R_{BB}. It is observed from the circuit diagram that

$$R_{BB} = R_1 + R_2 \qquad \textbf{(17-69)}$$

Assume that a voltage V_{BB} is connected between the two bases as shown in Figure 17-21. The voltage V_1 across R_1 will be

$$V_1 = \frac{R_1}{R_1 + R_2} V_{BB} = \eta V_{BB} \qquad \textbf{(17-70)}$$

where η is the voltage-divider ratio preceding V_{BB}. The quantity η is called the *intrinsic standoff ratio*. It is the fraction of the total voltage between the two bases that appears across R_1. A typical value of η is from about 0.5 to 0.8.

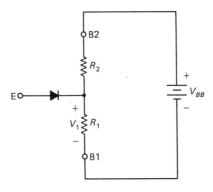

Figure 17-21 Connection of a voltage between the two bases of a UJT.

To illustrate the action of a UJT, consider the circuit of Figure 17-22. Along with a voltage V_{BB} connected between the two bases, an adjustable voltage V_{EE} is connected between emitter and the lower base. When V_E is less than ηV_{BB}, the emitter diode is reverse biased and the high resistance between the two bases is unchanged. However, when v_E is increased slightly above the level ηV_{BB}, holes from the heavily doped p region are injected into the lower half of the UJT and create a strong conducting path. The net effect is to decrease the resistance R_1 to a very low value. The voltage v_E likewise drops to a low value.

Once conduction is initiated, the UJT remains in a heavily conducting state even if V_{EE} is reduced over a reasonable range. The UJT will drop out of the strong conduction state if the emitter current is reduced below a specified *valley current*.

UJT Relaxation Oscillator

The schematic diagram of a sawtooth relaxation oscillator circuit utilizing a UJT is shown in Figure 17-23. Assume initially that the voltage across the capacitor is zero. The UJT will then be in the open state, and the voltage across the capacitor will charge exponentially toward V_{BB}. However, when this voltage slightly exceeds ηV_{BB}, strong conduction occurs between emitter and B1. This discharges the capacitor voltage rapidly down to a level at which the emitter current is reduced below the valley current. At this point, the resistance between emitter and B1 changes back to the value R_1. This completes one cycle and the process is repeated. The output voltage across the emitter then approximates a sawtooth waveform, as shown.

Figure 17-22 Circuit for illustrating UJT operations.

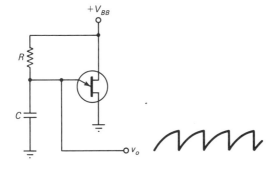

Figure 17-23 UJT sawtooth oscillator.

17-1. For the loop of Figure P17-1, determine the values of A and θ that will produce oscillations.

Amplifier

$A \underline{/\theta}$

Attenuator

0.02

Phase shift = +135°

Figure P17-1

17-2. For the loop of Figure P17-2, determine the values of the attenuation ratio A (output/input ratio) and phase shift θ that will produce oscillations.

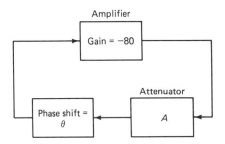

Amplifier

Gain = −80

Attenuator

A

Phase shift = θ

Figure P17-2

17-3. A certain Wien bridge oscillator employs the circuit of Figure 17-4 with $R = 20$ kΩ, $C = 0.005$ μF, and $R_f = 1$ kΩ. Determine **(a)** the frequency of oscillation and **(b)** the resistance of the lamp when stable oscillations are produced.

17-4. Repeat the analysis of Problem 17-3 if $R = 100$ kΩ, $C = 0.2$ μF, and $R_f = 2.4$ kΩ.

17-5. A certain phase shift oscillator has $R = 200$ kΩ and $C = 0.2$ μF (three values of each). Determine the frequency of oscillation.

17-6. A certain phase shift oscillator has $R = 5.1$ kΩ and $C = 0.001$ μF (three values of each). Determine the frequency of oscillation.

17-7. A Colpitts oscillator of the form shown in Figure 17-6 has $C_1 = 50$ pF, $C_2 = 400$ pF, and $L = 20$ μH. Determine the frequency of oscillation.

17-8. Repeat the analysis of Problem 17-7 if $C_1 = 220$ pF, $C_2 = 0.001$ μF, and $L = 250$ μH.

17-9. A Hartley oscillator of the form shown in Figure 17-9 has $L_1 = 50$ μH, $L_2 = 10$ μH, and $C = 68$ pF. Determine the frequency of oscillation.

17-10. Repeat the analysis of Problem 17-9 if $L_1 = 2$ mH, $L_2 = 400$ μH, and $C = 0.001$ μF.

17-11. An op-amp multivibrator of the form shown in Figure 17-11 has $R = 10$ kΩ, $C = 0.01$ μF, $R_1 = 3.9$ kΩ, and $R_2 = 1.5$ kΩ. Determine the period and frequency of oscillation.

17-12. An op-amp multivibrator of the form shown in Figure 17-11 has $R = 120$ kΩ, $C = 0.2$ μF, $R_1 = 4.7$ kΩ, and $R_2 = 2$ kΩ. Determine the period and frequency of oscillation.

17-13. A 555 timer connected for astable operation as shown in Figure 17-14 has $R_A = 10$ kΩ, $R_B = 62$ kΩ, and $C = 0.1$ μF. Determine the high and low timing intervals, the total period, the frequency, and the duty cycle.

17-14. Repeat the analysis of Problem 17-13 if $R_A = 200$ Ω, $R_B = 820$ Ω, and $C = 0.001$ μF.

17-15. A 555 timer connected for monostable operation as shown in Figure 17-17 has $R = 200$ kΩ and $C = 2$ μF. Determine the pulse width.

17-16. Repeat the analysis of Problem 17-15 if $R = 150$ Ω and $C = 0.001$ μF.

DERIVATION PROBLEMS

17-17. (Requires steady-state ac phasor analysis.) Consider the Wien bridge oscillator of Figure 17-3. Let $\overline{V}_o$ represent the feedback circuit input voltage phasor (which is also the amplifier output voltage), and let $\overline{V}_1$ represent the feedback circuit output voltage phasor. Show that

$$\overline{V}_1 = \frac{j\omega RC\overline{V}_o}{1 - \omega^2 R^2 C^2 + j3\omega RC}$$

17-18. (Requires steady-state ac phasor analysis.) **(a)** Use the results of Problem 17-17, and obtain an expression for the loop gain G_L for the Wien bridge oscillator. **(b)** Apply the Barkhausen criterion to the results of (a) to derive the requirements for oscillation [i.e., $A = 3$ and $f_o = 1/(2\pi RC)$].

17-19. Consider the op-amp square-wave generator of Figure 17-11 along with the waveforms of Figure 17-12. **(a)** Referring back to the process of capacitor charging between two states as discussed in Section 16-3, write an equation for v^- based on the given initial value and the final value that *would* result if the capacitor were allowed to completely charge to V_{sat}. **(b)** From the expression obtained in (a), set $v^- = V_f$ and determine the time t_1 corresponding to half of a cycle. **(c)** By the symmetry involved, determine the period T.

17-20. For the op-amp square-wave generator of Figure 17-11, show that if R_2 is selected to satisfy the condition that $R_2 = 0.86R_1$, the expression for the period reduces to

$$T = 2RC$$

DESIGN PROBLEMS

17-21. A Wien bridge oscillator is to be designed to operate at a fixed frequency of 1 kHz. If C is selected as 0.01 μF, determine the value of R required.

17-22. Assume that in the design of Problem 17-21, the value of R is selected as 10 kΩ. Determine the value of C required.

17-23. A certain Wien bridge oscillator is being designed to be tunable from 200 to 1200 Hz. Fixed values of 0.2 μF are employed for the two capacitors. Determine the range of resistance required.

17-24. A certain Wien bridge oscillator is being designed to be tunable from 100 Hz to 1 kHz using a dual potentiometer, from which the two sections can be varied from zero to 10 kΩ. Fixed values of resistance R_f will be connected in series with the sections to provide the minimum resistance required. Determine the values of R_f and C.

17-25. The resistance R (in ohms) of a certain incandescant lamp is assumed to increase linearly with the voltage across it and can be approximated as

$$R = 100 + 300V_p \ \Omega$$

where V_p is the peak voltage across the lamp. The lamp is used to stabilize a Wien bridge oscillator of the form shown in Figure 17-4. **(a)** Determine the value of R_f required in a certain design in order to produce an oscillator peak output voltage of 9 V. **(b)** In a different design, the feedback resistor R_f is set to 1.6 kΩ. Determine the peak value of the oscillator output voltage.

17-26. The resistance R (in ohms) of a certain incandescant lamp is assumed to be a quadratic function of the voltage across it and can be approximated as

$$R = 200 + 50V_p^2 \ \Omega$$

where V_p is the peak voltage across the lamp. The lamp is used to stabilize a Wien bridge oscillator of the form shown in Figure 17-4. **(a)** Determine the value of R_f required in a certain design in order to produce an oscillator peak output voltage of 6 V. **(b)** In a different design the feedback resistor R_f is set to 1.3 kΩ. Determine the peak value of the oscillator output voltage.

17-27. A phase shift oscillator is to be designed to operate at a fixed frequency of 1 kHz. If C is selected as 0.01 μF, determine the value of R required.

17-28. Assume that in the design of Problem 17-27, the value of R is selected as 10 kΩ. Determine the value of C required.

17-29. Design an op-amp square-wave generator of the form shown in Figure 17-11 to generate a 500-Hz square wave. In this particular design, select $R_1 = R_2$ and $C = 0.02$ μF.

17-30. Perform a new design of the circuit of Problem 17-29 by selecting the feedback voltage as $V_f = 0.1V_{\text{sat}}$. Choose $C = 0.02$ μF.

17-31. Design an op-amp square-wave generator of the form shown in Figure 17-11 to generate a 2-kHz square wave. In this particular design, select the feedback voltage as $V_f = 0.4V_{\text{sat}}$, and select $C = 0.01$ μF.

17-32. Perform a new design of the circuit of Problem 17-31 by selecting $R_1 = 4R_2$. Choose $C = 0.01$ μF.

17-33. Design a 555 astable circuit to produce a 1-kHz square wave with a duty cycle of 65%. Select $C = 0.02$ μF.

17-34. Perform a new design of the circuit of Problem 17-33 to achieve a duty cycle of 52%. Select $C = 0.02$ μF.

17-35. Design a 555 monostable circuit to produce a pulse width of 400 μs. Select $C = 0.01$ μF.

17-36. Design a 555 monostable circuit to produce a pulse width of 30 s. Select $C = 50$ μF.

TROUBLESHOOTING PROBLEMS

17-37. You are troubleshooting a Wien bridge oscillator of the type shown in Figure 17-4. The problem is that oscillation has stopped. Which *three* of the following *could be* the problem? **(a)** R too large; **(b)** R too small; **(c)** C too large; **(d)** C too small; **(e)** R_f too large; **(f)** R_f too small; **(g)** lamp shorted; **(h)** lamp open; **(i)** defective op-amp.

17-38. You are troubleshooting a Colpitts oscillator of the type shown in Figure 17-6. The problem is that the frequency is too low. Which *three* of the following *could be* the problem? **(a)** L too large; **(b)** L too small; **(c)** C_1 too large; **(d)** C_1 too small; **(e)** C_2 too small; **(f)** C_2 too large; **(g)** shorted RF choke; **(h)** R_1 too small; **(i)** R_2 too large.

COMPUTER PROBLEMS

17-39. Write a program to analyze an op-amp square-wave generator of the form shown in Figure 17-11.

Input data: R, C, R_1, R_2, V_{sat}
Output data: T, f_o, V_f

17-40. Write a program to analyze a 555 timer connected in the astable mode.

Input data: R_A, R_B, C
Output data: T, f_o, d

CHAPTER 18

High-Frequency Effects in Amplifiers

OBJECTIVES

After completing this chapter, the reader should be able to:

1. State the phasor forms for sinusoidal voltages and currents.
2. Construct phasor impedance models for circuit components.
3. Apply phasor analysis to determine steady-state voltages and currents in ac circuits.
4. Explain the Miller effect and determine the effective input capacitance of an inverting amplifier.
5. Analyze a series resonant circuit to determine resonant frequency, bandwidth, and Q.
6. Analyze a parallel resonant circuit to determine resonant frequency, bandwidth, and Q.
7. Analyze a simplified tuned amplifier and discuss its major advantages with respect to a low-pass amplifier.
8. Define and discuss the significance of the gain–bandwidth product of a transistor.
9. Draw the schematic diagram for and define the parameters contained in the hybrid-pi model for a BJT.
10. Draw the schematic diagram for and define the parameters contained in the hybrid-pi model for an FET.
11. Sketch the waveform of a transistor current pulse and identify and define the following parameters: delay time, rise time, turn-on time, storage time, fall time, and turn-off time.
12. Define and discuss the significance of the gain–bandwidth product of an operational amplifier.
13. Determine the closed-loop 3-dB bandwidth for either a noninverting or an inverting amplifier congifuration.
14. Discuss the significance of the slew rate parameter for an op-amp.
15. Determine an op-amp output rise time produced by the slew rate effect.
16. Determine the slew rate limiting frequency.
17. Compare the relative effects of finite gain–bandwidth product and finite slew rate.

A major portion of basic circuit theory is devoted to *steady-state ac circuit analysis*. This technique involves the representation of sinusoidal voltages and currents by *phasors*. Phasor voltages and currents associated with passive circuit elements are related to each other by *impedances* or *admittances*.

It is assumed by this point in the book that most readers will have studied basic ac circuit theory in some detail. The treatment here is intended to facilitate the use of that theory in analyzing certain types of electronic circuits. However, readers without a background in ac circuits should still be able to utilize the results given here in working with electronic circuits.

Phasor Form for Sinusoids

Consider the sinusoidal voltage v shown in Figure 18-1 with peak value V_p and a leading phase angle θ. This voltage may be expressed as

$$v = V_p \sin(\omega t + \theta) \qquad \text{(18-1)}$$

where $\omega = 2\pi f = 2\pi/T$. Reviewing the terminology introduced in Chapter 8, ω is the angular frequency in radians/second, f is the repetition frequency in hertz, and T is the period in seconds.

Consider now the sinusoidal current i of Figure 18-2 with peak value I_p and a leading phase angle ϕ. This current may be expressed as

$$i = I_p \sin(\omega t + \phi) \qquad \text{(18-2)}$$

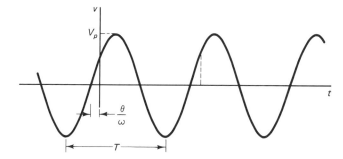

Figure 18-1 Instantaneous sinusoidal voltage with arbitrary phase angle.

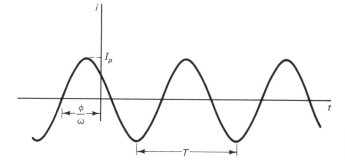

Figure 18-2 Instantaneous sinusoidal current with arbitrary phase angle.

A *phasor* is a vector quantity that represents a sinusoidal voltage or current in the so-called *complex plane*. This quantity has both length and angular orientation. The phasor representing the instantaneous voltage v is denoted as $\overline{V}$, and the phasor representing the instantaneous current i is denoted as $\overline{I}$.

The phasor form of $\overline{V}$ is expressed as

$$\overline{V} = V_p \underline{/\theta} \tag{18-3}$$

and is shown as one of the two quantities in Figure 18-3. The value V_p is the phasor length or *magnitude*, and θ is the angle measured in a counterclockwise sense from the positive x-axis.

The phasor form of $\overline{I}$ is expressed as

$$\overline{I} = I_p \underline{/\phi} \tag{18-4}$$

and is the other quantity of Figure 18-3. The value I_p is the length of this phasor, and ϕ is the angle measured from the positive x-axis. In the terminology of complex numbers, the x-axis is called the *real axis,* and the y-axis is called the *imaginary axis.*

Impedance

One of the most important properties of steady-state ac phasor analysis is that the algebraic properties of resistive circuits may be extended to circuits containing capacitance and inductance by use of the impedance concept. For any *passive* linear circuit consisting of combinations of resistance, capacitance, and inductance, the impedance $\overline{Z}$ is defined as

$$\overline{Z} = \frac{\overline{V}}{\overline{I}} \tag{18-5}$$

In general, impedance is a complex quantity having both magnitude and angle. However, it is best represented in *rectangular form.*

The rectangular form of an impedance may be expressed as

$$\overline{Z} = R + jX \tag{18-6}$$

The quantity R is the real part of $\overline{Z}$, and it is the net *resistance*. The quantity X is the

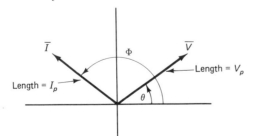

Figure 18-3 Phasor forms of sinusoids of Figures 18-1 and 18-2.

imaginary part of $\bar{Z}$, and it is the net *reactance*. The quantity $j = \sqrt{-1}$ is referred to as the *j*-operator. When displayed as a vector quantity, the real part represents the projection on the *x* or *real* axis, and the quantity preceded by *j* represents the projection on the *y* or *imaginary* axis.

An alternative way to characterize impedance is through the *admittance $\bar{Y}$*. The admittance is

$$\bar{Y} = \frac{\bar{I}}{\bar{V}} = \frac{1}{\bar{Z}}$$

(18-7)

Admittance can also be expressed in terms of real and imaginary parts in the form

$$\bar{Y} = G + jB$$

(18-8)

where *G* is called the *conductance* and *B* is called the *susceptance*.

Each of the three circuit parameters may be represented in terms of a complex impedance whenever ac circuit analysis is employed. Refer to Figure 18-4 in the discussion that follows. The instantaneous notational forms for voltage and current, along with the circuit elements, are shown in the left-hand column. The instantaneous voltage–current relationship for a resistance obeys Ohm's law as was seen in Chapter 1. However, the instantaneous voltage–current relationships for capacitance and inductance require differential and integral calculus, and these forms are not

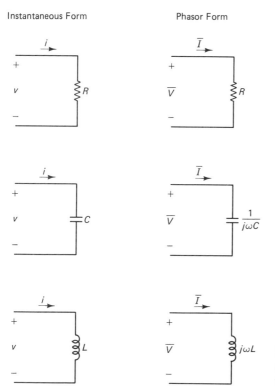

Figure 18-4 Resistance, capacitance, and inductance and their phasor impedances.

considered in this book. On the other hand, the steady-state phasor forms, which require only algebraic relationships with complex numbers, are given in the second column.

The first parameter shown is resistance R. The phasor relationship is

$$\bar{V} = R\bar{I} \qquad (18\text{-}9)$$

The resistive impedance $\bar{Z}_R$ is simply

$$\boxed{\bar{Z}_R = \frac{\bar{V}}{\bar{I}} = R} \qquad (18\text{-}10)$$

The next quantity is capacitance C. The phasor voltage–current relationship for a capacitance is

$$\bar{V} = \frac{1}{j\omega C}\bar{I} \qquad (18\text{-}11)$$

The capacitive impedance $\bar{Z}_C$ is

$$\boxed{\bar{Z}_C = \frac{1}{j\omega C}} \qquad (18\text{-}12)$$

A basic property of the j-operator is that $1/j = -j$. Thus (18-12) can be expressed in the alternative form

$$\boxed{\bar{Z}_C = \frac{-j}{\omega C}} \qquad (18\text{-}13)$$

Thus the *phasor impedance of a capacitor is a purely imaginary negative function*. Note that $\bar{Z}_C$ is simply $-j$ multiplied by the capacitive reactance as defined in Chapter 8.

The final quantity is inductance L. The phasor voltage–current relationship for an inductance is

$$\bar{V} = j\omega L\bar{I} \qquad (18\text{-}14)$$

The inductive impedance $\bar{Z}_L$ is

$$\boxed{\bar{Z}_L = j\omega L} \qquad (18\text{-}15)$$

Thus, *the phasor impedance of an inductor is a purely imaginary positive function*. Note that $\bar{Z}_L$ is simply j multiplied by the inductive reactance as defined in Chapter 8.

A positive j-operator corresponds to a positive (counterclockwise) rotation of $90°$, and a negative j-operator corresponds to a negative (clockwise) rotation of $90°$. This means that the sinusoidal voltage and current associated with an inductance or a capacitance have a $90°$ phase difference. The voltage and current for a resistance are, of course, in phase. Sinusoidal waveforms depicting these conditions, along with typical phasor forms, are shown in Figure 18-5. Thus *the voltage across a pure*

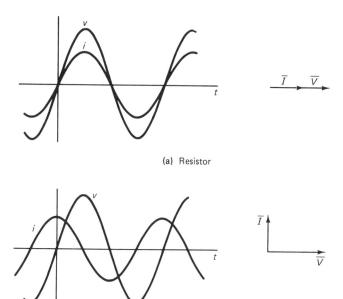

(a) Resistor

(b) Capacitor

(c) Inductor

Figure 18-5 Instantaneous and phasor forms for *R*, *C*, and *L*.

capacitor lags the current by 90°, and the voltage across a pure inductor leads the current by 90°.

Since the impedance of a pure capacitance or a pure inductance has only reactance, inductance and capacitance are referred to as *reactive elements*.

Example 18-1

Express the complex form of the capacitive impedance of a 2-μF capacitor at a frequency of 1 kHz.

⌐ **Solution**

The capacitive impedance $\bar{Z}_C$ is

$$\bar{Z}_C = \frac{-j}{\omega C} = \frac{-j}{2\pi \times 10^3 \times 2 \times 10^{-6}} = -j79.58 \ \Omega \qquad \textbf{(18-16)}$$

Example 18-2

Express the complex form of the inductive impedance of a 3-mH inductor at a frequency of 2 kHz.

Solution

The inductive impedance $\bar{Z}_L$ is

$$\bar{Z}_L = j\omega L = j2\pi \times 2 \times 10^3 \times 3 \times 10^{-3} = j37.70 \ \Omega \qquad \textbf{(18-17)}$$

18-2 MILLER EFFECT

An important phenomenon known as the *Miller effect* is derived and discussed in this section. This effect is most significant in a high-gain inverting amplifier, and it results in a severe degradation of the high-frequency response.

Consider the block shown in Figure 18-6(a) representing an inverting amplifier with gain $-|A|$. (The gain has been indicated in this manner to emphasize the fact that it is an inverting gain.) The amplifier may represent any type of circuit in which gain and inversion occur; that is, it could be a BJT amplifier, an FET amplifier, or an op-amp. The input and output phasor voltages with respect to ground are denoted as $\bar{V}_i$ and $\bar{V}_o$, respectively. The output voltage is related to the input voltage by

$$\bar{V}_o = -|A|\,\bar{V}_i \qquad \textbf{(18-18)}$$

It will be assumed for this development that the input impedance of the amplifier is sufficiently high that the input current is negligible. With this assumption, we need only concentrate on the current drawn by the capacitance. The assumed input current $\bar{I}_i$ must then flow into the capacitance.

To determine the capacitive current, it is necessary to know the voltage across

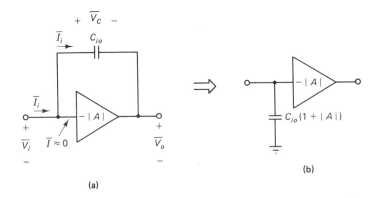

(a)

(b)

Figure 18-6 Inverting amplifier used to develop Miller effect and equivalent capacitance *as referred to input*.

the capacitance. Let $\bar{V}_C$ represent this phasor voltage. A loop is formed by starting from ground, moving up to the left-hand side of the capacitance, moving through the capacitance to $\bar{V}_o$, and then back to ground. The KVL equation reads

$$-\bar{V}_i + \bar{V}_C + \bar{V}_o = 0 \tag{18-19}$$

or

$$\bar{V}_C = \bar{V}_i - \bar{V}_o \tag{18-20}$$

Substitution of (18-18) in (18-20) results in

$$\bar{V}_C = \bar{V}_i - (-|A|\bar{V}_i) = (1 + |A|)\bar{V}_i \tag{18-21}$$

The voltage across the capacitance is thus $1 + |A|$ times the input signal voltage.
  The current $\bar{I}_i$ is expressed as

$$\bar{I}_i = \frac{\bar{V}_C}{1/j\omega C_{io}} = j\omega C_{io}\bar{V}_C \tag{18-22}$$

Substitution of (18-21) in (18-22) yields

$$\bar{I}_i = j\omega C_{io}(1 + |A|)\bar{V}_i \tag{18-23}$$

Since the input current is purely capacitive in nature, $\bar{I}_i$ must be related to $\bar{V}_i$ by an equation of the form

$$\bar{I}_i = j\omega C_{eq}\bar{V}_i \tag{18-24}$$

where C_{eq} is the equivalent capacitance seen at the input.
  Comparing (18-23) with (18-24), we see that the equivalent input capacitance is

$$\boxed{C_{eq} = (1 + |A|)C_{io}} \tag{18-25}$$

The capacitance between input and output is thus multiplied by $1 + |A|$. This is a direct result of the fact that the voltage across the capacitance is multiplied by $1 + |A|$.
  The equivalent circuit *referred to the input of the amplifier* is shown in Figure 18-6(b). An equivalent capacitance of value $(1 + |A|)C_{io}$ to ground can be considered to exist across the input of the amplifier.
  A typical amplifier will also have an actual shunt input capacitance C_{ig} to ground as shown in Figure 18-7(a). This capacitance simply adds to the Miller effect capacitance so that the total equivalent capacitance C_{eq} is now

$$\boxed{C_{eq} = C_{ig} + C_{io}(1 + |A|)} \tag{18-26}$$

An equivalent circuit *referred to the input* is shown in Figure 18-7(b).
  Since the 3-dB upper frequency of a circuit with shunt capacitance is inversely proportional to the value of the capacitance, the enlarged capacitance seen at the input acts to reduce significantly the upper operating frequency range. This phenomenon is a serious limitation of an inverting amplifier with capacitive coupling between input and output.

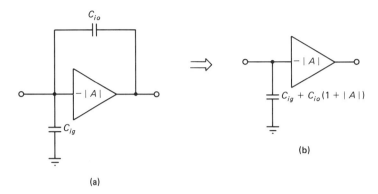

Figure 18-7 Combination of Miller effect capacitance and shunt input capacitance *as referred to input*.

Example 18-3

A certain inverting amplifier has a gain $A = -99$. The amplifier has an input shunt capacitance of 8 pF and a capacitance between input and output terminals of 2 pF. Determine the net equivalent input capacitance.

Solution

From the data provided $C_{ig} = 8$ pF and $C_{io} = 2$ pF. The equivalent capacitance is

$$C_{eq} = C_{ig} + (1 + |A|)C_{io} = 8 + (1 + 99)2 = 8 + 200 = 208 \text{ pF} \quad \textbf{(18-27)}$$

Note that the Miller effect capacitance alone is 200 pF, and this is 100 times as great as the actual input/output capacitance of 2 pF.

Example 18-4

Assume that the amplifier of Example 18-3 is driven from a source whose internal resistance is 2 kΩ. Determine the 3-dB break frequency f_b.

Solution

Since the net capacitance represents a shunt effect, the circuit model is a basic low-pass form. Recall from earlier work that the 3-dB break frequency for such a circuit is

$$f_b = \frac{1}{2\pi RC} \quad \textbf{(18-28)}$$

Substituting $R = 2$ kΩ and $C = C_{eq} = 208$ pF, we have

$$f_b = \frac{1}{2\pi \times 2 \times 10^3 \times 208 \times 10^{-12}} = 382.6 \text{ kHz} \qquad \textbf{(18-29)}$$

18-3 RESONANT CIRCUITS

As a necessary prelude to the introduction of tuned amplifier circuits, the basic properties of series and parallel resonant circuits are discussed in this section. The properties discussed are limited to those essential for a basic understanding of tuned amplifiers.

Series Resonant Circuit

In basic ac circuit texts, resonant circuits usually assume a single resistance. However, to make the work here applicable to some actual electronic circuits, the models assumed will contain several typical circuit resistances.

Consider the circuit of Figure 18-8(a) having a voltage source with internal resistance R_g and a load R_L. A series combination of a capacitance C and an inductance L is connected between the signal terminals and the load.

In addition to the source and load resistances, the inductance (or "coil") also contains a certain resistance R_{coil}. This coil resistance normally cannot be determined from a simple dc ohmmeter check since the actual resistance of the coil typically varies with frequency. The coil resistance may be determined indirectly from the "quality factor" of the inductance with a so-called Q-meter, as will be discussed later.

The phasor model of the circuit with the coil resistance indicated is shown in Figure 18-8(b). There are thus three series resistances in this circuit. For convenience in some later steps, we will define R_s as the net resistance, that is,

$$R_s = R_g + R_L + R_{coil} \qquad \textbf{(18-30)}$$

The phasor impedance Z of the series circuit is

$$\bar{Z} = R_g + \frac{1}{j\omega C} + j\omega L + R_{coil} + R_L = R_s + j\left(\omega L - \frac{1}{\omega C}\right) \qquad \textbf{(18-31)}$$

The phasor current $\bar{I}$ is

$$\bar{I} = \frac{\bar{V}_g}{\bar{Z}} = \frac{\bar{V}_g}{R_s + j(\omega L - 1/\omega C)} \qquad \textbf{(18-32)}$$

Assume that the desired output voltage is $\bar{V}_o$, which is given by

$$\bar{V}_o = R_L \bar{I} = \frac{R_L \bar{V}_g}{R_s + j(\omega L - 1/\omega C)} \qquad \textbf{(18-33)}$$

The magnitude of $\bar{V}_o$, indicated by $|\bar{V}_o|$, is given by

$$|\bar{V}_o| = \frac{R_L |\bar{V}_g|}{\sqrt{R_s^2 + (\omega L - 1/\omega C)^2}} \qquad \textbf{(18-34)}$$

where $|\bar{V}_g|$ is the magnitude of $\bar{V}_g$.

(a)

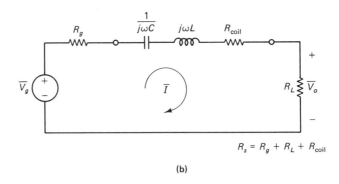

$$R_s = R_g + R_L + R_{coil}$$

(b)

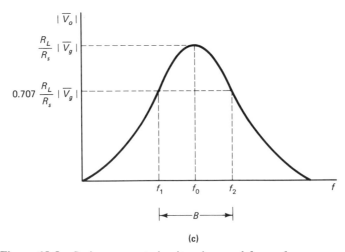

(c)

Figure 18-8 Series resonant circuit and general form of response.

The variation of $|\overline{V}_o|$ as a function of the cyclic frequency $f = \omega/2\pi$ is shown in Figure 18-8(c). At low frequencies, $|\overline{V}_o|$ is small due to the large series capacitive reactance. At high frequencies, $|\overline{V}_o|$ is small due to the large series inductive reactance. However, there is a frequency at which the reactive effects cancel. This is the frequency at which the quantity in parentheses in (18-34) is zero. Denoting this radian frequency as ω_o and the corresponding cyclic frequency as $f_o = \omega_o/2\pi$, we have

$$\omega_o L - \frac{1}{\omega_o C} = 0 \qquad (18\text{-}35)$$

which leads to

$$\omega_o = \frac{1}{\sqrt{LC}} \qquad (18\text{-}36)$$

and

$$f_o = \frac{1}{2\pi\sqrt{LC}} \qquad (18\text{-}37)$$

The frequency f_o is called the *series resonant frequency*. At the *series resonant frequency*, the impedance magnitude has its *minimum* value, the current has its *maximum* value, and the voltage across R_L has its *maximum* value. At f_o, the voltage across R_L from (18-34) is

$$|\overline{V}_o| = \frac{R_L}{R_s}|\overline{V}_g| \qquad (18\text{-}38)$$

The response curve for $|\overline{V}_o|$ thus has a *bandpass* form.

The relative width of the resonant curve is an important design parameter since it determines the *selectivity* of the circuit. The selectivity is a relative measure of the circuit's ability to pass a desired bandpass signal while rejecting others outside the desired passband. If a circuit is too selective, it will reject a portion of the desired signal, while if it is not sufficiently selective, it will pass undesired signals.

Let f_1 and f_2 represent frequencies on the lower and upper sides of resonance at which the response drops by 3 dB. The relative magnitude level at these points is $1/\sqrt{2}(= 0.707)$ times the peak value. Let B represent the 3-dB bandwidth defined by

$$B = f_2 - f_1 \qquad (18\text{-}39)$$

The selectivity of the circuit is often specified in terms of the Q (or "quality factor"). For this purpose, Q is defined as

$$Q = \frac{f_o}{B} \qquad (18\text{-}40)$$

As Q increases, the ratio of f_o to B increases and the relative selectivity increases.

To distinguish the Q of the series circuit from that of the parallel circuit to be studied later, let Q_s represent the Q of the series circuit. It is shown in basic ac circuit texts that Q_s may be expressed in either of the following three ways:

$$Q_s = \frac{\omega_o L}{R_s} = \frac{1}{\omega_o R_s C} = \frac{1}{R_s}\sqrt{\frac{L}{C}} \qquad \textbf{(18-41)}$$

where $\omega_o = 2\pi f_o$ and R_s was defined in (18-30).

Coil Q

In practice, the ac resistance of a coil is often determined indirectly from a measurement with a Q-meter. Let Q_{coil} represent a Q parameter for the coil itself, and it is defined by

$$Q_{\text{coil}} = \frac{\omega_o L}{R_{\text{coil}}} \qquad \textbf{(18-42)}$$

If (18-42) is compared with the first form of (18-41), it can be deduced that if R_{coil} were the only circuit resistance, Q_{coil} would be identical with Q_s. In practice, however, $R_s > R_{\text{coil}}$ and $Q_s < Q_{\text{coil}}$. Thus the circuit Q can never be greater than the coil Q but will almost always be smaller. Since R_{coil} is usually less predictable and adjustable than the other series resistances, the most desirable situation is to have the highest possible value of Q_{coil} and adjust the circuit Q with external resistances.

Parallel Resonant Circuit

Consider the circuit of Figure 18-9(a) having a voltage source with internal resistance R_g and a load R_L. A parallel combination of a capacitance C and an inductance L is connected across the load.

To simplify the analysis of the circuit, a source transformation is first applied to v_g and R_g, and this series combination is converted to a equivalent parallel combination of a current source $i_g = v_g/R_g$ and the resistance R_g. Incidentally, many electronic circuit models actually start with the current source parallel form, as we will see later, but the development here started with the voltage source form to show the equivalence.

In addition to the source and load resistances, the inductance also contains a resistance R_{coil} as shown in Figure 18-9(b). The natural form of this resistance is a series parameter, but it is awkward to analyze the circuit with this form since the remainder of the circuit is a parallel structure.

The key to this dilemma is to replace the series RL combination by an equivalent parallel RL combination. This approach is most meaningful if $Q_{\text{coil}} \geq 10$. For this condition, and to within an error not exceeding 1% in the vicinity of resonance, the inductance is unchanged, but the equivalent *parallel* coil resistance R'_{coil} is approximately

$$R'_{\text{coil}} \simeq Q^2_{\text{coil}} R_{\text{coil}} \qquad \textbf{(18-43)}$$

The net parallel equivalent circuit is shown in phasor form in Figure 18-9(c). For convenience, we will define the net parallel resistance R_p as

$$R_p = R_g \parallel R_L \parallel R'_{\text{coil}} \qquad \textbf{(18-44)}$$

(a)

(b)

$$R_p = R_g \parallel R_L \parallel R'_{coil}$$

(c)

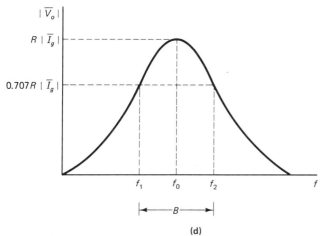

(d)

Figure 18-9 Parallel resonant circuit and general form of response.

The phasor admittance $\overline{Y}$ of the parallel circuit is

$$\overline{Y} = G_p + j\omega C + \frac{1}{j\omega L} = G_p + j\left(\omega C - \frac{1}{\omega L}\right) \qquad \textbf{(18-45)}$$

where

$$G_p = \frac{1}{R_p} \qquad \textbf{(18-46)}$$

The phasor voltage $\overline{V}_o$ is

$$\overline{V}_o = \frac{\overline{I}_g}{\overline{Y}} = \frac{\overline{I}_g}{G_p + j(\omega C - 1/\omega L)} \tag{18-47}$$

The magnitude of $\overline{V}_o$, indicated by $|\overline{V}_o|$, is given by

$$|\overline{V}_o| = \frac{|\overline{I}_g|}{\sqrt{G_p^2 + (\omega C - 1/\omega L)^2}} \tag{18-48}$$

where $|\overline{I}_g|$ is the magnitude of $\overline{I}_g$.

The variation of $|\overline{V}_o|$ as a function of f is shown in Figure 18-9(d). At low frequencies, $|\overline{V}_o|$ is small due to the shunting effect of the small inductive reactance. At high frequencies, $|\overline{V}_o|$ is small due to the shunting effect of the small capacitive reactance. However, there is a resonant frequency at which the reactive effects cancel. This frequency is determined by setting the quantity in parentheses in the denominator of (18-48) to zero. Denoting the resonant frequency as f_o, it is readily shown to be

$$\boxed{f_o = \frac{1}{2\pi\sqrt{LC}}} \tag{18-49}$$

Thus the expression for the parallel resonant frequency is the same as for the series resonant frequency.

At the *parallel resonant frequency*, the impedance magnitude has its *maximum* value, and the voltage across the resonant circuit has its *maximum* value. At f_o, the voltage across the resonant circuit from (18-48) is

$$|\overline{V}_o| = \frac{|\overline{I}_g|}{G_p} = R_p|\overline{I}_g| \tag{18-50}$$

As in the case of the series circuit, we may define B and Q as

$$\boxed{B = f_2 - f_1} \tag{18-51}$$

and

$$\boxed{Q = \frac{f_o}{B}} \tag{18-52}$$

However, the expressions for determining the Q from the circuit elements are different for the parallel resonant circuit. Let Q_p represent the parallel circuit Q. It is shown in basic ac circuit texts that Q_p may be expressed in either of the following three ways:

$$\boxed{Q_p = \frac{R_p}{\omega_o L} = \omega_o R_p C = R_p\sqrt{\frac{C}{L}}} \tag{18-53}$$

where $\omega_o = 2\pi f_o$ and R_p was defined in (18-44).

As in the case of the series RLC circuit, the maximum possible value of the net circuit Q is limited by Q_{coil}. If the driving source were an ideal current source ($R_g = \infty$), and if the load were an open circuit ($R_L = \infty$), $Q_p = Q_{coil}$. In general, however, $Q_p < Q_{coil}$.

Effect of Q on Response

For either the series or the parallel resonant circuits, the relative effect of the circuit Q on the response shape is illustrated in Figure 18-10. It is assumed here that Q_1 is the smallest of the three values and Q_3 is the largest. The smallest value Q_1 corresponds to a relatively broad bandpass response. At the other extreme, Q_3 corresponds to a very sharp, narrow bandpass response.

 Amplitude response

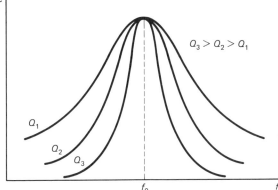

Figure 18-10 Effect of Q on resonance curves.

Example 18-5

A certain electronic amplifier utilizes a series resonant circuit whose form may be represented by the model of Figure 18-8(a). The generator resistance is 250 Ω and the load resistance is 500 Ω. The reactive elements are $L = 5$ mH and $C = 100$ pF. **(a)** Determine the resonant frequency. **(b)** At the resonant frequency, the Q of the coil is measured as 30. Determine the net circuit Q at that frequency. **(c)** Determine the 3-dB bandwidth B.

Solution

(a) The resonant frequency f_o is

$$f_o = \frac{1}{2\pi\sqrt{LC}} = \frac{1}{2\pi\sqrt{5 \times 10^{-3} \times 100 \times 10^{-12}}} = 225.1 \text{ kHz} \qquad (18\text{-}54)$$

(b) To determine the net circuit Q, we must first determine R_{coil}. The coil Q is related to its resistance by

$$Q_{\text{coil}} = \frac{\omega_o L}{R_{\text{coil}}} \qquad \text{(18-55)}$$

Solving for R_{coil}, we have

$$R_{\text{coil}} = \frac{\omega_o L}{Q_{\text{coil}}} = \frac{2\pi \times 225.1 \times 10^3 \times 5 \times 10^{-3}}{30} = 235.7 \ \Omega \qquad \text{(18-56)}$$

The net series resistance R_s is

$$R_s = R_g + R_L + R_{\text{coil}} = 250 + 500 + 235.7 = 985.7 \ \Omega \qquad \text{(18-57)}$$

The net series Q is

$$Q_s = \frac{\omega_o L}{R_s} = \frac{2\pi \times 225.1 \times 10^3 \times 5 \times 10^{-3}}{985.7} = 7.174 \qquad \text{(18-58)}$$

The net circuit Q is much less than the Q of the inductance alone.

(c) The net Q is related to the bandwidth B by

$$Q_s = \frac{f_o}{B} \qquad \text{(18-59)}$$

The 3-dB bandwidth B is

$$B = \frac{f_o}{Q_s} = \frac{225.1 \times 10^3}{7.174} = 31.38 \text{ kHz} \qquad \text{(18-60)}$$

Example 18-6

A certain electronic amplifier utilizes a parallel resonant circuit whose form may be represented by the model of Figure 18-9(a). The generator resistance is 5 kΩ, and the load resistance is 10 kΩ. The reactive elements are $L = 8 \ \mu H$ and $C = 125$ pF.

(a) Determine the resonant frequency. (b) At the resonant frequency, the Q of the coil is measured as 50. Determine the values of the series resistance and the equivalent parallel resistance. (c) Determine the net circuit Q at the resonant frequency. (d) Determine the 3-dB bandwidth B.

Solution

(a) The resonant frequency f_o is

$$f_o = \frac{1}{2\pi\sqrt{LC}} = \frac{1}{2\pi\sqrt{8 \times 10^{-6} \times 125 \times 10^{-12}}} = 5.033 \text{ MHz} \qquad \text{(18-61)}$$

(b) The coil Q is related to its series resistance R_{coil} by

$$Q_{\text{coil}} = \frac{\omega_o L}{R_{\text{coil}}} \qquad \text{(18-62)}$$

Solving for R_{coil}, we have

$$R_{coil} = \frac{\omega_o L}{Q_{coil}} = \frac{2\pi \times 5.033 \times 10^6 \times 8 \times 10^{-6}}{50} = 5.060\ \Omega \qquad \textbf{(18-63)}$$

The equivalent parallel resistance R'_{coil} is

$$R'_{coil} = Q^2_{coil} R_{coil} = (50)^2 \times 5.060 = 12.65\ k\Omega \qquad \textbf{(18-64)}$$

(c) The net parallel resistance R_p is

$$R_p = R_g \parallel R_L \parallel R'_{coil} = 5\ k\Omega \parallel 10\ k\Omega \parallel 12.65\ k\Omega = 2.638\ k\Omega \qquad \textbf{(18-65)}$$

The net circuit Q is

$$Q_p = \frac{R_p}{\omega_o L} = \frac{2638}{2\pi \times 5.033 \times 10^6 \times 8 \times 10^{-6}} = 10.43 \qquad \textbf{(18-66)}$$

(d) The 3-dB bandwidth is

$$B = \frac{f_o}{Q} = \frac{5.033 \times 10^6}{10.43} = 482.6\ kHz \qquad \textbf{(18-67)}$$

18-4 TUNED AMPLIFIER

Tuned amplifiers employ series or parallel resonant circuits to create a bandpass characteristic over the frequency range of interest. Shunt capacitive effects, which severely limit the upper frequency range of low-pass amplifiers, become an integral part of the resonant circuit and are thus used to advantage in tuned amplifiers. With careful design and adjustment, tuned amplifiers employing discrete components may be used to several hundred megahertz.

A full treatment of tuned amplifiers is beyond the scope of this text and is more appropriate for a text on communications or high-frequency electronics. The treatment here will be limited to the basic concepts involved. This should allow the reader insight into the more complex circuits encountered in the field.

A simplified representative tuned amplifer employing a BJT and a parallel resonant circuit is shown in Figure 18-11(a). The input of this particular circuit as shown is similar to any standard BJT amplifier, although the circuit could be driven by another tuned amplifier. The resistance R_L represents any load resistance, and C_{in} represents the net shunt capacitance of the circuit as viewed by the collector. This could represent the Miller effect capacitance looking into any additional stage plus all shunt capacitance across the circuit (including collector-to-ground capacitance of the BJT).

The inductance L and the capacitance C_t represent the discrete components constituting a parallel resonant circuit. Note that C_t is adjustable (i.e., it is a "tuning" capacitor). Seldom is it possible to predict with high accuracy the exact value of capacitance required for this capacitor due to the various stray circuit capacitances. Consequently, C_t is adjusted under dynamic circuit conditions to achieve

(a)

(b)

Figure 18-11 Typical tuned amplifier stage and a simplified equivalent circuit.

parallel resonance at the exact center frequency desired. An alternative approach is to employ a fixed capacitance and an adjustable inductance. In receivers, the process of tuning a capacitance or inductance to an exact frequency is known as *alignment*.

The capacitances C_i, C_E, and C_o are all assumed to be very large compared with C_t and C_{in}. This means that these first three capacitors may be assumed to be short circuits in the frequency range where resonance occurs.

With the preceding assumptions, a simplified small-signal phasor model for the circuit is shown in Figure 18-11(b). Note that the variables and components are re-placed by their phasor and impedance equivalents. The resistance R_{in} represents the parallel combination of R_1, R_2, and r_π; that is,

$$R_{in} = R_1 \parallel R_2 \parallel r_\pi \qquad (18\text{-}68)$$

The capacitance C represents the combination of the tuning capacitance C_t and the circuit shunt capacitance; that is,

$$C = C_t + C_{in} \qquad (18\text{-}69)$$

Finally, the resistance R represents the net parallel combination of the load resistance R_L and the parallel equivalent R'_{coil} of the coil resistance; that is,

$$R = R_L \parallel R'_{coil} \tag{18-70}$$

Analysis of this circuit is very similar to that of the parallel resonant circuit considered in the preceding section and differs only with respect to the fact that a gain can be achieved with this circuit. The admittance $\bar{Y}$ of the parallel circuit is

$$\bar{Y} = G + j\omega C + \frac{1}{j\omega L} \tag{18-71}$$

$$= G + j\left(\omega C - \frac{1}{\omega L}\right) \tag{18-72}$$

where $G = 1/R$. The voltage $\bar{V}_o$ is

$$\bar{V}_o = \frac{-g_m \bar{V}_i}{\bar{Y}}$$

$$= \frac{-g_m \bar{V}_i}{G + j(\omega C - 1/\omega L)} \tag{18-73}$$

The voltage gain has a frequency-dependent nature, so we will denote it as $\bar{A}$. It is

$$\bar{A} = \frac{\bar{V}_o}{\bar{V}_i} = \frac{-g_m}{G + j(\omega C - 1/\omega L)} \tag{18-74}$$

The resonant frequency f_o is readily shown to be

$$f_o = \frac{1}{2\pi \sqrt{LC}} \tag{18-75}$$

At resonance, the gain is

$$A = \frac{-g_m}{G} = -g_m R \tag{18-76}$$

A sketch of the gain magnitude as a function of frequency is shown in Figure 18-12. The form here is the same as that of the resonant circuits considered earlier. However, the magnitude of the gain at f_o is the same as that of a low-pass common-emitter amplifier. In effect, the use of the tuned circuit has resulted in the same gain at f_o as would occur in the midband for a low-pass type of amplifier.

Expressions for the Q for this amplifier are the same as those for the parallel resonant circuit of the preceding section, but they will be repeated here. Letting Q_p represent this Q, we have

$$Q_p = \frac{R}{\omega_o L} = \omega_o RC = R\sqrt{\frac{C}{L}} \tag{18-77}$$

While the concept shown here is generally applicable more or less to all tuned amplifiers, there are many variations in the circuit forms. Although this circuit em-

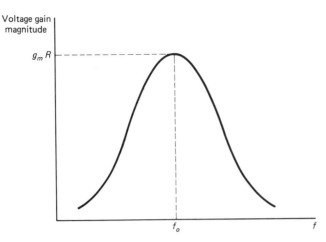

Figure 18-12 Voltage gain magnitude of tuned amplifier.

ployed parallel resonance, some employ series resonance, and some multistage tuned amplifiers employ a combination of both. In many cases the desired Q is easier to achieve with one form than the other.

An additional process used in many tuned amplifiers is that of impedance transformation. It turns out that the input and output resistance levels of many active devices, sources, and loads limit the value of Q directly attainable with realistic values of inductance and capacitance. By employing an impedance transformation, however, these resistance levels may be transformed to a range permitting the attainment of a desired Q with practical values of L and C.

Impedance transformation may be achieved with a transformer, a tapped inductance, or with a capacitive-divider circuit. A representative example of a tuned amplifier employing a tapped inductance and an FET is shown in Figure 18-13.

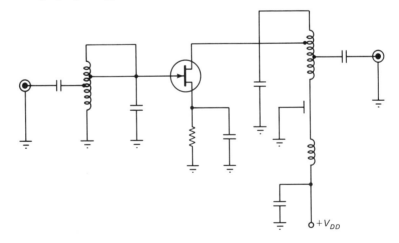

Figure 18-13 Typical high-frequency amplifier employing tapped inductors for impedance matching.

Example 18-7

An amplifier of the form shown in Figure 18-11(a) has a center frequency of 2 MHz and a 3-dB bandwidth of 200 kHz. The inductance is 50 μH, and the input capacitance C_{in} is estimated as 40 pF. The load resistance is 10 kΩ, and at the operating point, $g_m = 0.02$ S. Determine (a) the value of C_t required for correct tuning, (b) the net shunt resistance R, and (c) the gain A at resonance.

Solution

(a) Resonance results from an inductance of 50 μH and a net capacitance C. The resonance formula is

$$f_o = \frac{1}{2\pi \sqrt{LC}} \tag{18-78}$$

Solving for C and substituting values for f_o and L, we obtain

$$C = \frac{1}{4\pi^2 f_o^2 L} = \frac{1}{4\pi^2 (2 \times 10^6)^2 \times 50 \times 10^{-6}} = 126.7 \text{ pF} \tag{18-79}$$

Of this total capacitance, about 40 pF is supplied by input and transistor capacitance. The tuning capacitance C_t is

$$C_t = 126.7 - 40 = 86.7 \text{ pF} \tag{18-80}$$

(b) The net circuit Q is

$$Q_p = \frac{f_o}{B} = \frac{2 \times 10^6}{200 \times 10^3} = 10 \tag{18-81}$$

The net Q is related to R by

$$Q_p = R\omega_o C \tag{18-82}$$

Solving for R, we have

$$R = \frac{Q_p}{\omega_o C} = \frac{10}{2\pi \times 2 \times 10^6 \times 126.7 \times 10^{-12}} = 6281 \ \Omega \tag{18-83}$$

Note that the net shunt resistance is less than the load resistance of 10 kΩ. This is a result of the fact that R is a parallel combination of the load resistance, the equivalent parallel resistance of the coil, and any other shunt resistance (e.g., the dynamic output resistance of the BJT).

(c) The net gain A_{io} at resonance is

$$A_{io} = -g_m R = -0.02 \times 6281 = -126 \tag{18-84}$$

In earlier chapters the small-signal value of the short-circuit ac current gain β was assumed to be independent of frequency and was denoted as β_o. The actual behavior of β as a function of frequency follows a one-pole roll-off behavior quite closely.

Refer to the curve shown in Figure 18-14 in which logarithmic scales are assumed for both variables. At relatively low frequencies, $\beta = \beta_o$ as assumed earlier in the book. The value of β drops to 0.707 β_o at a frequency f_β, which corresponds to the break frequency of a Bode representation. The value of β continues to drop above f_β as shown.

At a frequency $f = f_T$, the value of β is $\beta = 1$ as shown. The frequency f_T is called the *transition frequency*. It can be shown from the one-pole roll-off model that

$$f_T = \beta_o f_\beta \qquad (18\text{-}85)$$

Since β_o is the low-frequency gain and f_β is a bandwidth parameter, the frequency f_T is also called the *gain–bandwidth* product.

Typical values of f_T for general-purpose transistors may be determined from inspecting the 2N3903/2N3904 specifications in Appendix D. On the third data page of Appendix D, it is noted that the minimum value for the 2N3903 is $f_T = 250$ MHz, and the minimum value for the 2N3904 is $f_T = 300$ MHz. Transistors designed specifically for microwave applications have values of f_T exceeding 1 GHz.

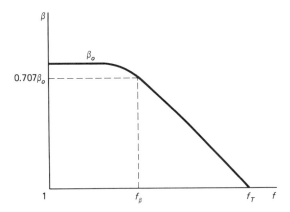

Figure 18-14 Variation of β with frequency for a BJT.

Example 18-8

The minimum value of f_T for a 2N3904 was noted to be 300 MHz. Determine the value of f_β for each of the following values of β_o: **(a)** 100; **(b)** 400.

Solution

An expression for f_β is determined from (18-85) to be

$$f_\beta = \frac{f_T}{\beta_o} \tag{18-86}$$

(a) For $\beta_o = 100$, we have

$$f_\beta = \frac{300 \times 10^6}{100} = 3 \text{ MHz} \tag{18-87}$$

(b) For $\beta_o = 400$, we have

$$f_\beta = \frac{300 \times 10^6}{400} = 0.75 \text{ MHz} \tag{18-88}$$

Note that for a given f_T, the bandwidth varies inversely with the low-frequency gain.

18-6 HYBRID-PI TRANSISTOR MODELS

When serious analysis of high frequency effects in a transistor is necessary or when a high-frequency amplifier is designed, a model displaying the high-frequency effects is required. The most convenient and widely employed circuit is the so-called *hybrid-pi model*. The hybrid-pi models for both a BJT and an FET are discussed in this section.

BJT Hybrid-Pi Model

The form of the hybrid-pi model of the BJT is shown in Figure 18-15. First, note that if $r_x = 0$, $C_\pi = 0$, $C_\mu = 0$, and $r_d = \infty$, the circuit reduces to the simple small-signal g_m-model for the BJT, which was used extensively in early chapters. For the frequency range we have considered thus far, the effects of these parameters are usually not significant.

A forward-biased *pn* junction exhibits a capacitive effect as a result of temporary charge storage in the semiconductor material. This capacitance for the base–emitter junction is referred to as *diffusion capacitance* and is denoted as C_π.

A reverse-biased *pn* junction exhibits a capacitive effect arising from the de-

Figure 18-15 Hybrid-pi model of bipolar junction transistor.

pletion layer between conducting regions. This capacitance for the base–collector junction is referred to as *depletion* or *junction* capacitance and is denoted as C_μ.

There is always some resistance between the active base region and the external contact. This resistance is referred to as the *base spreading resistance* and is denoted by r_x. Note that when the base spreading resistance is present, the control voltage v'_{be} for the dependent current source is not the same as v_{be} because of the drop across r_x.

The resistance r_d is shown as an optional quantity to use when it is necessary to consider the actual slope of the collector characteristics in the active region. In tuned amplifiers, this effective resistance will tend to lower the overall Q of the circuit.

FET Hybrid-Pi Model

As in the case of the BJT, there is a hybrid-pi model for an FET. The form of the hybrid-pi model for an FET is shown in Figure 18-16.

The hybrid-pi model of the FET contains three capacitances. The capacitance C_{gs} represents the *gate–source* capacitance, C_{gd} represents the *gate–drain* capacitance, and C_{ds} represents the *drain–source* capacitance. When the dynamic effects of the drain–source resistance are to be considered, r_d may be included. Note that this model reduces to the simpler FET model when the three capacitances are assumed to be zero.

Figure 18-16 Hybrid-pi model of field-effect transistor.

18-7 TRANSISTOR SWITCHING TIMES

In various transistor switching applications considered early in the text, the transistors were assumed to switch instantaneously between appropriate levels. In actual practice, a transistor requires some time to change between levels, and this limitation is important in high-speed pulse and switching applications. Various transistor capacitances act to slow down the switching process.

Consider the transistor circuit of Figure 18-17(a). Assume initially that $i_B = 0$ and $i_C = 0$. Assume that through some external switching arrangement (not shown), an ideal pulse of base current is applied to the transistor as shown in Figure 18-17(b). If the transistor were ideal, the collector current would be a pulse of the same shape as the base current, but would have a value $\beta_{dc} i_B$. The ideal collector current would start and stop at the same times as the base current.

The actual form of the collector current is as shown in Figure 18-17(c). On the

(a)

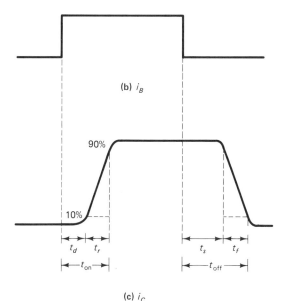

(b) i_B

90%

10%

t_d t_r t_s t_f

t_{on} t_{off}

(c) i_C

Figure 18-17 Transistor switching current waveforms.

leading edge of the collector current pulse are two time parameters. These are called the *delay time* and the *rise time* and are denoted by t_d and t_r, respectively. The *delay time* t_d is the time between the point where the input pulse is initiated and the point where the collector current reaches 10% of its final level. The *rise time* t_r is the time between the 10% and 90% points on the leading edge of the current pulse. The *turn-on time* t_{on} is the sum of the delay time and the rise time, that is,

$$\boxed{t_{on} = t_d + t_r}$$

(18-89)

On the trailing edge of the collector current pulse are two additional time parameters. These are called the *storage time* and the *fall time* and are denoted by t_s and t_f, respectively. The *storage time* t_s is the time between the point where the base current pulse drops to zero and the point where the collector current drops to 90% of its maximum value. The storage time is due to charge carriers being trapped in the depletion region. This time is usually longer than the other time parameters.

The *fall time* t_f is the time required for the collector to drop from 90% of its peak value down to 10% of its peak value. The *turn-off* time t_{off} is the sum of the storage time and the fall time, that is,

$$\boxed{t_{off} = t_s + t_f}$$ (18-90)

Typical values of these various parameters for general-purpose transistors may be noted by referring to the third and fourth pages of the 2N3903/2N3904 data in Appendix D. The maximum values of t_d and t_r are both listed as 35 ns. The maximum value of t_s for the 3904 is listed as 200 ns. (The value for the 3903 is 175 ns.) The maximum value of t_f is listed as 50 ns.

Example 18-9

Based on the data provided for the 2N3904 transistor, calculate **(a)** the turn-on time and **(b)** the turn-off time.

Solution

The values of t_d, t_r, t_s, and t_f were summarized in the last paragraph preceding this example.

(a) The turn-on time t_{on} is

$$t_{on} = t_d + t_r = 35 + 35 = 70 \text{ ns}$$ (18-91)

(b) The turn-off time t_{off} is

$$t_{off} = t_s + t_f = 200 + 50 = 250 \text{ ns}$$ (18-92)

Since the individual time parameters were given as maximum values, the computed turn-on and turn-off times are also maximum values.

18-8 OPERATIONAL AMPLIFIER HIGH-FREQUENCY LIMITATIONS

Operational amplifiers are generally most useful in the frequency range below about 1 MHz or so. While a few special-purpose op-amps are available with usable frequency capabilities well above 10 MHz, these units are relatively expensive.

In general, there are two separate effects that contribute to the high-frequency degradation of op-amps. These are (1) the *finite bandwidth effect* and (2) the *slew rate effect*. Each effect will be considered separately in the developments that follow.

Finite Bandwidth Effect

A discussion was given in Chapter 14 concerning the effect of feedback on the closed-loop bandwidth, and some results were stated at that time without proof. In this section, these results will be expanded, and more detailed considerations of the effects will be made.

We have seen that an op-amp has a very large open-loop dc gain, typically of the order of 10^5 or greater. However, the amplitude response of the open-loop response starts to decrease at a relatively low frequency, often less than 10 Hz.

The variation in open-loop gain as a function of frequency for a typical op-amp is shown by the Bode plot in Figure 18-18. The level A_o represents the dc (and very low frequency) open-loop gain, and f_o represents the 3-dB open loop or break frequency. If the op-amp were operated as an open-loop amplifier, the 3-dB bandwidth would be only f_o.

Although there are many frequency-limiting effects inside the op-amp, most modern op-amps are of the *internally compensated* variety. Such op-amps have included in their circuitry deliberate frequency shaping so as to attain a one-pole roll-off rate (-6 dB/octave) over the frequency range where the open-loop gain is greater than 1. It turns out that from more advanced feedback theory, this condition ensures that the op-amp will be stable for all closed-loop configurations. Unstabilized op-amps are available, in which the user adds a specifically tailored stabilizing circuit, but these will not be considered here.

The mathematical form of the open-loop differential gain $\bar{A}$ as a function of frequency for internally stabilized op-amps can be closely approximated by a one-pole roll-off function, which is of the form

$$\bar{A} = \frac{A_o}{1 + j(f/f_o)}$$

(18-93)

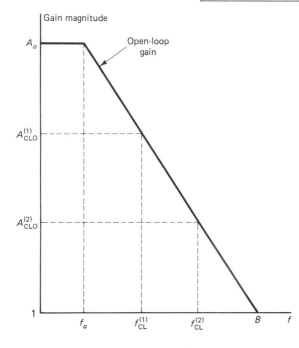

Figure 18-18 Manner in which closed-loop gain and break frequency depend on open-loop gain of an op-amp.

where A_o is the dc and very low frequency gain and f_o is the break frequency. The closed-loop gain A_{CL} expressed in terms of the open-loop differential gain is

$$\bar{A}_{CL} = \frac{\bar{A}}{1 + \bar{A}\beta} \tag{18-94}$$

When (18-93) is substituted in (18-94) and some rearrangement is made, the following result is obtained:

$$\bar{A}_{CL} = \frac{A_o/(1 + A_o\beta)}{1 + jf/f_o(1 + A_o\beta)} \tag{18-95}$$

This result may be expressed as

$$\bar{A}_{CL} = \frac{A_{CLO}}{1 + j(f/f_{CL})} \tag{18-96}$$

in which the new dc gain A_{CLO} is

$$A_{CLO} = \frac{A_o}{1 + A_o\beta} \tag{18-97}$$

and the new break frequency f_{CL} is

$$\boxed{f_{CL} = f_o(1 + A_o\beta)} \tag{18-98}$$

These results were, of course, stated in Chapter 14 without proof. Based on the one-pole roll-off model, the results can readily be derived.

The manner in which closed-loop gain and closed-loop bandwidth are related for an op-amp is illustrated in Figure 18-18. Consider first the closed-loop gain $A_{CLO}^{(1)}$. The point where this gain level intersects the open-loop gain magnitude curve is $f_{CL}^{(1)}$, and this value represents the 3-dB closed-loop bandwidth in this case. Consider next the smaller closed-loop gain $\bar{A}_{CLO}^{(2)}$. This level intersects the open-loop gain magnitude curve at $f_{CL}^{(2)}$, so the 3-dB closed-loop bandwidth is larger at this smaller gain.

Eventually, unity-gain level is reached, and the 3-dB frequency corresponding to this level is denoted as B. The value of B is called the *unity-gain frequency*, and it is the frequency corresponding to the intersection of the open-loop magnitude response and a gain level of unity.

It can be shown from the one-pole frequency response model that B is related to the dc gain A_o and 3-dB open-loop frequency f_o by

$$\boxed{B = A_o f_o} \tag{18-99}$$

From this relationship, an alternative definition of B is the *gain–bandwidth product*. Thus *unity-gain frequency* and *gain–bandwidth product* will be considered to be equivalent specifications.

Simplified Bandwidth Forms

The closed-loop bandwidth has been expressed in terms of the open-loop gain and the feedback factor. In practice, however, these quantities may not be readily apparent from a circuit diagram. Instead, some forms will now be given in which the closed-loop bandwidth is expressed in terms of the closed-loop gain and the unity-gain frequency.

The gain–bandwidth product is

$$A_{CLO}f_{CL} = A_o f_o = B \qquad (18\text{-}100)$$

Solving for f_{CL} and expressing A_{CLO} simply as a *desired* real gain A_{CL}, we have

$$f_{CL} = \frac{B}{A_{CL}} \qquad (\textit{noninverting} \text{ amplifier}) \qquad (18\text{-}101)$$

The value of β for an inverting amplifier is the same as for a noninverting amplifier, but the gain is different. Specifically,

$$\frac{1}{\beta} = \frac{R_i + R_f}{R_i} = 1 + \frac{R_f}{R_i} = 1 + |A_{CL}| \qquad (18\text{-}102)$$

where $|A_{CL}|$ is the magnitude of the *desired* real gain. The corresponding bandwidth relationship is

$$f_{CL} = \frac{B}{1 + |A_{CL}|} \qquad (\textit{inverting} \text{ amplifier}) \qquad (18\text{-}103)$$

For other types of op-amp circuits, the relationship between closed-loop bandwidth and open-loop unity-gain frequency involves a quantity called the "noise gain."* The two relationships given here will suffice for the purposes of this book. In all cases, the general inverse relationship between desired gain and attainable closed-loop bandwidth tends to be the case. In fact, some books simply state a formula such as

$$\text{closed-loop bandwidth} = \frac{\text{unity-gain frequency}}{\text{closed-loop gain}}$$

This formula may always be used qualitatively for estimation purposes, but it is strictly correct only for the case of the noninverting amplifier.

---

* W. D. Stanley, *Operational Amplifiers with Linear Integrated Circuits* (Columbus, Ohio: Charles E. Merrill, 1984).

Example 18-10

A certain op-amp has a unity-gain frequency B = 2 MHz. Determine the value of closed-loop 3-dB bandwidth B_{CL} for each of the following values of *closed-loop noninverting gain*: **(a)** 1000; **(b)** 100; **(c)** 10; **(d)** 1.

Solution

From (18-101), the relationship used is

$$f_{CL} = \frac{B}{A_{CL}} = \frac{2 \times 10^6}{A_{CL}} \qquad \textbf{(18-104)}$$

The four values of A_{CL} are then substituted in (18-104), and the four values of f_{CL} are determined. The resulting values of f_{CL} are summarized in the second column of the table in Example 18-11. Comments will be made at the conclusion of Example 18-11.

Example 18-11

For the op-amp of Example 18-10, determine the closed-loop 3-dB bandwidth for the following values of *closed-loop inverting gain*: **(a)** −1000; **(b)** −100; **(c)** −10; **(d)** −1.

Solution

From (18-103), the relationship used is

$$f_{CL} = \frac{B}{1 + |A_{CL}|} = \frac{2 \times 10^6}{1 + |A_{CL}|} \qquad \textbf{(18-105)}$$

The four values of A_{CL} are substituted in (18-105), and the four values of f_{CL} are determined. The resulting values of f_{CL} are summarized in the third column of the table that follows.

	$|A_{CL}|$	Non-inverting f_{CL}	Inverting f_{CL}
(a)	1000	2 kHz	1998 Hz
(b)	100	20 kHz	19.8 kHz
(c)	10	200 kHz	181.9 kHz
(d)	1	2 MHz	1 MHz

Observe the manner in which the 3-dB frequency increases as the desired gain level decreases. For moderate and large gains, there is negligible difference between the closed-loop 3-dB frequencies for the noninverting and inverting configurations. However, at very low gain, there is a significant difference.

Slew Rate Effect

To provide the internal compensation discussed earlier, a capacitor is added to the op-amp at a critical point in the circuitry. This capacitor in conjunction with the current capability of the stage driving it will limit the rate at which the output of the op-amp can change. This limitation is given in specifications as a *slew rate* parameter.

The slew rate, denoted as S, is the maximum rate of change of the op-amp *output* voltage. The basic units are volts per second. However, because of the large value involved, it is customary to specify the slew rate in volts per microsecond ($V/\mu s$). A typical value of slew rate is 0.5 $V/\mu s$ for a general-purpose op-amp.

The manner in which the slew rate effect exhibits itself in pulse-type situations is illustrated in Figure 18-19. It is assumed in this development that rise time effects due to finite bandwidth are negligible.

Assume an input pulse V_i shown in (a) in which there is a rise time t_i. (Only the leading edge is shown.) The input height is V_i. The closed-loop gain of the amplifier is A_{CL}, so the desired output is a pulse whose height should be $A_{CL} V_i$. The required rate of change of the output voltage v_o is $A_{CL} V_i/t_i$. If the slew rate of the op-amp equals or exceeds this rate of change, the pulse will be reproduced without distortion. This situation is shown in Figure 18-19(b).

Next, assume an input voltage having a shorter rise time t_i as shown in Figure 18-19(c). The rate of change of the desired output voltage is much greater than for the first voltage and is assumed to exceed the slew rate of the op-amp. Since the op-amp output cannot change faster than S, the output will now be of the form shown in Figure 18-19(d). Consequently, the rise time will be degraded.

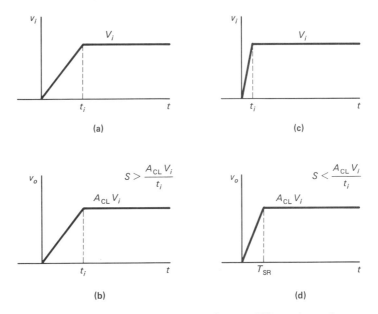

Figure 18-19 Effects of slew rate for two different input rise times.

In Figure 18-19(d), the actual time required for the pulse to reach full output is a function of the slew rate and is denoted as T_{SR}. For a given *output voltage change* V_o, the value of T_{SR} is

$$T_{SR} = \frac{V_o}{S}$$

(18-106)

If V_o is expressed in volts and S is expressed in volts per microsecond, T_{SR} will be expressed in microseconds.

Slew Rate Limiting Frequency

The slew rate concept has been developed on the basis of a pulse-type waveform. However, for complex waveforms (e.g., music, speech, etc.), the effect of the slew rate is to limit the usable frequency range to those frequencies in which the output can follow the maximum rate of change required.

Let f_{SR} represent the frequency of a sinusoidal waveform at which the op-amp output can track the required signal. Using calculus, it can be shown that

$$f_{SR} = \frac{S}{2\pi V_o}$$

(18-107)

where V_o is the peak value of the output sinusoid. In using (18-107), S should be expressed in the basic units of volts per second instead of volts per microsecond. The relationship of (18-107) provides an estimate of the highest operating frequency for complex signals. The frequency will be denoted as the *slew rate limiting frequency*.

Finite Bandwidth and Slew Rate

We have discussed finite bandwidth and slew rate as two separate isolated phenomena. In practice, of course, both effects occur simultaneously, and the combined frequency limitation of the op-amp will result from both effects.

Although both effects are always present, the following general guidelines should be noted: (1) At *very low output signal levels,* the *finite bandwidth effect* is *usually* the dominating effect; and (2) at *very high output signal levels,* the *slew rate* effect is *usually* the dominating effect. In general, however, calculations should be performed to ascertain the degree of each effect.

Example 18-12

From 741 data in Appendix D, it can be determined that the slew rate is 0.5 V/μs. Determine the time required for an output pulse to reach its full level for each of the following op-amp output voltages: **(a)** 0.1 V; **(b)** 1 V; **(c)** 10 V.

Solution

The pertinent relationship is (18-106), and since S is given in V/μs, the time T_{SR} will be expressed directly in microseconds. We have

$$T_{SR}(\mu s) = \frac{V_o}{0.5} \tag{18-108}$$

The three values of V_o may be substituted in (18-108), and the results are summarized below.

	V_o (V)	T_{SR} (μs)
(a)	0.1	0.2
(b)	1	2
(c)	10	20

The output rise time is thus directly proportional to the peak value of the output voltage. Note that the gain does not enter into these results since we are given the output voltage in each case. Thus, for slew rate purposes, it is immaterial whether an output voltage of 10 V arose from an input voltage of 10 V and a gain of 1 or an input voltage of 1 V and a gain of 10.

Example 18-13

For the 741 op-amp with a slew rate of 0.5 V/μs, determine the highest operating frequency with analog-type signals for each of the following peak output voltages: **(a)** 0.1 V; **(b)** 1 V; **(c)** 10 V.

Solution

The pertinent relationship is (18-107). In this case we should first express the slew rate as $S = 0.5 \times 10^6$ V/s. When this value is substituted in (18-107), we have

$$f_{SR} = \frac{0.5 \times 10^6}{2\pi V_o} = \frac{79,577}{V_o} \tag{18-109}$$

The three values of V_o may be substituted in (18-109), and the results are summarized below.

	V_o (V)	f_{SR} (kHz)
(a)	0.1	796
(b)	1	79.6
(c)	10	7.96

The highest operating frequency is thus inversely proportional to the peak value of the output voltage.

DRILL PROBLEMS

18-1. Express the complex form of the capacitive impedance of a 0.01-μF capacitor at a frequency of 5 kHz.

18-2. Express the complex form of the capacitive impedance of a 200-pF capacitor at a frequency of 3 MHz.

18-3. Express the complex form of the inductive impedance of a 50-mH inductor at a frequency of 8 kHz.

18-4. Express the complex form of the inductive impedance of a 200-μH inductor at a frequency of 6 MHz.

18-5. A certain inverting amplifier has a gain $A = -50$. The amplifier has an input shunt capacitance of 20 pF and a capacitance between input and output terminals of 3 pF. Determine the net equivalent input capacitance.

18-6. A certain inverting amplifier has a gain $A = -80$. The amplifier has an input shunt capacitance of 40 pF and a capacitance between input and output terminals of 5 pF. Determine the net equivalent input capacitance.

18-7. Assume that the amplifier of Problem 18-5 is driven from a source whose internal resistance is 1 kΩ. Determine the 3-dB break frequency f_b.

18-8. Assume that the amplifier of Problem 18-6 is driven from a source whose internal resistance is 500 Ω. Determine the 3-dB break frequency f_b.

18-9. Determine the resonant frequency corresponding to $L = 80$ mH and $C = 0.001$ μF.

18-10. Determine the resonant frequency corresponding to $L = 50$ μH and $C = 30$ pF.

18-11. A certain resonant circuit has a resonant frequency of 5 MHz and a 3-dB bandwidth of 200 kHz. Determine the circuit Q.

18-12. A certain resonant circuit has a resonant frequency of 12 MHz and the circuit Q is 50. Determine the 3-dB bandwidth.

18-13. The Q of a certain 20-mH inductance is 50 at a frequency of 250 kHz. Determine the series resistance of the coil.

18-14. The Q of a certain 40-μH inductance is 80 at a frequency of 5 MHz. Determine the series resistance of the coil.

18-15. A certain coil having a Q of 12 at a certain frequency has a series resistance of 20 Ω. Determine the equivalent parallel resistance.

18-16. A certain coil having a Q of 20 at a certain frequency has an equivalent parallel resistance of 2-kΩ. Determine the series resistance.

18-17. A certain electronic amplifier utilizes a series resonant circuit whose form may be represented by the model of Figure 18-8(a). Both the generator and load resistances are 50 Ω. The reactive elements are $L = 40$ μH and $C = 50$ pF. **(a)** Determine the resonant frequency. **(b)** At the resonant frequency, the Q of the coil is measured as 50. Determine the net circuit Q at that frequency. **(c)** Determine the 3-dB bandwidth B.

18-18. A certain electronic amplifier utilizes a parallel resonant circuit whose form may be represented by the model of Figure 18-9(a). The generator resistance is 8 kΩ, and the load resistance is 16 kΩ. The reactive elements are $L = 2$ mH and $C = 0.003$ μF. **(a)** Determine the resonant frequency. **(b)** At the resonant frequency, the Q of the coil is measured as 30. Determine the values of the series resistance and the equivalent parallel resistance. **(c)** Determine the net circuit Q at the resonant frequency. **(d)** Determine the 3-dB bandwidth B.

18-19. An amplifier of the form shown in Figure 18-11(a) has a center frequency of 5 MHz and a 3-dB bandwidth of 250 KHz. The inductance is 10 μH, and resonance at the center frequency is achieved when the tuning capacitance is $C_t = 80$ pF. The load resistance is 12 kΩ, and at the operating point, $g_m = 0.03$ S. **(a)** Estimate the circuit shunt capacitance C_{in}. **(b)** Determine the net shunt resistance R. **(c)** Determine the gain A at resonance.

18-20. An amplifier of the form shown in Figure 18-11(a) has an inductance of 20 μH, and the circuit shunt capacitance C_{in} is estimated as 12 pF. **(a)** Determine the resonant frequency when the tuning capacitance is set at $C_t = 40$ pF. **(b)** For the condition of (a), the 3-dB bandwidth is measured as 200 kHz. Determine the net shunt resistance R. **(c)** If $g_m = 5$ mS, determine the gain A at resonance.

18-21. The value of f_T for a certain transistor is 500 MHz. Determine the value of f_B for each of the following values of β_0: **(a)** 50; **(b)** 200.

18-22. The value of f_B for a certain transistor is 4 MHz and the value of β_0 is 200. Determine the value of f_T.

18-23. A certain BJT has the following parameters:

> Base–emitter diffusion capacitance = 18 pF
> Base–collector depletion capacitance = 2 pF
> Base spreading resistance = 50 Ω
> Dynamic base–emitter resistance = 2 kΩ
> Dynamic collector–emitter resistance = 100 kΩ

For $I_C = 2$ mA, draw a complete hybrid-pi equivalent circuit model.

18-24. A certain FET has the following parameters:

Gate–source capacitance = 8 pF
Gate–drain capacitance = 4 pF
Drain–source capacitance = 1 pF
Dynamic drain resistance = 200 kΩ
g_m = 8 mS

Draw a complete hybrid-pi equivalent circuit model.

18-25. A certain BJT has the following switching parameters:

Delay time = 40 ns
Rise time = 30 ns
Storage time = 120 ns
Fall time = 45 ns

Determine **(a)** the turn-on time and **(b)** the turn-off time.

18-26. A certain BJT has the following parameters:

Turn-on time = 100 ns
Turn-off time = 240 ns

The rise time is 50% longer than the delay time, and the storage time is three times as long as the fall time. Determine the **(a)** delay time, **(b)** rise time, **(c)** storage time, and **(d)** fall time.

18-27. A certain op-amp has a unity-gain frequency B = 1.2 MHz. Determine the value of closed-loop 3-dB bandwidth for each of the following values of *closed-loop noninverting gain:* **(a)** 500; **(b)** 50; **(c)** 5; **(d)** 1.

18-28. A certain op-amp has a unity-gain frequency B = 4 MHz. Determine the value of closed-loop 3-dB bandwidth for each of the following values of *closed-loop noninverting gain:* **(a)** 1000; **(b)** 100; **(c)** 10; **(d)** 1.

18-29. For the op-amp of Problem 18-27, determine the closed-loop 3-dB bandwidth for the following values of *closed-loop inverting gain:* **(a)** −500; **(b)** −50; **(c)** −5; **(d)** −1.

18-30. For the op-amp of Problem 18-28, determine the closed-loop 3-dB bandwidth for the following values of *closed-loop inverting gain:* **(a)** −1000; **(b)** −100; **(c)** −10; **(d)** −1.

18-31. A certain op-amp has a slew rate of 2 V/μs. Determine the time required for an output pulse to reach its full level for each of the following op-amp output voltages: **(a)** 0.2 V; **(b)** 4 V; **(c)** 12 V.

18-32. A certain op-amp has a slew rate of 40 V/μs. Determine the time required for an output pulse to reach its full level for each of the following op-amp output voltages: **(a)** 0.1 V; **(b)** 1 V; **(c)** 10 V.

18-33. For the op-amp of Problem 18-31, determine the highest operating frequency with analog-type signals for each of the following peak output voltages: **(a)** 0.2 V; **(b)** 4 V; **(c)** 12 V.

18-34. For the op-amp of Problem 18-32, determine the highest operating frequency with analog-type signals for each of the following peak output voltages: **(a)** 0.1 V; **(b)** 1 V; **(c)** 10 V.

DERIVATION PROBLEMS

18-35. Consider the series combination of an inductance L and resistance R_{coil} as shown in a portion of the circuit of Figure 18-9(b). **(a)** Write an expression for the impedance $\bar{Z}$ of the series combination. **(b)** Write an expression for the admittance $\bar{Y}$. Rationalize the denominator by multiplying both numerator and denominator by the complex conjugate of the denominator. **(c)** From the result of (b), draw an equivalent parallel circuit showing the values of the equivalent parallel inductance L' and the parallel resistance R'. **(d)** Show that if $Q_{coil} \gg 1$, $L' \approx L$ and $R' \approx Q_{coil}^2 \mathbf{R}_{coil}$.

18-36. Consider the equivalent parallel resonant circuit of Figure 18-9(c). Show that the net Q of the circuit may be determined from the expression

$$\frac{1}{Q} = \frac{1}{Q_g} + \frac{1}{Q_L} + \frac{1}{Q_{coil}}$$

where $Q_g = \omega_o R_g C$, $Q_L = \omega_o R_L C$, and Q_{coil} is the coil Q.

18-37. Starting with (18-94), substitute the expression of (18-93) and carry out the details leading to (18-98).

18-38. (Requires calculus.) Assume that the output v_o of an op-amp is expressed as

$$v_o = V_o \sin \omega t$$

(a) Applying differential calculus, determine the rate of change of the output, that is,

$$\frac{dv_o}{dt}$$

(b) Determine the maximum value of the result of (a), that is, $dv_o/dt]_{max}$. **(c)** To determine the highest frequency at which the op-amp can track the required output, set the result of (b) equal to the slew rate S. Let f_{SR} represent the frequency involved, and show that the result of (18-107) is obtained.

18-39. A certain tuned amplifier utilizing series resonance is being designed. The basic circuit can be modeled by the circuit form of Figure 18-8(b). The inductance is selected as $L = 8$ mH. Determine the value of capacitance required for a center frequency of 150 kHz.

18-40. A certain tuned amplifier utilizing parallel resonance is being designed. The basic circuit can be modeled by the circuit form of Figure 18-9(c). The inductance is selected as $L = 12$ μH. Determine the net value of capacitance required for a center frequency of 4 MHz.

18-41. An op-amp is being selected for a noninverting amplifier application requiring a closed-loop gain of 30. If the closed-loop 3-dB bandwidth is required to be at least 50 kHz, determine the minimum acceptable value of the gain–bandwidth product of the op-amp selected.

18-42. An op-amp is being selected for an inverting amplifier application requiring a closed-loop gain of -7. If the closed-loop 3-dB bandwidth is required to be at least 100 kHz, determine the minimum acceptable value of the gain–bandwidth product of the op-amp selected.

18-43. For the op-amp of Problem 18-41, assume that the slew rate limiting frequency is also specified as 50 kHz. If the maximum input voltage is 0.2 V, determine the minimum acceptable slew rate of the op-amp selected.

18-44. For the op-amp of Problem 18-42 assume that the slew rate limiting frequency is also specified as 100 kHz. If the maximum input voltage is 0.5 V, determine the minimum acceptable slew rate of the op-amp selected.

TROUBLESHOOTING PROBLEMS

18-45. You are troubleshooting a tuned amplifier of the type shown in Figure 18-11(a) and find that the inductor L is open. The original inductor had a value of 40 μH but the only readily available inductor has a value of 20 μH. While waiting for an exact replacement, you decide to temporarily replace the faulty inductor with the available one. To compensate for the smaller inductance, the capacitance of the tuning capacitor will be changed, and the circuit will be realigned. Indicate which one of the following represents the expected required change in C_t. (*Hint:* Remember that a portion of the capacitance is C_{in}.) (**a**) C_t must be more than doubled; (**b**) C_t must be increased some, but not doubled; (**c**) C_t must be decreased to less than half of the original value; (**d**) C_t must be decreased some, but not to half of the original value.

18-46. In the situation of Problem 18-45, assume that the net shunt resistance has the same value at the center frequency as in the original situation. Indicate what change (if any) you would expect in the bandwidth B. (**a**) Increase; (**b**) remain the same; (**c**) decrease.

18-47. Write a program to evaluate the gain magnitude of a tuned amplifier as a function of frequency over the range from f_a to f_b in steps of Δf [see equation (18-74)].

> Input data: g_m, R, L, C, f_a, f_b, Δf
> Output data: f, $|\bar{A}|$

18-48. Write a program to evaluate the closed-loop gain magnitude of an op-amp for the *noninverting* amplifier configuration as a function of frequency over the range from f_a to f_b in steps of Δf.

> Input data: A_{CL}, B, f_a, f_b, Δf
> Output data: f, $|\bar{A}_{CL}|$

18-49. Repeat Problem 18-49 for an *inverting* amplifier configuration.

CHAPTER 19

Power Amplifiers

OBJECTIVES

After completing this chapter, the reader should be able to:

1. Explain how a power amplifier differs from a small-signal amplifier.
2. Express the mathematical form of and sketch the hyperbola of maximum power dissipation.
3. Define class A, B, and C amplifiers and discuss their relative efficiencies.
4. Discuss the balance between power supply input power, signal output power, and internal power dissipation.
5. Analyze a transformer-coupled class A power amplifier for input power, output power, internal power dissipation, and efficiency.
6. Analyze a class B push-pull amplifier for input power, output power, internal power dissipation, and efficiency.
7. Discuss the operation of a class C amplifier.
8. Draw an electrical analog of the heat flow process in a transistor.
9. State and apply the equation relating temperature difference, power dissipation, and thermal resistance of a transistor.
10. Apply the derating factor to determine maximum power dissipation at a particular temperature.
11. Discuss the significance of various power transistor specifications.

19-1 OVERVIEW

A *power amplifier* is an amplifier circuit in which the primary objective is to deliver power to an external load. There is no precise boundary between the so-called "small-signal" amplifiers that we have studied thus far and power amplifiers since all amplifiers are capable of delivering some output power. However, a characterizing feature of most power amplifiers is that a specified power level must be delivered to some external load. For example, in an audio amplifier, the power amplifier stage delivers electrical power to the speakers, and the speakers in turn convert the electrical power to sound. Other examples of power amplifiers are the amplifiers used to provide drive for motors and the output stages of broadcast transmitters, which supply power to the antennas.

A second characterizing feature of most power amplifiers is the level of the required power. Earlier in the book, the level of 0.5 W was indicated as a common dividing line between small-signal amplifiers and power amplifiers. Actually, there are some power amplifiers that operate below this level (e.g., small portable radios with miniature speakers). However, the output power levels of some power amplifiers can be as high as hundreds of kilowatts.

Two factors that are of major importance in power amplifiers are (1) efficiency and (2) heat transfer or thermal conductivity. Efficiency may not be important in a small-signal amplifier operating at 10 mW, but it is extremely important in a power amplifier operating at 10 kW. Similarly, if adequate heat transfer is not provided, power amplifier transistors may be quickly destroyed.

19-2 HYPERBOLA OF MAXIMUM POWER DISSIPATION

Since power dissipation is a major factor in a power transistor, an important concept concerning its operation may be depicted graphically. For this purpose, a BJT will be used, although an FET could be used as well.

Refer to Figure 19-1 in which the coordinate axes of I_C and V_{CE} are shown. The actual transistor output characteristic curves are not shown here since they would add confusion to the curve to be developed. However, their presence is understood.

For this purpose, assume a dc collector–emitter voltage V_{CE} and a dc collector

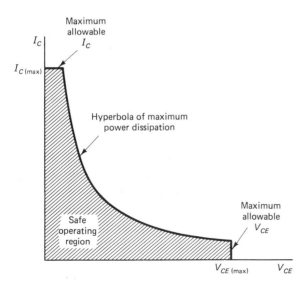

Figure 19-1 Hyperbola of maximum power dissipation.

current I_C. Neglecting any small base power, the total power P_D dissipated in the transistor is given by

$$\boxed{P_D = V_{CE}I_C} \qquad \text{(19-1)}$$

For a given transistor, heat transfer conditions, and temperature, there is a maximum value of P_D specified for operation. Indicating this value as $P_{D(max)}$, we can express (19-1) as

$$V_{CE}I_C = P_{D(max)} \qquad \text{(19-2)}$$

If $P_{D(max)}$ is a constant, then V_{CE} and I_C may be considered as variables. This means that as V_{CE} increases, the maximum allowable value of I_C must decrease, and vice versa. The curve is of the form $xy = constant$ and this is a *hyperbola*. The resulting curve is called the *hyperbola of maximum power dissipation*, and its form is shown in Figure 19-1.

The hyperbola is not the only limit that must be observed. There is a maximum value of V_{CE}, indicated here as $V_{CE(max)}$, and there is also a maximum value of I_C, indicated here as $I_{C(max)}$. Thus the safe operating region for a transistor is that region bounded by the hyperbola of maximum power dissipation and the maximum respective values of voltage and current. For any specified operating point, it is absolutely essential that voltage, current, and power all be below their maximum respective values.

Example 19-1

A certain power transistor has a maximum allowable power dissipation $P_{D(max)} = 30$ W at 25°C. The maximum allowable collector–emitter voltage is $V_{CE(max)} = 50$ V, and the maximum allowable collector current is $I_{C(max)} = 2$ A. **(a)** Determine the maximum allowable collector current when $V_{CE} = 25$ V. **(b)** Determine the maximum allowable collector current when $V_{CE} = 6$ V. **(c)** Determine the maximum allowable collector–emitter voltage when $I_C = 1.5$ A. **(d)** Determine the maximum allowable collector–emitter voltage when $I_C = 0.25$ A. All calculations are to be based on 25°C.

Solution

The first point to note is that the product $V_{CE(max)}I_{C(max)} = 50$ V $\times$ 2 A $=$ 100 W, and this exceeds $P_{D(max)}$ by a factor of $100/30 = 3.33$! This means that the transistor cannot operate at maximum collector–emitter voltage and maximum collector current simultaneously. Referring back to Figure 19-1, note that when V_{CE} is near $V_{CE(max)}$, I_C is quite small. Similarly, when I_C is near $I_{C(max)}$, V_{CE} is quite small.

For a given value of V_{CE}, we will first calculate I_C based on the maximum power rating. If $I_C \leq I_{C(max)}$, the calculated current is the maximum allowable value for that particular value of V_{CE}. If $I_C > I_{C(max)}$, we must use $I_{C(max)}$ rather than the calculated current as the acceptable value. An analogous process is used with V_{CE} when I_C is given.

(a) For $V_{CE} = 25$ V, the value of I_C corresponding to a dissipation of 30 W is

$$I_C = \frac{30 \text{ W}}{25 \text{ V}} = 1.2 \text{ A} \tag{19-3}$$

Since this value is less than 2 A, the maximum current allowed for $V_{CE} = 25$ V is 1.2 A.

(b) For $V_{CE} = 6$ V, the value of I_C corresponding to 30 W dissipation is

$$I_C = \frac{30 \text{ W}}{6 \text{ V}} = 5 \text{ A} \tag{19-4}$$

However, this value *exceeds* $I_{C(\text{max})}$. Consequently, the maximum current allowed for $V_{CE} = 6$ V is $I_{CE(\text{max})}$ (i.e., 2 A).

(c) For $I_C = 1.5$ A, the value of V_{CE} corresponding to 30 W dissipation is

$$V_{CE} = \frac{30 \text{ W}}{1.5 \text{ A}} = 20 \text{ V} \tag{19-5}$$

Since this value is less than 50 V, the maximum voltage allowed for $I_C = 1.5$ A is 20 V.

(d) For $I_C = 0.25$ A, the value of V_{CE} corresponding to a dissipation of 30 W is

$$V_{CE} = \frac{30 \text{ W}}{0.25 \text{ A}} = 120 \text{ V} \tag{19-6}$$

However, this value *exceeds* $V_{CE(\text{max})}$. Consequently, the maximum voltage allowed for $I_C = 0.25$ A is $V_{CE(\text{max})}$ (i.e., 50 V).

Reviewing the four parts of the problem, the conditions of (a) and (c) represent *power-limiting* situations, since it was the power rating rather than the voltage or current rating that established the limit. The condition of (b) represented a *current-limiting* situation since it was the maximum current rating that established the limit. Finally, the condition of (d) represented a *voltage-limiting* condition.

19-3 CLASSIFICATION OF AMPLIFIERS

We will now consider a particular classification scheme for amplifiers that has been widely used since the early days of vacuum tube amplifiers. This classification scheme, which is very important in dealing with power amplifiers, relates to the *conduction angle* of active operation of the device.

It is customary for this purpose to assume an input sinusoidal signal. Although a sinusoidal signal rarely represents an actual useful signal in an application of an amplifier, it is widely used as a standard test signal. The response of the amplifier to a sinusoidal test signal is a meaningful way to provide specifications for the unit, and the properties resulting from other waveforms may be inferred from the sinusoidal response.

A *full cycle* of the input signal refers to the sinusoid over a time duration equal to one period. The angle involved is 360°. A *half-cycle* of the input signal refers to the sinusoid over a time duration equal to one half-cycle. The angle involved is 180°.

Class A Amplifier

A *class A* amplifier is one in which the active device conducts for a full-cycle (or 360°) of the input signal. All amplifiers considered in the earlier chapters on transistors and op-amps have been assumed to be of the class A type. Most small-signal amplifiers are operated as class A amplifiers.

Class B Amplifier

A class B amplifier is one in which each active device conducts for one-half cycle (or 180°) of the input signal. Normally, class B amplifiers employ two active devices, with each device conducting on alternate half-cycles. The combination of the two devices then results in full angle conduction. Strictly speaking, there is no conduction in a class B amplifier without a signal applied so that power dissipation occurs only when the amplifier is excited.

Class AB Amplifier

Some amplifiers deviate slightly from the class B condition and provide a slight forward bias with no signal applied. In this case, conduction for each active device occurs for somewhat more than a half-cycle. Such amplifiers are classified as class AB. Except for the slight forward bias, class AB amplifiers are very close to class B amplifiers in their various properties.

Class C Amplifier

A class C amplifier is one in which the active device conducts for less than one-half cycle. Conduction thus occurs in short pulses. A class C amplifier is basically a nonlinear circuit and imparts significant distortion to an input signal in its basic form. However, by utilizing one or more tuned circuits, a useful output may be derived for specialized applications, such as in radio-frequency circuits.

Efficiency

The efficiency of amplifiers increases with the alphabetical listing of the amplifier type. Thus a class A amplifier has the lowest efficiency and a class C amplifier has the highest efficiency. Certain idealized values of efficiencies will be stated in subsequent sections.

It is essential to understand the manner in which power is distributed in a power amplifier. Actually, power is a time-varying function, and it is possible to describe all power relationships on an instantaneous basis. However, it is more customary to assume a standard signal waveform and develop the various power relationships on an average basis. The standard waveform used for this basis is the sinusoidal function. Thus if an amplifier is rated at 20 W output, the specification usually means that the average power supplied to some specified load is 20 W based on an assumed sinusoidal signal.

The following quantities are defined:

P_S: average power delivered by power supply to amplifier (also called the *dc input power*)

P_O: output signal power delivered to load (also called the *ac output power*)

P_D: power dissipated by amplifier

The conservation of power for the circuit requires that

$$P_S = P_O + P_D \qquad \text{(19-7)}$$

Stated in words, the power delivered by the power supply must equal the sum of output signal power delivered to the load plus any power dissipated in the amplifier.

The equation as given applies only to the active device. Obviously, any bias circuitry involved will consume power. Similarly, small-signal amplifiers used to amplify the signal to a level sufficient to drive the power amplifier stage will also consume power. Our focus now is on the power output stage only.

Efficiency

The efficiency η of the power amplifier can be expressed as

$$\eta = \frac{P_O}{P_S} \times 100\% \qquad \text{(19-8)}$$

Greater efficiency is thus obtained as more of the dc input power is converted to ac output power and less of the power is dissipated in the amplifier. Depending on the amplifier type and the manner in which it is operated, efficiency may range from a few percent to a theoretical level approaching 100% for very ideal conditions. Most practical power amplifiers, however, operate well below 100% efficiency.

Example 19-2

A certain power amplifier delivers 12 W to an external load. The power delivered by the power supply is 20 W. Determine **(a)** the internal power dissipated by the power amplifier and **(b)** the efficiency.

Solution

(a) Since the dc input power is 20 W and the ac output power is 12 W, the internal power dissipation P_D is

$$P_D = P_S - P_O = 20 - 12 = 8 \text{ W} \tag{19-9}$$

(b) The efficiency is

$$\eta = \frac{P_O}{P_S} \times 100\% = \frac{12}{20} \times 100\% = 60\% \tag{19-10}$$

19-5 CLASS A POWER AMPLIFIERS

For reasons of operational comparison and historical perspective, a transformer-coupled class A power amplifier will be discussed even though it is limited in modern usage. The circuit form is shown in Figure 19-2. As shown, the circuit does not have a stabilized bias circuit and is thus impractical. However, this simplification enhances the load-line analysis and the analytical analysis to be given later in this section.

The transformer serves two purposes: (1) it allows the full dc supply voltage to be applied to the collector (in practice, there is some resistance in the transformer windings, but it will be neglected in this analysis), and (2) the turns ratio $N_1 : N_2$ of the transformer can be selected to provide an optimum impedance transformation between the transistor and the load.

Assuming negligible resistance in the transformer, the quiescent dc collector-emitter voltage is simply $V_{CEQ} = V_{CC}$. A *dc load line* may be constructed as shown in Figure 19-3. Since the assumed resistance of the transformer is zero, the slope of the dc load line is infinite and is a vertical line as shown.

When an input signal is applied, the action of the transformer becomes

Note: This circuit is simplified for analysis and is not intended to be a practical circuit.

Figure 19-2 Simplified transformer-coupled class A power amplifier.

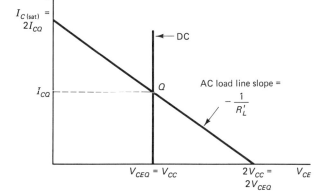

Figure 19-3 Load lines for class A power amplifier based on optimum conditions.

significant. An effective dynamic resistance R'_L is coupled back to the primary and is given by

$$R'_L = \left(\frac{N_1}{N_2}\right)^2 R_L \qquad (19\text{-}11)$$

The primary voltage of the transformer will then vary up and down about V_{CEQ} due to transformer action. This process may be illustrated with an *ac load line* as shown in Figure 19-3. Thus a transformer-coupled amplifier may be characterized by the combination of a *dc load line and an ac load line*.

The optimum primary condition for class A operation is when V_{CE} reaches a level of $2V_{CC}$ on a positive half cycle and 0 on a negative half-cycle. Similarly, I_C should reach $2I_{CQ} = I_{C(\text{sat})}$ on a positive half-cycle and 0 on a negative half-cycle. The inverse magnitude of the slope of the ac load line is given by

$$\frac{2V_{CC}}{2I_{CQ}} = \frac{V_{CC}}{I_{CQ}} = \frac{V_{CEQ}}{I_{CQ}} = R'_L \qquad (19\text{-}12)$$

From the maximum power transfer theorem, maximum output power is obtained when

$$\left(\frac{N_1}{N_2}\right)^2 R_L = R'_L = \frac{V_{CEQ}}{I_{CQ}} \qquad (19\text{-}13)$$

Power and Efficiency

Assume now that the instantaneous collector current i_C is varying sinusoidally about a Q-point according to

$$i_C = I_{CQ} + kI_{CQ} \sin \omega t \qquad (19\text{-}14)$$

The quantity k is a *drive factor* and lies in the range $0 \le k \le 1$. When $k = 0$, there is no signal present, and when $k = 1$, the signal has the maximum allowable variation without distortion, assuming idealized characteristics.

Assume that the collector–emitter voltage is of the form

$$v_{CE} = V_{CQ} - kV_{CQ} \sin \omega t$$
$$= V_{CC} - kV_{CC} \sin \omega t \tag{19-15}$$

Note that the signal component of V_{CE} is chosen to be $180°$ out of phase with the signal component of i_C as required.

The instantaneous power p_S delivered by the power supply is

$$p_S = V_{CC}i_C = V_{CC}(I_{CQ} + kI_{CC} \sin \omega t)$$
$$= V_{CC}I_{CQ} + kV_{CC}I_{CQ} \sin \omega t \tag{19-16}$$

The average or dc power P_S delivered by the power supply is simply the constant part of (19-16) since the sinusoidal portion has an average value of zero. Hence

$$P_S = V_{CC}I_{CQ} \tag{19-17}$$

This average power is independent of the drive, so the class A power amplifier consumes the same dc power with or without a signal.

The instantaneous output power p_O is also the power accepted by the transformer primary (assuming a loseless transformer), and it is given by

$$p_O = i_C(V_{CC} - v_{CE}) = kV_{CEQ}I_{CQ} \sin \omega t + k^2 V_{CEQ}I_{CQ} \sin^2 \omega t \tag{19-18}$$

It can be shown that the average output power P_O can be determined from (19-18) to be

$$P_O = \frac{k^2 V_{CEQ}I_{CQ}}{2} = \frac{k^2 V_{CC}I_{CQ}}{2} = \frac{k^2 V_{CC}^2}{2R_L'} \tag{19-19}$$

The maximum output power $P_{O(\max)}$ occurs when $k = 1$ and is

$$P_{O(\max)} = \frac{V_{CEQ}I_{CQ}}{2} = \frac{V_{CC}I_{CQ}}{2} = \frac{V_{CC}^2}{2R_L'} \tag{19-20}$$

The efficiency η is

$$\eta = \frac{P_O}{P_S} \times 100\% = \frac{k^2 V_{CC}I_{CQ}/2}{V_{CC}I_{CQ}} \times 100\% = \frac{k^2}{2} \times 100\% \tag{19-21}$$

The maximum efficiency $\eta_{\max}$ occurs when $k = 1$ and is

$$\boxed{\eta_{\max} = 50\%} \tag{19-22}$$

To simplify the analysis, recall that the form of the amplifier chosen was highly simplified. When stable bias is added, the maximum peak output will be reduced, and this lowers the efficiency. Consequently, the 50% value should be considered as an ideal upper bound. Practical class A amplifiers operate with somewhat lower efficiencies.

Internal Dissipation

The average power P_D dissipated in the transistor is

$$P_D = P_S - P_O = V_{CC}I_{CQ} - \frac{k^2 V_{CC}I_{CQ}}{2} = V_{CC}I_{CQ}\left(1 - \frac{k^2}{2}\right) \qquad \text{(19-23)}$$

The maximum dissipation $P_{D(\text{max})}$ occurs for $k = 0$ and is

$$\boxed{P_{D(\text{max})} = V_{CC}I_{CQ} = 2P_{O(\text{max})}} \qquad \text{(19-24)}$$

Thus the maximum internal dissipation in a class A transformer-coupled amplifier is twice the maximum output power. The major results of this section, along with results to be developed in Section 19-6, are summarized in Table 19-1 in Section 19-6.

We have only considered a transformer-coupled class A amplifier in this section. It is possible to utilize an RC-coupled amplifier as a class A power amplifier, but the efficiency is even less. The theoretical maximum efficiency of a class A RC-coupled amplifier is 25%, and practical amplifiers of this type have efficiencies well below this value.

Example 19-3

A certain idealized class A transformer-coupled power amplifier has $V_{CC} = 24$ V, $I_{CQ} = 0.5$ A, and it is used to drive a speaker whose dynamic resistance is $R_L = 8\ \Omega$. Based on the simplified assumptions of the model of Figure 19-2, determine **(a)** V_{CEQ}, **(b)** R_L', **(c)** N_1/N_2, **(d)** P_S, **(e)** $P_{O(\text{max})}$, and **(f)** $P_{D(\text{max})}$.

Solution

(a) Assuming a lossless transformer, the Q-point collector–emitter voltage is

$$V_{CEQ} = V_{CC} = 24 \text{ V} \qquad \text{(19-25)}$$

(b) The dynamic resistance R_L' reflected to the primary should satisfy

$$R_L' = \frac{V_{CEQ}}{I_{CQ}} = \frac{25 \text{ V}}{0.5 \text{ A}} = 48\ \Omega \qquad \text{(19-26)}$$

(c) The turns ratio should satisfy the equation

$$\left(\frac{N_1}{N_2}\right)^2 R_L = R_L' \qquad \text{(19-27)}$$

or

$$\frac{N_1}{N_2} = \sqrt{\frac{R_L'}{R_L}} = \sqrt{\frac{48}{8}} = 2.449 \qquad \text{(19-28)}$$

(d) The average power delivered by the power supply is

$$P_S = V_{CC}I_{CQ} = 24 \times 0.5 = 12 \text{ W} \qquad (19\text{-}29)$$

(e) The maximum value of the output power is

$$P_{O(\text{max})} = \frac{V_{CC}I_{CQ}}{2} = \frac{24 \times 0.5}{2} = 6 \text{ W} \qquad (19\text{-}30)$$

(f) The maximum internal dissipation is

$$P_{D(\text{max})} = 2P_{O(\text{max})} = 2 \times 6 = 12 \text{ W} \qquad (19\text{-}31)$$

19-6 CLASS B AMPLIFIER

A class B amplifier is one in which conduction in each transistor occurs for 50% of a cycle. To obtain full-cycle coverage without tuned circuits, therefore, it is necessary that two transistors be employed.

Earlier class B amplifiers employed transformers to couple from the output stages to the load. Most modern circuits utilize direct coupling between a low-impedance output stage and the load. The treatment here will focus on a very popular circuit of this type, namely the *emitter-follower push-pull amplifier*. This circuit

TABLE 19-1 SUMMARY OF CERTAIN POWER RELATIONSHIPS[a]

	Class A (with transformer)	Class B
P_O	$\dfrac{k^2 V_{CC} I_{CQ}}{2}$	$\dfrac{k^2 V_{CC} I_{C(\text{sat})}}{2}$
P_S	$V_{CC}I_{CQ}$	$\dfrac{2}{\pi} k V_{CC} I_{C(\text{sat})}$
P_D	$2P_{O(\text{max})}\left(1 - \dfrac{k^2}{2}\right)$	$P_{O(\text{max})}\left(\dfrac{4k}{\pi} - k^2\right)$
η	$\dfrac{k^2}{2} \times 100\%$	$\dfrac{\pi}{4}k \times 100\%$
$P_{O(\text{max})}$[b]	$\dfrac{V_{CC}I_{CQ}}{2}$	$\dfrac{V_{CC}I_{C(\text{sat})}}{2}$
$P_{S(\text{max})}$	$V_{CC}I_{CQ}$	$\dfrac{4}{\pi}P_{O(\text{max})}$
$P_{D(\text{max})}$	$2P_{O(\text{max})}$	$\dfrac{4}{\pi^2}P_{O(\text{max})}$
η_{max}	50%	78.5%

[a] k is a "drive" factor $0 \leq k \leq 1$.
[b] $P_{D(\text{max})}$ occurs at a different value of k than $P_{O(\text{max})}$ and $P_{S(\text{max})}$.

is widely employed, both in discrete form and as the output stage of many integrated circuits.

Emitter-Follower Push-Pull Amplifier

The basic form of the circuit is shown in Figure 19-4. Some refinements will be made on this circuit later to make it better, but the basic form shown here will suffice for the discussion that follows.

Note that one transistor is an *npn* unit, and the other is a *pnp* unit. To minimize distortion, their characteristics should be identical (except, of course, for opposite polarities). A pair of transistors with matched characteristics and opposite polarities is said to be a *complementary-symmetry matched pair*.

Assume initially that $v_i = 0$. Neither base–emitter diode has any forward bias, so there is no conduction in either transistor and $v_o = 0$. Thus, in this basic circuit, there is no power dissipation with no signal applied.

Now assume that v_i is a positive voltage well above 0.7 V. The *npn* transistor T1 is driven into conduction, but the *pnp* transistor T2 is reverse biased. The input signal is thus coupled to the load, but suffers a loss of about 0.7 V due to the base–emitter drop of T1.

Next, assume that v_i is a negative voltage well below −0.7 V. The base–emitter junction of T2 is now forward biased, but T1 is reverse biased. Conduction between source and load is now through T2. Once again, the output tracks the input, but with a loss of about 0.7 V in magnitude.

Power amplification is achieved by the fact that the required power at the input is very small, but the transistors are capable of delivering considerable power to a load R_L. Thus, although the voltage is not amplified, the power level can be increased dramatically.

The basic circuit as given suffers from a phenomenon called *crossover distortion*. This distortion arises from the *deadband* between about −0.7 and +0.7 V when neither transistor is conducting. The manner in which crossover distortion affects a sinusoidal input signal is illustrated in Figure 19-5.

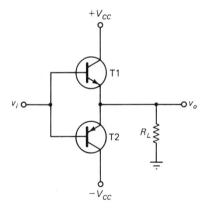

Figure 19-4 Basic form of a class B emitter-follower push-pull amplifier. (Circuit as given is subject to crossover distortion.)

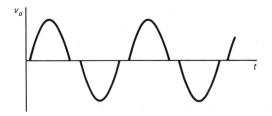

Figure 19-5 Crossover distortion in basic push-pull amplifier of Figure 19-4.

Crossover distortion is eliminated by placing a forward bias on each of the two base–emitter junctions. There are several ways to do this, one of which employs two diodes and dual power supplies and is shown in Figure 19-6. Assume first that $v_i = 0$. The base of T1 will then be at 0.7 V above ground and the base of T2 will be at 0.7 V below ground. The two transistors will barely be starting to conduct at $v_o = 0$.

Full conduction for T1 is now established at a very low positive signal level, and full conduction for T2 is now established at a very low negative signal level. In effect, the base of T1 tracks v_i, but it is at a level of 0.7 V above v_i. Conversely, the base of T2 tracks v_i, but it is at a level of 0.7 V below v_i. Crossover distortion is virtually eliminated and $v_o = v_i$ without the 0.7-V loss.

Maximum effectiveness in this circuit is achieved when the characteristics of

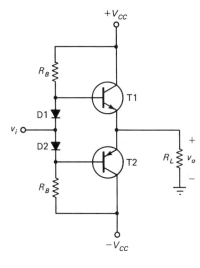

Figure 19-6 Class B emitter-follower push-pull amplifier with circuit used to eliminate crossover distortion.

the diodes match those of the base–emitter transistor junctions. On integrated circuit chips, this is achieved by using transistors as diodes. This process is illustrated in Figure 19-7.

Incidentally, the push-pull emitter-follower amplifier may be biased with a single power supply if ac coupling is employed at the output. This form of the circuit is shown in Figure 19-8. Depending on the manner in which the circuit is driven, ac coupling may or may not be required at the input.

Power and Efficiency

The various power and efficiency relationships of the class B amplifier will now be developed. A sinusoidal signal will be assumed, and the circuit form of Figure 19-6 employing dual power supplies will be used.

If the saturation drops of the two transistors are neglected, the output voltage v_o can swing all the way down to $-V_{CC}$ and all the way up to $+V_{CC}$. The output voltage v_o can then be expressed as

$$v_O = kV_{CC} \sin \omega t \tag{19-32}$$

where k is a "drive factor" in the range $0 \leq k \leq 1$. The instantaneous output current i_O can be expressed in a similar form as

$$i_O = kI_{C(\text{sat})} \sin \omega t \tag{19-33}$$

where $I_{C(\text{sat})}$ is the transistor saturation current occurring at the peak voltage V_{CC}. The peak load voltage and peak load current are related by

$$\frac{V_{CC}}{I_{C(\text{sat})}} = R_L \tag{19-34}$$

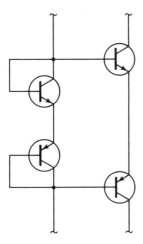

Figure 19-7 Integrated-circuit power amplifier stage with transistors used for diodes.

Figure 19-8 Class B amplifier operated with single power supply.

The rms values of the output voltage and current are $kV_{CC}/\sqrt{2}$ and $kI_{C(\text{sat})}/\sqrt{2}$, respectively. The output load power P_O can then be expressed as

$$P_O = \frac{kV_{CC}}{\sqrt{2}}\frac{kI_{C(\text{sat})}}{\sqrt{2}} = \frac{k^2 V_{CC}I_{C(\text{sat})}}{2} \tag{19-35}$$

The maximum value of this power corresponds to $k = 1$ and is denoted as $P_{O(\text{max})}$. We have

$$P_{O(\text{max})} = \frac{V_{CC}I_{C(\text{sat})}}{2} \tag{19-36}$$

The instantaneous source power p_S from *either* one of the two sources over an interval of *one-half cycle* is

$$p_S = V_{CC}i_O = kV_{CC}I_{C(\text{sat})} \sin \omega t \tag{19-37}$$

It can be shown using calculus that the average value P_S of the source power determined over a half-cycle interval, which is then the total average power due to the symmetry, is

$$P_S = \frac{2}{\pi}kV_{CC}I_{C(\text{sat})} \tag{19-38}$$

A significant point to note is that $P_S = 0$ when $k = 0$. This means that the amplifier consumes no power in the absence of a signal. (Of course, power is consumed in the bias circuit and any class A driver circuits, etc.)

The maximum value of P_S is denoted as $P_{S(\text{max})}$, and it corresponds to $k = 1$ in (19-38).

$$P_{S(\text{max})} = \frac{2}{\pi}V_{CC}I_{C(\text{sat})} \tag{19-39}$$

The efficiency η is

$$\eta = \frac{P_O}{P_S} \times 100\% = \frac{k^2 V_{CC}I_{C(\text{sat})}/2}{2kV_{CC}I_{C(\text{sat})}/\pi} \times 100\% = \frac{\pi}{4}k \times 100\% \tag{19-40}$$

The maximum efficiency η_{max} occurs at $k = 1$ and is

$$\boxed{\eta_{\text{max}} = \frac{\pi}{4} \times 100\% = 78.5\%} \tag{19-41}$$

The class B amplifier is thus theoretically capable of achieving an efficiency of 78.5% at full output. This compares with the theoretical maximum of 50% for a class A amplifier. In addition, the ideal class B amplifier consumes no power with no input signal while the class A amplifier consumes power on a continual basis.

Practical class B amplifiers, like class A amplifiers, operate at efficiencies below the theoretical maximum level.

Internal Dissipation

The average internal power dissipation P_D may be expressed as

$$P_D = P_S - P_O = \frac{2}{\pi} k V_{CC} I_{C(\text{sat})} - \frac{k^2}{2} V_{CC} I_{C(\text{sat})} \tag{19-42}$$

It can be shown using calculus that the maximum value of the power dissipation occurs when $k = 2/\pi = 0.637$. The maximum value of dissipation $P_{D(\text{max})}$ corresponding to this value of k is

$$P_{D(\text{max})} = \frac{2}{\pi^2} V_{CC} I_{C(\text{sat})} \tag{19-43}$$

Using (19-36), the maximum power dissipation may be expressed in terms of the maximum output power as

$$P_{D(\text{max})} = \frac{4}{\pi^2} P_{O(\text{max})} \approx 0.405\, P_{O(\text{max})} \tag{19-44}$$

Thus the maximum power dissipation for the circuit is about 40% of the maximum output power. This is a considerable improvement over the class A circuit.

The major results for this section, along with the results of Section 19-5, are summarized in Table 19-1.

Example 19-4

A certain class B push-pull emitter-follower power amplifier is biased by two power supplies with values ± 12 V, and it is used to drive a speaker whose dynamic resistance is $R_L = 8\ \Omega$. Based on the assumptions in this section, determine (a) $I_{C(\text{sat})}$; (b) $P_{S(\text{max})}$; (c) $P_{O(\text{max})}$; (d) $P_{D(\text{max})}$.

Solution

(a) From (19-34), the peak current $I_{C(\text{sat})}$ is determined as

$$I_{C(\text{sat})} = \frac{V_{CC}}{R_L} = \frac{12\ \text{V}}{8\ \Omega} = 1.5\ \text{A} \tag{19-45}$$

(b) The peak power delivered by the power supply is

$$P_{S(\text{max})} = \frac{2}{\pi} V_{CC} I_{C(\text{sat})} = \frac{2}{\pi} \times 12 \times 1.5 = 11.46\ \text{W} \tag{19-46}$$

(c) The maximum output power is

$$P_{O(\text{max})} = \frac{V_{CC} I_{C(\text{sat})}}{2} = \frac{12 \times 1.5}{2} = 9\ \text{W} \tag{19-47}$$

(d) The maximum internal power dissipated is

$$P_{D(\text{max})} = \frac{4}{\pi^2}P_{O(\text{max})} = \frac{4}{\pi^2} \times 9 = 3.648 \text{ W} \qquad \textbf{(19-48)}$$

19-7 CLASS C AMPLIFIER

In a class C power amplifier, conduction occurs for less than 50% of a full cycle. Because of the low conduction cycle, the use of class C amplifiers is primarily limited to narrow-band signals in which a high-Q resonant circuit can be used to restore the output signal to a full conduction cycle. Primary application of class C amplifiers is in communications circuits such as radio transmitters and radio receivers. Signals in these systems tend to have all their frequency components in a narrow portion of the frequency spectrum, and the resonant effect can be used to advantage in high-efficiency power applications.

A simplified class C amplifier employing a BJT is shown in Figure 19-9. The output of the circuit is very similar to that of a tuned amplifier as described in Section 18-4. An LC resonant circuit is employed with the Q selected for the appropriate bandwidth. However, the feature that establishes the circuit as a class C amplifier is the presence of a negative voltage $-V_{BB}$ in the base circuit. The presence of this negative voltage means that the base of the BJT is at a negative voltage with respect to ground, and conduction in the transistor can only occur whenever the input voltage exceeds this negative voltage by about 0.7 V.

In the discussion that follows, refer to the waveforms of Figure 19-10 as a supplement to the schematic diagram. For discussion purposes, assume a sinusoidal input v_i as shown in Figure 19-10(a). The transistor begins conducting when the input signal exceeds the dc base bias by about 0.7 V. Conduction continues until the base voltage drops below the cut-in voltage. The result is a sequence of collector current pulses as shown in Figure 19-10(b).

Although collector current flow consists of short pulses, the high-Q resonant

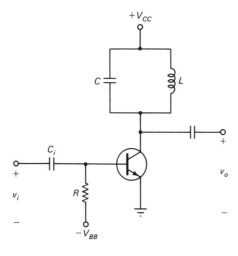

Figure 19-9 Basic form of a class C amplifier.

(a)

(b)

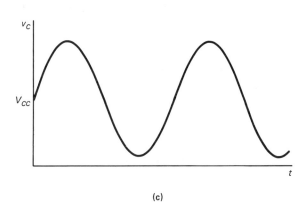

(c)

Figure 19-10 Waveforms in a class C amplifier.

circuit produces a "flywheel effect." Energy stored in the capacitor on one half-cycle is released to the inductor on the next half-cycle, and the resulting voltage across the tuned circuit is sinusoidal in form. The result is that the collector voltage v_C is centered at V_{CC} but oscillates above and below this level as shown in Figure 19-10(c). After coupling through the output capacitor, the output waveform is a sinusoid oscillating about zero voltage.

Optimum efficiency conditions occur when the transistor is driven to full saturation on the peak of the input. Under this condition, the voltage v_C varies all the way from zero to $2V_{CC}$, and the output voltage v_O is a sinusoid with peak value V_{CC}. Efficiency in this limiting case approaches 100%.

19-8 THERMAL ANALYSIS

Because of the large power levels in power amplifiers, a considerable amount of heat may have to be dissipated. Some important insight into this process may be obtained from certain thermal models and the associated analysis.

Figure 19-11 Basic form of thermal model for heat flow analysis.

Consider the model shown in Figure 19-11. Assume a physical situation in which two temperatures T_A and T_B are separated by some thermal path. Assume that $T_A > T_B$. Heat will then flow from T_A in the direction of T_B. The amount of heat flow is directly proportional to the difference in temperatures and inversely proportional to a quantity called the *thermal resistance*.

The situation may be analyzed by an analogy to Ohm's law. The quantities involved, units, and the corresponding electrical variables are summarized in Table 19-2. The form of the heat flow equation is

$$P = \frac{T_A - T_B}{\Theta_{AB}}$$ (19-49)

where Θ_{AB} is the thermal resistance between points A and B. This equation is analogous to $I = V/R$. Note that the "flow variable" for the thermal system is power, representing the amount of heat flow in watts. The units for thermal resistance are degrees Celsius per watt (°C/W).

A thermal model describing heat flow in a transistor is shown in Figure 19-12. The following quantities are defined:

T_J = transistor junction temperature

T_C = transistor case temperature

T_A = ambient or air temperature outside transistor

Θ_{JC} = thermal resistance between junction and case

Θ_{CA} = thermal resistance between case and air

Θ_{JA} = net thermal resistance between junction and air = $\Theta_{JC} + \Theta_{CA}$

P_D = power dissipated by transistor

The power P_D dissipated by the transistor is converted to heat, and this heat must be

TABLE 19-2 THERMAL QUANTITIES AND THE ASSOCIATED ELECTRICAL ANALOGIES

Quantity	Symbol	Units	Abbreviation of units	Electrical analog
Temperature	T	degrees Celsius	°C	Voltage
Heat flow	P	watts	W	Current
Thermal resistance	Θ	degrees Celsius per watt	°C/W	Resistance

Junction   Case   Ambient

Figure 19-12 Thermal model for transistor.

transferred to the outside. By the Ohm's law analogy, this power must satisfy the equation

$$P_D = \frac{T_J - T_A}{\Theta_{JA}} \tag{19-50}$$

When (19-50) is solved for T_J, the following result is obtained:

$$T_J = T_A + \Theta_{JA} P_D \tag{19-51}$$

For a given transistor, there is a maximum value permitted for the junction temperature T_J. From (19-51), for a given thermal resistance, T_J increases as either the ambient temperature T_A increases or as the power dissipation P_D increases. The specified maximum value of power dissipation $P_{D(max)}$ is normally based on $T_A = 25°C$, and it represents the value that would result in the maximum allowable junction temperature.

If $T_A > 25°C$, the only way to maintain a balance on the right-hand side of (19-51) is to reduce P_D since Θ_{JA} is fixed. This is accomplished through a derating factor, for which the concept was briefly introduced back in Chapter 4.

To keep the notation from getting too bulky, the subscript "max" in the power equation that follows will be omitted, but the power level will be understood to be the maximum value. Let $P_D(T_A = 25°C)$ represent the maximum transistor power dissipation at an ambient temperature of 25°C. The value of maximum power dissipation at a temperature T_A will be denoted as $P_D(T_A)$ and it is

$$\boxed{P_D(T_A) = P_D(T_A = 25°C) - (T_A - 25)D_{JA}} \tag{19-52}$$

where D_{JA} is the junction-to-air derating factor measured in watts/°C. Inspection of (19-52) reveals that there is a simple relationship between the thermal resistance and the derating factor, which is

$$\Theta_{JA} = \frac{1}{D_{JA}} \tag{19-53}$$

Heat Sinks

Power transistors are almost always operated with *heat sinks*. A heat sink is a quantity of metal having a relatively large surface area and high heat conductivity. It facilitates heat transfer to the outside air and allows the transistor to operate at a cooler case temperature. This results in the ability of the transistor to operate at higher values of power dissipation, as we will see shortly.

Heat Flow with Heat Sink

We now consider the effect of a heat sink on the thermal circuit. Refer to Figure 19-13. The heat sink provides a separate parallel path for power to flow to the ambient environment. If the heat sink had infinite area, the shunt path would be a short circuit and the case and ambient temperatures would be the same. In practice, however, T_C will always be larger than T_A, but a good heat sink reduces the difference significantly.

A separate specification for power transistors is usually given based on $T_C = 25°C$. The junction temperature may be expressed as a function of the case temperature as

$$T_J = T_C + \Theta_{JC} P_D \tag{19-54}$$

The maximum power dissipation at a temperature T_C may now be expressed in terms of the power dissipation when $T_C = 25°C$ and a junction-to-case derating factor D_{JC}. As in the previous case, the subscript "max" will be omitted from $P_D(T_C)$, but the value is understood to be the maximum value. We have

$$\boxed{P_D(T_C) = P_D(T_C = 25°C) - (T_C - 25)D_{JC}} \tag{19-55}$$

The thermal resistance Θ_{JC} is related to D_{JC} by

$$\Theta_{JC} = \frac{1}{D_{JC}} \tag{19-56}$$

Figure 19-13 Effect of heat sink on thermal model.

Example 19-5

The specifications of some actual power transistors will be studied in Section 19-9. The thermal specifications of these particular transistors may be summarized as follows:

Total power dissipation at $T_C = 25°C$: 30 W

Derate above 25°C: 0.24W/°C

Total power dissipation at $T_A = 25°C$ 2 W

Derate above 25°C: 0.016 W/°C

Solution

(a) The value of Θ_{JC} is

$$\Theta_{JC} = \frac{1}{D_{JC}} = \frac{1}{0.24} = 4.167°C/W \qquad (19\text{-}57)$$

The value of Θ_{JA} is

$$\Theta_{JA} = \frac{1}{D_{JA}} = \frac{1}{0.016} = 62.5°C/W \qquad (19\text{-}58)$$

From the circuit model analogy of Figure 19-12, the value of Θ_{CA} is

$$\Theta_{CA} = \Theta_{JA} - \Theta_{JC} = 62.5 - 4.167 = 58.3°C/W \qquad (19\text{-}59)$$

Note that Θ_{CA} is considerably larger than Θ_{JC}. This is a common characteristic of power transistors.

(b) For $T_C = 35°C$, the maximum allowable power dissipation is

$$P_D(T_C) = P_D(T_C = 25) - (35 - 25)D_{JC}$$
$$= 30 - 10 \times 0.24 = 27.6 \text{ W} \qquad (19\text{-}60)$$

Although this transistor is considered as a 30-W transistor, it should be noted that the allowable power dissipation will be less than 30 W unless the case temperature can be maintained at 25°C.

19-9 POWER TRANSISTOR SPECIFICATIONS

Earlier in the book, some typical specifications of small-signal transistors were studied. Now that we have studied the operation of power amplifiers, it seems appropriate to consider some typical specifications of power transistors.

A wide variety of power transistors are available, with power levels ranging from less than 1 W to thousands of watts. The transistors selected for our review are the TIP29 and TIP30 units along with several variations. The specifications are given in Appendix D, so the reader should refer to the two-page set for the discussion that follows.

The TIP29 and TIP30 are a complementary pair, with the first being an *npn* unit and the second being a *pnp* unit. Several variations of each are indicated with the suffix A, B, or C following the transistor type.

First, study the transistor package arrangement as shown. In addition to the normal emitter, base, and collector terminals, there is also a larger tab shown at the top of the transistor and internally connected to the collector. The purpose of this connection is to provide an effective means for heat transfer out of the transistor.

Refer now to the "maximum ratings." The maximum collector–emitter voltage rating with base open is indicated as V_{CEO}, and the value varies from 40 to 100 V, depending on the suffix. The value of the collector–base voltage V_{CB} is the same in each case as V_{CEO}. Strictly speaking, these values are correct as given for the *npn* units, but they must be interpreted as magnitudes for the *pnp* units, since the actual reverse-biased values are negative in that sense. The maximum reverse voltage V_{EB} between emitter and base is 5 V for all units. Again, this is the magnitude for the *pnp* units.

The continuous maximum collector current rating is 1 A, and the peak current rating is 3 A. The maximum base current is 0.4 A.

It is customary to provide two sets of power dissipation data for power transistors. One set is based on $T_C = 25°C$ and the other is based on $T_A = 25°C$. In each case, the maximum power dissipation and the derating factor are provided. To achieve the maximum power dissipation rating, it would be necessary to maintain the case temperature at 25°C, and this is difficult to achieve in practice. The power rating is then an upper bound in a practical situation. The particular data given here were utilized in Example 19-5.

Some data concerning "thermal characteristics" are provided in a separate table. The notation there differs slightly from the notation of Section 19-10, but the meanings should be clear. In fact, these particular values were determined from the derating factors in Example 19-15.

Refer now to the "electrical characteristics" data. The collector–emitter maximum voltage values indicated earlier are now repeated as "sustaining voltages" based on a pulse test. The quantity I_{CEO} represents the collector-to-emitter current with the base open, and I_{CES} represents the collector-to-emitter current with the base shorted to the emitter. The maximum value of I_{CEO} is indicated as 0.3 mA, but the maximum value of I_{CES} is 200 μA = 0.2 mA. The quantity I_{EBO} is the emitter-to-base current with the collector open. The maximum value of this quanitity is 1 mA. All of the preceding current values should be interpreted as magnitudes for the *pnp* units.

Some data concerning $h_{FE}(\beta_{dc})$ are provided in the table, but more extensive data on this parameter are provided from the curves on the second page. The typical range of h_{FE} at 25°C is around 100 to 150 or so for collector currents less than 1 A.

Returning to the table, note that the maximum value of $V_{CE(sat)}$ is 0.7 V. The maximum value of $V_{BE(on)}$ is 1.3 V, which is considerably larger than the "standard" value of 0.7 V assumed for small-signal units.

DRILL PROBLEMS

19-1. A certain power transistor has a maximum allowable power dissipation of 50 W at 25°C. The maximum allowable collector–emitter voltage is 80 V, and the maximum allowable collector current is 4 A. Determine the maximum allowable collector current **(a)** when $V_{CE} = 10$ V and **(b)** when $V_{CE} = 25$ V. Determine the maximum allowable collector–emitter voltage **(c)** when $I_C = 2$ A and **(d)** when $I_C = 0.5$ A. All calculations are to be based on 25°C.

19-2. A certain power transistor has the following specifications at 25°C: $P_{D(max)} = 10$ W, $V_{CE(max)} = 40$ V, $I_{C(max)} = 2$ A. Determine the maximum allowable collector current **(a)** when $V_{CE} = 4$ V and **(b)** when $V_{CE} = 20$ V. Determine the maximum allowable collector–emitter voltage **(c)** when $I_C = 0.2$ A and **(d)** when $I_C = 0.5$ A.

19-3. A certain idealized class A transformer-coupled power amplifier has $V_{CC} = 32$ V, $I_{CQ} = 0.5$ A, and it is used to drive a load $R_L = 4$ Ω. Based on the simplified assumptions in the text, determine **(a)** V_{CEQ}, **(b)** R'_L, **(c)** N_1/N_2, **(d)** P_S, **(e)** $P_{O(max)}$, and **(f)** $P_{D(max)}$.

19-4. Repeat the analysis of Problem 19-3 if $V_{CC} = 12$ V, $I_{CQ} = 2$ A, and $R_L = 96$ Ω.

19-5. A certain class B push-pull emitter-follower power amplifier is biased by two power supplies with values ± 16 V, and it is used to drive a load $R_L = 4$ Ω. Based on the assumptions of this section, determine **(a)** $I_{C(sat)}$, **(b)** $P_{S(max)}$, **(c)** $P_{O(max)}$, and **(d)** $P_{D(max)}$.

19-6. Repeat the analysis of Problem 19-5 if the two power supply voltages are ± 6 V and $R_L = 96$ Ω.

19-7. A certain power transistor has the following specifications:

> Total power dissipation at $T_C = 25$°C: 50 W
> Derate above 25°C: 0.4 W/°C
> Total power dissipation at $T_A = 25$°C: 4 W
> Derate above 25°C: 20 mW/°C

(a) Determine Θ_{JC}, Θ_{JA}, and Θ_{CA}. **(b)** If the transistor is used without a heat sink and the ambient temperature is 30°C, determine the maximum allowable power dissipation. **(c)** If the transistor is used with a heat sink and the case temperature is 40°C, determine the maximum allowable power dissipation.

19-8. A certain power transistor has the following specifications:

> Total power dissipation at $T_C = 25$°C: 10 W
> Derate above 25°C: 0.1 W/°C
> Total power dissipation at $T_A = 25$°C: 1.5 W
> Derate above 25°C: 10 mW/°C

(a) Determine Θ_{JC}, Θ_{JA}, and Θ_{CA}. **(b)** If the transistor is used without a heat sink and $T_A = 32°C$, determine the maximum allowable power dissipation. **(c)** If the transistor is used with a heat sink and $T_C = 38°C$, determine the maximum allowable power dissipation.

DERIVATION PROBLEMS

19-9. (Requires calculus.) Consider the expression for the source power p_S as given by (19-37). Determine the average value of this function over the interval from $t = 0$ to $t = T/2$ and show that the result is given by (19-38).

19-10. (Requires calculus.) Consider the expression for the internal power dissipation P_D as given by (19-42). **(a)** Show that the maximum value occurs for $k = 2/\pi$. **(b)** Show that the maximum value of P_D is the quantity of (19-44).

DESIGN PROBLEMS

19-11. Some preliminary design on a class B push-pull emitter-follower power amplifier is to be performed. The desired maximum output power is 4 W into a resistive load of 8 Ω. Based on a sinusoidal waveform, determine **(a)** the minimum value of the power dissipation rating of the transistors to be selected, **(b)** the maximum value of collector current, and **(c)** the maximum value of power delivered by the power supplies.

19-12. Repeat Problem 19-11 if the desired maximum output power is 25 W into a resistive load of 10 Ω.

TROUBLESHOOTING PROBLEMS

19-13. You are troubleshooting a push-pull amplifier of the type shown in Figure 19-6. Some dc voltages are measured as follows:

> Positive power supply: $+15$ V
> Negative power supply: -15 V
> Base of T1: 0.7 V
> Base of T2: -0.7 V

With a sine wave applied to the input, the output is a half-wave rectified version of the input (i.e., the negative segments are missing). Which one of the following is the most likely trouble? **(a)** T1 defective; **(b)** T2 defective; **(c)** upper bias resistor open; **(d)** lower bias resistor open; **(e)** lower bias resistor shorted.

19-14. You are troubleshooting a class C amplifier of the type shown in Figure 19-9. All dc voltages appear to be normal, and an oscilloscopic display of v_i appears normal. However, the transistor collector voltage, while periodic, is not sinusoidal. Which *two* of the following could represent a possible trouble or troubles? **(a)** Defective transistor; **(b)** insufficient negative bias; **(c)** incorrect value of C; **(d)** incorrect value of L; **(e)** R shorted; **(e)** incorrect value of C_i.

COMPUTER PROBLEMS

19-15. Write a program to perform an analysis of an idealized class A transformer-coupled power amplifier.

Input data: V_{CC}, I_{CQ}, R_L
Output data: R_L', N_1/N_2 (optimum value), P_S, $P_{O(max)}$, $P_{D(max)}$

19-16. Write a program to perform an analysis of an idealized class B push-pull emitter-follower power amplifier with power supplies $\pm V_{CC}$.

Input data: V_{CC}, R_L
Output data: $I_{C(sat)}$, $P_{S(max)}$, $P_{O(max)}$, $P_{D(max)}$

CHAPTER 20

Power Supplies

OBJECTIVES

After completing this chapter, the reader should be able to:

1. Apply the voltage and current relationships for an ideal transformer.
2. Draw the schematic diagram for and analyze the operation of a half-wave rectifier.
3. Draw the schematic diagram for and anayze the operation of a full-wave rectifier with a center-tapped transformer.
4. Draw the schematic diagram for and analyze the operation of a full-wave bridge rectifier.
5. Discuss the need for and analyze the operation of a ripple filter in a power supply.
6. Discuss the importance of the peak reverse voltage for a diode, and determine the value for each of the rectifier circuits studied.
7. Explain the operation of an IC regulator circuit.
8. Discuss the operation of a silicon-control rectifier (SCR), a diac, and a triac.

20-1 OVERVIEW

As we have seen throughout the text, dc voltages of various levels are required to properly bias various electronic devices such as transistors and op-amps. However, the commercial power distribution system is an ac system. The commercial power system in the United States has as its standard level close to 120 V rms at a frequency of 60 Hz. This voltage level must be changed and converted to dc before it can be used for electronic device operation.

A circuit that provides the required level change and converts the ac to dc is called a *power supply*. Virtually all electronic equipment operated from the ac line (e.g., radios, television sets, VCRs, etc.) has one or more power supply circuits.

A block diagram displaying the major components of a power supply is given in Figure 20-1. The circuit shows a grounded plug, a switch, and a fuse, all of which

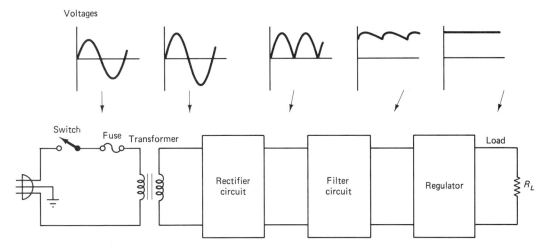

Figure 20-1 Block diagram of complete power supply circuit.

are essential for safety. To simplify subsequent diagrams, these components will be omitted, but they should be considered as requirements in a practical power supply.

The transformer is used to change the level of the voltage. Depending on the final power supply requirements, the voltage level may be increased or decreased. Note the ac waveforms shown above the transformer on both sides.

The *rectifier circuit* is used to convert the alternating transformer output voltage to a voltage having only one polarity. However, this voltage will still vary considerably over the range of a cycle as shown. This widely varying voltage of one polarity is called *pulsating dc*.

The *filter circuit* provides a substantial smoothing effect on the pulsating dc. The result is a voltage having a relatively small variation as shown. This variation is called *ripple*.

Most modern power supplies employ a *regulator*. The regulator provides further smoothing on the voltage, and it establishes and maintains a predetermined output voltage level. This last process is quite important since there are many causes of possible output voltage variation in a typical power supply.

Much older equipment designed before the advent of the modern regulator circuit is still in service. In those units, the filter circuit was designed to provide all the smoothing required. Consequently, filter circuits in such older equipment were much more elaborate than is required with modern regulator circuits.

20-2 IDEAL TRANSFORMERS

At several earlier points in the text, transformers have appeared in certain amplifier circuits. A brief discussion of the property utilized was given in each situation, but a lengthy discussion of transformer properties was not provided. However, before considering the operation of power supplies, it is necessary to take a more detailed look at the properties of transformers.

A transformer is a circuit consisting of two or more coils wound on a common magnetic core. The coils are usually electrically isolated from each other, but the magnetic fields are coupled together. Because of the magnetic coupling, energy may be transferred between the coils through the magnetic field. In the process, the voltage may be increased ("stepped up") or decreased ("stepped down"). The coils of the transformers may be tightly coupled or loosely coupled depending on the application. Virtually all power supply transformers are tightly coupled. Tight coupling is achieved through a ferromagnetic core and close windings.

Tightly coupled transformers such as used in power supplies may be closely approximated by the *ideal transformer* model. The schematic symbol for an ideal transformer is shown in Figure 20-2. The coil on the left is called the *primary winding,* and the coil on the right is called the *secondary winding*. The primary winding is the one normally connected to the ac line, and the secondary winding is the one from which the working voltage for the power supply is obtained.

The voltage–current relationships for the ideal transformer to be given here apply to either instantaneous values, peak values, or rms values provided that *all* variables are expressed in the *same forms*. For the moment, we leave the exact forms unspecified. The primary voltage and current are then denoted simply as v_1 and i_1, and the secondary voltage and current are denoted as v_2 and i_2 in Figure 20-2. Obviously, the transformer must be connected to other parts of a circuit for all the variables to be present.

The ideal transformer is characterized by a *turns ratio,* denoted as $N_1 : N_2$ on Figure 20-2. Ideally, N_1 is the number of primary turns and N_2 is the number of secondary turns. In a practical transformer, it is usually impossible to determine the actual number of turns, so it is the *ratio* that is important. For example, assume that $N_1 = 500$ and $N_2 = 2000$. The turns ratio of such a transformer would be $N_1/N_2 = 1 : 4$. In actual practice, transformer specifications are often stated in terms of the intended working voltages. For example, a transformer might be specified as operating at 120 V primary/24 V secondary. In this case the turns ratio could be assumed to be $N_1/N_2 = 120/24 = 5/1$ and $N_1 : N_2 = 5 : 1$.

The voltage–current relationships of the ideal transformer are as follows:

$$\boxed{\frac{v_2}{v_1} = \frac{N_2}{N_1}} \qquad \textbf{(20-1)}$$

and

$$\boxed{\frac{i_2}{i_1} = \frac{N_1}{N_2}} \qquad \textbf{(20-2)}$$

Figure 20-2 Schematic symbol for ideal transformer with primary and secondary labeled.

Thus the voltage changes in direct proportion to the turns ratio, while the current changes in reverse proportion to the turns ratio. This means a "stepup" from primary to secondary requires that $N_2 > N_1$, while a "stepdown" requires that $N_2 < N_1$.

If (20-1) and (20-2) are multiplied together, we obtain

$$\frac{v_2 i_2}{v_1 i_1} = \frac{N_2}{N_1} \frac{N_1}{N_2} = 1 \qquad \textbf{(20-3)}$$

Equation (20-3) simply tells us that the total power is not changed by the transformer. If the voltage is increased, the current is decreased, but the product remains the same. In an actual transformer, some power is always dissipated internally, so the output power is less than the input power.

Example 20-1

An ideal transformer has a turns ratio of 4:1 from primary to secondary. The primary voltage is the ac line voltage whose rms value will be assumed to be 120 V. For a particular secondary load, the primary rms current is 2 A. Determine the secondary rms voltage and current.

Solution

Since the primary variables are specified as rms values, the corresponding secondary values will also be determined as rms values. By convention, uppercase symbols will be used for rms values.

The voltages are related by

$$\frac{V_2}{V_1} = \frac{N_2}{N_1} = \frac{1}{4} \qquad \textbf{(20-4)}$$

from which we obtain

$$V_2 = \frac{1}{4} V_1 = \frac{1}{4} \times 120 = 30 \text{ V rms} \qquad \textbf{(20-5)}$$

The currents are related by

$$\frac{I_2}{I_1} = \frac{N_1}{N_2} = 4 \qquad \textbf{(20-6)}$$

from which we determine that

$$I_2 = 4I_1 = 4 \times 2 = 8 \text{ A rms} \qquad \textbf{(20-7)}$$

Thus, when the voltage is decreased from 120 V to 30 V, the current is increased from 2 A to 8 A.

In working with small-signal diodes early in the book, a forward drop of 0.7 V was always assumed. Power supply levels vary considerably, and for large currents, typical of many power supplies, diode drops may be somewhat greater. In addition, transformer losses will also be present, and these are difficult to predict exactly.

In view of more uncertain diode and transformer voltage losses, a simplified approach will be taken in the analysis of this rectifier circuit and others to follow. The simplified approach will be to neglect all voltage drops in the discussion of the basic operation of the circuit, and this should allow the reader to focus on the principle of operation without undue "clutter" resulting from losses and unwieldy notation. Bear in mind, however, that the actual output voltages will be somewhat smaller than the values predicted from the idealized assumptions, and it will be necessary to allow for this in a practical application.

The simplest rectifier power supply circuit is the unfiltered half-wave rectifier circuit, whose basic form is shown in Figure 20-3. Let $V_{1(rms)}$ represent the rms value of the primary voltage v_1, and let $V_{2(rms)}$ represent the rms value of the secondary voltage v_2. These voltages are related by

$$V_{2(rms)} = \frac{N_2}{N_1} V_{1(rms)} \tag{20-8}$$

Let V_{2p} represent the peak value of the secondary voltage, which is related to the secondary rms voltage by

$$V_{2p} = \sqrt{2} \, V_{2(rms)} \tag{20-9}$$

Refer to the waveforms of Figure 20-4 in the discussion that follows. The voltage v_2 is shown as (a).

When the secondary voltage v_2 is positive at the top of the transformer secondary winding with respect to the bottom, the diode will conduct. However, when the secondary voltage has the opposite polarity, the diode will be open, and no conduction will occur. The load voltage v_L thus has the form of Figure 20-4(b). If there were no losses, the peak load voltage *would be* V_{2p}.

The load voltage as shown is pulsating dc. It is certainly not pure dc, but it does have a dc component. It has only limited value as shown, but there are some industrial and commercial processes for which pulsating dc is used. Let $V_{L(dc)}$ represent the dc value of this waveform. It can be shown with calculus that this voltage has the value

$$\boxed{V_{L(dc)} = \frac{V_{2p}}{\pi} = 0.318 V_{2p}} \tag{20-10}$$

In practice, the value will always be smaller than (20-10) due to transformer and diode losses.

For a resistive load R_L, the dc load current $I_{L(dc)}$ is

$$I_{L(dc)} = \frac{V_{L(dc)}}{R_L} \tag{20-11}$$

Figure 20-3 Schematic diagram of basic half-wave rectifier circuit.

(a)

(b)

Figure 20-4 Waveforms of half-wave rectifier circuit.

Example 20-2

A certain unfiltered half-wave rectifier circuit of the form shown in Figure 20-3 is powered by the 120-V rms ac power system, and the turns ratio of the transformer is $N_1 : N_2 = 6 : 1$. Determine **(a)** the rms secondary voltage, **(b)** the peak secondary voltage, **(c)** the dc load voltage, and **(d)** the dc load current if $R_L = 5\ \Omega$. Neglect any diode and transformer losses.

Solution

(a) The rms secondary voltage $V_{2(\text{rms})}$ is

$$V_{2(\text{rms})} = \frac{N_2}{N_1} V_{1(\text{rms})} = \frac{1}{6} \times 120 = 20\ \text{V} \qquad (20\text{-}12)$$

(b) The peak secondary voltage V_{2p} is

$$V_{2p} = \sqrt{2}\ V_{2(\text{rms})} = \sqrt{2} \times 20 = 28.28\ \text{V} \qquad (20\text{-}13)$$

(c) The dc load voltage $V_{L(\text{dc})}$ is

$$V_{L(\text{dc})} = \frac{V_{2p}}{\pi} = \frac{28.28}{\pi} = 9.00\ \text{V} \qquad (20\text{-}14)$$

(d) The dc load current $I_{L(\text{dc})}$ is

$$I_{L(\text{dc})} = \frac{V_{L(\text{dc})}}{R_L} = \frac{9.00}{5} = 1.8 \text{ A} \tag{20-15}$$

20-4 FULL-WAVE RECTIFIER (UNFILTERED)

The basic form of a full-wave rectifier circuit is shown in Figure 20-5. The secondary winding of the transformer is center tapped in this case. As will be seen shortly, only one-half of the secondary will conduct at a given time. For convenience, the total number of secondary turns will be denoted as $2N_2$, with N_2 representing the number of turns on half of the secondary.

Refer to Figures 20-6 and 20-7 in the discussion that follows. Let v_2 represent

Figure 20-5 Schematic diagram of basic full-wave rectifier circuit.

(a)

(b)

Figure 20-6 Manner in which conduction occurs on alternate half-cycles in full-wave rectifier circuit.

(a)

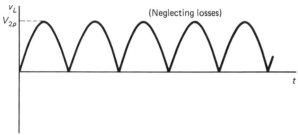

(b)

Figure 20-7 Waveforms in full-wave rectifier circuit.

the voltage across the upper half of the secondary. The rms value of this voltage is

$$V_{2(\text{rms})} = \frac{N_2}{N_1} V_{1(\text{rms})} \tag{20-16}$$

The peak value is

$$V_{2p} = \sqrt{2}\ V_{2(\text{rms})} \tag{20-17}$$

When v_2 is positive as defined, the upper diode D1 conducts and the lower diode D2 is open as shown in Figure 20-6(a). The first half-cycle of v_2 as shown in Figure 20-7(a) is then transferred to the load, and the load current returns through ground and the center tap of of the transformer.

When v_2 as defined is negative, the voltage on the lower half of the transformer is positive at the bottom with respect to the center tap. Under this condition, D2 conducts and D1 is open. From Figure 20-6(b), note that the direction of this voltage is still positive at the top of the load with respect to ground as shown. Current again flows to ground and back to the center tap of the transformer.

The form of v_L is shown in Figure 20-7(b). Once again, rectification has taken place, but now it is *full-wave rectification*.

As in the case of the half-wave rectifier, this voltage is pulsating dc. However, its dc value $V_{L(\text{dc})}$ has twice the value as for the half-wave rectifier and is

$$\boxed{V_{L(\text{dc})} = \frac{2}{\pi} V_{2p} = 0.637 V_{2p}} \tag{20-18}$$

Example 20-3

A certain unfiltered full-wave rectifier circuit of the form of Figure 20-5 is powered by the 120-V rms ac power system, and the turns ratio of primary to *one-half* of the secondary is $N_1 : N_2 = 6 : 1$. (Equivalently, $N_1 : 2N_2 = 3 : 1$.) Determine **(a)** the rms secondary voltage across one-half of the transformer, **(b)** the peak secondary voltage, **(c)** the dc load voltage, and **(d)** the dc load current if $R_L = 5 \ \Omega$. Neglect any diode and transformer losses.

9-13-88 AC2

(a) The rms voltage across one-half the secondary is

$$V_{2(\text{rms})} = \frac{N_2}{N_1} V_{1(\text{rms})} = \frac{1}{6} \times 120 = 20 \text{ V} \qquad (20\text{-}19)$$

(b) The peak voltage across one-half of the secondary is

$$V_{2p} = \sqrt{2} \ V_{2(\text{rms})} = \sqrt{2} \times 20 = 28.28 \text{ V} \qquad (20\text{-}20)$$

(c)

$$V_{L(\text{dc})} = \frac{2}{\pi} V_{2p} = \frac{2}{\pi} \times 28.28 = 18.00 \text{ V} \qquad (20\text{-}21)$$

(d) The dc load current is

$$I_{L(\text{dc})} = \frac{V_{L(\text{dc})}}{R_L} = \frac{18.00}{5} = 3.6 \text{ A} \qquad (20\text{-}22)$$

20-5 BRIDGE RECTIFIER (UNFILTERED)

Another type of full-wave rectifier circuit is the so-called bridge rectifier, whose basic form is shown in Figure 20-8. Although four diodes are required, the circuit has the advantage that no center tap is required on the transformer, and the entire secondary voltage is available for rectification. Another advantage is the lower peak reverse voltage requirement, and this will be discussed in Section 20-7. Because of the popularity of the circuit, diode bridge packages containing the four required diodes are available as stock items.

Operation of the circuit can be explained with the diagram of Figure 20-9. Assume first that the top of the transformer is positive with respect to the bottom. As shown in Figure 20-9(a), diodes D1 and D3 are forward biased, while D2 and D4 are reverse biased. Current flows out of the top of the transformer through D1, through the load to ground, up from ground through D3, and back to the lower terminal of the transformer.

Assume next that the bottom of the transformer secondary is positive with respect to the top as shown in Figure 20-9(b). Diodes D2 and D4 are now forward biased, while diodes D1 and D3 are reverse biased. Current flows out of the bottom of

Power Supplies   Chap. 20

Figure 20-8 Schematic diagram of full-wave bridge rectifier circuit.

(a)

(b)

Figure 20-9 Manner in which conduction occurs on alternate half-cycles in bridge rectifier circuit.

the transformer through D2, through the load to ground, up from ground through D4, and back to the upper terminal of the transformer.

As can be seen, current flows through the load in the same direction on both half-cycles. The waveforms in this case are essentially the same as in Figure 20-7.

The various relationships for the bridge circuit will now be given. The secondary rms voltage is

$$V_{2(\text{rms})} = \frac{N_2}{N_1} V_{1(\text{rms})} \tag{20-23}$$

The peak secondary voltage is

$$V_{2p} = \sqrt{2} \, V_{2(\text{rms})} \qquad \text{(20-24)}$$

The dc load voltage is

$$\boxed{V_{L(\text{dc})} = \frac{2}{\pi} = 0.637 V_{2p}} \qquad \text{(20-25)}$$

It should be noted that when losses are considered, there are *two* forward-biased diode drops on each half-cycle to be considered in the bridge rectifier.

Example 20-4

A certain unfiltered bridge rectifier circuit of the form of Figure 20-8 is powered by the 120-V rms ac power system, and the turns ratio is $N_1 : N_2 = 6 : 1$. Determine **(a)** the rms secondary voltage, **(b)** the peak secondary voltage, **(c)** the dc load voltage, and **(d)** the dc load current if $R_L = 5 \, \Omega$. Neglect any diode and transformer losses.

Solution

(a) The rms secondary voltage is

$$V_{2(\text{rms})} = \frac{1}{6} \times 120 = 20 \text{ V} \qquad \text{(20-26)}$$

(b) The peak secondary voltage is

$$V_{2p} = \sqrt{2} \times 20 = 28.28 \text{ V} \qquad \text{(20-27)}$$

(c) The dc load voltage is

$$V_{L(\text{dc})} = \frac{2}{\pi} \times 28.28 = 18.00 \text{ V} \qquad \text{(20-28)}$$

(d) The dc load current is

$$I_{L(\text{dc})} = \frac{18.00}{5} = 3.6 \text{ A} \qquad \text{(20-29)}$$

The reader has likely observed a great deal of redundancy between Examples 20-2, 20-3, and 20-4 because of the similarities. However, the values should help the reader in understanding the differences between the different circuit forms.

20-6 FILTER CIRCUITS

The rectifier circuits considered thus far convert ac to pulsating dc. As noted earlier, pulsating dc has only limited value in practice. The large output variations in pulsating dc can be converted to very small variations by means of a *filter circuit*. The resulting small variations are referred to as *ripple*. The action of filter circuits will be explained qualitatively in this section and a detailed quantitative analysis will be given in Section 20-8.

Capacitor Filter

Since modern power supplies use IC regulators, which provide additional smoothing, a very widely used power supply filter is a *single-capacitor filter* shown in Figure 20-10. The resistance R_L represents any load. This could be an actual load or an equivalent load seen from a regulator input. The particular rectifier circuit shown in this case is the basic full-wave rectifier.

Refer to the waveforms of Figure 20-11. The dashed curve in (a) shows the pulsating dc that *would* appear across the load *if* the capacitor were not present. The

Figure 20-10   Full-wave rectifier with capacitor filter.

(a)

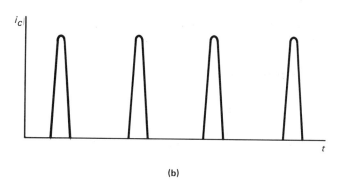

(b)

Figure 20-11   Waveforms of full-wave rectifier with capacitor filter.

capacitor, however, charges to the peak value of this voltage very soon after the circuit is energized. If there were no load present, this voltage would remain at the peak value.

The presence of a load causes the voltage on the capacitor to decay slowly over the interval of a half-cycle for a full-wave rectifier (or a full-cycle in the case of a half-wave rectifier). The decay has been exaggerated here for clarity.

At some point in each half-cycle, the rectified voltage exceeds the decaying voltage v_L. At that point, and depending on the appropriate half-cycle, one of the two diodes turns on again, and the capacitor is recharged quickly to the peak voltage. The form of the charging current is shown in Figure 20-11(b). The peak value of this charging current is rather large and is an important factor in selecting the diode ratings for a power supply.

Other Filters

As noted earlier, the use of modern IC regulators has reduced the requirements for filtering with reactive elements. However, in special designs and in earlier equipment, more elaborate filters are used.

One filter that found widespread usage in earlier equipment is the *inductive-input LC filter,* for which a single section is shown in Figure 20-12. Additional sections could be added for further filtering.

A second earlier popular filter was the *capacitive input LC filter,* for which a single section is shown in Figure 20-13. As in the previous case, additional sections may be added.

Another filter was the *RC filter,* for which a typical circuit form is shown in Figure 20-14. In a sense, the single capacitor filter is a special case of the *RC* filter in which there is one capacitance, and the load serves as the resistance.

Figure 20-12   Single section of an inductive-input *LC* filter.

Figure 20-13   Single section of a capacitive-input *LC* filter.

Figure 20-14   Typical *RC* filter.

A very important rating that must be considered in all power supply rectifier circuits is the *peak reverse voltage* (PRV) rating of the diodes. The peak reverse voltage of a diode is the peak voltage across the diode when it is reverse biased. As we will see shortly, this voltage may be larger than any of the obvious circuit voltages. Each of the major circuits previously considered will now be analyzed to determine the peak reverse voltage.

Half-Wave Rectifiers

First consider the unfiltered half-wave rectifier circuit of Figure 20-15(a). Operation is assumed to be at the point in the cycle where the diode is reverse biased ("off") and the secondary voltage is at a peak negative value $-V_{2p}$. Since no current is flowing in the load at this point, there is no voltage across the resistor, and the value of the peak reverse voltage V_{pr} is simply

$$\boxed{V_{pr} = V_{2p}} \qquad\qquad \textbf{(20-30)}$$

Thus, with no filter, *the peak reverse diode voltage of a half-wave rectifier is equal to the peak voltage across the transformer secondary.*

Next, consider the case where there is a capacitor filter as shown in Figure 20-15(b). The voltage across the capacitor is the load voltage V_L. A direct application of KVL to the loop yields

$$V_{pr} = V_{2p} + V_L \qquad\qquad \textbf{(20-31)}$$

(a)

(b)

Figure 20-15 Analysis of diode peak reverse voltage in two half-wave rectifier circuits.

Although V_L varies slightly during a cycle due to the discharge, for very low ripple and for the purpose of this analysis, it will be assumed that $V_L = V_{2p}$. Thus the maximum peak reverse voltage is simply

$$V_{pr} = 2V_{2p} = 2V_L \qquad (20\text{-}32)$$

The presence of the capacitor thus results in a peak reverse voltage across the diode equal to twice the dc output voltage.

Full-Wave Rectifier

Consider now the full-wave rectifier with no filter of Figure 20-16. Operation is assumed to be at the point in the cycle where D1 is reverse biased, D2 is forward biased, and the secondary voltage across each transformer half is V_{2p} with the polarities shown. Neglecting the small forward drop across D2, a KVL loop will be formed around the secondary. We have

$$V_{2p} + V_{2p} - V_{pr} = 0 \qquad (20\text{-}33)$$

or

$$\boxed{V_{pr} = 2V_{2p}} \qquad (20\text{-}34)$$

A similar analysis for the opposite half-cycle results in the same reverse voltage for the other diode.

Before commenting on this result, it can readily be demonstrated that the presence of a filter across the load does not affect the value in this case. *Thus the peak reverse voltage of a full-wave rectifier is twice the peak voltage across each half or, equivalently, the voltage across the entire transformer secondary.*

Bridge Rectifier

Finally, consider the bridge rectifier circuit of Figure 20-17. Operation is assumed to be at the point in the cycle where D1 and D3 are reverse biased and D2 and D4 are forward biased. The secondary voltage is at a peak V_{2p} with the positive terminal at the bottom.

Two reverse voltages, V_{pr1} and V_{pr2}, must be considered in this case. A KVL loop through the transformer, down through D1 and D2, and back to the transformer

Figure 20-16 Analysis of diode peak reverse voltage in full-wave rectifier circuit.

Figure 20-17 Analysis of diode peak reverse voltage in bridge rectifier circuit.

will be formed to determine V_{pr1}. Neglecting the small forward drop of D2, this loop equation reads

$$V_{2p} - V_{pr1} = 0 \tag{20-35}$$

or

$$V_{pr1} = V_{2p} \tag{20-36}$$

From the symmetry of the circuit, it can be readily inferred that

$$\boxed{V_{pr2} = V_{pr1} = V_{2p}} \tag{20-37}$$

A similar analysis for the opposite half-cycle results in the same reverse voltage for all four diodes.

As in the case of the basic full-wave rectifier, the presence of a filter on the load does not affect the value in this case. *Thus the peak reverse voltage of a bridge rectifier is simply the peak voltage across the secondary of the transformer.*

Let us now compare the bridge rectifier with the basic full-wave rectifier circuit. *For a given dc output voltage, the peak reverse voltage rating of the diodes in a full-wave rectifier must be twice as large as for a bridge rectifier.* This fact provides another distinct advantage of a bridge rectifier.

The various peak reverse voltage relationships developed in this section are summarized in Table 20-1.

TABLE 20-1 SUMMARY OF PEAK REVERSE VOLTAGES FOR DIFFERENT RECTIFIER CONFIGURATIONS

Rectifier type	Diode peak reverse voltage[a]
Half-wave (unfiltered)	V_{2p}
Half-wave (capacitor filter)	$2V_{2p}$
Full-wave rectifier with secondary center tap (with or without filtering)	$2V_{2p}$
Full-wave bridge rectifier (with or without filtering)	V_{2p}

[a]V_{2p} is the peak voltage across the complete tranformer secondary for half-wave rectifiers and the full-wave bridge rectifier. It is the voltage across half of the transformer secondary for the full-wave rectifier with center tap.

Example 20-5

Determine the peak reverse voltages for the diodes in each of the four following rectifier circuits: **(a)** unfiltered half-wave rectifier with $V_{2p} = 30$ V; **(b)** filtered half-wave rectifier with $V_L = 30$ V; **(c)** filtered full-wave rectifier (using a center-tapped transformer) with $V_L = 30$ V; **(d)** filtered bridge rectifier with $V_L = 30$ V.

Solution

The relationships of this section are readily applied, and the following short table can be constructed.

	Circuit	V_{pr} (V)
(a)	Unfiltered half-wave rectifier	30
(b)	Filtered half-wave rectifier	60
(c)	Filtered full-wave rectifier	60
(d)	Filtered bridge rectifier	30

20-8 ANALYSIS OF RIPPLE

As discussed earlier, the output of a filter circuit does contain some ripple. We will now investigate the nature of this ripple to determine how it relates to various circuit parameters. The analysis will be limited to the case of a single capacitor across a resistive load since this is the most important case.

An enlarged view of the ripple is shown in Figure 20-18(a). The exact form of the discharge for a constant resistance load is an exponential function. In practice, however, the actual ripple level of realistic filters is so low that some simplifying approximations may be made.

The following approximations have been found to greatly simplify the analysis of the ripple without creating significant errors (assuming that the ripple level is relatively low): (1) The ripple waveform is approximated by a triangular waveform as shown in Figure 20-18(b). (2) The load current is assumed to be constant for the purpose of calculating the change in voltage across the capacitor. This change in voltage may then be used to compute the ripple level and the actual dc output. (Note that the actual dc level is at the midpoint of the triangular waveform.) (3) The short charging time at the beginning of a half-cycle will be neglected in comparison with the discharge time.

A point to note from the waveforms is that the period T of the ripple for a full-wave rectifier is one-half the period of the ac line voltage. In contrast, the period of the ripple for a half-wave rectifier is the actual period of the ac line voltage. In order not to derive two sets of formulas, we will employ the symbol $f = 1/T$ as the fre-

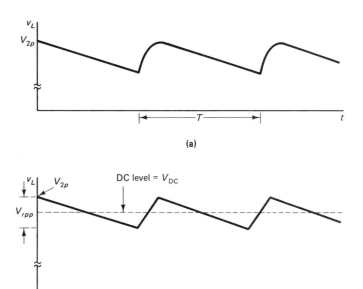

Figure 20-18 Waveforms used in analysis of ripple.

quency of the ripple. Thus we have the following relationships for the commercial 60-Hz power system:

$$f = 60 \text{ Hz (half-wave)} \qquad \textbf{(20-38a)}$$

$$= 120 \text{ Hz (full-wave)} \qquad \textbf{(20-38b)}$$

First, a fundamental circuit principle must be established. Assume that a capacitor C is being charged from a constant current source I. Let Δv represent the change in voltage across the capacitor in a time Δt. It is established in basic circuit theory that

$$\Delta v = \frac{1}{C} I \, \Delta t \qquad \textbf{(20-39)}$$

Let V_{rpp} represent the peak-to-peak ripple voltage and let I_L represent the dc load current at the output of the filter. Referring to Figure 20-18(b) and applying (20-39), we have

$$\boxed{V_{rpp} = \frac{I_L T}{C} = \frac{I_L}{fC}} \qquad \textbf{(20-40)}$$

From (20-40), the following points are readily deduced: (1) The ripple level increases with an increase in the load current and (2) the ripple level decreases with increasing frequency or with increasing capacitance.

From the frequency dependency, it should be clear that a full-wave rectifier is easier to filter than a half-wave rectifier since its frequency is twice as great. Generally, the full-wave rectifier is almost always preferred over the half-wave rectifier.

Since the ripple decreases with increased capacitance, one might be tempted to suggest that the capacitance could be increased to any size required for a given ripple level. In practice, however, there is an important tradeoff that must be considered. As the capacitance increases, the initial surge of charging current when the power supply is first turned on increases. This surge is limited only by diode and transformer resistance and can be very large. In some circuits, a surge resistance is placed in the circuit to limit the value of this current.

The actual dc output voltage V_{dc} can be determined by locating the midpoint of the ripple. For a peak secondary voltage V_{2p}, the dc output can be expressed as

$$V_{dc} = V_{2p} - \frac{V_{rpp}}{2} = V_{2p} - \frac{I_L}{2fC} \qquad \text{(20-41)}$$

Ripple Factor

Frequently, the rms ripple voltage is used to express the ripple level. Let V_r represent this value. It can be shown that this rms value is related to the peak-to-peak ripple by

$$V_r = \frac{V_{rpp}}{2\sqrt{3}} = \frac{I_L}{2\sqrt{3}\,fC} \qquad \text{(20-42)}$$

The ripple factor r is defined as

$$r = \frac{\text{rms ripple voltage}}{\text{dc output voltage}} \qquad \text{(20-43a)}$$

$$= \frac{V_r}{V_{dc}} \qquad \text{(20-43b)}$$

Example 20-6

A certain full-wave power supply with a single capacitor filter of the type shown in Figure 20-10 is powered by the 120-V rms, 60-Hz power system. The peak value of the output voltage under loaded conditions is 30 V. (The peak transformer output voltage across half the secondary is somewhat greater to compensate for losses.) The capacitance is $C = 2000\ \mu\text{F}$. For a load current of 2 A, determine **(a)** the peak-to-peak ripple voltage, **(b)** the rms ripple voltage, **(c)** the dc load voltage, and **(d)** the ripple factor.

Solution

(a) The peak-to-peak ripple voltage is

$$V_{rpp} = \frac{I_L}{fC} = \frac{2}{120 \times 2000 \times 10^{-6}} = 8.33 \text{ V} \qquad \textbf{(20-44)}$$

(b) The rms ripple voltage is

$$V_r = \frac{V_{rpp}}{2\sqrt{3}} = \frac{8.33}{2\sqrt{3}} = 2.41 \text{ V} \qquad \textbf{(20-45)}$$

(c) The dc output voltage is

$$V_{sc} = V_{2p} - \frac{V_{rpp}}{2} = 30 - \frac{8.33}{2} = 25.83 \text{ V} \qquad \textbf{(20-46)}$$

(d) The ripple factor is

$$r = \frac{V_r}{V_{dc}} = \frac{2.41}{25.83} = 0.093 \qquad \textbf{(20-47)}$$

20-9 REGULATED POWER SUPPLIES

The various power supply circuits considered thus far all suffer from the fact that the dc output voltage is not regulated. At the front end, the commercial ac supply voltage is subject to a wide variation, and this variation is in turn transferred to the output dc voltage. Coupled with this are transformer losses and diode drops. As load requirements change, voltage variations also occur. Consequently, these basic power supply circuits are used alone only in applications where the expected variation of the output dc voltage is not critical.

The modern approach is to use a voltage regulator circuit to establish a nearly precision dc output voltage. A *voltage regulator* is an electronic control circuit capable of establishing and maintaining a nearly constant dc output voltage as load requirements vary and as the input voltage varies.

For moderate degrees of regulation and where variations are not too great, zener diodes can be used directly for regulation. The approach for analyzing and designing zener circuits was considered back in Chapter 3, and that material is applicable to the present discussion. However, the approach here will be to focus on the more comprehensive voltage regulator circuit. As a part of such a circuit, zener diodes are often used. However, the variation of the current in the zener will be far less than expected for the complete power supply, and zeners can be used to establish references within regulator circuits.

Basic Regulator Circuit

The operation of a basic regulator circuit implemented with an op-amp is illustrated in Figure 20-19. The op-amp is connected in the noninverting configuration with the feedback resistor connected to the dc output voltage. A zener reference voltage

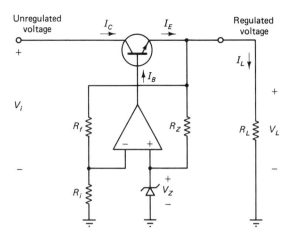

Unregulated voltage

Regulated voltage

V_i

I_C I_E

I_B

I_L

R_f R_z R_L V_L

R_i V_z

+

Figure 20-19 Voltage regulator implemented with an op-amp and a pass transistor.

V_Z is also obtained from the output voltage using a dropping resistor R_Z. The reference voltage V_Z is less than the required load voltage V_L. For a zener current I_Z, the resistance R_Z must satisfy

$$R_Z = \frac{V_L - V_Z}{I_Z} \tag{20-48}$$

Assume for the moment that the transistor operates in the active linear region. The nonlinear base–emitter junction is a part of the forward block of a feedback loop. The feedback network consisting of R_f and R_i forces the output voltage V_L to assume the value

$$V_L = \left(1 + \frac{R_f}{R_i}\right)V_Z \tag{20-49}$$

The output of the op-amp assumes a voltage equal to V_L plus the transistor base–emitter drop. For the current levels involved, the base–emitter drop is typically greater than the 0.7-V drop assumed for small-signal transistors.

The op-amp need only supply the base current required to drive the transistor. However, the transistor and collector currents will be considerably greater. Because of the action of the transistor in allowing a much greater load current than its base drive current, it is referred to as a *pass transistor*.

In a typical application with moderate to large load currents, the current through the feedback circuit and the zener bias current will both be considerably smaller than the load current I_L. Thus a reasonable approximation in many cases is

$$I_E \simeq I_L \tag{20-50}$$

The collector current I_C can then be expressed as

$$I_C = I_E - I_B \tag{20-51}$$

As a further approximation, I_B will be neglected in comparison with I_E in (20-51), which results in the simplification

$$I_C \simeq I_E \simeq I_L \tag{20-52}$$

Thus, in subsequent computations, it will be assumed that collector current, emitter current, and load current are all essentially the same. Although these assumptions are not always necessary, they do simplify the subsequent analysis.

Based on the preceding assumptions, the base current I_B can be expressed as

$$I_B \approx \frac{I_L}{\beta_{dc}} \tag{20-53}$$

where β_{dc} is the dc transistor current gain. The op-amp must be capable of providing this value of base drive. Where the required load current is considerably greater than the available op-amp drive current, a transistor compound connection may be used to achieve a very large value of β_{dc}.

Efficiency

Let V_i represent the value of the unregulated voltage at the input to the regulator. In order to ensure linear active region operation for the pass transistor, it is necessary that $V_i > V_L$. Because of the typical larger saturation voltage of a power transistor and the necessity to provide some leeway for line variations, ripple, and other factors, V_i is often chosen to be at least 2 or 3 V greater than V_L.

The input power P_i can be closely approximated as

$$P_i \simeq V_i I_L \tag{20-54}$$

The output load power P_L is

$$P_L = V_L I_L \tag{20-55}$$

The pass transistor must dissipate a power P_D given by the difference between the preceding values, that is,

$$P_D = P_i - P_L = V_i I_L - V_L I_L = (V_i - V_L)I_L \tag{20-56}$$

Let η represent the efficiency as a percentage. It is given by

$$\eta = \frac{P_L}{P_i} \times 100\% \tag{20-57}$$

Substitution of (20-54) and (20-55) in (20-57) results in

$$\eta = \frac{V_L}{V_i} \times 100\% \tag{20-58}$$

Current Limiting

To prevent the possibility of drawing excessive load current, a current-limiting transistor may be added to the regulator as shown in Figure 20-20. The combination of the pass transistor and the current-limiting transistor in Figure 20-20 replace the pass transistor in Figure 20-19. All connections to the left, right, and below are the same as before.

The concept behind the limiting action is as follows: When the current I_L is

Figure 20-20 Addition of current-limiting transistor (T2) to regulator circuit.

small, the base–emitter voltage of T2, which is given by R_sI_L, is insufficient to forward bias T2, and thus $I_{C2} = 0$. The limiter T2 has no effect for this condition. As I_L increases, the voltage R_sI_L becomes large enough to forward bias the base–emitter junction of T2, and base current flows. This results in collector current for T2, as indicated by I_{C2} on Figure 20-20. This current has to be supplied from the op-amp and is "robbed" from the base supply of T1. This effect tends to reduce the collector and emitter currents of T1, and the result is to effectively limit further increases in the load current. The resistance R_s is selected so that the product of R_s and the maximum load current is the expected cut-in voltage for T2.

Integrated-Circuit Regulators

This development has explained the basic operation of the regulator utilizing an op-amp. In actual practice, a wide variety of integrated-circuit chips are available, in which the op-amp, reference zener, and pass transistor are all part of the given circuit. Some are designed for fixed dc output voltages such as +5 V, +15 V, and −15 V. Others are "programmable" by providing suitable resistors (i.e., R_i and R_f in Figure 20-19).

The wiring layout of a typical IC regulator used for a fixed dc output voltage is shown in Figure 20-21. This particular regulator is a three-terminal device having an input, an output, and a common ground. It is assumed that a reasonable amount of filtering has been provided for the input since this voltage must not drop below some minimum level for regulation to be sustained. Most regulators also recommend additional capacitors at input and output as shown.

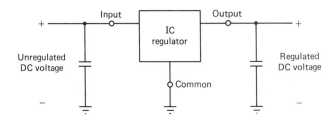

Figure 20-21 Connections for fixed positive voltage regulator.

Switching Regulator Power Supply

One of the newest concepts in regulated power supplies is that of the *switching regulator power supply*. The circuit is quite complex and will be discussed here from a block diagram point of view.

Refer to Figure 20-22, representing a typical switching power supply block diagram. The input ac line voltage is first rectified with a conventional rectifier circuit, and some filtering is employed. This voltage is applied to a high-frequency inverter which is controlled by a control circuit. The output of this inverter is a series of high-frequency pulses, for which a typical frequency (20 kHz) is indicated. The width of the pulses is controlled by the modulator in the control circuit. This higher-frequency voltage is rectified and filtered to produce the dc output voltage as shown.

The regulation action of the circuit is a function of the feedback signal applied to the control voltage. If the output dc voltage starts to drop, the sense of the comparator is such as to increase the width of the modulated pulses. Since these pulses control the high-frequency inverter pulses, the result is to increase the dc output to

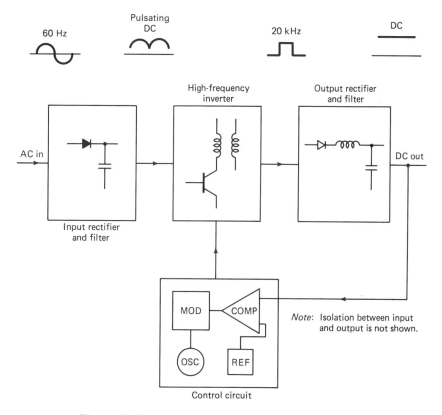

Figure 20-22 Block diagram of switching power supply.

compensate. The opposite situation occurs if the dc output decreases. The output signal from the modulator is called a *pulse-wide modulated* (PWM) signal.

The primary advantage of switching regulator power supplies is their relatively high efficiencies. Efficiencies may run as high as 80% in many cases, which is considerably higher than the typical efficiencies of standard regulator circuits.

Example 20-7

Assume an op-amp regulator of the type shown in Figure 20-19 with $R_i = 2$ kΩ, $R_f = 3.357$ kΩ, and $V_z = 5.6$ V. The unregulated voltage is approximately 20 V.
   (a) Determine V_L. (b) If the maximum value of load current is 1 A and $\beta_{dc} = 100$ for T1, determine the maximum op-amp output current required for base drive I_B. (c) Determine the load power, the approximate input power, and the efficiencies under peak-load conditions. (d) Determine the maximum power dissipation $P_{D(max)}$ for the pass transistor.

Solution

(a) The value of the load voltage is

$$V_L = \left(1 + \frac{R_f}{R_i}\right)V_z = \left(1 + \frac{3357}{2000}\right)5.6 = 15 \text{ V} \qquad \textbf{(20-59)}$$

(b) For $I_L = 1$ A and $\beta_{dc} = 100$, the op-amp must supply a base drive of

$$I_B = \frac{I_L}{\beta_{dc}} = \frac{1 \text{ A}}{100} = 10 \text{ mA} \qquad \textbf{(20-60)}$$

(c) The load power is

$$P_L = V_L I_L = 15 \times 1 = 15 \text{ W} \qquad \textbf{(20-61)}$$

The approximate input power is

$$P_i = V_i I_L = 20 \times 1 = 20 \text{ W} \qquad \textbf{(20-62)}$$

The efficiency is

$$\eta = \frac{15 \text{ W}}{20 \text{ W}} \times 100\% = 75\% \qquad \textbf{(20-63)}$$

(d) The maximum power dissipation occurs under full-load conditions and is

$$P_{D(max)} = 20 - 15 = 5 \text{ W} \qquad \textbf{(20-64)}$$

20-10 THYRISTORS

Thyristors are semiconductor devices that use internal feedback to produce latching action. Thyristors operate only in the switching mode. Generally, they are used to control large amounts of power, such as in lighting systems, heaters, motors, and so on. Because of their importance in power systems, a discussion of their properties has been included in this chapter.

Silicon-Control Rectifier

A *silicon-control rectifier* is a thyristor employing four layers of semiconducting materials. The construction is suggested by the diagram of Figure 20-23(a). The three external connections are denoted as the *cathode,* the *anode,* and the *gate*.

The SCR may be visualized in the manner shown in Figure 20-23(b). It is equivalent to the connection of a *pnp* transistor and an *npn* transistor as shown in Figure 20-23(c). The corresponding schematic symbol is shown in Figure 20-23(d).

The SCR will act as an open circuit unless either one of two possible conditions are satisfied: (1) A positive trigger pulse is received at the gate, or (2) the anode–cathode voltage exceeds a certain minimum voltage called the *breakover voltage*. Most SCRs are designed to function with a trigger input.

Assume that the voltage at the anode is positive with respect to the cathode, but there is no base current flowing in either transistor. This is the initial condition encountered with the SCR, and it acts as an open circuit under this condition as long as the anode–cathode voltage is less than the breakover voltage.

Assume now that a positive pulse is received at the gate terminal. The base–emitter junction of the *npn* transistor will be forward biased, and current will flow in its collector through the base–emitter junction of the *pnp* transistor. This will result in current flow in the *pnp* transistor collector, and this current flow adds to the initial base current of the *npn* transistor. The result is a regenerative action that rapidly changes the SCR to a nearly short-circuit condition. The SCR thus acts as a closed switch under this condition.

Even though the base pulse is removed, the SCR will remain in the closed con-

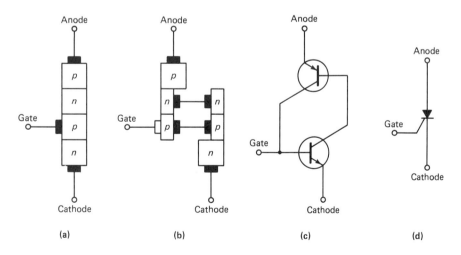

Figure 20-23 Construction of silicon-control rectifier (SCR) and the schematic symbol.

dition until the current through the SCR drops to a small value. The transistors will come out of saturation when the current is reduced below the *holding current*. It then remains in the open-circuit condition until a new pulse is received.

Half-Wave Control

A typical control circuit utilizing an SCR to control the voltage over a half-cycle is shown in Figure 20-24. The resistance R_1 is used to limit the current, and R_2 is used to establish the trigger level for the SCR. If the SCR is turned on near the beginning of a positive half-cycle, conduction in the load R_L will occur for nearly 180°. Conversely, if the SCR is turned on only near the peak of the half-cycle, conduction will occur for only about 90°. By adjustment of the variable arm for R_2, the conduction angle can thus be varied, and the power delivered to the load can be adjusted. During the negative half-cycle of the input, the SCR is turned off. The diode prevents this negative ac voltage from being applied to the SCR gate.

Diac

A *diac* is equivalent to the parallel connection of two four-layer semiconductor devices in opposite directions, and the schematic diagram is shown in Figure 20-25(a). It is a *bidirectional* thyristor in that current can flow in either direction.

The diac acts like an open circuit until the voltage in either direction across it tries to exceed the breakover voltage. In that event one of two parallel diodes changes state and conducts strongly, approximating a short circuit in the ideal case. Once the diac starts to conduct, the only way it can be stopped is for the current through it to drop below the rated holding current of the device. Operation can be visualized as two switches in parallel as shown in Figure 20-25(b).

Triac

The *triac* is a bidirectional device with a control terminal. The schematic symbol is shown in Figure 20-26(a). It can be visualized as a parallel connection of two SCRs, as shown in Figure 20-26(b). The normal mode of operation for the triac is to turn it on with a trigger at the gate. When the voltage across the triac has one polarity, a positive trigger is required, but when the voltage has the opposite polarity, a negative trigger is required.

Figure 20-24 Half-wave SCR control circuit.

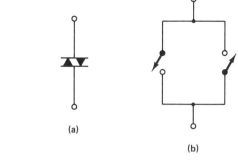

(a)

(b)

Figure 20-25 Schematic symbol for diac and equivalent switching circuit.

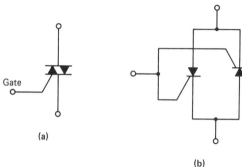

Gate

(a)

(b)

Figure 20-26 Schematic symbol for triac and its equivalence as two SCRs.

DRILL PROBLEMS

20-1. An ideal transformer has a turns ratio of 5 : 1 from primary to secondary. Assume a primary ac line voltage of 120 V rms. For a particular secondary load, the primary current is 3 A. Determine the secondary rms voltage and current.

20-2. An ideal transformer has a turns ratio of 1 : 4 from primary to secondary. Assume a primary ac line voltage of 120 V rms. For a particular secondary load, the primary current is 8 A. Determine the secondary rms voltage and current.

20-3. An ideal transformer has a turns ratio of 8 : 1 from primary to secondary. Assume a primary ac line voltage of 120 V rms. A certain load draws a secondary current of 4 A rms. Determine the rms values of the primary current and the secondary voltage.

20-4. The primary of an ideal transformer is connected to the 120-V rms ac line voltage. A secondary 20-Ω load draws an rms current of 2 A. Determine the rms secondary voltage, the rms primary current, and the turns ratio.

20-5. A certain unfiltered half-wave rectifier circuit of the form of Figure 20-3 is powered by the 120-V rms ac power system. The turns ratio is $N_1 : N_2 = 8 : 1$. Determine (a) the rms secondary voltage, (b) the peak secondary voltage, (c) the dc load voltage, and (d) the dc load current if $R_L = 6 \ \Omega$. Neglect any diode and transformer losses.

20-6. Repeat the analysis of Problem 20-5 if $N_1 : N_2 = 1 : 4$ and $R_L = 50 \ \Omega$.

20-7. A certain unfiltered full-wave rectifier circuit of the form of Figure 20-5 is powered by the 120-V rms ac power system. The turns ratio is $N_1 : N_2 = 8 : 1$. Determine **(a)** the rms secondary voltage, **(b)** the peak secondary voltage, **(c)** the dc load voltage, and **(d)** the dc load current if $R_L = 6 \ \Omega$. Neglect any diode and transformer losses.

20-8. Repeat the analysis of Problem 20-7 if $N_1 : N_2 = 1 : 4$ and $R_L = 50 \ \Omega$.

20-9. A certain unfiltered bridge rectifier circuit of the form of Figure 20-8 is powered by the 120-V rms ac power system. The turns ratio is $N_1 : N_2 = 8 : 1$. Determine **(a)** the rms secondary voltage, **(b)** the peak secondary voltage, **(c)** the dc load voltage, and **(d)** the dc load current if $R_L = 6 \ \Omega$. Neglect any diode and transformer losses.

20-10. Repeat the analysis of Problem 20-9 if $N_1 : N_2 = 1 : 4$ and $R_L = 50 \ \Omega$.

20-11. Determine the peak reverse voltages for the diodes in the rectifier circuits of the following problems: **(a)** 20-5; **(b)** 20-7; **(c)** 20-9.

20-12. Determine the peak reverse voltages for the diodes in the rectifier circuits of the following problems: **(a)** 20-6; **(b)** 20-8; **(c)** 20-10.

20-13. Determine the peak reverse voltages for the diodes in each of the four following rectifier circuits: **(a)** unfiltered half-wave rectifier with $V_{2P} = 24$ V; **(b)** filtered half-wave rectifier with $V_L = 24$ V; **(c)** unfiltered full-wave rectifier (using a center-tapped transformer) with $V_{2p} = 24$ V; **(d)** unfiltered bridge rectifier with $V_{2p} = 24$ V.

20-14. Determine the peak reverse voltages for the diodes in each of the four following rectifier circuits: **(a)** unfiltered half-wave rectifier with $V_{2P} = 400$ V; **(b)** filtered half-wave rectifier with $V_L = 400$ V; **(c)** filtered full-wave rectifier (using a center-tapped transformer) with $V_L = 400$ V; **(d)** filtered bridge rectifier with $V_L = 400$ V.

20-15. A certain half-wave rectifier circuit employs a single capacitor filter and $C = 600 \ \mu F$. It is powered by the 120-V rms, 60-Hz ac power system. The dc load current is 180 mA. Determine **(a)** the peak-to-peak ripple voltage and **(b)** the rms ripple voltage.

20-16. Repeat the analysis of Problem 20-15 if the circuit is a full-wave rectifier. All other conditions are unchanged.

20-17. A certain full-wave power supply with a single capacitor filter of the type shown in Figure 20-10 is powered by the 120-V rms, 60-Hz ac power system. The peak value of the output voltage under loaded conditions is 36 V and $C = 2000 \ \mu F$. For a load current of 3 A, determine **(a)** the peak-to-peak ripple voltage, **(b)** the rms ripple voltage, **(c)** the dc load voltage, and **(d)** the ripple factor.

20-18. A certain full-wave power supply with a single capacitor filter utilizes a bridge circuit and is powered by the 120-V, 60-Hz ac power system. The peak value of the output voltage under loaded conditions is 600 V and $C = 100 \ \mu F$. For a load current of 200 mA, determine **(a)** the peak-to-peak ripple voltage, **(b)** the rms ripple voltage, **(c)** the dc load voltage, and **(d)** the ripple factor.

20-19. An op-amp regulator of the type shown in Figure 20-19 has $R_i = 10 \ k\Omega$, $R_f = 14.24 \ k\Omega$, and $V_Z = 3.3$ V. The unregulated voltage is approximately 12 V. **(a)** Determine V_L. **(b)** If the maximum value of load current is 0.5 A and $\beta_{dc} = 80$ for the transistor, determine the maximum op-amp output current required for base drive I_B. **(c)** Determine the load power, the approximate input power, and the efficiency under peak load conditions. **(d)** Determine the maximum power dissipation $P_{D(max)}$ for the pass transistor.

20-20. An op-amp regulator of the type shown in Figure 20-19 has $R_i = 10 \ k\Omega$, $V_Z = 5.1$ V, and $V_L = 20$ V. The unregulated voltage is approximately 28 V. **(a)** Determine R_f. **(b)** If the maximum value of load current is 250 mA and $\beta_{dc} = 50$ for the transistor, determine the maximum op-amp output current required for base drive I_B. **(c)** Determine the load power, the approximate input power, and the efficiency under peak load conditions. **(d)** Determine the maximum power dissipation $P_{D(max)}$ for the pass transistor.

DERIVATION PROBLEMS

20-21. (Requires calculus.) Consider a half-wave rectified voltage as defined by

$$v(t) \begin{cases} = V_p \sin \omega t & 0 < t < \dfrac{T}{2} \\ = 0 & \dfrac{T}{2} < t < T \end{cases}$$

Prove that the dc value V_{dc} is given by

$$V_{dc} = \frac{V_p}{\pi}$$

20-22. (Requires calculus.) Consider a full-wave rectified voltage as defined by

$$v(t) = |V_p \sin \omega t| \qquad 0 < t < T$$

Prove that the dc value V_{dc} is given by

$$V_{dc} = \frac{2V_{dc}}{\pi}$$

[*Hint:* In the first half-cycle, $v(t)$ is the same as in Problem 20-21, and the integral for the complete cycle is double the value for the first half-cycle.]

20-23. In a certain industrial process, an unfiltered half-wave rectifier circuit of the form of Figure 20-3 is to be employed and is to be powered by the 120-V ac power system. In this initial design stage, idealized components will be assumed, and losses will be neglected. A dc current of 5 A in a 4-Ω load is required. Determine the turns ratio $N_1 : N_2$.

20-24. Repeat the design of Problem 20-23 if an unfiltered bridge rectifier of the form of Figure 20-8 is employed. (In this case, N_2 is the number of turns on the complete secondary.)

20-25. In a certain rectifier circuit employing a single-capacitor filter, the dc output voltage is to be 30 V at a load current of 800 mA. At that load current, the ripple factor is not to exceed 0.04. Determine the minimum value of capacitance required if a full-wave rectifier is used. The circuit is powered by the 120-V rms, 60-Hz ac power system.

20-26. Determine the minimum value of capacitance required in Problem 20-25 if a half-wave rectifier is used.

20-27. An unregulated full-wave rectifier circuit employing a center-tapped transformer and a single capacitor filter is to be designed for a particular application. In this initial design step, idealized components will be assumed, and losses will be neglected. The design specifications are

Dc load voltage = 20 V at a load current of 0.5 A

Ripple factor = 0.05

Determine **(a)** the capacitance required and **(b)** the rms voltage required across each half of transformer secondary (assuming the 120-V, 60-Hz ac power system).

20-28. An unregulated full-wave bridge rectifier circuit employing a single-capacitor filter is to be designed for a particular application. In this initial design step, idealized components will be assumed, and losses will be neglected. The design specifications are

Dc load voltage = 300 V at a load current of 100 mA

Ripple factor = 0.02

Determine **(a)** the capacitance required and **(b)** the rms voltage required across transformer secondary (assuming the 120-V, 60-Hz ac power system).

20-29. You are to perform a preliminary design of a voltage regulator circuit of the form shown in Figure 20-19. The desired dc load voltage is 12 V, and a 5.1-V zener diode is to be used as a reference voltage. Maximum load current is specified as 1 A. Zener bias current is to be established as 20 mA. (a) Determine the value of R_Z required. (b) Determine possible values of R_i and R_f. (c) Assume that the value of β_{dc} for the pass transistor is subject to a range from 50 to 200. Determine the maximum output drive current for the op-amp. (d) If the maximum unregulated voltage is 16 V, determine the maximum power dissipation of the pass transistor.

20-30. Based on the circuit form of Figure 20-19, draw a new schematic diagram of a regulator circuit for regulating a negative dc voltage.

TROUBLESHOOTING PROBLEMS

20-31. You are troubleshooting a circuit of the form shown in Figure 20-3. The measured ac value for v_2 is within the acceptable range as given by the manual, but there is no dc load voltage. The following voltages are measured on separate dc and ac scales of a voltmeter:

v_L (V)	DC	AC (rms)
v_2 (V)	0	23.5
v_L (V)	0	23.5

Which one of the following is a possible trouble? (a) Diode open; (b) diode shorted; (c) shorted load; (d) open transformer.

20-32. You are troubleshooting a circuit of the form shown in Figure 20-5. The specified value of v_L is about 24 V dc, but the dc value read by a voltmeter is only about 11.8 V. The ac rms voltage read across half the secondary is measured as 26.2 V and across the full secondary as 52.4 V. Which one of the following is a possible trouble? (a) One diode open; (b) two diodes open; (c) one diode shorted; (d) two diodes shorted.

20-33. A power supply of the form of Figure 20-10 is used to power an amplifier circuit. There is an excessive level of hum accompanying the signal. Which one of the following could be the cause of the difficulty? (a) One or more shorted diodes; (b) one or more open diodes; (c) half of transformer secondary shorted; (d) open C; (e) shorted C.

20-34. You are troubleshooting a circuit of the form of Figure 20-10, for which there is no dc output voltage. After performing ohmmeter checks on the diodes, you discover that both are open and appear to have experienced excessive current flow. Which one of the following could be the possible trouble? (a) load disconnected; (b) half of transformer secondary open; (c) half of transformer secondary shorted; (d) open C; (e) shorted C.

APPENDIX A

Resistor Values

A-1 TYPICAL 5% AND 10% VALUES

The most commonly used resistors in electronic circuits are the low-power, axial-lead, carbon-composition, and film resistors with tolerances of $\pm 5\%$ and $\pm 10\%$. The values in the range 10 Ω to 10 MΩ are listed below. All values listed are available in $\pm 5\%$ tolerances, but only the *boldface* values are available in $\pm 10\%$ tolerances.

Ω	Ω	kΩ	kΩ	kΩ	MΩ
10	**100**	**1**	**10**	**100**	**1**
11	110	1.1	11	110	1.1
12	**120**	**1.2**	**12**	**120**	**1.2**
13	130	1.3	13	130	1.3
15	**150**	**1.5**	**15**	**150**	**1.5**
16	160	1.6	16	160	1.6
18	**180**	**1.8**	**18**	**180**	**1.8**
20	200	2.0	20	200	2.0
22	**220**	**2.2**	**22**	**220**	**2.2**
24	240	2.4	24	240	2.4
27	**270**	**2.7**	**27**	**270**	**2.7**
30	300	3.0	30	300	3.0
33	**330**	**3.3**	**33**	**330**	**3.3**
36	360	3.6	36	360	3.6
39	**390**	**3.9**	**39**	**390**	**3.0**
43	430	4.3	43	430	4.3
47	**470**	**4.7**	**47**	**470**	**4.7**
51	510	5.1	51	510	5.1
56	**560**	**5.6**	**56**	**560**	**5.6**
62	620	6.2	62	620	6.2
68	**680**	**6.8**	**68**	**680**	**6.8**
75	750	7.5	75	750	7.5
82	**820**	**8.2**	**82**	**820**	**8.2**
91	910	9.1	91	910	9.1

A-2 TYPICAL 1% VALUES

Values in the range 100 to 1000 Ω are listed below. Values in other ranges are obtained by adding zeros or a decimal point to these values.

100	147	215	316	464	681
102	150	221	324	475	698
105	154	226	332	487	715
107	158	232	340	499	732
110	162	237	348	511	750
113	165	243	357	523	768
115	169	249	365	536	787
118	174	255	374	549	806
121	178	261	383	562	825
124	182	267	392	576	845
127	187	274	402	590	866
130	191	280	412	604	887
133	196	287	422	619	909
137	200	294	432	634	931
140	205	301	442	649	953
143	210	309	453	665	976

APPENDIX B

Capacitor Values*

The patterns of capacitor values tend to track those of 5% and 10% resistor values as far as the digits are concerned. However, these patterns are somewhat more complex and depend on capacitor types and tolerances. A few digit combinations used for capacitors are different than for resistors. The values listed below should be interpreted as *typical* rather than *general*.

pF	pF	pF	pF	μF	μF	μF	μF	μF	μF	μF
5	50	500	5000		0.05	0.5	5	50	500	5000
—	51	510	5100		—	—	—	—	—	—
—	56	560	5600		0.056	0.56	5.6	56	—	5600
—	—	—	6000		0.06	—	6	—	—	6000
—	62	620	6200		—	—	—	—	—	—
—	68	680	6800		0.068	0.68	6.8	—	—	—
—	75	750	7500		—	—	—	75	—	—
—	—	—	8000		—	—	8	80	—	—
—	82	820	8200		0.082	0.82	8.2	82	—	—
—	91	910	9100		—	—	—	—	—	—
10	100	1000		0.01	0.1	1	10	100	1000	10000
—	110	1100		—	—	—	—	—	—	
12	120	1200		0.012	0.12	1.2	—	—	—	
—	130	1300		—	—	—	—	—	—	
15	150	1500		0.015	0.15	1.5	15	150	1500	
—	160	1600		—	—	—	—	—	—	
18	180	1800		0.018	0.18	1.8	18	180	—	
20	200	2000		0.02	0.2	2	20	200	2000	
22	220	2200		—	0.22	2.2	22	—	—	
24	240	2400		—	—	—	—	240	—	
—	250	2500		—	0.25	—	25	250	2500	
27	270	2700		0.027	0.27	2.7	27	270	—	
30	300	3000		0.03	0.3	3	30	300	3000	
33	330	3300		0.033	0.33	3.3	33	330	3300	
36	360	3600		—	—	—	—	—	—	
39	390	3900		0.039	0.39	3.9	39	—	—	
—	—	4000		0.04	—	4	—	400	—	
43	430	4300		—	—	—	—	—	—	
47	470	4700		0.047	0.47	4.7	47	—	—	

*The element values in this table were extracted from *Electronic Devices and Circuits,* 3rd. Ed., David A. Bell. (Englewood Cliffs, N.J.: Reston/Prentice Hall, 1986).

APPENDIX C
Hybrid Parameters

Within the text, the modeling of small-signal transistor behavior has been achieved primarily with the hybrid-pi equivalent circuit because of its simplicity in the midfrequency range and because it can be easily adapted to high-frequency models. However, many books and data manuals use references to the classical hybrid parameter equivalent circuit. This model was heavily utilized in much of the earlier transistor work.

It turns out that three separate hybrid parameter equivalent circuits can actually be developed, corresponding to common emitter, common base, and common collector. However, all cases can be developed with the common-emitter configuration, and most data manuals provide numerical values based solely on that configuration. Consequently, the treatment here will be limited to the common-emitter situation.

Consider the common-emitter configuration shown in Figure C-1. The dc circuit used to establish an operating point is not shown but is obviously understood. Only the small-signal or ac variables are shown. The pertinent dynamic variables are as follows:

v_{be}: base–emitter signal voltage

i_b: base signal current

v_{ce}: collector–emitter signal voltage

i_c: collector signal current

The hybrid parameters as commonly applied to BJTs are based on assuming i_b and v_{ce} as the *independent* variables and v_{be} and i_c as the *dependent* variables. The resulting equations are

$$v_{be} = h_{ie}i_b + h_{re}v_{ce} \qquad \text{(C-1)}$$

Figure C-1 BJT common-emitter configuration with signal variables labeled.

$$\boxed{i_c = h_{fe}i_b + h_{oe}v_{ce}} \qquad \text{(C-2)}$$

The quantities h_{ie}, h_{re}, h_{fe}, and h_{oe} are the *hybrid parameters* for the common-emitter configuration. The second subscript e in each case refers to common-*emitter*. The subscripts i, r, f, and o refer to *input, reverse, forward,* and *output,* respectively.

To understand the physical significance of the parameters, assume in (C-1) and (C-2) that the total collector–emitter voltage is held constant so that the signal voltage is $v_{ce} = 0$. The two equations then reduce to

$$v_{be} = h_{ie}i_b \qquad \text{for } v_{ce} = 0 \qquad \text{(C-3)}$$

$$i_c = h_{fe}i_b \qquad \text{for } v_{ce} = 0 \qquad \text{(C-4)}$$

The quantities h_{ie} and h_{fe} may then be expressed as

$$\boxed{h_{ie} = \left.\frac{v_{be}}{i_b}\right]_{v_{ce}=0}} \qquad \text{(C-5)}$$

and

$$\boxed{h_{fe} = \left.\frac{i_c}{i_b}\right]_{v_{ce}=0}} \qquad \text{(C-6)}$$

From these expressions, it is observed that h_{ie} is an input resistance, and h_{fe} is a forward current gain. Both are determined under conditions of an "ac short" (but not a total short, obviously) on the output.

Next, assume that the total input base current is held constant so that the signal current is $i_b = 0$. Equations (C-1) and (C-2) then reduce to

$$v_{be} = h_{re}v_{ce} \qquad \text{for } i_b = 0 \qquad \text{(C-7)}$$

$$i_c = h_{oe}v_{ce} \qquad \text{for } i_b = 0 \qquad \text{(C-8)}$$

The quantities h_{re} and h_{oe} may then be expressed as

$$\boxed{h_{re} = \left.\frac{v_{be}}{v_{ce}}\right]_{i_b = 0}} \qquad \text{(C-9)}$$

$$\boxed{h_{oe} = \left.\frac{i_c}{v_{ce}}\right]_{i_b = 0}} \qquad \text{(C-10)}$$

The parameter h_{re} is observed to be a reverse feedback ratio, and h_{oe} is an output conductance. Both are determined under conditions of an "ac open" (but not a total open) at the input.

C-1 EQUIVALENT CIRCUIT

Equation (C-1) expresses v_{be} as the sum of two voltages, while (C-2) expresses i_c as the sum of two currents. The first case can be interpreted as a series connection, while the second can be interpreted as a parallel connection.

A hybrid parameter equivalent circuit based on the preceding interpretations is shown in Figure C-2. From the preceding discussion and the circuit diagram, it can be inferred that h_{ie} has the dimensions of ohms, h_{oe} has the dimensions of siemens, and h_{re} and h_{fe} are dimensionless. Note that the resistance at the output is labeled as $1/h_{oe}$ (which would be in ohms), as is customary.

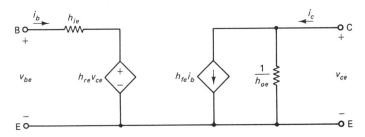

Figure C-2 Hybrid-parameter equivalent circuit based on common-emitter configuration.

C-2 RELATIONSHIP TO HYBRID-PI

By comparing the circuit of Figure C-2 with the low-frequency hybrid-pi model and other parameters of Chapter 10, some relationships between certain quantities may be established. First, h_{ie} is the input dynamic resistance, which was defined as r_π in Chapter 10. Second, h_{fe} is the same as the flat-band current gain, which was defined as β_o in Chapter 10.

In the midfrequency hybrid-pi model of Chapter 10, the following two assumptions were made: (1) Reverse feedback from collector back to base was ignored, and (2) the dependent output current source was assumed to be ideal. Because of these assumptions, there are no parameters in Chapter 10 that correspond directly with h_{re} and h_{oe}. The hybrid-pi model can be modified to take these effects into consideration, but considering the approximate nature of the parameters, such efforts are mandated in only the most exacting situations.

A summary of the comparison is as follows:

Hybrid parameters	Chapter 10 and later
h_{ie}	r_π
h_{fe}	β_o
h_{ie}	Effect is neglected
h_{oe}	Assumed to be zero

C-3 AMPLIFIER ANALYSIS

The use of the hybrid parameters for analyzing a typical amplifier circuit will now be illustrated. Consider the common-emitter circuit of Figure C-3(a). Assume operation in the midfrequency range where external capacitances are represented as shorts.

The net small-signal equivalent circuit based on replacing the transistor by its hybrid equivalent circuit is shown in Figure C-3(b). Note that the input voltage is $v_i = v_{be}$, and the output voltage is $v_o = v_{ce}$.

A KVL loop equation around the input reads

$$-v_i + h_{ie}i_b + h_{re}v_{ce} = 0 \qquad\qquad \textbf{(C-11)}$$

Solving for i_b, there results

$$i_b = \frac{v_i - h_{re}v_{ce}}{h_{ie}} = \frac{v_i - h_{re}v_o}{h_{ie}} \qquad\qquad \textbf{(C-12)}$$

where the substitution $v_{ce} = v_o$ was made.

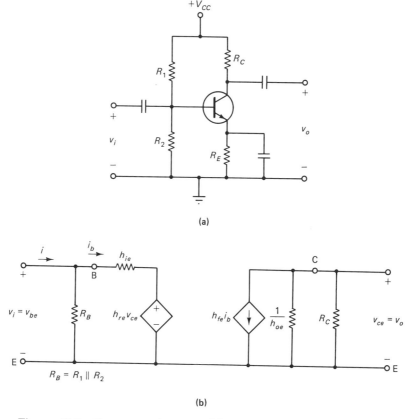

(a)

(b)

Figure C-3 Common-emitter amplifier and small-signal equivalent circuit using hybrid parameters.

The output voltage v_o can be determined from the output portion of the circuit as

$$v_o = -\left(R_C \| \frac{1}{h_{oe}}\right) h_{fe} i_b = -\frac{R_C h_{fe} i_b}{1 + R_C h_{oe}} \tag{C-13}$$

The result of (C-12) is substituted in (C-13) and the equation is manipulated to obtain an expression for v_o. The result is

$$v_o = \frac{-h_{fe} R_C v_i}{h_{ie} + R_C(h_{ie} h_{oe} - h_{re} h_{fe})} \tag{C-14}$$

The gain A is obtained as

$$A = \frac{v_o}{v_i} = \frac{-h_{fe} R_C}{h_{ie} + R_C(h_{ie} h_{oe} - h_{re} h_{fe})} \tag{C-15}$$

APPENDIX D

Data Sheets

Data sheets for the particular devices discussed in the text are reproduced in this appendix. Appreciation is expressed to the following organizations for permission to reproduce the specific data sheets indicated:

1. Pages 760 through 771: Courtesy of Motorola, Inc.
2. Pages 772 through 777: Reprinted with permission of GE Semiconductor, Research Triangle Park, N. C.
3. Pages 778 through 789: Copyright, Fairchild Semiconductor Corporation; Used by permission.

1N4001
thru
1N4007

GENERAL-PURPOSE RECTIFIERS

. . . subminiature size, axial lead mounted rectifiers for general-purpose low-power applications.

*MAXIMUM RATINGS

Rating	Symbol	1N4001	1N4002	1N4003	1N4004	1N4005	1N4006	1N4007	Unit
Peak Repetitive Reverse Voltage Working Peak Reverse Voltage DC Blocking Voltage	V_{RRM} V_{RWM} V_R	50	100	200	400	600	800	1000	Volts
Non-Repetitive Peak Reverse Voltage (halfwave, single phase, 60 Hz)	V_{RSM}	60	120	240	480	720	1000	1200	Volts
RMS Reverse Voltage	$V_{R(RMS)}$	35	70	140	280	420	560	700	Volts
Average Rectified Forward Current (single phase, resistive load, 60 Hz, see Figure 8, $T_A = 75^oC$)	I_O				1.0				Amp
Non-Repetitive Peak Surge Current (surge applied at rated load conditions, see Figure 2)	I_{FSM}				30 (for 1 cycle)				Amp
Operating and Storage Junction Temperature Range	T_J, T_{stg}				−65 to +175				^oC

*ELECTRICAL CHARACTERISTICS

Characteristic and Conditions	Symbol	Typ	Max	Unit
Maximum Instantaneous Forward Voltage Drop ($i_F = 1.0$ Amp, $T_J = 25^oC$) Figure 1	v_F	0.93	1.1	Volts
Maximum Full-Cycle Average Forward Voltage Drop ($I_O = 1.0$ Amp, $T_L = 75^oC$, 1 inch leads)	$V_{F(AV)}$	−	0.8	Volts
Maximum Reverse Current (rated dc voltage) $T_J = 25^oC$ $T_J = 100^oC$	I_R	0.05 1.0	10 50	μA
Maximum Full-Cycle Average Reverse Current ($I_O = 1.0$ Amp, $T_L = 75^oC$, 1 inch leads	$I_{R(AV)}$	−	30	μA

*Indicates JEDEC Registered Data.

MECHANICAL CHARACTERISTICS

CASE: Transfer Molded Plastic

MAXIMUM LEAD TEMPERATURE FOR SOLDERING PURPOSES: 350^oC, 3/8" from case for 10 seconds at 5 lbs. tension

FINISH: All external surfaces are corrosion-resistant, leads are readily solderable

POLARITY: Cathode indicated by color band

WEIGHT: 0.40 Grams (approximately)

NOTES:
1. POLARITY DENOTED BY CATHODE BAND.

2. LEAD DIAMETER NOT CONTROLLED WITHIN "F" DIMENSION.

DIM	MILLIMETERS		INCHES	
	MIN	MAX	MIN	MAX
A	5.97	6.60	0.235	0.260
B	2.79	3.05	0.110	0.120
D	0.76	0.86	0.030	0.034
K	27.94	−	1.100	

CASE 59-04
(Does not meet DO-41 outline)

(Courtesy of Motorola Inc.)

1.5KE6.8,A thru 1.5KE200,A
See Page 4-74

Designers Data Sheet

1N746 thru 1N759
1N957A thru 1N986A
1N4370 thru 1N4372

500-MILLIWATT HERMETICALLY SEALED GLASS SILICON ZENER DIODES

- Complete Voltage Range — 2.4 to 110 Volts
- DO-35 Package — Smaller than Conventional DO-7 Package
- Double Slug Type Construction
- Metallurgically Bonded Construction
- Nitride Passivated Die

Designer's Data for "Worst Case" Conditions

The Designer's Data sheets permit the design of most circuits entirely from the information presented. Limit curves — representing boundaries on device characteristics — are given to facilitate "worst case" design.

GLASS ZENER DIODES
500 MILLIWATTS
2.4–110 VOLTS

MAXIMUM RATINGS

Rating	Symbol	Value	Unit
DC Power Dissipation @ $T_L \leqslant 50^oC$, Lead Length = 3/8''	P_D		
*JEDEC Registration		400	mW
*Derate above $T_L = 50^oC$		3.2	mW/^oC
Motorola Device Ratings		500	mW
Derate above $T_L = 50^oC$		3.33	mW/^oC
Operating and Storage Junction Temperature Range	T_J, T_{stg}		^oC
*JEDEC Registration		−65 to +175	
Motorola Device Ratings		−65 to +200	

*Indicates JEDEC Registered Data.

MECHANICAL CHARACTERISTICS

MAXIMUM LEAD TEMPERATURE FOR SOLDERING PURPOSES: 230^oC, 1/16'' from case for 10 seconds

FINISH: All external surfaces are corrosion resistant with readily solderable leads.

POLARITY: Cathode indicated by color band. When operated in zener mode, cathode will be positive with respect to anode.

MOUNTING POSITION: Any

NOTES:
1. PACKAGE CONTOUR OPTIONAL WITHIN A AND B. HEAT SLUGS, IF ANY, SHALL BE INCLUDED WITHIN THIS CYLINDER, BUT NOT SUBJECT TO THE MINIMUM LIMIT OF B.
2. LEAD DIAMETER NOT CONTROLLED IN ZONE F TO ALLOW FOR FLASH, LEAD FINISH BUILDUP AND MINOR IRREGU-LARITIES OTHER THAN HEAT SLUGS.
3. POLARITY DENOTED BY CATHODE BAND.
4. DIMENSIONING AND TOLERANCING PER ANSI Y14.5, 1973.

DIM	MILLIMETERS		INCHES	
	MIN	MAX	MIN	MAX
A	3.05	5 08	0.120	0.200
B	1.52	2.29	0.060	0.090
D	0.46	0.56	0.018	0.022
F	–	1.27	–	0.050
K	25.40	38.10	1.000	1.500

All JEDEC dimensions and notes apply.

CASE 299-02
DO-204AH
(DO-35)

STEADY STATE POWER DERATING

(Courtesy of Motorola Inc.)

1N746 thru 1N759, 1N957A thru 1N986A, 1N4370 thru 1N4372

ELECTRICAL CHARACTERISTICS ($T_A = 25°C$, $V_F = 1.5$ V max at 200 mA for all types)

Type Number (Note 1)	Nominal Zener Voltage V_Z @ I_{ZT} (Note 2) Volts	Test Current I_{ZT} mA	Maximum Zener Impedance Z_{ZT} @ I_{ZT} (Note 3) Ohms	*Maximum DC Zener Current I_{ZM} (Note 4) mA		Maximum Reverse Leakage Current $T_A = 25°C$ I_R @ $V_R = 1$ V µA	$T_A = 150°C$ I_R @ $V_R = 1$ V µA
1N4370	2.4	20	30	150	190	100	200
1N4371	2.7	20	30	135	165	75	150
1N4372	3.0	20	29	120	150	50	100
1N746	3.3	20	28	110	135	10	30
1N747	3.6	20	24	100	125	10	30
1N748	3.9	20	23	95	115	10	30
1N749	4.3	20	22	85	105	2	30
1N750	4.7	20	19	75	95	2	30
1N751	5.1	20	17	70	85	1	20
1N752	5.6	20	11	65	80	1	20
1N753	6.2	20	7	60	70	0.1	20
1N754	6.8	20	5	55	65	0.1	20
1N755	7.5	20	6	50	60	0.1	20
1N756	8.2	20	8	45	55	0.1	20
1N757	9.1	20	10	40	50	0.1	20
1N758	10	20	17	35	45	0.1	20
1N759	12	20	30	30	35	0.1	20

Type Number (Note 1)	Nominal Zener Voltage V_Z (Note 2) Volts	Test Current I_{ZT} mA	Maximum Zener Impedance (Note 3) Z_{ZT} @ I_{ZT} Ohms	Z_{ZK} @ I_{ZK} Ohms	I_{ZK} mA	*Maximum DC Zener Current I_{ZM} (Note 4) mA		Maximum Reverse Current I_R Maximum µA	Test Voltage Vdc 5% V_R	10%
1N957A	6.8	18.5	4.5	700	1.0	47	61	150	5.2	4.9
1N958A	7.5	16.5	5.5	700	0.5	42	55	75	5.7	5.4
1N959A	8.2	15	6.5	700	0.5	38	50	50	6.2	5.9
1N960A	9.1	14	7.5	700	0.5	35	45	25	6.9	6.6
1N961A	10	12.5	8.5	700	0.25	32	41	10	7.6	7.2
1N962A	11	11.5	9.5	700	0.25	28	37	5	8.4	8.0
1N963A	12	10.5	11.5	700	0.25	26	34	5	9.1	8.6
1N964A	13	9.5	13	700	0.25	24	32	5	9.9	9.4
1N965A	15	8.5	16	700	0.25	21	27	5	11.4	10.8
1N966A	16	7.8	17	700	0.25	19	37	5	12.2	11.5
1N967A	18	7.0	21	750	0.25	17	23	5	13.7	13.0
1N968A	20	6.2	25	750	0.25	15	20	5	15.2	14.4
1N969A	22	5.6	29	750	0.25	14	18	5	16.7	15.8
1N970A	24	5.2	33	750	0.25	13	17	5	18.2	17.3
1N971A	27	4.6	41	750	0.25	11	15	5	20.6	19.4
1N972A	30	4.2	49	1000	0.25	10	13	5	22.8	21.6
1N973A	33	3.8	58	1000	0.25	9.2	12	5	25.1	23.8
1N974A	36	3.4	70	1000	0.25	8.5	11	5	27.4	25.9
1N975A	39	3.2	80	1000	0.25	7.8	10	5	29.7	28.1
1N976A	43	3.0	93	1500	0.25	7.0	9.6	5	32.7	31.0
1N977A	47	2.7	105	1500	0.25	6.4	8.8	5	35.8	33.8
1N978A	51	2.5	125	1500	0.25	5.9	8.1	5	38.8	36.7
1N979A	56	2.2	150	2000	0.25	5.4	7.4	5	42.6	40.3
1N980A	62	2.0	185	2000	0.25	4.9	6.7	5	47.1	44.6
1N981A	68	1.8	230	2000	0.25	4.5	6.1	5	51.7	49.0
1N982A	75	1.7	270	2000	0.25	1.0	5.5	5	56.0	54.0
1N983A	82	1.5	330	3000	0.25	3.7	5.0	5	62.2	59.0
1N984A	91	1.4	400	3000	0.25	3.3	4.5	5	69.2	65.5
1N985A	100	1.3	500	3000	0.25	3.0	4.5	5	76	72
1N986A	110	1.1	750	4000	0.25	2.7	4.1	5	83.6	79.2

NOTE 1. TOLERANCE AND VOLTAGE DESIGNATION

Tolerance Designation

The type numbers shown have tolerance designations as follows:

1N4370 series: ±10%, suffix A for ±5% units.

1N746 series: ±10%, suffix A for ±5% units.

1N957 series: suffix A for ±10% units, suffix B for ±5% units.

Voltage Designation

To designate units with zener voltages other than those listed, the Motorola type number should be modified as shown below. Unless otherwise specified, the electrical characteristics other than the nominal voltage (V_Z) and test voltage for leakage current will conform to the characteristics of the next higher voltage type shown in the table.

EXAMPLE: 1N746 series, 1N4370 series variations

EXAMPLE: 1N957 series variations

Matched Sets for Closer Tolerances or Higher Voltages

Series matched sets make zener voltages in excess of 100 volts or tolerances of less than 5% possible as well as providing lower temperature coefficients, lower dynamic impedance and greater power handling ability.

For Matched Sets or other special circuit requirements, contact your Motorola Sales Representative.

(Courtesy of Motorola Inc.)

1N746 thru 1N759, 1N957A thru 1N986A, 1N4370 thru 1N4372

NOTE 2. ZENER VOLTAGE (V$_Z$) MEASUREMENT

Nominal zener voltage is measured with the device junction in thermal equilibrium at the lead temperature of 30°C ±1°C and 3/8" lead length.

NOTE 3. ZENER IMPEDANCE (Z$_Z$) DERIVATION

Z$_{ZT}$ and Z$_{ZK}$ are measured by dividing the ac voltage drop across the device by the ac current applied. The specified limits are for I$_Z$(ac) = 0.1 I$_Z$(dc) with the ac frequency = 60 Hz.

NOTE 4. MAXIMUM ZENER CURRENT RATINGS (I$_{ZM}$)

Maximum zener current ratings are based on the maximum voltage of a 10% 1N746 type unit or a 20% 1N957 type unit. For closer tolerance units (10% or 5%) or units where the actual zener voltage (V$_Z$) is known at the operating point, the maximum zener current may be increased and is limited by the derating curve.

APPLICATION NOTE

Since the actual voltage available from a given zener diode is temperature dependent, it is necessary to determine junction temperature under any set of operating conditions in order to calculate its value. The following procedure is recommended:

Lead Temperature, T$_L$, should be determined from:

$$T_L = \theta_{LA}P_D + T_A$$

θ_{LA} is the lead-to-ambient thermal resistance (°C/W) and P$_D$ is the power dissipation. The value for θ_{LA} will vary and depends on the device mounting method. θ_{LA} is generally 30-40°C/W for the various clips and tie points in common use and for printed circuit board wiring.

The temperature of the lead can also be measured using a thermocouple placed on the lead as close as possible to the tie point. The thermal mass connected to the tie point is normally large enough so that it will not significantly respond to heat surges generated in the diode as a result of pulsed operation once steady-state conditions are achieved. Using the measured value of T$_L$, the junction temperature may be determined by:

$$T_J = T_L + \Delta T_{JL}$$

ΔT_{JL} is the increase in junction temperature above the lead temperature and may be found from Figure 1 for dc power.

$$\Delta T_{JL} = \theta_{JL}P_D$$

For worst-case design, using expected limits of I$_Z$, limits of P$_D$ and the extremes of T$_J$(ΔT$_J$) may be estimated. Changes in voltage, V$_Z$, can then be found from:

$$\Delta V = \theta_{VZ}\Delta T_J$$

θ_{VZ}, the zener voltage temperature coefficient, is found from Figures 3 and 4.

Under high power-pulse operation, the zener voltage will vary with time and may also be affected significantly by the zener resistance. For best regulation, keep current excursions as low as possible.

Surge limitations are given in Figure 6. They are lower than would be expected by considering only junction temperature, as current crowding effects cause temperatures to be extremely high in small spots, resulting in device degradation should the limits of Figure 6 be exceeded.

FIGURE 1 — TYPICAL THERMAL RESISTANCE

FIGURE 2 — TYPICAL LEAKAGE CURRENT

FIGURE 3 — TEMPERATURE COEFFICIENTS
(-55°C to +150°C temperature range; 90% of the units are in the ranges indicated.)

a — RANGE FOR UNITS TO 12 VOLTS

b — RANGE FOR UNITS 12 TO 100 VOLTS

FIGURE 4 — EFFECT OF ZENER CURRENT

FIGURE 5 — TYPICAL CAPACITANCE

FIGURE 6 — MAXIMUM SURGE POWER

This graph represents 90 percentil data points.
For worst case design characteristics, multiply surge power by 2/3.

(Courtesy of Motorola Inc.)

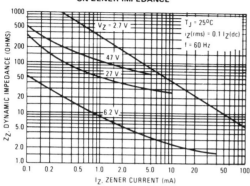

FIGURE 7 – EFFECT OF ZENER CURRENT ON ZENER IMPEDANCE

FIGURE 8 – EFFECT OF ZENER VOLTAGE ON ZENER IMPEDANCE

FIGURE 9 – TYPICAL NOISE DENSITY

FIGURE 10 – NOISE DENSITY MEASUREMENT METHOD

FIGURE 11 – TYPICAL FORWARD CHARACTERISTICS

(Courtesy of Motorola Inc.)

FIGURE 12 — ZENER VOLTAGE versus ZENER CURRENT — V_Z = 1 THRU 16 VOLTS

FIGURE 13 — ZENER VOLTAGE versus ZENER CURRENT — V_Z = 15 THRU 30 VOLTS

FIGURE 14 — ZENER VOLTAGE versus ZENER CURRENT — V_Z = 30 THRU 105 VOLTS

(Courtesy of Motorola Inc.)

2N4220
thru
2N4222

2N4220,A
thru
2N4222,A

CASE 20-03, STYLE 3
TO-72 (TO-206AF)

JFET
LOW-FREQUENCY, LOW NOISE

N-CHANNEL — DEPLETION

MAXIMUM RATINGS

Rating	Symbol	Value	Unit
Drain-Source Voltage	V_{DS}	30	Vdc
Drain-Gate Voltage	V_{DG}	30	Vdc
Gate-Source Voltage	V_{GS}	-30	Vdc
Drain Current	I_D	15	mAdc
Total Device Dissipation @ T_A = 25°C Derate above 25°C	P_D	300 2	mW mW/°C
Junction Temperature Range	T_J	175	°C
Storage Channel Temperature Range	T_{stg}	-65 to $+200$	°C

ELECTRICAL CHARACTERISTICS (T_A = 25°C unless otherwise noted.)

Characteristic		Symbol	Min	Typ	Max	Unit
OFF CHARACTERISTICS						
Gate-Source Breakdown Voltage ($I_G = -10~\mu Adc$, $V_{DS} = 0$)		$V_{(BR)GSS}$	-30	—	—	Vdc
Gate Reverse Current ($V_{GS} = -15$ Vdc, $V_{DS} = 0$) ($V_{GS} = -15$ Vdc, $V_{DS} = 0$, $T_A = 150°C$)		I_{GSS}	 — —	 — —	 -0.1 -100	nAdc
Gate Source Cutoff Voltage ($I_D = 0.1$ nAdc, $V_{DS} = 15$ Vdc)	2N4220,A 2N4221,A 2N4222,A	$V_{GS(off)}$	 — — —	 — — —	 -4 -6 -8	Vdc
Gate Source Voltage ($I_D = 50~\mu Adc$, $V_{DS} = 15$ Vdc) ($I_D = 200~\mu Adc$, $V_{DS} = 15$ Vdc) ($I_D = 500~\mu Adc$, $V_{DS} = 15$ Vdc)	2N4220,A 2N4221,A 2N4222,A	V_{GS}	 -0.5 -1.0 -2.0	 — — —	 -2.5 -5.0 -6.0	Vdc
ON CHARACTERISTICS						
Zero-Gate-Voltage Drain Current* ($V_{DS} = 15$ Vdc, $V_{GS} = 0$)	2N4220,A 2N4221,A 2N4222,A	I_{DSS}	 0.5 2.0 5.0	 — — —	 3.0 6.0 15	mAdc
Static Drain-Source On Resistance ($V_{DS} = 0$, $V_{GS} = 0$)	2N4220,A 2N4221,A 2N4222,A	$r_{DS(on)}$	 — — —	 500 400 300	 — — —	Ohms
SMALL-SIGNAL CHARACTERISTICS						
Forward Transfer Admittance Common Source* ($V_{DS} = 15$ Vdc, $V_{GS} = 0$, $f = 1.0$ kHz)	2N4220,A 2N4221,A 2N4222,A	$\|Y_{fs}\|$	 1000 2000 2500	 — — —	 4000 5000 6000	$\mu mhos$
Output Admittance Common Source ($V_{DS} = 15$ Vdc, $V_{GS} = 0$, $f = 1.0$ kHz)	2N4220,A 2N4221,A 2N4222,A	$\|Y_{os}\|$	 — — —	 — — —	 10 20 40	$\mu mhos$
Input Capacitance ($V_{DS} = 15$ Vdc, $V_{GS} = 0$, $f = 1.0$ MHz)		C_{iss}	—	4.5	6.0	pF
Reverse Transfer Capacitance ($V_{DS} = 15$ Vdc, $V_{GS} = 0$, $f = 1.0$ MHz)		C_{rss}	—	1.2	2.0	pF
Common-Source Output Capacitance ($V_{DS} = 15$ Vdc, $V_{GS} = 0$, $f = 30$ MHz)		C_{osp}	—	1.5	—	pF

MOTOROLA SEMICONDUCTORS

(Courtesy of Motorola Inc.)

SMALL-SIGNAL DEVICES

767

2N4220 thru 2N4222, 2N4220A thru 2N4222A

ELECTRICAL CHARACTERISTICS (continued) (T_A = 25°C unless otherwise noted.)

Characteristic		Symbol	Min	Typ	Max	Unit
FUNCTIONAL CHARACTERISTICS						
Noise Figure		NF				dB
(V_{DS} = 15 Vdc, V_{GS} = 0, R_S = 1.0 megohm,	2N4220A		—	—	2.5	
f = 100 Hz)	2N4221A		—	—	2.5	
	2N4222A		—	—	2.5	

*Pulse Test: Pulse Width = 630 ms, Duty Cycle = 10%.

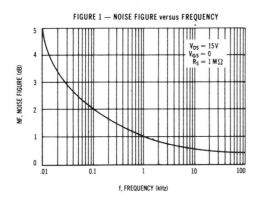

FIGURE 1 — NOISE FIGURE versus FREQUENCY

FIGURE 2 — NOISE FIGURE versus SOURCE RESISTANCE

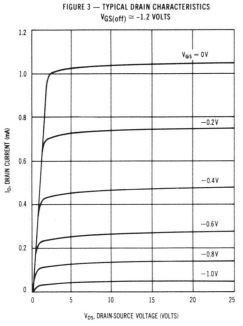

FIGURE 3 — TYPICAL DRAIN CHARACTERISTICS
$V_{GS(off)} \cong$ -1.2 VOLTS

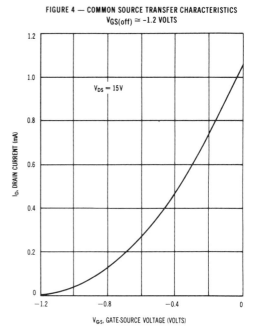

FIGURE 4 — COMMON SOURCE TRANSFER CHARACTERISTICS
$V_{GS(off)} \cong$ -1.2 VOLTS

(Courtesy of Motorola Inc.)

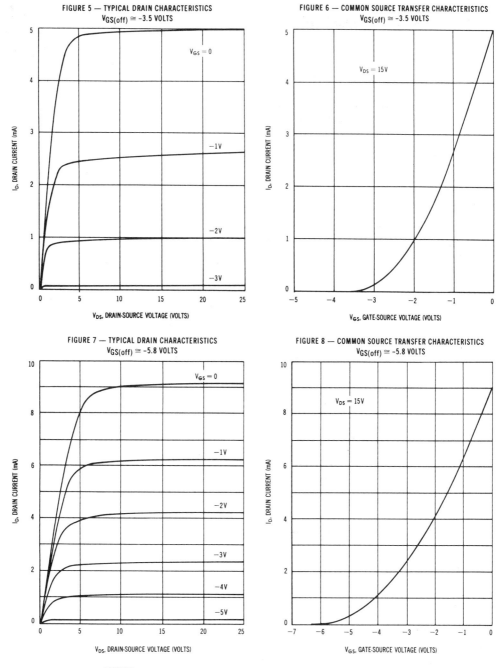

FIGURE 5 — TYPICAL DRAIN CHARACTERISTICS
$V_{GS(off)} \cong -3.5$ VOLTS

FIGURE 6 — COMMON SOURCE TRANSFER CHARACTERISTICS
$V_{GS(off)} \cong -3.5$ VOLTS

FIGURE 7 — TYPICAL DRAIN CHARACTERISTICS
$V_{GS(off)} \cong -5.8$ VOLTS

FIGURE 8 — COMMON SOURCE TRANSFER CHARACTERISTICS
$V_{GS(off)} \cong -5.8$ VOLTS

NOTES: 1. Graphical data is presented for dc conditions. Tabular data is given for pulsed conditions (Pulse Width = 630 ms, Duty Cycle = 10%). Under dc conditions, self heating in higher I_{DSS} units reduces I_{DSS} (See Figure 10).

2. Figures 8, 9, 10: Data taken in a standard printed circuit with a TO-18 type socket mounting and 1/4" lead length.

(Courtesy of Motorola Inc.)

MOTOROLA

NPN   PNP

TIP29   TIP30
TIP29A TIP30A
TIP29B TIP30B
TIP29C TIP30C

COMPLEMENTARY SILICON PLASTIC POWER TRANSISTORS

... designed for use in general purpose amplifier and switching applications. Compact TO-220 AB package. TO-66 leadform also available.

1 AMPERE

POWER TRANSISTORS COMPLEMENTARY SILICON

40-60-80-100 VOLTS
30 WATTS

MAXIMUM RATINGS

Rating	Symbol	TIP29 TIP30	TIP29A TIP30A	TIP29B TIP30B	TIP29C TIP30C	Unit
Collector-Emitter Voltage	V_{CEO}	40	60	80	100	Vdc
Collector-Base Voltage	V_{CB}	40	60	80	100	Vdc
Emitter-Base Voltage	V_{EB}	5.0				Vdc
Collector Current — Continuous Peak Peak	I_C	1.0 3.0				Adc
Base Current	I_B	0.4				Adc
Total Power Dissipation @ T_C = 25°C Derate above 25°C	P_D	30 0.24				Watts W/°C
Total Power Dissipation @ T_A = 25°C Derate above 25°C	P_D	2.0 0.016				Watts W/°C
Unclamped Inductive Load Energy (See Note 3)	E	32				mJ
Operating and Storage Junction Temperature Range	T_J, T_{stg}	-65 to +150				°C

THERMAL CHARACTERISTICS

Characteristic	Symbol	Max	Unit
Thermal Resistance, Junction to Case	$R_{\theta JC}$	4.167	°C/W
Thermal Resistance, Junction to Ambient	$R_{\theta JA}$	62.5	°C/W

ELECTRICAL CHARACTERISTICS (T_C = 25°C unless otherwise noted)

Characteristic		Symbol	Min	Max	Unit
OFF CHARACTERISTICS					
Collector-Emitter Sustaining Voltage (1) (I_C = 30 mAdc, I_B = 0)	TIP29, TIP30 TIP29A, TIP30A TIP29B, TIP30B TIP29C, TIP30C	$V_{CEO(sus)}$	40 60 80 100	– – – –	Vdc
Collector Cutoff Current (V_{CE} = 30 Vdc, I_B = 0) TIP29, TIP29A, TIP30, TIP30A (V_{CE} = 60 Vdc, I_B = 0) TIP29B, TIP29C, TIP30B, TIP30C		I_{CEO}		0.3 0.3	mAdc
Collector Cutoff Current (V_{CE} = 40 Vdc, V_{EB} = 0) TIP29, TIP30 (V_{CE} = 60 Vdc, V_{EB} = 0) TIP29A, TIP30A (V_{CE} = 80 Vdc, V_{EB} = 0) TIP29B, TIP30B (V_{CE} = 100 Vdc, V_{EB} = 0) TIP29C, TIP30C		I_{CES}		200 200 200 200	µAdc
Emitter Cutoff Current (V_{BE} = 5.0 Vdc, I_C = 0)		I_{EBO}		1.0	mAdc
ON CHARACTERISTICS (1)					
DC Current Gain (I_C = 0.2 Adc, V_{CE} = 4.0 Vdc) (I_C = 1.0 Adc, V_{CE} = 4.0 Vdc)		h_{FE}	40 15	– 75	
Collector-Emitter Saturation Voltage (I_C = 1.0 Adc, I_B = 125 mAdc)		$V_{CE(sat)}$	–	0.7	Vdc
Base-Emitter On Voltage (I_C = 1.0 Adc, V_{CE} = 4.0 Vdc)		$V_{BE(on)}$	–	1.3	Vdc
DYNAMIC CHARACTERISTICS					
Current Gain — Bandwidth Product (2) (I_C = 200 mAdc, V_{CE} = 10 Vdc, f_{test} = 1 MHz)		f_T	3.0	–	MHz
Small-Signal Current Gain (I_C = 0.2 Adc, V_{CE} = 10 Vdc, f = 1 kHz)		h_{fe}	20	–	

(1) Pulse Test: Pulse Width ≤ 300 µs, Duty Cycle ≤ 2.0%.
(2) f_T = |h_{fe}| • f_{test}
(3) This rating based on testing with L_C = 20 mH, R_{BE} = 100 Ω, V_{CC} = 10 V, I_C = 1.8 A, P.R.F. = 10 Hz.

STYLE 1
PIN 1 BASE
2 COLLECTOR
3 EMITTER
4 COLLECTOR

DIM	MILLIMETERS		INCHES	
	MIN	MAX	MIN	MAX
A	14.60	15.75	0.575	0.620
B	9.65	10.29	0.380	0.405
C	4.06	4.82	0.160	0.190
D	0.64	0.89	0.025	0.035
F	3.61	3.73	0.142	0.147
G	2.41	2.67	0.095	0.105
H	2.79	3.93	0.110	0.155
J	0.36	0.56	0.014	0.022
K	12.70	14.27	0.500	0.562
L	1.14	1.39	0.045	0.055
N	4.83	5.33	0.190	0.210
Q	2.54	3.04	0.100	0.120
R	2.04	2.79	0.080	0.110
S	1.14	1.39	0.045	0.055
T	5.97	6.48	0.235	0.255
U	0.00	1.27	0.000	0.050
V	1.14		0.045	
Z		2.03		0.080

CASE 221A-02
TO-220AB

(Courtesy of Motorola Inc.)

TIP29, TIP29A, TIP29B, TIP29C, NPN, TIP30, TIP30A, TIP30B, TIP30C, PNP

FIGURE 1 – DC CURRENT GAIN

FIGURE 2 – TURN-OFF TIME

FIGURE 3 – SWITCHING TIME EQUIVALENT CIRCUIT

FIGURE 4 – TURN-ON TIME

FIGURE 5 – ACTIVE REGION SAFE OPERATING AREA

There are two limitations on the power handling ability of a transistor: average junction temperature and second breakdown. Safe operating area curves indicate I_C-V_{CE} operation; i.e., the transistor must not be subjected to greater dissipation than the curves indicate.

The data of Figure 5 is based on $T_{J(pk)} = 150^{\circ}C$; T_C is variable depending on conditions. Second breakdown pulse limits are valid for duty cycles to 10% provided $T_{J(pk)} \leqslant 150^{\circ}C$. At high case temperatures, thermal limitations will reduce the power that can be handled to values less than the limitations imposed by second breakdown.

(Courtesy of Motorola Inc.)

2N3903
2N3904

Silicon Transistors

The General Electric 2N3903 and 2N3904 are silicon NPN planar epitaxial transistors designed for general purpose switching and amplifier applications.

absolute maximum ratings: (T$_A$ = 25°C unless otherwise specified)

VOLTAGES

Collector to Emitter	V$_{CEO}$	40	Volts
Collector to Base	V$_{CBO}$	60	Volts
Emitter to Base	V$_{EBO}$	6	Volts

CURRENT

Collector	I$_C$	200	mA

DISSIPATION

Total Power T$_A$ ≤ 25°C	P$_T$	350	m Watts
Derate Factor T$_A$ > 25°C	P$_T$	2.8	mW/°C

TEMPERATURE

Operating	T$_J$	-55°C to +135°C	°C
Storage	T$_{STG}$	-55°C to +135°C	°C
Lead (1/16" ± 1/32" from case for 10 sec.)	T$_L$	+230°C	°C

1. EMITTER
2. BASE
3. COLLECTOR

SEATING PLANE

TO-92

SYMBOL	MILLIMETERS MIN	MAX	INCHES MIN	MAX	NOTES
A	4.320	5.330	.170	.210	
φb	.407	.550	.016	.022	1,3
φb2	.407	.482	.016	.019	3
φD	4.450	5.200	.175	.205	3
E	3.180	4.190	.125	.165	
e	2.410	2.670	.095	.105	
e1	1.150	1.395	.045	.055	
J	3.430	4.320	.135	.170	
L	12.700	—	.500	—	1,3
L1	—	1.270	—	.050	3
L2	6.350	—	.250	—	3
Q	2.920	—	.115	—	2
s	2.030	2.670	.080	.105	

NOTES:
1. THREE LEADS
2. CONTOUR OF PACKAGE UNCONTROLLED OUTSIDE THIS SIDE.
3. (THREE LEADS) φb2 APPLIES BETWEEN L1 AND L2. φb APPLIES BETWEEN L2 AND 12.70 MM (.500") FROM THE SEATING PLANE. DIAMETER IS UNCONTROLLED IN L1 AND BEYOND 12.70 MM (.500") FROM SEATING PLANE.

* electrical characteristics: $(T_A = 25°C$ unless otherwise specified)

STATIC CHARACTERISTICS		SYMBOL	MIN.	MAX.	UNITS
Collector-Emitter Breakdown Voltage					
$(I_C = 1mA, I_B = 0)$		$V_{(BR)CEO}$	40	—	Volts
Collector-Base Breakdown Voltage					
$(I_C = 10\mu A, I_E = 0)$		$V_{(BR)CBO}$	60	—	Volts
Emitter-Base Breakdown Voltage					
$(I_E = 10\mu A, I_C = 0)$		$V_{(BR)EBO}$	6	—	Volts
Collector Cutoff Current					
$(V_{CE} = 30V, V_{EB}$ (off) $= 3V)$		I_{CEV}	—	50	nA
Base Cutoff Current					
$(V_{CE} = 30V, V_{EB}$ (off) $= 3V)$		I_{BEV}	—	50	nA
Forward Current Transfer Ratio					
$(V_{CE} = 1V, I_C = 100\mu A)$	2N3903	h_{FE}	20	—	
	2N3904	h_{FE}	40	—	
$(V_{CE} = 1V, I_C = 1mA)$	2N3903	h_{FE}	35	—	
	2N3904	h_{FE}	70	—	
$(V_{CE} = 1V, I_C = 10mA)$	2N3903	$\dagger h_{FE}$	50	150	
	2N3904	$\dagger h_{FE}$	100	300	
$(V_{CE} = 1V, I_C = 50mA)$	2N3903	$\dagger h_{FE}$	30	—	
	2N3904	$\dagger h_{FE}$	60	—	
$(V_{CE} = 1V, I_C = 100mA)$	2N3903	$\dagger h_{FE}$	15	—	
	2N3904	$\dagger h_{FE}$	30	—	

(Reprinted with permission of GE Semiconductor Research, Triangle Park, N.C.)

2N3903

2N3904

STATIC CHARACTERISTICS (Continued)

		SYMBOL	MIN.	MAX.	UNITS
Collector-Emitter Saturation Voltage					
(I_C = 10mA, I_B = 1mA)		†$V_{CE(sat)}$	—	.200	Volts
(I_C = 50mA, I_B = 5mA)		†$V_{CE(sat)}$	—	.300	Volts
Base-Emitter Saturation Voltage					
(I_C = 10mA, I_B = 1mA)		†$V_{BE(sat)}$	.65	.85	Volts
(I_C = 50mA, I_B = 5mA)		†$V_{BE(sat)}$	—	.95	Volts

DYNAMIC CHARACTERISTICS

		SYMBOL	MIN.	MAX.	UNITS
Collector-Base Capacitance					
(V_{CB} = 5V, I_E = O,f = 1 MHz)		C_{cb}	—	4	pF
Emitter-Base Capacitance					
(V_{EB} = .5V, I_C = O,f = 1 MHz)		C_{eb}	—	8	pF
Current – Gain – Bandwidth Product					
(V_{CE} = 20V, I_E = 10mA,f = 100 MHz)	2N3903	f_T	250	—	MHz
	2N3904	f_T	300	—	MHz
Noise Figure					
(I_E = 100μA, V_{CE} = 5V, R_G = 1 kHz)	2N3903	NF	—	6	dB
BW = 15.7 kHz	2N3904	NF	—	5	dB
Turn-On Delay		t_d	—	35	ns
Collector Current Rise Time		t_r	—	35	ns
(I_C = 10mA, I_{B1} = 1mA, V_{BE} (off) = .5V)					
(R_L = 275Ω)					
Storage Delay Time	2N3903	t_s	—	175	ns
	2N3904	t_s	—	200	ns

774

Collector Current Fall Time				
($I_C = 10mA$, $I_{B1} = I_{B2} = 1mA$)				
($R_L = 275\Omega$, $V_{CC} = 3V$)				
t_f		—	50	ns
Hybrid Parameters				
($I_E = 1mA$, $V_{CE} = 10V$, f = 1KHz)				
h_{fe}	2N3903	50	200	
h_{fe}	2N3904	100	400	
h_{ie}	2N3903	.5	8	kΩ
h_{ie}	2N3904	1	10	kΩ
h_{re}	2N3903	.1	5	X10^{-4}
h_{re}	2N3904	.5	.8	X10^{-4}
h_{oe}		1.0	40	μmhos

†Pulse width ≤ 300 μsec., Duty Cycle ≤ 2%.
*JEDEC Registered Parameters.

SWITCHING TIME EQUIVALENT TEST CIRCUITS

1. TURN-ON TIME TEST CIRCUIT t_d AND t_r

2. TURN-OFF TIME TEST CIRCUIT t_s AND t_f

C_s = Total shunt capacitance of test jig and connectors

(Reprinted with permission of GE Semiconductor Research, Triangle Park, N.C.)

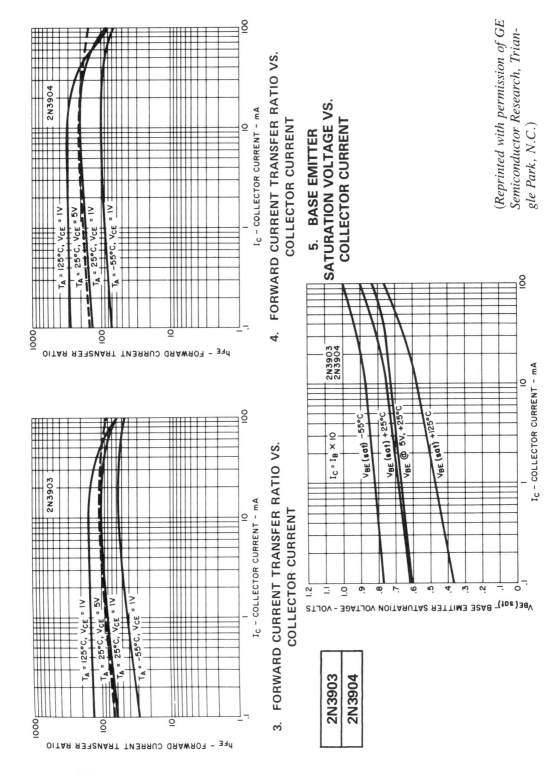

3. FORWARD CURRENT TRANSFER RATIO VS.
COLLECTOR CURRENT

4. FORWARD CURRENT TRANSFER RATIO VS.
COLLECTOR CURRENT

5. BASE EMITTER
SATURATION VOLTAGE VS.
COLLECTOR CURRENT

| 2N3903 |
| 2N3904 |

(Reprinted with permission of GE Semiconductor Research, Triangle Park, N.C.)

776

6. COLLECTOR EMITTER SATURATION
   VOLTAGE VS. BASE CURRENT

7. COLLECTOR EMITTER SATURATION
   VOLTAGE VS. COLLECTOR CURRENT

8. COLLECTOR EMITTER
   SATURATION VOLTAGE VS.
   COLLECTOR CURRENT

(Reprinted with permission of GE Semiconductor Research, Triangle Park, N.C.)

| 2N3903 |
| 2N3904 |

µA741
Operational Amplifier

Linear Division Operational Amplifiers

Description

The µA741 is a high performance monolithic operational amplifier constructed using the Fairchild Planar Epitaxial process. It is intended for a wide range of analog applications. High common mode voltage range and absence of latch up tendencies make the µA741 ideal for use as a voltage follower. The high gain and wide range of operating voltage provide superior performance in integrator, summing amplifier, and general feedback applications.

- **No Frequency Compensation Required**
- **Short Circuit Protection**
- **Offset Voltage Null Capability**
- **Large Common Mode And Differential Voltage Ranges**
- **Low Power Consumption**
- **No Latch Up**

Absolute Maximum Ratings

Storage Temperature Range
Metal Can and Ceramic DIP	−65°C to +175°C
Molded DIP and SO-8	−65°C to +150°C

Operating Temperature Range
Extended (µA741AM, µA741M)	−55°C to +125°C
Commercial (µA741EC, µA741C)	0°C to +70°C

Lead Temperature
Metal Can and Ceramic DIP (soldering, 60 s)	300°C
Molded DIP and SO-8 (soldering, 10 s)	265°C

Internal Power Dissipation[1, 2]
8L-Metal Can	1.00 W
8L-Molded DIP	0.93 W
8L-Ceramic DIP	1.30 W
SO-8	0.81 W

Supply Voltage
µA741A, µA741, µA741E	± 22 V
µA741C	± 18 V

Differential Input Voltage	± 30 V
Input Voltage[3]	± 15 V
Output Short Circuit Duration[4]	Indefinite

Notes

1. $T_{J\ Max}$ = 150°C for the Molded DIP and SO-8, and 175°C for the Metal Can and Ceramic DIP.
2. Ratings apply to ambient temperature at 25°C. Above this temperature, derate the 8L-Metal Can at 6.7 mW/°C, the 8L-Molded DIP at 7.5 mW/°C, the 8L-Ceramic DIP at 8.7 mW/°C, and the SO-8 at 6.5 mW/°C.
3. For supply voltages less than ± 15 V, the absolute maximum input voltage is equal to the supply voltage.
4. Short circuit may be to ground or either supply. Rating applies to 125°C case temperature or 75°C ambient temperature.

Connection Diagram
8-Lead Metal Package
(Top View)

CD00751F

Lead 4 connected to case.

Order Information

Device Code	Package Code	Package Description
µA741HM	5W	Metal
µA741HC	5W	Metal
µA741AHM	5W	Metal
µA741EHC	5W	Metal

Connection Diagram
8-Lead DIP and SO-8 Package
(Top View)

CD00761F

Order Information

Device Code	Package Code	Package Description
µA741RM	6T	Ceramic DIP
µA741RC	6T	Ceramic DIP
µA741SC	KC	Molded Surface Mount
µA741TC	9T	Molded DIP
µA741ARM	6T	Ceramic DIP
µA741ERC	6T	Ceramic DIP
µA741ETC	9T	Molded DIP

Equivalent Circuit

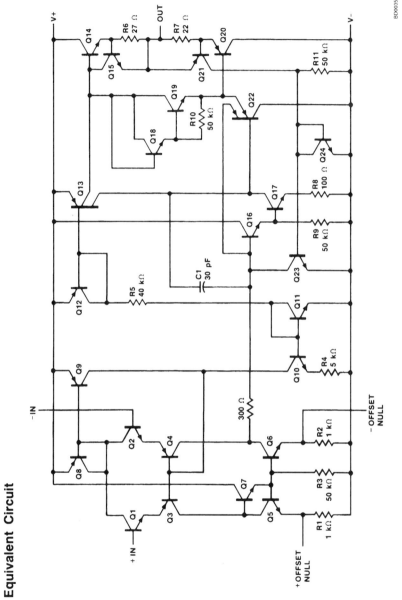

(Copyright Fairchild Semiconductor Corporation, 1987. Used by permission.)

779

μA741 and μA741C
Electrical Characteristics $T_A = 25°C$, $V_{CC} = \pm 15$ V, unless otherwise specified.

Symbol	Characteristic	Condition	μA741 Min	μA741 Typ	μA741 Max	μA741C Min	μA741C Typ	μA741C Max	Unit
V_{IO}	Input Offset Voltage	$R_S \leqslant 10$ kΩ		1.0	5.0		2.0	6.0	mV
$V_{IO\ adj}$	Input Offset Voltage Adjustment Range			±15			±15		mV
I_{IO}	Input Offset Current			20	200		20	200	nA
I_{IB}	Input Bias Current			80	500		80	500	nA
Z_I	Input Impedance		0.3	2.0		0.3	2.0		MΩ
I_{CC}	Supply Current			1.7	2.8		1.7	2.8	mA
P_c	Power Consumption			50	85		50	85	mW
CMR	Common Mode Rejection		70			70	90		dB
V_{IR}	Input Voltage Range		±12	±13		±12	±13		V
PSRR	Power Supply Rejection Ratio	$V_{CC} = \pm 5.0$ V to ±18 V		30	150		30	150	μV/V
I_{OS}	Output Short Circuit Current			25			25		mA
A_{VS}	Large Signal Voltage Gain	$R_L \geqslant 2.0$ kΩ, $V_O = \pm 10$ V	50	200		20	200		V/mV
V_{OP}	Output Voltage Swing	$R_L = 10$ kΩ	±12			±12	±14		V
		$R_L = 2.0$ kΩ	±10			±10	±13		V
TR	Transient Response — Rise time	$V_I = 20$ mV, $R_L = 2.0$ kΩ, $C_L = 100$ pF, $A_V = 1.0$		0.3			0.3		μs
	Transient Response — Overshoot			5.0			5.0		%
BW	Bandwidth			1.0			1.0		MHz
SR	Slew Rate	$R_L \geqslant 2.0$ kΩ, $A_V = 1.0$		0.5			0.5		V/μs

(Copyright Fairchild Semiconductor Corporation, 1987. Used by permission.)

μA741 and μA741C (Cont.)

Electrical Characteristics Over the range of −55°C ≤ T_A ≤ +125°C for μA741, 0°C ≤ T_A ≤ +70°C for μA741C, unless otherwise specified.

Symbol	Characteristic	Condition	μA741 Min	μA741 Typ	μA741 Max	μA741C Min	μA741C Typ	μA741C Max	Unit
V_{IO}	Input Offset Voltage	R_S ≤ 10 kΩ		1.0	6.0			7.5	mV
$V_{IO\ adj}$	Input Offset Voltage Adjustment Range			±15			±15		mV
I_{IO}	Input Offset Current							300	nA
		T_A = +125°C		7.0	200				nA
		T_A = −55°C		85	500				nA
I_{IB}	Input Bias Current							800	nA
		T_A = +125°C		0.03	0.5				μA
		T_A = −55°C		0.3	1.5				μA
I_{CC}	Supply Current	T_A = +125°C		1.5	2.5				mA
		T_A = −55°C		2.0	3.3				mA
P_c	Power Consumption	T_A = +125°C		45	75				mW
		T_A = −55°C		60	100				mW
CMR	Common Mode Rejection	R_S ≤ 10 kΩ	70	90					dB
V_{IR}	Input Voltage Range		±12	±13					V
PSRR	Power Supply Rejection Ratio			30	150				μV/V
A_{VS}	Large Signal Voltage Gain	R_L ≥ 2.0 kΩ, V_O = ±10 V	25			15			V/mV
V_{OP}	Output Voltage Swing	R_L = 10 kΩ	±12	±14					V
		R_L = 2.0 kΩ	±10	±13		±10	±13		V

781

μA741A and μA741E

Electrical Characteristics $T_A = 25°C$, $V_{CC} = \pm 15$ V, unless otherwise specified.

Symbol	Characteristic		Condition	Min	Typ	Max	Unit
V_{IO}	Input Offset Voltage		$R_S \leqslant 50\ \Omega$		0.8	3.0	mV
I_{IO}	Input Offset Current				3.0	30	nA
I_{IB}	Input Bias Current				30	80	nA
Z_I	Input Impedance		$V_{CC} = \pm 20$ V	1.0	6.0		MΩ
P_c	Power Consumption		$V_{CC} = \pm 20$ V		80	150	mW
PSRR	Power Supply Rejection Ratio		$V_{CC} = +10$ V, -20 V to $V_{CC} = +20$ V, -10 V, $R_S = 50\ \Omega$		15	50	μV/V
I_{OS}	Output Short Circuit Current			10	25	40	mA
A_{VS}	Large Signal Voltage Gain		$V_{CC} = \pm 20$ V, $R_L \geqslant 2.0$ kΩ, $V_O = \pm 15$ V	50	200		V/mV
TR	Transient Response	Rise time	$A_V = 1.0$, $V_{CC} = \pm 20$ V, $V_I = 50$ mV, $R_L = 2.0$ kΩ, $C_L = 100$ pF		0.25	0.8	μs
		Overshoot			6.0	20	%
BW	Bandwidth			0.437	1.5		MHz
SR	Slew Rate		$V_I = \pm 10$ V, $A_V = 1.0$	0.3	0.7		V/μs

The following specifications apply over the range of $-55°C \leqslant T_A \leqslant +125°C$ for the μA741A, and $0°C \leqslant T_A \leqslant +70°C$ for the μA741E.

Symbol	Characteristic	Condition	Min	Typ	Max	Unit
V_{IO}	Input Offset Voltage				4.0	mV
$\Delta V_{IO}/\Delta T$	Input Offset Voltage Temperature Sensitivity				15	μV/°C
$V_{IO\ adj}$	Input Offset Voltage Adjustment Range	$V_{CC} = \pm 20$ V	10			mV
I_{IO}	Input Offset Current				70	nA

Symbol	Parameter	Conditions	Min	Typ	Max	Unit
$\Delta I_{IO}/\Delta T$	Input Offset Current Temperature Sensitivity				0.5	nA/°C
I_{IB}	Input Bias Current				210	nA
Z_I	Input Impedance		0.5			MΩ
P_c	Power Consumption	$V_{CC} = \pm 20$ V, μA741A, −55°C			165	mW
		μA741A, +125°C			135	
		μA741E			150	
CMR	Common Mode Rejection	$V_{CC} = \pm 20$ V, $V_I = \pm 15$ V, $R_S = 50\ \Omega$	80	95		dB
I_{OS}	Output Short Circuit Current		10		40	mA
A_{VS}	Large Signal Voltage Gain	$V_{CC} = \pm 20$ V, $R_L \geqslant 2.0$ kΩ, $V_O = \pm 15$ V	32			V/mV
		$V_{CC} = \pm 5.0$ V, $R_L \geqslant 2.0$ kΩ, $V_O = \pm 2.0$ V	10			
V_{OP}	Output Voltage Swing	$V_{CC} = \pm 20$ V, $R_L = 10$ kΩ	±16			V
		$R_L = 2.0$ kΩ	±15			

(Copyright Fairchild Semiconductor Corporation, 1987. Used by permission.)

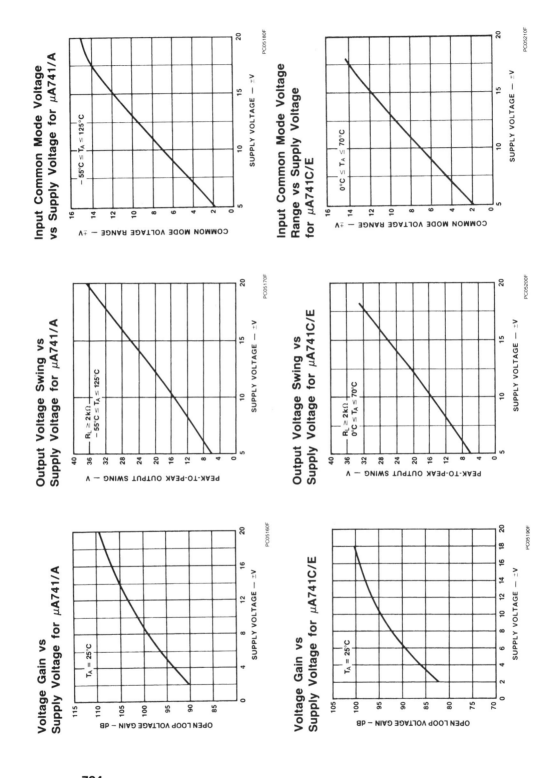

784

Common Mode Rejection Ratio vs Frequency for µA741C/E

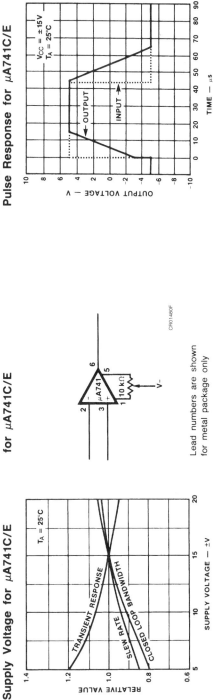

Transient Response Test Circuit for µA741C/E

Transient Response for µA741C/E

Voltage Follower Large Signal Pulse Response for µA741C/E

Voltage Offset Null Circuit for µA741C/E

Frequency Characteristics vs Supply Voltage for µA741C/E

785

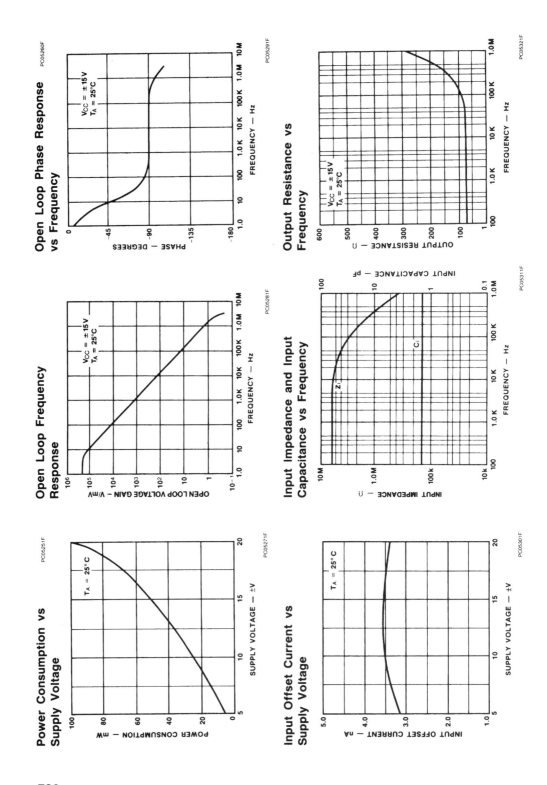

Open Loop Phase Response vs Frequency

PHASE — DEGREES

$V_{CC} = \pm 15V$
$T_A = 25°C$

FREQUENCY — Hz

PC05260F

Output Resistance vs Frequency

OUTPUT RESISTANCE — Ω

$V_{CC} = \pm 15V$
$T_A = 25°C$

FREQUENCY — Hz

PC05321F

Open Loop Frequency Response

OPEN LOOP VOLTAGE GAIN — V/mV

$V_{CC} = \pm 15V$
$T_A = 25°C$

FREQUENCY — Hz

PC05281F

Input Impedance and Input Capacitance vs Frequency

INPUT IMPEDANCE — Ω

INPUT CAPACITANCE — pF

C_i

Z_i

FREQUENCY — Hz

PC05311F

Power Consumption vs Supply Voltage

POWER CONSUMPTION — mW

$T_A = 25°C$

SUPPLY VOLTAGE — ±V

PC05251F

PC05271F

Input Offset Current vs Supply Voltage

INPUT OFFSET CURRENT — nA

$T_A = 25°C$

SUPPLY VOLTAGE — ±V

PC05301F

786

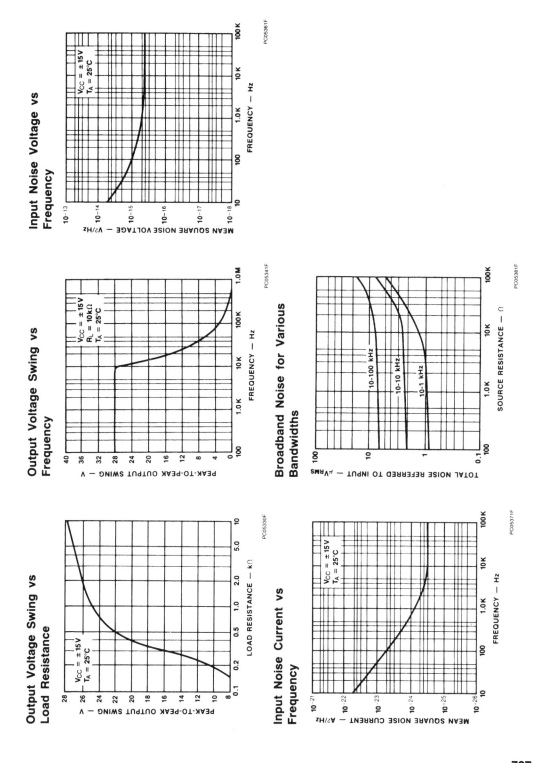

Input Noise Voltage vs Frequency

Output Voltage Swing vs Frequency

Output Voltage Swing vs Load Resistance

Broadband Noise for Various Bandwidths

Input Noise Current vs Frequency

787

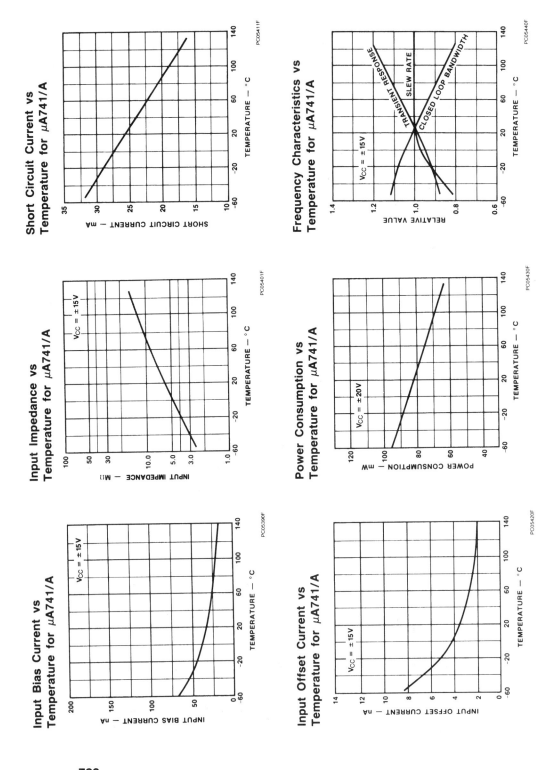

Short Circuit Current vs Temperature for μA741/A

Input Impedance vs Temperature for μA741/A

Input Bias Current vs Temperature for μA741/A

Frequency Characteristics vs Temperature for μA741/A

Power Consumption vs Temperature for μA741/A

Input Offset Current vs Temperature for μA741/A

788

Input Bias Current vs Temperature for μA741C/E

PC05450F

Input Impedance vs Temperature for μA741C/E

PC05461F

Input Offset Current vs Temperature for μA741C/E

PC05470F

Power Consumption vs Temperature for μA741C/E

PC05480F

Short Circuit Current vs Temperature for μA741C/E

PC05491F

Frequency Characteristics vs Temperature for μA741C/E

PC05500F

789

Answers to Selected Odd-numbered Problems

CHAPTER 1

1-1. (a) 15 W delivered, (b) 24 W absorbed

1-3. 4.667 mA, 58.8 mW

1-5. 36 V, 432 mW

1-7. 15 kΩ, 60 mW

1-9. 0.5 mS

1-11. 1 kΩ

1-13. 6 V, 20 mA, 11 mA, 1.636 kΩ

1-15. 15 V, 3 Ω

1-17. 50 kΩ

1-19. (a) 3 A, (b) 12 V, 9 V, 15 V, (c) total power = 150 W

1-21. (a) 2 A, (b) 4 V, 8 V, 10 V, (c) −16 V

1-23. (a) 24 V, (b) 1 A, 2 A, 3 A

1-25. (a) 20 V, (b) 4 A, 10 A, (c) 5 A

1-27. 12 V, 8 V

1-29. 24 mA, 12 mA

1-31. R: 30 kΩ   60 kΩ   18 kΩ   12 kΩ
       I: 2 mA 0.667 mA 1.333 mA 1.333 mA
       V: 60 V   40 V   24 V   16 V

1-33. 18 V

1-35. 12 V

1-37. 3 A in parallel with 8 Ω

1-39. 8 V in series with 4 Ω

1-41. Ideal 6-V voltage source

1-43. Ideal 40-μA current source

1-45. 9-V source in series with 4 kΩ

1-47. 14-V source in series with 7 Ω

1-49. 8-V source in series with 5 kΩ

1-59. 5.556 Ω

1-61. $R_1 = 450\ \Omega$, $R_2 = 55.56\ \Omega$

1-63. (c)

CHAPTER 2

2-1. **(a)** 8 mA, 0, 16 V   **(b)** 7.65 mA, 0.7 V, 15.3 V
2-3. -16 V
2-5. 7.662 mA, 0.677 V, 15.323 V
2-7. 7.68 V
2-9. 7.7 V
2-11. 2.9 V
2-13. 20.77 mA
2-15. **(a)** A, B, G, E, D,   **(b)** 100 mA
2-17. **(a)** 150 Ω,   **(b)** 23.33 mA to 16.67 mA
2-19. Diode open (or possibly connected backwards)

CHAPTER 3

3-1. **(a)** 19.3 V,   **(b)** 19.3 mA,   **(c)** 20 V
3-3. **(a)** 24.3 V,   **(b)** 24.3 mA,   **(c)** 30 V
3-5. **(a)** 20 V,   **(b)** 19.3 mA,   **(c)** 20 V
3-7. **(a)** 25 V,   **(b)** 24.3 mA,   **(c)** 25 V
3-9. **(a)** $v_o = 0$ for $v_s < 0.7$ V
$\qquad = 0.6667v_s - 0.4667$ for $v_s > 0.7$ V
$\quad$ **(b)** 19.53 V, 19.53 mA, 30 V
3-11. $v_o = -0.7$ V for $v_i < -0.933$ V
$\qquad = 0.75v_i$ for $v_i > -0.933$ V
3-13. **(a)** $v_o = v_i$ for $v_i < 6.7$ V
$\qquad = 6.7$ V for $v_i > 6.7$ V
$\quad$ **(b)** -12 V, -5 V, 5 V, 6.7 V,   **(c)** 23.3 mA
3-15. **(a)** 12.7 V, -12.7 V
$\quad$ **(b)** $v_o = v_i$ for -12.7 V $< v_i <$ 12.7 V
$\qquad = -12.7$ V for $v_i < -12.7$ V
$\qquad = 12.7$ V for $v_i > 12.7$ V
$\quad$ **(c)** 17.3 mA,   **(d)** 24.7 V
3-17. **(a)** $v_o = v_i$ for -0.7 V $< v_i <$ 6.2 V
$\qquad = -0.7$ V for $v_i < -0.7$ V
$\qquad = 6.2$ V for $v_i > 6.2$ V
$\quad$ **(b)** 6.2 V, 69 mA
3-19. **(a)** $v_o = -6.9$ V for $v_i < -6.9$ V
$\qquad = v_i$ for -6.9 V $< v_i <$ 6.9 V
$\qquad = 6.9$ V for $v_i > 6.9$ V
$\quad$ **(b)** 6.9 V, 65.5 mA
3-21. **(a)** 100 mA, 175 mA, 75 mA,   **(b)** 125 mA, 175 mA, 50 mA
3-23. **(a)** 100 mA, 111.1 mA, 11.1 mA,   **(b)** 100 mA, 129.6 mA, 29.6 mA,
$\quad$ **(c)** 88.2 mA, 129.6 mA, 41.4 mA
3-25. 1465 Ω, 30 V

3-27. 857.5 Ω, 30.7 V

3-29. Use 5.1-V zener diodes; $R = 1105\ \Omega$

3-31. $R = 1.72\ k\Omega$

3-33. 141.4 Ω, 70 mA

3-35. Either the diode or the 1-kΩ resistor open

3-37. Either the 500-Ω resistor shorted or the 1500-Ω resistor open

CHAPTER 4

4-1. (a) Cutoff, (b) active, (c) saturation, (d) active, (e) cutoff, (f) saturation

4-3. (a) -14 V, (b) -14.3 V, (c) 0.7 V, (d) 14.3 V, (e) 13 V, (f) -0.7 V

4-5. 250, 10.04 mA

4-7. 3.6 mA, 3.63 mA

4-9. 200

4-13. 1.9 mA

4-15. 20 μA

4-17. 336 mW

4-19. Calculated value ≈ 0.47 V, which compares with 0.48 V on curve

4-21. (a) 0.98, (b) 0.99, (c) 0.995, (d) 0.9998

4-27. (a) Cutoff, (b) active, (c) saturation, (d) cutoff, (e) saturation, (f) active

4-29. Transistor is saturated.

CHAPTER 5

5-1. 6 mA, 75 μA, 7.525 V

5-3. (a) 0, 0, 0, 12 V, (b) 25.3 μA, 2.022 mA, 2.047 mA, 7.956 V, (c) 58.2 μA, 4.659 mA, 4.718 mA, 2.681 V, (d) 91.2 μA, 6 mA, 6.091 mA, 0

5-5. (a) 5.295 mA, 13.41 V, (b) 10.59 mA, 2.818 V, (c) 12 mA, 0

5-7. (a) 5 mA, 0, (b) 3.667 mA, 4 V, (c) 2.554 mA, 7.339 V

5-9. (a) 9.32 mA, 14.68 V, (b) 9.32 mA, 5.36 V, (c) 8 mA, 0

5-11. (a) 2 mA, (b) 4850 Ω

5-13. (a) $v_i - 0.7$, (b) 0.7 V $< v_i <$ 24.7 V, (c) 1.582 mA

5-15. (a) 1.90 mA, 1.90 mA, 8.18 V, 6.28 V, 9.54 V, (b) 38.1 μA

5-17. 1.19 mA, 1.19 mA, 5.10 V, 4.62 V, 5.28 V

5-19. (a) Load-line equation: $4000I_C + V_{CE} = 20$; $V_{CEQ} = 12$ V, $I_{CQ} = 2$ mA (b) 5 mA, 25 μA

5-21. (a) 4.9 V, (b) 0.1 V

5-23. (a) 5 V, (b) 0, (c) NOR (combination of OR and NOT)

5-27. 4708 Ω

5-29. Load is connected between 15-V supply and collector. Emitter is biased through R_s to 15-V supply. Required value of emitter resistance is $R_E = 14.3$ kΩ. R_s is not critical as long as $R_s i_B \ll 0.7$. A reasonable value is $R_s = 2$ kΩ.

5-31. $R_C = 10 \text{ k}\Omega$, $R_E = 4 \text{ k}\Omega$, $R_1 = 48.25 \text{ k}\Omega$, $R_2 = 12.37 \text{ k}\Omega$

5-33. $R_C = 10 \text{ k}\Omega$, $R_E = 4 \text{ k}\Omega$, $R_1 = 12.06 \text{ k}\Omega$, $R_2 = 2975 \text{ }\Omega$

5-35. (b)

5-37. **(a)** Increase, **(b)** decrease, **(c)** decrease, **(d)** increase, **(e)** decrease, **(f)** decrease

CHAPTER 6

6-1. **(a)** 12 mS, 83.3 Ω, **(b)** 8 mS, 125 Ω, **(c)** 4 mS, 250 Ω

6-3. **(a)** -3 V, 3 V, **(b)** 3 V, 2 V, 1 V

6-5. **(a)** 18 mA, **(b)** $18 \times 10^{-3}\left(1 - \dfrac{V_{GS}}{-3}\right)^2$

6-7. **(a)** -1.419 V, **(b)** -0.764 V, **(c)** -0.261 V

6-11. **(a)** $12 \times 10^{-3}\left(1 - \dfrac{V_{GS}}{-3}\right)$

6-13. **(a)** 347 Ω, **(b)** 4.32 mA

6-17. **(a)** 24.5 mA, **(b)** 32 mA

6-19. **(a)** 3 V, **(b)** 0, 1 V, 2 V, 3 V

6-21. **(a)** 18×10^{-3} A, **(b)** $18 \times 10^{-3}\left(\dfrac{V_{GS}}{3} - 1\right)^2$

6-23. **(a)** 4.581 V, **(b)** 5.449 V, **(c)** 6.162 V

6-27. 20 mA

6-31. **(a)** -2.3 V, **(b)** 6 mA

6-33. 2.8 V

6-35. **(a)** beyond pinchoff, **(b)** ohmic, **(c)** cutoff, **(d)** beyond pinchoff

6-37. **(a)** Cutoff, **(b)** beyond pinchoff, **(c)** ohmic

6-41. Cutoff

CHAPTER 7

7-1. 8 mA, 6 V

7-3. Operation moves to the ohmic region.

7-7. $I_D \approx 4.7$ mA, $V_{DS} \approx 0.4$ V

7-9. 8.164 mA, -0.9796 V, 0.9796 V, 6.123 V, 7.897 V

7-13. **(b)** $v_i = 0$ and -3 V, **(c)** 0.3 V

7-15. $R_S = 97.6 \text{ }\Omega$, $R_D = 1285 \text{ }\Omega$

7-17.

	I_{DQ}	V_{R_S}	V_{R_D}	V_{DSQ}
Lower bound:	6 mA	0.586 V	7.71 V	15.70 V
Upper bound:	12 mA	1.171 V	15.42 V	7.409 V

7-19. **(a)** Nearly the same, **(b)** increase, **(c)** decrease, **(d)** decrease, **(e)** increase

7-21. (b)

CHAPTER 8

8-1. (a) 0.5 ms,   (b) $5 \sin 4000\pi t$
8-3. (a) 25 V,   (b) $10,000\pi$ rad/s,   (c) 5 kHz,   (d) 0.2 ms
8-5. (a) 3.536 V,   (b) 69.44 mW
8-7. (a) 17.678 V,   (b) 142.0 mW
8-9. (a) 7.958 MΩ,   (b) 15.92 kΩ,   (c) 198.9 Ω
8-11. (a) 50.27 mΩ,   (b) 25.13 Ω,   (c) 1257 Ω
8-13. 15 V
8-15. 10 V, 6 V
8-17. 6 mA, 12 V
8-19. $V_{C1} = 6.211$ V, $V_{C2} = 19.25$ V, $V_{C_E} = 5.51$ V
8-21. $0.4v_g$
8-23. (a) $V_{C1} = V_{C2} = V_A = 24$ V, $V_B = 0$,   (b) $v_{c1} = 0$, $v_{c2} = 0$, $v_a = v_b = 0.2v_g$,   (c) $v_{C1} = 24$ V, $v_{C2} = 24$ V, $v_A = 24 + 0.2v_g$, $v_B = 0.2v_g$
8-25. 15.92 kHz
8-27. 3979 Hz
8-29. 795.8 Hz
8-31. 442.1 Hz
8-33. 1.056 μs
8-35. 22 μs
8-37. 88 μs
8-39. 17.5 kHz
8-45. 31.83 μF
8-47. 3.979 mH
8-49. 3183 Ω
8-51. 265.3 Ω
8-53. (e), (f)
8-55. (a), (c)

CHAPTER 9

9-1. (a) 0,   (b) -1 V,   (c) 5 V
9-3. -40
9-5. -56.57
9-9. 450 Ω, 75 Ω, 120
9-11. 100 kΩ, 800 Ω, -40
9-13. 80
9-15. 64
9-17. 60
9-19. 48
9-21. 240

9-23. 160

9-25. 39.03 dB

9-27. 63.10×10^3

9-29. 400, 26.02 dB

9-31. 0.04, −13.98 dB, 25, 13.98 dB

9-33. 41.58 dB

9-35. 31.62

9-37. 47.60 dB

9-39. **(a)** 30 dBm, **(b)** 36.99 dBm, **(c)** 14.47 dBm, **(d)** −12.22 dBm, **(e)** −106.99 dBm

9-41. Output level = 37.78 dBm

9-43. Break frequency = 3979 Hz

9-45. Break frequency = 3979 Hz

9-51. 102.33, 4318 Ω, 23.16 Ω

9-53. Circuit form of Figure 9-27; $R = 7.958$ kΩ

9-55. Circuit form of Figure 8-19(a) with $R_1 = R_2 = 31.83$ kΩ

9-57. (b), (d)

CHAPTER 10

10-1. −50

10-3. 250 Ω

10-5. 60

10-7. **(a)** 0.32 S, **(b)** 468.75 Ω

10-11. **(a)** 1.974 mA, 5.923 V, 11.057 V, 13.020 V, **(c)** −442.3, 2532 Ω, ∞, 1713 Ω, 5600 Ω

10-13. **(b)** −17.91, 62.52 kΩ, ∞, 4885 Ω, 5600 Ω

10-15. **(b)** 442.3, 12.60 Ω, ∞, 12.55 Ω, 5600 Ω

10-17. **(a)** 1.852 mA, 15.182 V, 14.818 V, **(b)** 0.998, 1.643 MΩ, 77.03 Ω, 12.608 kΩ, 76.32 Ω

10-21. $R_C = 10$ kΩ, $R_1 = 48.25$ kΩ, $R_2 = 28.16$ kΩ, $R_{E1} = 808$ Ω, $R_{E2} = 9192$ Ω

10-23. (d)

CHAPTER 11

11-1. 9 mS

11-3. 12 mS

11-5. **(a)** 12 mS, **(b)** 12 mS, 8 mS, 4 mS, 0

11-9. **(a)** $12 \times 10^{-3}\left(\dfrac{V_{GS}}{3} - 1\right)$, **(b)** 0, 4 mS, 8 mS, 12 mS

11-11. **(b)** −24, ∞, ∞, 123.2 kΩ, 2 kΩ

11-13. -8.571, ∞, ∞, 123.2 kΩ, 2 kΩ

11-15. 45, 66.67 Ω, ∞, 50 Ω, 3 kΩ

11-17. 0.923, ∞, 83.33 Ω, 142.1 kΩ, 76.92 Ω

11-21. 2 kΩ

11-23. 126.7 Ω

11-25. (b)

CHAPTER 12

12-1. -88.89, -50.70, 10, 233

12-3. 36.32, 0.5822, 105.7

12-5. -9.311, -671.8, 8794

12-7. 5.714, 0.5455, 6.545

12-9. 270.8

12-11. 346.5

12-13. (a) -24, (b) -24 mV, -0.48 V

12-15. (d)

CHAPTER 13

13-1. (a) 1 mA, (b) 0.5 mA, 0.5 mA, 7 V, 7 V, 0

13-3. (a) 200, (b) 100, (c) -0.4425, (d) 226

13-5. (a) 12 mV, 6 mV, (b) 2.4 V, 1.2 V, -2.66 mV, (c) 6 mV

13-7. (a) 12 mV, 2.006 V, (b) 2.4 V, 1.2 V, -0.8877 V, (c) 2.006 V

13-9. (a) -10 mV, -1.795 V, (b) -2 V, -1 V, 0.7943 V, (c) -1.795 V

13-11. 2.26 mA

13-13. 3860 Ω

13-15. (a) $\pm$ 14 V, (b) 400,000

13-17. 55 μV

13-19. 105 nA

13-21. 10 nA

13-23. 4000, 72.04 dB

13-25. 1.2

CHAPTER 14

14-1. (a) 20, (b) 11.91

14-3. (a) 49.67, (b) 50

14-5. (a) 16.67, (b) 19.61, (c) 19.96, (d) 20

14-7. 0.119 %

Answers to Selected Odd-numbered Problems

14-9. **(a)** 0.02, **(b)** 2400, **(c)** 49.98, **(d)** 50

14-11. **(a)** 10, **(b)** 20, **(c)** 50, **(d)** 200

14-13. **(a)** 960 MΩ, **(b)** 0.025 Ω

14-15. **(a)** 2.4 MHz, **(b)** 48.02 kHz

14-17. **(a)** 2 kHz, **(b)** 20 kHz, **(c)** 200 kHz, **(d)** 2 MHz

14-21. 0.009

14-23. 31

CHAPTER 15

15-1. **(a)** 120 μV, **(b)** 600 μV, **(c)** 2.4 mV

15-3. 8 V

15-5. **(a)** -20, **(b)** 7.5 kΩ, **(c)** 0.65 V, **(d)** 0, -4 V, 4 V, -10 V, -13 V

15-7. $-2v_1 - 1.5v_2$

15-9. **(a)** -9 V, **(b)** 0, **(c)** 2 V, **(d)** 13 V

15-11. $3(v_1 - v_2)$

15-13. **(a)** 0, **(b)** 12 V, **(c)** -12 V, **(d)** 13 V

15-15. **(a)** 2 kΩ, **(b)** 0, -4 V, 4 V, -13 V

15-17. **(a)** 0.5 mS, **(b)** 13 kΩ, **(c)** 2.6 V

15-19. **(a)** 0.5 mS, **(b)** 6.5 kΩ, **(c)** 1.3 V

15-21. **(a)** 5, **(b)** 1 mA

15-23. 28.6 dB

15-33. $R_i = 15$ kΩ, $R_f = 180$ kΩ

15-35. $R_i = 10$ kΩ, $R_f = 240$ kΩ

15-37. $R_1 = 24$ kΩ, $R_2 = 10$ kΩ, $R_f = 120$ kΩ

15-39. $R = 1.5$ kΩ

15-41. $R = 1$ kΩ

15-43. 3 poles

15-45. $C_1 = 0.05644$ μF, $C_2 = 0.02215$ μF, $C_3 = 0.003221$ μF

15-47. $R = 16.60$ kΩ, $C_2 = 0.02669$ μF, $C_3 = 0.003881$ μF

15-49. (a), (d)

CHAPTER 16

16-1. **(a)** 8.7 V, 8 V, **(b)** -13.5 V, 0

16-3. **(a)** -0.7 V, 0, **(b)** 8.7 V, 8 V

16-5. 1.1 μs

16-7. 2 min

16-9. **(a)** 100 V, **(b)** 36.79 V, **(c)** 13.53 V, **(d)** 0.674 V, **(e)** 4.54×10^{-3} V

16-11. **(a)** 5.3 V, **(b)** 11.3 V

16-13. **(a)** 4 V, 4.7 V, **(b)** 4 V, -1.7 V

16-15. (a) 15.3 V,   **(b)** 15.3 V

16-17. (a) 16 V,   **(b)** 29 V

16-19. (a) 13 V,   **(b)** −13 V

16-21. $v_o = v_i$ for $v_i < 0$

$\qquad = 0$ for $v_i > 0$

16-25. (a) 10 kΩ

16-27. (b)

CHAPTER 17

17-1. 50, −135°

17-3. (a) 1592 Hz,   **(b)** 500 Ω

17-5. 1.625 Hz

17-7. 5.338 MHz

17-9. 2.492 MHz

17-11. 114.1 μs, 8764 Hz

17-13. 4.99 ms, 4.30 ms, 9.29 ms, 108 Hz, 53.7%

17-15. 0.440 s

17-21. 15.92 kΩ

17-23. 3979 Ω to 663.1 Ω

17-25. (a) 2 kΩ,   **(b)** 7 V

17-27. 6497 Ω

17-29. 45.51 kΩ

17-31. $R_2 = 10$ kΩ, $R_1 = 15$ kΩ, $R = 29.51$ kΩ

17-33. $R_A = 21.65$ kΩ, $R_B = 25.25$ kΩ

17-35. 36.36 kΩ

17-37. (f), (h), (i)

CHAPTER 18

18-1. $-j3183$ Ω

18-3. $j2513$ Ω

18-5. 173 pF

18-7. 920.0 kHz

18-9. 17.79 kHz

18-11. 25

18-13. 628.3 Ω

18-15. 2880 Ω

18-17. (a) 3.559 MHz,   **(b)** 7.587 Ω,   **(c)** 469.1 kHz

18-19. (a) 21.32 pF,   **(b)** 6283 Ω,   **(c)** −188.5
18-21. (a) 10 MHz,   **(b)** 2.5 MHz
18-25. (a) 70 ns,   **(b)** 165 ns
18-27. (a) 2.4 kHz,   **(b)** 24 kHz,   **(c)** 240 kHz,   **(d)** 1.2 MHz
18-29. (a) 2395 Hz,   **(b)** 23.53 kHz,   **(c)** 200 kHz,   **(d)** 600 kHz
18-31. (a) 0.1 μs,   **(b)** 2 μs,   **(c)** 6 μs
18-33. (a) 1.592 MHz,   **(b)** 79.58 kHz,   **(c)** 26.53 kHz
18-39. 140.7 pF
18-41. 1.5 MHz
18-43. 1.885 V/μs
18-45. (a)

CHAPTER 19

19-1. (a) 4 A,   **(b)** 2 A,   **(c)** 25 V,   **(d)** 80 V
19-3. (a) 32 V,   **(b)** 64 Ω,   **(c)** 4,   **(d)** 16 W,   **(e)** 8 W,   **(f)** 16 W
19-5. (a) 4 A,   **(b)** 40.74 W,   **(c)** 32 W,   **(d)** 12.97 W
19-7. (a) 2.5°C/W, 50°C/W, 47.5°C/W,   **(b)** 3.9 W,   **(c)** 44 W
19-11. (a) 0.8106 W for each transistor,   **(b)** 1 A,   **(c)** 5.093 W
19-13. (b)

CHAPTER 20

20-1. 24 V, 15 A
20-3. 0.5 A, 15 V
20-5. (a) 15 V,   **(b)** 21.21 V,   **(c)** 6.752 V,   **(d)** 1.125 A
20-7. (a) 15 V,   **(b)** 21.21 V,   **(c)** 13.50 V   **(d)** 2.251 A
20-9. Same answers as Problem 20-7
20-11. (a) 21.21 V,   **(b)** 42.43 V,   **(c)** 21.21 V
20-13. (a) 24 V,   **(b)** 48 V,   **(c)** 48 V,   **(d)** 24 V
20-15. (a) 5 V,   **(b)** 1.443 V
20-17. (a) 12.5 V,   **(b)** 3.608 V,   **(c)** 29.75 V,   **(d)** 0.1213
20-19. (a) 8 V,   **(b)** 6.25 mA,   **(c)** 4 W, 6 W, 66.7%,   **(d)** 2 W
20-23. 2.707
20-25. 1604 μF
20-27. (a) 1203 μF,   **(b)** 15.37 V
20-29. (a) 345 Ω,   **(b)** $R_i = 10$ kΩ, $R_f = 13.53$ kΩ,   **(c)** 20 mA,   **(d)** 4 W
20-31. (b)
20-33. (d)

Index